D1479474

Springer Series in Statistics

Perspectives in Statistics

Springer Series in Statistics

Andrews/Herzberg: Data: A Collection of Problems from Many Fields for the Student and Research Worker.
Anscombe: Computing in Statistical Science through APL.
Berger: Statistical Decision Theory and Bayesian Analysis, 2nd edition.
Brémaud: Point Processes and Queues: Martingale Dynamics.
Brockwell/Davis: Time Series: Theory and Methods, 2nd edition.
Daley/Vere-Jones: An Introduction to the Theory of Point Processes.
Dzhaparidze: Parameter Estimation and Hypothesis Testing in Spectral Analysis of Stationary Time Series.
Farrell:: Multivariate Calculation.
Fienberg/Hoaglin/Kruskal/Tanur (Eds.): A Statistical Model: Frederick Mosteller's Contributions to Statistics, Science, and Public Policy.
Goodman/Kruskal: Measures of Association for Cross Classifications.
Grandell: Aspects of Risk Theory.
Härdle: Smoothing Techniques: With Implementation in S.
Hartigan: Bayes Theory.
Heyer: Theory of Statistical Experiments.
Jolliffe: Principal Component Analysis.
Kotz/Johnson: Breakthroughs in Statistics Volume I: Foundations and Basic Theory
Kotz/Johnson: Breakthroughs in Statistics Volume II: Methodology and Distribution
Kres: Statistical Tables for Multivariate Analysis.
Leadbetter/Lindgren/Rootzén: Extremes and Related Properties of Random Sequences and Processes.
Le Cam: Asymptotic Methods in Statistical Decision Theory.
Le Cam/Yang: Asymptotics in Statistics: Some Basic Concepts.
Manoukian: Modern Concepts and Theorems of Mathematical Statistics.
Miller, Jr.: Simultaneous Statistical Inference, 2nd edition.
Mosteller/Wallace: Applied Bayesian and Classical Inference: The Case of *The Federalist Papers.*
Pollard: Convergence of Stochastic Processes.
Pratt/Gibbons: Concepts of Nonparametric Theory.
Read/Cressie: Goodness-of-Fit Statistics for Discrete Multivariate Data.
Reiss: Approximate Distributions of Order Statistics: With Applications to Nonparametric Statistics.
Ross: Nonlinear Estimation.
Sachs: Applied Statistics: A Handbook of Techniques, 2nd edition.
Seneta: Non-Negative Matrices and Markov Chains.
Siegmund: Sequential Analysis: Tests and Confidence Intervals.
Tong: The Multivariate Normal Distribution.
Vapnik: Estimation of Dependences Based on Empirical Data.
West/Harrison: Bayesian Forecasting and Dynamic Models.
Wolter: Introduction to Variance Estimation.
Yaglom: Correlation Theory of Stationary and Related Random Functions I: Basic Results.
Yaglom: Correlation Theory of Stationary and Related Random Functions II: Supplementary Notes and References.

Samuel Kotz Norman L. Johnson
Editors

Breakthroughs in Statistics Volume II

Methodology and Distribution

Springer-Verlag
New York Berlin Heidelberg London Paris
Tokyo Hong Kong Barcelona Budapest

Samuel Kotz
College of Business
and Management
University of Maryland
at College Park
College Park, MD 20742
USA

Norman L. Johnson
Department of Statistics
Phillips Hall
The University of North Carolina
at Chapel Hill
Chapel Hill, NC 27599
USA

Library of Congress Cataloging-in-Publication Data
Breakthroughs in statistics / Samuel Kotz, Norman L. Johnson, editors.
p. cm.
Includes bibliographical references and indexes.
Contents: v. 2. Methodology—v. 1. Foundations and basic theory v.2. Methodology and distribution.
ISBN 0-387-97572-1
1. Statistics. I. Kotz, Samuel. II. Johnson, Norman Lloyd.
QA276.A12B74 1991
519.5—dc20 91-28183

Printed on acid-free paper.

Production managed by Karen Phillips.
Typeset by Asco Trade Typesetting Ltd., Hong Kong.
Printed and bound by Edwards Brothers, Inc., Ann Arbor, MI.
Printed in the United States of America.

9 8 7 6 5 4 3 2 1

ISBN 0-387-97572-1 Springer-Verlag New York Berlin Heidelberg
ISBN 3-540-97572-1 Springer-Verlag Berlin Heidelberg New York

To the memory of Guta S. Kotz
1901–1989

Preface

McCrimmon, having gotten Grierson's attention, continued: "A breakthrough, you say? If it's in economics, at least it can't be dangerous. Nothing like gene engineering, laser beams, sex hormones or international relations. That's where we don't want any breakthroughs." (Galbraith, J.K. (1990) *A Tenured Professor*, Houghton Mifflin; Boston.)

To judge [*astronomy*] in this way [*a narrow utilitarian point of view*] demonstrates not only how poor we are, but also how small, narrow, and indolent our minds are; it shows a disposition always to calculate the payoff before the work, a cold heart and a lack of feeling for everything that is great and honors man. One can unfortunately not deny that such a mode of thinking is not uncommon in our age, and I am convinced that this is closely connected with the catastrophes which have befallen many countries in recent times; do not mistake me, I do not talk of the general lack of concern for science, but of the source from which all this has come, of the tendency to everywhere look out for one's advantage and to relate everything to one's physical well-being, of the indifference towards great ideas, of the aversion to any effort which derives from pure enthusiasm: I believe that such attitudes, if they prevail, can be decisive in catastrophes of the kind we have experienced. [Gauss, K.F.: *Astronomische Antrittsvorlesung* (cited from Buhler, W.K. (1981) *Gauss*: *A Biographical Study*, Springer: New York)].

This collection of papers (reproduced in whole or in part) is an indirect outcome of our activities, during the decade 1979–88, in the course of compiling and editing the *Encyclopedia of Statistical Sciences* (nine volumes and a Supplementary volume published by John Wiley and Sons, New York). It is also, and more directly, motivated by a more recent project, a systematic rereading and assessment of Presidential Addresses delivered to the Royal

Statistical Society, the International Statistical Institute, and the American Statistical Association during the last 50 years.

Our studies revealed a growing, and already embarrassingly noticeable, diversification among the statistical sciences that borders on fragmentation. Although our belief in the unified nature of statistics remains unshaken, we must recognize certain dangers in this steadily increasing diversity accompanying the unprecedented penetration of statistical methodology into many branches of the social, life, and natural sciences, and engineering and other applied fields.

The initial character of statistics as the "science of state" and the attitudes summed up in the Royal Statistical Society's original motto (now abandoned) of *aliis exterendum* ("let others thresh")—reflecting the view that statisticians are concerned solely with the collection of data—have changed dramatically over the last 100 years and at an accelerated rate during the last 25 years.

To trace this remarkably vigorous development, it seemed logical (to us) to search for "growth points" or "breakthrough" publications that have initiated fundamental changes in the development of statistical methodology. It also seemed reasonable to hope that the consequences of such a search might result in our obtaining a clearer picture of likely future developments. The present collection of papers is an outcome of these thoughts.

In the selection of papers for inclusion, we have endeavored to identify papers that have had lasting effects, rather than search to establish priorities. However, there are introductions to each paper that do include references to important precursors, and also to successor papers elaborating on or extending the influence of the chosen papers.

We were fortunate to have available S.M. Stigler's brilliant analysis of the history of statistics up to the beginning of the 20th century in his book, *The History of Statistics: The Measurement of Uncertainty* (Belknap Press, Cambridge, Mass., 1986), which, together with Claire L. Parkinson's *Breakthroughs: A Chronology of Great Achievements in Science and Mathematics* 1200–1930 (G.K. Hall, Boston, Mass., 1985), allowed us to pinpoint eleven major breakthroughs up to and including F. Galton's *Natural Inheritance*. These are, in chronological order, the following:

C. Huyghens (1657). *De Ratiociniis in Aleae Ludo* (Calculations in Games of Dice), in *Exercitationum Mathematicarum* (F. van Schooten, ed.). Elsevier, Leiden, pp. 517–534.
(The concept of *mathematical expectation* is introduced, as well as many examples of combinatorial calculations.)

J. Graunt (1662). *Natural and Political Observations Mentioned in a Following Index and Made upon the Bills of Mortality*. Martyn and Allestry, London.
(Introduced the idea that vital statistics are capable of scientific analysis.)

E. Halley (1693). An estimate of the degrees of mortality of mankind, drawn from the curious "*Tables of the Births and Funerals* at the City of Breslaw;

with an attempt to ascertain the price of *annuities* upon *lives*," *Philos. Trans. Roy. Soc., Lon.*, **17**, 596–610, 654–656.
[Systematized the ideas in Graunt (1662).]
J. Arbuthnot (1711). An argument for Divine Providence, taken from the constant regularity observed in the births of both sexes, *Philos. Trans. Roy. Soc., Lon.*, **27**, 186–190.
(This is regarded as the first use of a test of significance, although not described as such explicitly.)
J. Bernoulli (1713). *Ars Conjectandi* (The Art of Conjecture). Thurnisorium, Basel.
(Development of combinatorial methods and concepts of statistical inference.)
A. De Moivre (1733, 1738, 1756). *The Doctrine of Chances*, 1st-3rd eds. Woodfall, London.
(In these three books, the normal curve is obtained as a limit and as an approximation to the binomial.)
T. Bayes (1763). Essay towards solving a problem in the doctrine of chances, *Philos. Trans. Roy. Soc., Lon.*, **53**, 370–418.
(This paper has been the source of much work on inverse probability. Its influence has been very widespread and persistant, even among workers who insist on severe restrictions on its applicability.)
P.S. Laplace (1812). *Théorie Analytique des Probabilités.* Courcier, Paris.
(The originating inspiration for much work in probability theory and its applications during the 19th century. Elaboration of De Moivre's work on normal distributions.)
K.F. Gauss (1823). *Theoria Combinationis Observationum Erroribus Minimis Obnoxiae.* Dieterich, Gottingen.
(The method of least squares and associated analysis have developed from this book, which systematized the technique introduced by A.M. Legendre in 1805. Also, the use of "optimal principles" in choosing estimators.)
L.A.J. Quetelet (1846). *Lettres à S.A.R. le Duc Regnant de Saxe-Cobourg et Gotha, sur la Théorie des Probabilités, appliquée aux Sciences Morales et Politiques.* Hayez, Brussels. [English Translation, Layton: London 1849.]
(Observations on the stability of certain demographic indices provided empirical evidence for applications of probability theory.)
F. Galton (1889). *Natural Inheritance.* Macmillan, London.
[This book introduces the concepts of correlation and regression; also mixtures of normal distributions and the bivariate normal distribution. Its importance derives largely from the influence of Karl Pearson. In regard to correlation, an interesting precursor, by the same author, is 'Co-relations and their measurement, chiefly from anthropometric data,' *Proc. Roy. Soc., Lon.*, **45**, 135–145 (1886).]

In our efforts to establish subsequent breakthroughs in our period of study (1890–1989), we approached some 50 eminent (in our subjective evaluation) statisticians, in various parts of the world, asking them if they would supply us with "at least five (a few extra beyond five is very acceptable) possibly suitable references...". We also suggested that some "explanations of reasons for choice" would be helpful.

The response was very gratifying. The requests were sent out in June–July 1989; during July–August, we received over 30 replies, with up to 10 references each, the modal group being 8. There was remarkable near-unanimity recommending the selection of the earlier work of K. Pearson, "Student," R.A. Fisher, and J. Neyman and E.S. Pearson up to 1936. For the years following 1940, opinions became more diverse, although some contributions, such as A. Wald (1945), were cited by quite large numbers of respondents. After 1960, opinions became sharply divergent. The latest work cited by a substantial number of experts was B. Efron (1979). A number of replies cautioned us against crossing into the 1980s, since some time needs to elapse before it is feasible to make a sound assessment of the long-term influence of a paper. We have accepted this viewpoint as valid.

Originally, we had planned to include only 12 papers (in whole or in part). It soon became apparent, especially given the diversity of opinions regarding the last 50 years, that the field of statistical sciences is now far too rich and heterogeneous to be adequately represented by 12 papers over the last 90 years. In order to cover the field satisfactorily, it was decided that at least 30 references should be included. After some discussion, the publisher generously offered to undertake two volumes, which has made it possible to include 39 references! Assignment to the two volumes is on the basis of broad classification into "Foundations and Basic Theory" (Vol. 1) and "Methodology and Distribution" (Vol. 2). Inevitably, there were some papers that could reasonably have appeared in either volume. When there was doubt, we resolved it in such a way as to equalize the size of the two volumes, so far as possible. There are 19 introductions in the first volume and 20 in the second. In addition, we have included Gertrude Cox's 1956 Presidential Address "Frontiers of Statistics" to the American Statistical Association in Vol. 1, together with comments from a number of eminent statisticians indicating some lines on which statistical thinking and practice have developed in the succeeding years.

Even with the extension to two volumes, in order to keep the size of the books within reasonable limits, we found it necessary to reproduce only those parts of the papers that were relevant to our central theme of recording "breakthroughs," points from which subsequent growth can be traced. The necessary cutting caused us much "soul-searching," as did also the selection of papers for inclusion. We also restricted rather severely the lengths of the introductions to individual items. We regret that practical requirements made it necessary to enforce these restrictions. We also regret another consequence of the need to reduce size—namely, our inability to follow much of the advice of our distinguished correspondents, even though it was most cogently advocated. In certain instances the choice was indeed difficult, and a decision was reached only after long discussions. At this point, we must admit that we have included two or three choices of our own that appeared only sparsely among the experts' suggestions.

The division between the two volumes is necessarily somewhat arbitrary. Some papers could equally appear in either. However, papers on fundamental concepts such as probability and mathematical foundation of statistical inference are clearly more Vol. I than Vol. II material (though not entirely so, because concepts can influence application).

There have been laudable and commendable efforts to put the foundations of statistical inference, and more especially probability theory on a sound footing, according to the viewpoint of mathematical self-consistency. Insofar as these may be regarded as attempts to reconcile abstract mathematical logic with phenomena observed in the real world—via interpretation (subjective or objective) of data—we feel that the aim may be too ambitious and even doomed to failure. We are in general agreement with the following remarks of the physicist H.R. Pagels:

> "Centuries ago, when some people suspended their search for absolute truth and began instead to ask how things worked, modern science was born. Curiously, it was by abandoning the search for absolute truth that science began to make progress, opening the material universe to human exploration. It was only by being provisional and open to change, even radical change, that scientific knowledge began to evolve. And ironically, its vulnerability to change is the source of its strength." (From *Perfect Symmetry: The Search for the Beginning of Time*, Simon and Schuster, New York 1985, p. 370).

It is evident that this work represents the fruits of collaboration among many more individuals than the editors. Our special thanks go to the many distinguished statisticians who replied to our inquiries, in many cases responding to further "follow-up" letters requiring additional effort in providing more details that we felt were desirable; we also would like to thank those who have provided introductions to the chosen papers. The latter are acknowledged at the appropriate places where their contributions occur.

We take this opportunity to express our gratitude to Dean R.T. Lamone of the College of Business and Professor B.L. Golden, Chairman of the Department of Management Science and Statistics at the University of Maryland at College Park, and to Professor S. Cambanis, Chairman of the Department of Statistics at the University of North Carolina at Chapel Hill, for their encouragement and the facilities they provided in support of our work on this project.

We are also grateful to the various persons and organizations who have given us reprint permission. They are acknowledged, together with source references, in the section "Sources and Acknowledgments."

We welcome constructive criticism from our readers. If it happens that our first sample proves insufficiently representative, we may be able to consider taking another sample (perhaps of similar size, *without replacement*).

Samuel Kotz
College Park, Maryland
November 1990

Norman L. Johnson
Chapel Hill, North Carolina
November 1990

Contents

Contents

Volume I: Foundations and Basic Theory

Contributors

BARNARD, G.A. Mill House, 54 Hurst Green Brightlingsea, Essex C07 0EH, UK.

LEHMANN, E.L. Department of Statistics, University of California, Berkeley, CA 94720 USA.

PEARCE, S.C. Applied Statistics Research Unit Mathematical Institute, Cornwallis Building, The University of Canterbury, Kent CT2 7NF, UK.

SPEED, T.P. Department of Statistics, University of California, Berkeley, CA 94720 USA.

STEPHENS, M.A. Simon Fraser University, Burnaby, British Columbia, Canada V5A 1S6.

DALENIUS, T. 109 Benevolent Street, Providence, RI 02906 USA.

ANDERSON, T.W. Department of Statistics, Sequoia Hall, Stanford University, Stanford, CA 94305 USA.

NOETHER, G.E. 988 Boulevard of the Arts, Apt. #1812, Sarasota, FL 34236 USA.

DAVID, H.A. Department of Statistics, Snedecor Hall, Iowa State University, Ames, IA 50011 USA.

KING, M.L. Department of Econometrics, Monash University, Clayton, Melbourne, Victoria 3168, Australia.

DRAPER, N.R. Department of Statistics, University of Wisconsin-Madison, 1210 W. Dayton Street, Madison, WI 53706-1593 USA.

BRESLOW, N.E. Department of Biostatistics, School of Public Health and Community Medicine, University of Washington, Seattle, WA 98195 USA.

SEN, P.K. Department of Statistics, University of North Carolina, Chapel Hill, NC 27599-3260 USA.

WEGMAN, E.J. Center for Computational Statistics and Probability, George Mason University, 4400 University Drive, Fairfax, VA 22030-4444 USA.

JONES, LYLE V. Department of Psychology, University of North Carolina, Chapel Hill, NC 27599-3270 USA.

FIENBERG, S.E. College of Humanities and Social Sciences, Carnegie Mellon University, Pittsburgh, PA 15213-3890 USA.

HAMPEL, F.R. Seminar für Statistik, ETH-Zentrum, CH-8092 Zürich, Switzerland.

PRENTICE, R.L. Division of Public Health Sciences, Fred Hutchinson Cancer Research Center, 1124 Columbia Street MP-665, Seattle, WA 98104 USA.

MCCULLAGH, P. Department of Statistics, 5734 University Avenue, University of Chicago, IL 60637 USA.

BERAN, R.J. Department of Statistics, University of California, Berkeley, CA 94720 USA.

Sources and Acknowledgments

Pearson K. (1900) On the criterion that a given system of deviations from the probable in the case of a correlated system of variables is such that it can reasonably be supposed to have arisen from random sampling, *Philos. Mag. Ser.*, **5**, 157–175. Reproduced by the kind permission of Taylor and Francis.

Student (1908) The probable error of a mean. *Biometrika*, **6**, 1–25. Reproduced by the kind permission the Biometrika Trustees.

Fisher R.A. (1925) *Statistical Methods for Research Workers.* Oliver & Boyd: London and Edinburgh (§4 and §42 (Ex. 41)). Reproduced by the kind permission of Oxford University Press.

Fisher R.A. (1926) The arrangement of field experiments. *J. Min. Agric. G. Br.*, **33**, 503–515. Reproduced by the kind permission of Oxford University Press.

Kolmogorov A.N. (1933) On the empirical determination of a distribution. *G. Ist. Ital. Attuari*, **4**, 83–91, (english translation by Dr. Quirino Menegheni).

Neyman J. (1934) On the two different aspects of the representative method: the method of stratified sampling and the method of purposive selection. *J. R. Statist. Soc. Ser. A.*, **97**, 558–606. Reproduced by kind permission of the Royal Statistical Society and Basil Blackwell, Publishers.

Hotelling H. (1936) Relations between two sets of variaties. *Biometrika*, **28**, 321–377. Reproduced by the kind permission of the Biometrika Trustees.

Wilcoxon F. (1945) Individual comparisons by ranking methods. *Biometrics Bull.*, **1**, 80–83. Reproduced by the kind permission of The Biometrics Society.

Mosteller F. (1946) On some useful "inefficient" statistics. *Ann. Math. Statist.*, **17**, 377–408. Reproduced by the kind permission of the Institute of Mathematical Statistics.

Durbin J. and Watson G.S. (1950–51) Testing for serial correlation in least squares regression, I/II. *Biometrika*, **37**, 409–, 428/**38**, 159–178. Reproduced by the kind permission of the Biometrika Trustees.

Box G.E.P. and Wilson K.B. (1951) On the experimental attainment of optimal conditions. *J.R. Statist. Soc., Ser. B.*, **13**, 1–38. Reproduced by the kind permission of the Royal Statistical Society and Basil Blackwell, Publishers.

Kaplan E.L. and Meier P. (1958) Nonparametric estimation from incomplete observations. *J. Amer. Statist. Assoc.*, **53**, 456–481. Reproduced by the kind permission of the American Statistical Association.

Chernoff H. (1959) Sequential design of experiments. *Ann. Math. Statist.*, **29**, 755–770. Reproduced by the kind permission of the Institute of Mathematical Statistics.

Box G.E.P. and Jenkins G.M. (1962) Some statistical aspects of adaptive optimization and control. *J. R. Statist. Soc., Ser. B*, **24**, 297–331. Reproduced by kind permission of the Royal Statistical Society and Basil Blackwell, Publishers.

Tukey J.W. (1962) The future of data analysis. *Ann. Math. Statist.*, **33**, 1–67. Reproduced by the kind permission of the Institute of Mathematical Statistics.

Birch M.W. (1963) Maximum likelihood in three-way contingency tables. *J. R. Statist. Soc. Ser. B*, **25**, 220–233. Reproduced by the kind permission of the Royal Statistical Society.

Huber P.J. (1964) Robust estimation of a location parameter. *Ann. Math. Statist.*, **35**, 73–101. Reproduced by the kind permission of the Institute of Mathematical Statistics.

Cox D.R. (1972) Regression models and life tables. *J. R. Statist. Soc., Ser. B*, **34**, 187–202. Reproduced by the kind permission of the Royal Statistical Society.

Nelder J.A. and Wedderburn, R.W.M. (1972) Generalized linear models. *J. R. Statist. Soc., Ser. A.*, **135**, 370–384. Reproduced by the kind permission of the Royal Statistical Society and Basil Blackwell, Publishers.

Efron. B. (1979) Bootstrap methods: another look at the jackknife. *Ann. Statist.*, **7**, 1–26. Reproduced by the kind permission of the Institute of Mathematical Statistics.

Introduction to Pearson (1900) On the Criterion that a Given System of Deviations from the Probable in the Case of a Correlated System of Variables is Such that it Can be Reasonably Supposed to have Arisen from Random Sampling

G.A. Barnard

1. Introduction

> Before 1900 [the date of this paper by Karl Pearson] we see many scientists of different fields developing and using techniques we now recognise as belonging to modern statistics. After 1900 we begin to see identifiable statisticians developing such techniques into a unified logic of empirical science that goes far beyond its component parts.
>
> —Stephen J. Stigler; *The History of Statistics.*

In 1984, the journal *Science* published a series of articles listing 20 important scientific breakthroughs of the 20th century: relativity, the quantum theory, and so on. One article in the series was devoted to the paper we are about to discuss. Written by Ian Hacking, it is referenced below. Karl Pearson's paper is the fifth in a series of his early papers, most of which are concerned with the mathematical problems of biological evolution. Insofar as this paper is the first of the series that is not primarily concerned with biological problems, it could be taken to represent the break *into* modern, 20th century statistics.

Because it applies to such a wide variety of statistical problems, we can illustrate the use of "chi-squared" by a simple imaginary example. Suppose we have a bag containing a lot of tickets, each carrying one of the 7 letters P,E,A,R,S,O,N, and someone suggests that all vowels occur with equal frequency in the bag, and similarly all consonants, and that each vowel occurs with twice the frequency of each consonant. This would mean that the percentages of the successive letters in the bag would be 10, 20, 20, 10, 10, 20, and 10, and a series of 100 random drawings, with replacement, would give expected frequencies as in the following table:

Ticket carrying	P	E	A	R	S	O	N	Total
Expected frequency	10	20	20	10	10	20	10	100
Observed frequency	7	25	23	6	7	24	8	100

If the observed frequencies are as in the last line, we can ask whether the observed frequencies tend to confirm or deny the hypothesis H leading to the suggested percentages. To answer such a question, Karl Pearson proposed calculating χ, the square root of the sum over all the cells of

$$\frac{(\text{Observed frequency minus expected frequency})^2}{\text{Expected frequency}}, \tag{1}$$

which in the example above is

$$\sqrt{\left\{\frac{3^2}{20}+\frac{5^2}{20}+\frac{3^2}{10}+\frac{4^2}{20}+\frac{3^2}{10}+\frac{4^2}{10}+\frac{2^2}{10}\right\}}=\sqrt{6.30}=2.51.$$

He showed how to calculate the probability P, given H, of getting a value of χ larger than this. If χ was small, the fit of the observed to the expected frequencies would be good and P would be near 1. If, on the other hand, χ was large, the fit would be poor and P would be small. For sufficiently small P, we could conclude that the hypothesis H was wrong.

Later on, it became clear that taking the square root of (1) was not only unnecessary but even undesirable, because the properties of χ^2 are simpler than those of χ itself. For example, if the hypothesis H was true, the average value of χ^2 would be one less than the number of cells—in our example, $7 - 1 = 6$. The standard deviation of χ^2 would be the square root of twice the average value—$\sqrt{(2 \times 6)} = 3.46$ in our case. Thus, the figure of 6.30 is well within the limits that chance would suggest. Indeed, the closeness to the expected value might give rise to a slight suspicion that the data might have been faked. The probability, on H, of getting by chance a value larger than 6.30 can be found from tables to be 0.3905. These tables were calculated by Pearson's student, William Elderton, later known as Sir Willlam, a senior actuary.

Why calculate (1), instead of, for instance, the sum over all cells of (difference between observed and expected frequency)/$\sqrt{}$(expected frequency) where the difference is always taken to be positive? There is no fully compelling reason. In fact, Pearson himself at first thought of doing something similar. He claimed only that his (1), or equivalently, its square root, which he denoted by

$$\chi = \sqrt{S\left\{\frac{(\text{Observed-expected})^2}{(\text{expected})}\right\}} \tag{2}$$

(where S denotes summation over all values in the sample), is a "fairly reasonable" measure of "goodness of fit." It has since transpired that when the hypothetical expectations are correct, not only does χ^2 have the simple expressions for its average value and its standard deviation, but it also has properties similar to those possessed by the variance of a measured quantity: Two independent χ^2's can be added to give another χ^2, and conversely, a given χ^2 can sometimes be meaningfully split into two or more "components." Using χ without squaring would lose these advantages. Thus, it has come about that χ^2 has joined other statistical quantities such as R^2 (multiple correlation) in almost always being referred to in their squared form.

For illustration, we refer to the example of the bag of tickets. The conjecture about the expected values of the numbers of tickets can be expressed in three independent propositions: (1) All vowels are equally frequent; (2) all consonants are equally frequent; (3) any vowel is twice as frequent as any consonant. Since there are three vowels and four consonants, (3) would mean that the total chance of a vowel was $3 \times 0.2 = 0.6$, whereas the chance of a consonant was $4 \times 0.1 = 0.4$. Among 100 tickets, the expected frequency V of vowels would, if (3) were true, be 60, whereas the expected frequency of consonants C would be 40. We have

	V	C
Expected	60	40
Observed	72	38

giving

$$\chi^2 = \frac{12^2}{60} + \frac{12^2}{40} = 144\left(\frac{1}{60} + \frac{1}{40}\right) = \frac{(144 \times 100)}{2400} = 6.$$

We now have two boxes, so the expected χ^2 is $2 - 1 = 1$, and the standard error is only $\sqrt{(2 \times 1)} = 1.414$. The observed value is 6, much too far from 1 to be ascribed to chance. We conclude that (3) cannot be true. On the other hand, the remaining χ^2, corresponding to (1) and (2), is $6.3 - 6 = 0.3$, while its expected value is 5, so this value is suspiciously small. Elderton's tables give the probability of such a small value as 0.0005, strongly suggesting faking. In fact, the figures were faked to make the arithmetic convenient.

2. Background

Karl Pearson, originator of the chi-squared test, was born in London, England on March 27, 1857, the second son of William Pearson, a barrister of the Inner Temple, and his wife, Fanny Smith. Both parents were of Yorkshire stock, and the wider Pearson family continues in 1990 to be well represented in that ancient county. At the age of nine, Carl (as he then spelled his name) went to University College School in London. In 1875, he obtained a scholarship to King's College, Cambridge. He sat the Mathematics Tripos in 1879 and placed third in the class list. After a year in Germany, studying physics, metaphysics, Darwinism, and mediaeval German language and folklore at Heidelberg and Berlin, he returned to London, taking chambers in the Inner Temple to read law. He was called to the bar in 1881, but he never practiced. In 1884, he was persuaded to apply for and was appointed to the Goldsmid Professorship of Engineering and Applied Mathematics at University College in London. In 1900 when he published his paper on χ^2, Karl Pearson had long been well known in Britain as the author of *The Ethic of Free Thought*, a collection of essays mainly on social questions, and he had also become well known throughout the world with *The Grammar of Science*. This book, which continues to be reprinted, propounded the view that the essence of science is its method and there are no areas of human experience inaccessible to study by this method. Along with Olive Schreiner, the South African novelist, feminist, and early advocate of racial equality, he had been a member of the "Men's and Women's Club" for the improvement of relations between the sexes. In 1890, he had married another member of the club, Maria Sharpe, niece of the poet Samuel Rogers. He proclaimed himself a socialist, and a prominent socialist contemporary, the playwright George Bernard Shaw, used to say he relied on Pearson for his statistical arguments. While Pearson's support of women's rights would nowadays be seen as progressive, and in those days socialism was regarded as radical, he also shared and promulgated the social Darwinist views that were common currency in his day but that presently would be deemed racist.

He came late to his life's work in statistics. Influenced by Francis Galton and his close friend, the zoologist W.F.R. Weldon, he began in 1894 the long series of "Contributions to the Mathematical Theory of Evolution," the second of which introduced the system of seven types of frequency curve now known by his name.

Given that we have fitted a curve of the Pearson type to a set of observations, the question naturally arises of whether the fit can be considered satisfactory. This is the problem addressed by the paper here disscussed.

3. The Paper

The "contributions" mentioned above were published in the *Philosophical Transactions* of the Royal Society of London, of which Pearson was elected a Fellow in 1896. The present paper was published on pp. 157–175 of Volume 50 of the 5th series of the *London, Edinburgh and Dublin Philosophical Magazine and Journal of Science* "conducted" by Lord Kelvin, George Francis Fitzgerald, and William Francis. This journal was at the time largely devoted to geology, chemistry, and experimental physics. It carried, for example, much of the pioneer work then being done by J. J. Thomson, the discoverer of the electron. It continued into the 1920s to carry an occasional paper on probability, but it is now almost exclusively devoted to physics

Pearson begins:

> The object of this paper is to investigate a criterion of the probability on any theory of an observed system of errors, and to apply it to the determination of goodness of fit in the case of frequency curves.

He then gives the expression for the normal density function of an n vector $\mathbf{x}$ of variables with mean $\boldsymbol{\mu}$ and variance-covariance matrix $\mathbf{V}$. In modern notation, this is

$$K \exp^{-\frac{1}{2}}(\mathbf{x} - \boldsymbol{\mu})'\mathbf{V}^{-1}(\mathbf{x} - \boldsymbol{\mu}),$$

where $'$ denotes transpose and K is determined by the condition that the density integrates to 1. Pearson sets

$$\chi^2 = (\mathbf{x} - \boldsymbol{\mu})'\mathbf{V}^{-1}(\mathbf{x} - \boldsymbol{\mu}). \tag{3}$$

Then χ^2 = constant, is the equation to a generalized "ellipsoid" all over the surface, of which the frequency of the system of errors or deviations $x_1, x_2, \ldots, x_n$ is constant. The values that χ^2 must be given to cover the whole of space range from 0 to ∞. Now supposing the ellipsoid referred to its principal axes, and then by squeezing reduced to a sphere, $\mathbf{X}$ now being the coordinates: then the chances of a system of errors with as great or greater frequency than that denoted by χ are given by

$$P = \frac{[\iiiint \ldots e^{\chi^2/2}\, dX_1\, dX_2 \ldots dX_n]_\chi^\infty}{[\iiiint \ldots e^{\chi^2/2}\, dX_1\, dX_2 \ldots dX_n]_0^\infty},$$

the numerator being an n-fold integral from the ellipsoid χ to the ellipsoid ∞, and the denominator an n-fold integral from the ellipsoid 0 to the ellipsoid ∞. A common constant factor divides out. Transforming to polar coordinates, Pearson finds

$$P = \frac{\int_\chi^\infty e^{\chi^2/2}\chi^{n-1}\, d\chi}{\int_0^\infty e^{\chi^2/2}\chi^{n-1}\, d\chi} \tag{4}$$

and says,

> This is the measure of the probability of a complex system of n errors occurring with a frequency as great or greater than that of the observed system.... (It) gives us what appears to be a fairly reasonable criterion of the probability of such an error occurring on a random selection being made.

After pointing out that (4) can be evaluated by integration by parts, Pearson refers to a table given at the end of his paper, where $P(\chi^2, n')$ for $\chi^2 = 1(1)10(5)30(10)70$ and $n' = n + 1 = 3(1)20$. He says,

> Suppose we have n correlated variables and we desire to ascertain whether an outlying observed set is really anomalous.... The row χ^2 and the column $n + 1$ will give the value of P, the probability of a system of deviations as great or greater than the outlier in question.... The rough interpolation which this table affords will enable us to ascertain the general order of probability or improbability of the observed result, and this is usually what we want.

He then applies these results to the problem of the fit of an observed to a theoretical frequency distribution. Let there be an $(n + 1)$-fold grouping, and let the observed frequencies of the groups be

$$m'_1, m'_2, m'_3, \ldots, m'_n, m'_{n+1},$$

and the theoretical frequencies supposedly known a priori be

$$m_1, m_2, m_3, \ldots, m_n, m_{n+1}.$$

Then $S(m) = S(m') = N =$ total frequency. (Here, S denotes summation over the sample values. Fisher later used a similar notation.)

Furthermore, if $e = m' - m$ gives the error, we have

$$e_1 + e_2 + e_3 + \cdots + e_{n+1} = 0. \tag{5}$$

Hence only n of the $n + 1$ errors are variables; the $n + 1$th is determined when the first n are known and in using formula (1), we treat only n variables. Pearson states without proof that what would now be called the variance of e_p and the covariance of e_p with e_q are $N(1 - m_p/N)m_p/N$ and $-m_p m_q/N$, respectively. Inserting these values into the expression (1), he arrives, after two and a half pages of algebra, at a "result of great simplicity, and very easily applicable." The quantity

$$\chi = \sqrt{S}\left(\frac{e^2}{m}\right)$$

"is a measure of goodness of fit and the stages of our investigation are pretty clear."

He next considers the case where the m's are not known a priori, but must be derived from estimates of population parameters, giving values m_* with

$$m = m_* + \varepsilon,$$

where the ratio ε/m_* "will, as a rule, be small." The true m's will give a value χ^2 somewhat different from the value χ_*^2 obtained from the m_*'s. Pearson then gives a page and a half of vague argument leading to the conclusion that the difference between χ_* and χ can be neglected in practice, provided the samples are sufficiently large.

The final seven pages of the paper are devoted to illustrations. Pearson fits a binomial distribution with $p = 1/3$ to Weldon' s results of 26,306 throws of 12 dice and finds $\chi = 6.623{,}625$, giving $P = 0.000016$, "or the odds are 62,499 to 1 against such a system of deviations on a random selection." He then fits with p estimated from the data as 0.3377, finds $\chi^2 = 17.775{,}7555$, $P = 0.1227$, and

> The odds are now only 8 to 1 against a system of deviations as improbable or more improbable than this one. It may be said accordingly that the dice experiments of Professor Weldon are consistent with the chance of five or six points being thrown by a single die being 0.3377, but they are excessively improbable, if the chance of all the faces is alike and equal to 1/6.

His third illustration shows it to be excessively unlikely that the fortnight of runs at roulette in July 1892 at Monte Carlo was a random result of a true roulette. His fourth shows that the type-I curve fitted to data on buttercup petals in his second "mathematical contribution" is a satisfactory fit, and his fifth illustration does the same for some data of Thiele's.* His sixth and seventh illustrations show that data produced by Sir George Airy and Mansfield Merriman and alleged by them to demonstrate that the normal distribution fits observational errors, in fact, demonstrate the opposite. Finally, he produces a set of observational errors that do happen to give a good fit to a normal curve.

4. Comment

It is a pity that in 1900 Pearson was not yet acquainted with the Poisson distribution, for he might have seen the argument that was put in the 1920s to his son Egon by W.S. Gosset (Student of "Student's *t*"): If we estimate a Poisson mean in the usual way, we not only have (5), with m_p denoting the frequency of the value $r - 1$, but also

$$e_2 + 2e_3 + 3e_4 + \cdots + ne_{n+1} = 0,$$

so that only $n - 1$ of the $n + 1$ errors are variables in this case. It was left to Greenwood and Yule to show that for the 2×2 table, χ^2 has only one degree of freedom, not three as Pearson's argument would suggest. And Fisher later argued more generally that, *provided* the r parameters were *efficiently* estimated, the asymptotic distribution of χ^2 for a frequency distribution with $n + 1$ classes would have $n - r$ degrees of freedom. This proposition the elder Pearson never appears to have accepted, although the vast majority of statisticians nowadays have accepted Fisher's argument.

Before we attribute Karl Pearson's unwillingness to acknowledge his error to the stubbornness of man well into his 60s, however, we need to consider a point made by Fisher in Sec. 5, Chap. IV of his "Statistical Methods and Scientific Inference" concerned with "excluding a composite hypothesis." While accepting that some examples have the exceptional feature that a single test can be found in which, were any of this class of hypothesis true, the criterion of rejection would be satisfied with the same frequency as the appropriate level of significance (i.e., the test involved would be "similar," in the sense of Neyman and Pearson), Fisher says that, in general, to exclude a composite hypothesis, it is necessary to exclude each and every hypothesis in the class that makes it up. This would seem to imply that, in general, to exclude the composite hypothesis that a set of observations fits a curve of the form $\varphi(x; \theta)$ with θ an unknown parameter, it would be necessary to exclude the particular hypothesis in the family corresponding to x having the distribution $\varphi(x; \hat{\theta})$, where $\hat{\theta}$ is the estimated θ. But to exclude this particular hypothesis, it might seem we should use Pearson's degrees of freedom.

* *Editor's note*: The fourth and fifth illustrations have been omitted.

But things are not so simple. Reverting to our tickets example and using p,e,a,r,s,o,n to denote any possible set of observed frequencies, we can partition the overall χ^2 into the sum of six quantities, corresponding to parts (1), (2), and (3) of H as follows: Let

$$\begin{aligned}
&(\text{Ia}) &&= 1\text{p} && && - 1\text{r} - 1\text{s} && && + 1\text{n}\\
&(\text{Ib}) &&= 1\text{p} && && + 1\text{r} - 1\text{s} && && - 1\text{n}\\
&(\text{Ic}) &&= 1\text{p} && && - 1\text{r} + 1\text{s} && && - 1\text{n}\\
&(\text{IIa}) &&= && 1\text{e} - 2\text{a} && && + 1\text{o} &&\\
&(\text{IIb}) &&= && 1\text{e} && && - 1\text{o} &&\\
&(\text{III}) &&= 3\text{p} - 4\text{e} - 4\text{a} + 3\text{r} + 3\text{s} - 4\text{o} + 3\text{n}.
\end{aligned}$$

The observed values of these quantities are +2, −2, 0, 3, 1, and 60. If H were true, their expected values would all be zero, and their variances could be calculated by summing the squares of their coefficients' multiplied by the expected values of the respective letters. For example, the variance of (IIa) is

$$1^2 \times 20 + 2^2 \times 20 + 1^2 \times 20 = 120,$$

whereas that of (III) is 600. The contribution to χ^2 from (IIa) is the square of its observed value, divided by its variance, i.e., $3^2/120 = 3/40$. The contribution from (III) is $60^2/600 = 6$. The total contribution from (Ia, b, c,) is $8/40 = 0.2$, that from (IIa,b) is $4/40 = 0.1$, whereas that from (III) is 6, making up the original total of 6.30. The covariance, for example, of (IIa) with (III) can be calculated by summing the products of the coefficients: $-4 \times 1 - 4 \times -2 + -4 \times 1 = 0$. It wlll be found that all the covariances vanish. This is the reason why the total χ^2 for H can be split into three "degrees of freedom" corresponding to (1) being true, two degrees of freedom corresponding to (2), and one degree of freedom for (3).

Not all hypotheses about the bag of tickets could be split up in this way. For example, an hypothesis H^* that asserted that the frequencies of P tickets and of R tickets were not both greater than 0.1 could not be so split. To rule out H^* at a given level of significance, we would need to rule out the possibility that the frequency of P tickets was not greater than 0.1 and the corresponding possibility for R tickets.

Conclusion

Disputes over χ^2 were not the only differences that led to notoriously sharp controversies between Karl Pearson and Ronald Fisher, although it should be noted that after Fisher succeeded Pearson as Galton Professor at University College in London, Fisher would make a point of respectful conversation with "KP" whenever the latter came into the common room for tea. The idea of partitioning χ^2 into components was Fisher's major contribution and their combined efforts have made χ^2 into one of the most useful items in today's statistician's toolbox.

Appendix

We here give an account using modern matrix notation, of the pages of algebra used by Pearson to derive his expression for χ^2. We modify his notation, replacing his n by k, his $m'_1, \ldots, m'_{n+1}$ by $N_0, \ldots, N_k$, and his $m_1, m_2, \ldots, m_{n+1}$ by $nq, np_1, \ldots, np_k$.

Suppose there are $k + 1$ cells, labeled 0, 1, 2, ... k, with probabilities $q, p_1, p^2, \ldots, p_k$. Denote by $\mathbf{p}$ the vector with the ith component p_i, $i = 1, 2, \ldots k$. We shall denote by $\sqrt{\mathbf{p}}$, $1/\mathbf{p}$ the vectors with the ith components $\sqrt{p_i}$, $1/p_i$, etc., and by $\mathbf{p}^\delta$ the diagonal (matrix) form of $\mathbf{p}$, etc. $\mathbf{p}'$ denotes the transpose of $\mathbf{p}$. The k vector $\mathbf{I}_j$ denotes the "indicator" of the jth item, $j = 1, 2, \ldots, n$, defined, as the k vector whose ith component is 1 if the jth item falls in cell i, otherwise, 0. Thus, if the jth item falls in cell 0, $\mathbf{I}_j$ is 0. Note that for all j, $\mathbf{I}_j\mathbf{I}'_j = \mathbf{I}_j^\delta$.

We must have $q + 1'\mathbf{p} = 1$. Also, $E\mathbf{I}_j = \mathbf{p}$, where E denotes expectation. The variance-covariance matrix of $\mathbf{I}_j$ is $\text{var}\mathbf{I}_j = E\mathbf{I}'_{j\mathbf{I}j} - E\mathbf{I}_jE\mathbf{I}'_j = \mathbf{p}^\delta - \mathbf{pp}'$.

If N_i denotes the observed frequency in the ith cell, $\mathbf{N} = \sum \mathbf{I}_j$, so that $E\mathbf{N} = n\mathbf{p}$, and since the items are presumed to be uncorrelated,

$$\mathbf{V} = \text{var } \mathbf{N} = \sum \text{var } \mathbf{I}_j = n(\mathbf{p}^\delta - \mathbf{pp}') = \sqrt{(n\mathbf{p})}(1^\delta - \sqrt{\mathbf{p}}\sqrt{\mathbf{p}'})\sqrt{n\mathbf{p}}.$$

It is easily verified by direct multiplication that for any vector $\mathbf{x}$, if $a = 1 - \mathbf{x}'\mathbf{x}$, then $(1^\delta - \mathbf{xx}')[1^\delta + (\mathbf{xx}')/a] = 1^\delta$. So noting that $1 - \sqrt{\mathbf{p}'}\sqrt{\mathbf{p}} = 1 - 1'\mathbf{p} = q$, we obtain

$$\mathbf{V}^{-1} = \frac{1}{\sqrt{(n\mathbf{p})}} \frac{(1^\delta + \sqrt{\mathbf{p}}\sqrt{\mathbf{p}})1}{\sqrt{(n\mathbf{p})}} = \left(\frac{1}{n}\right)\mathbf{p}^\delta + (1/qn)1^\delta,$$

so that

$$(\mathbf{N} - E\mathbf{N})'\mathbf{V}^{-1}(\mathbf{N} - E\mathbf{N}) = \frac{\sum(N_i - np_i)^2}{np_i} + \left(\frac{1}{nq}\right)[\sum(N_i - np_i)]^2.$$

But if N_0 denotes the observed frequency in the cell labeled 0, $N_0 = n - \sum N_i$, whereas $nq = n - \sum p_i$. Thus, the second term on the right-hand side is equal to $(N_0 - nq)^2/nq$, so that the left-hand side is equal to the standard expression for

$$\chi^2 = \sum \frac{\text{Observed frequency} - \text{expected frequency})^2}{(\text{expected frequency})},$$

where the summation is now taken over all $k + 1$ cells.

For large n, the distribution of $\mathbf{N}$ converges to normality, and so by the first part of Pearson's paper, the distribution of χ^2 converges to the expression he gives.

Fisher's expressions for the means and variances of χ^2 components can be obtained similarly, by using a $k + 1$ vector $\mathbf{I}_{j*}$ as the indicator of the jth item, $j = 0, 1, 2, \ldots, k$, with a "1" where the jth item falls and 0 elsewhere. We still have $\mathbf{I}^*\mathbf{I}^{*\prime} = \mathbf{I}^{*\delta}$, and total frequencies can still be regarded as the sum of independent individual $\mathbf{I}^{*\prime}$s, although $\mathbf{V}$ is no longer invertible.

Reference

Hacking, I. (1984). Trial by Number. In *Science 84*, pp. 69–70.

Stigler, Stephen J. (1986). *The History of Statistics: The Measurement of Uncertainty*, Belknap, Harvard University Press, p. 361.

On the Criterion that a Given System of Deviations from the Probable in the Case of a Correlated System of Variables is Such that it Can be Reasonably Supposed to have Arisen from Random Sampling

Karl Pearson
University College, London

The object of this paper is to investigate a criterion of the probability on any theory of an observed system of errors, and to apply it to the determination of goodness of fit in the case of frequency curves.

1. Preliminary Proposition

Let $x_1, x_2 \ldots x_n$ be a system of deviations from the means of n variables with standard deviations $\sigma_1, \sigma_2 \ldots \sigma_n$ and with correlations $r_{12}, r_{13}, r_{23} \ldots r_{n-1,n}$.

Then the frequency surface is given by

$$Z = Z_0 e^{-\frac{1}{2}\left\{S_1\left(\frac{R_{pp}}{R}\frac{x_p^2}{\sigma_p^2}\right) + 2S_2\left(\frac{R_{pq}}{R}\frac{x_p}{\sigma_p}\frac{x_q}{\sigma_q}\right)\right\}}, \qquad \text{(i)}$$

where R is the determinant

$$\begin{vmatrix} 1 & r_{12} & r_{13} & \ldots & r_{1n} \\ r_{21} & 1 & r_{23} & \ldots & r_{2n} \\ r_{31} & r_{32} & 1 & \ldots & r_{3n} \\ \ldots & \ldots & \ldots & \ldots & \ldots \\ \ldots & \ldots & \ldots & \ldots & \ldots \\ r_{n1} & r_{n2} & r_{n3} & \ldots & 1 \end{vmatrix}$$

and R_{pp}, R_{pq} the minors obtained by striking out the pth row and pth column, and the pth row and qth column. S_1 is the sum for every value of p, and S_2 for every pair of values of p and q.

Now let

$$\chi^2 = S_1\left(\frac{R_{pp}}{R}\frac{x_p^2}{\sigma_p^2}\right) + 2S_2\left(\frac{R_{pq}}{R}\frac{x_p x_q}{\sigma_p \sigma_q}\right) \tag{ii}$$

Then: $\chi^2 =$ constant, is the equation to a generalized "ellipsoid," all over the surface of which the frequency of the system of errors or deviations $x_1, x_2 \ldots x_n$ is constant. The values which χ must be given to cover the whole of space are from 0 to ∞. Now suppose the "ellipsoid" referred to its principal axes, and then by squeezing reduced to a sphere, $X_1, X_2, \ldots X_n$ being now the coordinates; then the chances of a system of errors with as great or greater frequency than that denoted by χ is given by

$$P = \frac{[\iiiint \ldots e^{\chi^2/2}\, dX_1\, dX_2 \ldots dX_n]_\chi^\infty}{[\iiiint \ldots e^{\chi^2/2}\, dX_1\, dX_2 \ldots dX_n]_0^\infty},$$

the numerator being an n-fold integral from the ellipsoid χ to the ellipsoid ∞, and the denominator an n-fold integral from the ellipsoid 0 to the ellipsoid ∞. A common constant factor divides out. Now suppose a transformation of coordinates to generalized polar coordinates, in which χ may be treated as the ray, then the numerator and denominator will have common integral factors really representing the generalized "solid angles" and having identical limits. Thus we shall reduce our result to

$$P = \frac{\int_\chi^\infty e^{\chi^2/2}\chi^{n-1}d\chi}{\int_0^\infty e^{\chi^2/2}\chi^{n-1}d\chi}. \tag{iii}$$

This is the measure of the probability of a complex system of n errors occurring with a frequency as great or greater than that of the observed system.

(2) So soon as we know the observed deviations and the probable errors (or σ's) and correlations of errors in any case we can find χ from (ii), and then an evaluation of (iii) gives us what appears to be a fairly reasonable criterion of the probability of such an error occurring on a random selection being made.

For the special purpose we have in view, let us evaluate the numerator of P by integrating by parts; we find

$$\begin{aligned}\int_\chi^\infty e^{\chi^2/2}\chi^{n-1}d\chi &= [\chi^{n-2} + (n-2)\chi^{n-4} + (n-2)(n-4)\chi^{n-6} + \cdots \\ &\quad + (n-2)(n-4)(n-6)\ldots(n-2r-2)\chi^{n-2r}]e^{-1/2\chi^2} \\ &\quad + (n-2)(n-4)(n-6)\ldots(n-2r)\int_\chi^\infty e^{-1/2}\chi^{n-2r-1}d\chi \\ &= (n-2)(n-4)(n-6)\ldots(n-2r)\left[\int_\chi^\infty e^{-1/2\chi^2}\chi^{n-2r-1}d\chi\right.\end{aligned}$$

$$+ e^{\chi^2/2}\left\{\frac{\chi^{n-2r}}{n-2r} + \frac{\chi^{n-2r+2}}{(n-2r)(n-2r+2)}\right.$$

$$+ \frac{\chi^{n-2r+4}}{(n-2r)+(n-2r+2)(n-2r+4)}\cdots$$

$$\left.\left. + \frac{\chi^{n-2}}{(n-2r)(n-2r+2)\dots(n-2)}\right\}\right].$$

Further,

$$\int_0^\infty e^{\chi^2/2}\chi^{n-1}d\chi = (n-2)(n-4)(n-6)\dots(n-2r)\int_0^\infty e^{-1/2\chi^2}\chi^{n-2r-1}d\chi.$$

Now n will either be even or odd, or if n be indefinitely great we may take it practically either.

Case (i) n odd. Take $r = \dfrac{n-1}{2}$. Hence

$$P = \frac{\int_\chi^\infty e^{\chi^2/2}d\chi + e^{\chi^2/2}\left\{\frac{\chi}{1} + \frac{\chi^3}{1\cdot 3} + \frac{\chi^5}{1\cdot 3\cdot 5} + \cdots + \frac{\chi^{n-2}}{1\cdot 3\cdot 5\dots\overline{n-2}}\right\}}{\int_0^\infty e^{\chi^2/2}d\chi}. \quad \text{(iv)}$$

But

$$\int_0^\infty e^{\chi^2/2}d\chi = \sqrt{\frac{\pi}{2}}.$$

Thus

$$P = \sqrt{\frac{2}{\pi}}\int_\chi^\infty e^{\chi^2/2}d\chi$$

$$+ \sqrt{\frac{2}{\pi}}e^{\chi^2/2}\left(\frac{\chi}{1} + \frac{\chi^3}{1\cdot 3} + \frac{\chi^5}{1\cdot 3\cdot 5} + \cdots + \frac{\chi^{n-2}}{1\cdot 3\cdot 5\dots\overline{n-2}}\right). \quad \text{(v)}$$

As soon as χ is known this can be at once evaluated.

Case (ii) n even. Take $r = 1/2\,n - 1$. Hence

$$P = \frac{\int_\chi^\infty e^{\chi^2/2}\chi d\chi + e^{\chi^2/2}\left\{\frac{\chi^2}{2} + \frac{\chi^4}{2\cdot 4} + \frac{\chi^6}{2\cdot 4\cdot 6} + \cdots + \frac{\chi^{n-2}}{2\cdot 4\cdot 6\dots\overline{n-2}}\right\}}{\int_0^\infty e^{\chi^2/2}\chi d\chi}$$

$$= e^{\chi^2/2}\left(1 + \frac{\chi^2}{2} + \frac{\chi^4}{2\cdot 4} + \frac{\chi^6}{2\cdot 4\cdot 6} + \cdots + \frac{\chi^{n-2}}{2\cdot 4\cdot 6\dots\overline{n-2}}\right). \quad \text{(vi)}$$

The series (v) and (vi) both admit of fairly easy calculation, and give sensibly the same results if n be even moderately large. If we put $P = 1/2$ in (v) and (vi) we have equations to determine $\chi = \chi_0$, the value giving the "probability ellipsoid." This ellipsoid has already been considered by Bertrand for $n = 2$ (probability ellipse) and Czuber for $n = 3$. The table which concludes this paper gives the values of P for a series of values of χ^2 in a slightly different case. We can, however, adopt it for general purposes, when we only want a rough approximation to the probability or improbability of a given system of deviations. Suppose we have n correlated variables and we desire to ascertain whether an outlying observed set is really anomalous. Then we calculate χ^2 from (ii); next we take $n' = n + 1$ to enter our table, i.e. if we have 7 correlated quantities we should look in the column marked 8. The row χ^2 and the column $n + 1$ will give the value of P, the probability of a system of deviations as great or greater than the outlier in question. For many practical purposes, the rough interpolation which this table affords will enable us to ascertain the general order of probability or improbability of the observed result, and this is usually what we want.

If n be very large, we have for the series in (v) the value $e^{1/2\chi^2} \int_0^\chi e^{\chi^2/2} d\chi$*, and accordingly

$$P = \sqrt{\frac{2}{\pi}} \int_0^\infty e^{-(1/2)\chi^2} d\chi = 1.$$

Again, the series in (vi) for n very large becomes $e^{1/2\chi^2}$, and thus again $P = 1$. These results show that if we have only an indefinite number of groups, each of indefinitely small range, it is practically certain that a system of errors as large or larger than that defined by any value of χ will appear.

Thus, if we take a very great number of groups our test becomes illusory. We must confine our attention in calculating P to a finite number of groups, and this is undoubtedly what happens in actual statistics. n will rarely exceed 30, often not be greater than 12.

(3) Now let us apply the above results to the problem of the fit of an observed to a theoretical frequency distribution. Let there be an $(n + 1)$-fold grouping, and let the observed frequencies of the groups be

$$m'_1, m'_2, m'_3, \ldots m'_n, m'_{n+1},$$

and the theoretical frequencies supposed known *a priori* be

$$m_1, m_2, m_3 \ldots m_n, m_{n+1};$$

then $S(m) = S(m') = N =$ total frequency.

Further, if $e = m' - m$ give the error, we have

$$e_1 + e_2 + e_3 + \cdots + e_{n+1} = 0.$$

Hence only n of the $n + 1$ errors are variables; the $n + 1$th is determined when

* Write the series as F, then we easily find $dF/d\chi = 1 + \chi F$, whence by integration the above result follows. Geometrically, $P = 1$ means that if n be indefinitely large, the nth moment of the tail of the normal curve is equal to the nth moment of the whole curve, however much or however little we cut off as "tail."

the first n are known, and in using formula (ii) we treat only of n variables. Now the standard deviation for the random variation of e_p is

$$\sigma_p = \sqrt{N\left(1 - \frac{m_p}{N}\right)\frac{m_p}{N}}, \tag{vii}$$

and if r_{pq} be the correlation of random error e_p and e_q,

$$\sigma_p \sigma_q r_{pq} = -\frac{m_p m_q}{N}. \tag{viii}$$

Now let us write $m_q/N = \sin^2 \beta_q$, where β_q is an auxiliary angle easily found. Then we have

$$\sigma_q = \sqrt{N} \sin \beta_q \cos \beta_q, \tag{ix}$$

$$r_{pq} = -\tan \beta_q \tan \beta_p. \tag{x}$$

We have from the value of R in §1

$$R = \begin{vmatrix} 1 & -\tan \beta_2 \tan \beta_1 & -\tan \beta_3 \tan \beta_1 & \ldots & -\tan \beta_n \tan \beta_1 \\ -\tan \beta_1 \tan \beta_2 & 1 & -\tan \beta_2 \tan \beta_2 & \ldots & -\tan \beta_n \tan \beta_2 \\ -\tan \beta_1 \tan \beta_3 & -\tan \beta_2 \tan \beta_3 & 1 & \ldots & -\tan \beta_n \tan \beta_3 \\ \cdots & \cdots & \cdots & \cdots & \cdots \\ \cdots & \cdots & \cdots & \cdots & \cdots \\ \cdots & \cdots & \cdots & \cdots & \cdots \\ \cdots & \cdots & \cdots & \cdots & \cdots \\ -\tan \beta_1 \tan \beta_n & -\tan \beta_2 \tan \beta_n & -\tan \beta_3 \tan \beta_n & & 1 \end{vmatrix}$$

$$= (-1)^n \tan^2 \beta_1 \tan^2 \beta_2 \tan^2 \beta_3 \ldots \tan^2 \beta_n \times$$

$$\begin{vmatrix} -\cot^2 \beta_1 & 1 & 1 & \ldots & 1 \\ 1 & -\cot^2 \beta_2 & 1 & \ldots & 1 \\ 1 & 1 & -\cot^2 \beta_3 & \ldots & 1 \\ \cdots & \cdots & \cdots & \cdots & \cdots \\ \cdots & \cdots & \cdots & \cdots & \cdots \\ \cdots & \cdots & \cdots & \cdots & \cdots \\ 1 & 1 & 1 & & -\cot^2 \beta_n \end{vmatrix}$$

$$= \tan^2 \beta_1 \tan^2 \beta_2 \tan^2 \beta_3 \ldots \tan^2 \beta_n \times J, \text{ say.}$$

Similarly,

$$R_{11} = (-1)^{n-1} \tan^2 \beta_2 \tan^2 \beta_3 \ldots \tan^2 \beta_n \times J_{11},$$

$$R_{12} = (-1)^{n-1} \tan \beta_1 \tan \beta_2 \tan^2 \beta_3 \ldots \tan^2 \beta_n \times J_{12}.$$

Hence the problem reduces to the evaluation of the determinant J and its minors.

If we write

$$\eta_q = \cot^2 \beta_q = \frac{N}{m_q} - 1. \tag{xi}$$

$$J = \begin{vmatrix} -\eta_1 & 1 & 1 & \ldots & 1 \\ 1 & -\eta_2 & 1 & \ldots & 1 \\ 1 & 1 & -\eta_3 & \ldots & 1 \\ \ldots & \ldots & \ldots & \ldots & \ldots \\ \ldots & \ldots & \ldots & \ldots & \ldots \\ 1 & 1 & 1 & \ldots & -\eta_n \end{vmatrix}$$

Clearly,

$$J_{12} = \begin{vmatrix} 1 & 1 & 1 & \ldots & 1 \\ 1 & -\eta_3 & 1 & \ldots & 1 \\ 1 & 1 & -\eta_4 & \ldots & 1 \\ \ldots & \ldots & \ldots & \ldots & \ldots \\ \ldots & \ldots & \ldots & \ldots & \ldots \\ 1 & 1 & 1 & \ldots & -\eta_n \end{vmatrix}$$

$$= (-1)^{n-1}(\eta_3 + 1)(\eta_4 + 1)\ldots(\eta_n + 1).$$

Generally, if $\lambda = (\eta_1 + 1)(\eta_2 + 1)(\eta_3 + 1)\ldots(\eta_n + 1)$,

$$J_{pq} = (-1)^{n-1}\frac{\lambda}{(\eta_p + 1)(\eta_q + 1)}. \tag{xii}$$

But

$$J_{11} - \eta_2 J_{12} + J_{13} + J_{14} + \cdots + J_{1n} = 0.$$

Hence

$$\begin{aligned} J_{11} &= (1 + \eta_2)J_{12} - J_{12} - J_{13} - J_{14} \cdots - J_{1n} \\ &= \frac{(-1)^{n-1}\lambda}{1 + \eta_1}\left(1 - \frac{1}{1 + \eta_2} - \frac{1}{1 + \eta_3} - \frac{1}{1 + \eta_4} \cdots - \frac{1}{1 + \eta_n}\right). \end{aligned}$$

Whence, comparing J with J_{11}, it is clear that:

$$J = (-1)^n \lambda\left(1 - \frac{1}{1 + \eta_1} - \frac{1}{1 + \eta_2} - \frac{1}{1 + \eta_3} - \frac{1}{1 + \eta_4} \cdots - \frac{1}{1 + \eta_n}\right).$$

Now

$$S\left(\frac{1}{1 + \eta}\right) = S\left(\frac{m}{N}\right), \quad \text{by (xi)}, \quad = \frac{N - m_{n+1}}{N} = 1 - \frac{m_{n+1}}{N}.$$

Thus:

$$J = (-1)^n \lambda \frac{m_{n+1}}{N}.$$

Similarly:

$$J_{pp} = (-1)^{n-1} \frac{\lambda}{1+\eta_p}\left(\frac{m_p}{N} + \frac{m_{n+1}}{N}\right).$$

Thus:

$$\frac{R_{pp}}{R} = -\frac{J_{pp}}{J}\cot^2\beta_p = \cot^2\beta_p \frac{m_p}{N}\left(1 + \frac{m_p}{m_{n+1}}\right);$$

or from (vii)

$$\frac{R_{pp}}{R}\frac{1}{\sigma_p^2} = \frac{1}{m_p} + \frac{1}{m_{n+1}}. \qquad \text{(xiii)}$$

Again:

$$\frac{R_{pq}}{R} = -\cot\beta_p \cot\beta_q \frac{J_{pq}}{J} = \cot\beta_p \cot\beta_q \frac{m_p m_q}{N m_{n+1}}.$$

and:

$$\frac{R_{pq}}{R}\frac{1}{\sigma_p\sigma_q} = \frac{1}{m_{n+1}}. \qquad \text{(xiv)}$$

Thus by (ii):

$$\begin{aligned}\chi^2 &= S_1\left\{\left(\frac{1}{m_p} + \frac{1}{m_{n+1}}\right)e_p^2\right\} + 2S_2\left\{\frac{1}{m_{n+1}}e_p e\right\} \\ &= S_1\left(\frac{e_p^2}{m_p}\right) + \frac{1}{m_{n+1}}\{S_1(e_p)\}^2.\end{aligned}$$

But

$$S_1(e_p^2) = -e_{n+1},$$

hence:

$$\chi^2 = S\left(\frac{e^2}{m}\right), \qquad \text{(xv)}$$

where the summation is now to extend to all $(n + 1)$ errors, and not merely to the first n.

(4) This result is of very great simplicity, and very easily applicable. The quantity

$$\chi = \sqrt{S\left(\frac{e^2}{m}\right)}$$

is a measure of the goodness of fit, and the stages of our investigation are pretty clear. They are:

(i) Find χ from Equation (xv):

(ii) If the number of errors, $n' = n + 1$, be odd, find the improbability of the system observed from

$$P = e^{\chi^2/2}\left(1 + \frac{\chi^2}{2} + \frac{\chi^4}{2\cdot 4} + \frac{\chi^6}{2\cdot 4\cdot 6} + \cdots + \frac{\chi^{n'-3}}{2\cdot 4\cdot 6\ldots n'-3}\right).$$

If the number of errors, $n' = n + 1$, be even, find the probability of the system observed from

$$P = \sqrt{\frac{2}{\pi}}\int_{\chi}^{\infty} e^{\chi^2/2} d\chi$$

$$+ \sqrt{\frac{2}{\pi}} e^{\chi^2/2}\left(\frac{\chi}{1} + \frac{\chi^3}{1\cdot 3} + \frac{\chi^5}{1\cdot 3\cdot 5} + \cdots + \frac{\chi^{n'-3}}{1\cdot 3\cdot 5\ldots n'-3}\right).$$

(iii) If n be less than 13, then the Table at the end of this paper will often enable us to determine the general probability or improbability of the observed system without using these values for P at all.

(5) Hitherto we have been considering cases in which the theoretical probability is known *a priori*. But in a great many cases this is not the fact; the theoretical distribution has to be judged from the sample itself. The question we wish to determine is whether the sample may be reasonably considered to represent a random system of deviations from the theoretical frequency distribution of the general population, but this distribution has to be inferred from the sample itself. Let us look at this somewhat more closely. If we have a fairly numerous series, and assume it to be really a random sample, then the theoretical number m for the whole population falling into any group and the theoretical number m_δ as deduced from the data for the sample will only differ by terms of the order of the probable errors of the constants of the sample, and these probable errors will be small, as the sample is supposed to be fairly large. We may accordingly take:

$$m = m_\delta + \mu,$$

where the ratio of μ to m_δ, will, as a rule, be small. It is only at the "tails" that μ/m_δ, may become more appreciable, but here the errors or deviations will be few or small*.

Now let χ_δ be the value found for the sample, and χ the value required marking the system of deviations of the observed quantities from a group-system of the same number *accurately* representing the general population.

* A theoretical probability curve without limited range will never at the extreme tails exactly fit observation. The difficulty is obvious where the observations go by units and the theory by fractions. We ought to take our final theoretical groups to cover as much of the tail area as amounts to at least a unit of frequency in such cases.

Then:

$$\chi^2 = S\left\{\frac{(m'-m)^2}{m}\right\} = S\left\{\frac{(m'-m_\delta-\mu)^2}{m_\delta+\mu}\right\}$$

$$= S\left\{\frac{(m'-m_\delta)^2}{m_\delta}\right\} - S\left\{\frac{\mu(m'^2-m_\delta^2)}{m_\delta}\right\} + S\left\{\left(\frac{\mu}{m_\delta}\right)^2\frac{m'^2}{m_\delta}\right\},$$

if we neglect terms of the order $(\mu/m_\delta)^3$.

Hence:

$$\chi^2 - \chi_\delta^2 = -S\left\{\frac{\mu}{m_\delta}\frac{m'^2-m_\delta^2}{m_\delta}\right\} + S\left\{\left(\frac{\mu}{m_\delta}\right)^2\frac{m'^2}{m_\delta}\right\}.$$

Now χ_δ must, I take it, be less than χ, for otherwise the general population distribution or curve would give a better fit than the distribution or curve actually fitted to the sample. But we are supposed to fit a distribution or curve to the sample so as to get the "best" values of the constants. Hence the right-hand side of the above equation must be positive. If the first term be negative then it must be less than the second, or the difference of χ and χ_δ is of the order, not of the first but of the *second* power of quantities depending on the probable errors of the sample. On the other hand, if the first term be positive, it means that there is negative correlation between μ/m_δ and $m'^2 - m_\delta^2/m_\delta$, or that when the observed frequency exceeds the theoretical distribution given by the sample ($m' > m_\delta$), then the general population would fall below the theoretical distribution given by the sample ($m < m_\delta$), and *vice versa.* In other words the general population and the observed population would always tend to fall on opposite sides of the sample theoretical distribution. Now this seems impossible; we should rather expect, when the observations exceeded the sample theoretical distribution, that the general population would have also excess, and *vice versa.* Accordingly, we should either expect the first term to be negative, or to be very small (or zero) if positive. In either case I think we may conclude that χ only differs from χ_δ by terms of the order of the squares of the probable errors of the constants of the sample distribution. Now our argument as to goodness of fit will be based on the general order of magnitude of the probability P, and not on slight differences in its value. Hence, if we reject the series as a random variation from the frequency distribution determined from the sample, we must also reject it as a random variation from a theoretical frequency distribution differing by quantities of the order of the probable errors of the constants from the sample theoretical distribution. On the other hand, if we accept it as a random deviation from the sample theoretical distribution, we may accept it as a random variation from a system differing by quantities of the order of the probable errors of the constants from this distribution.

Thus I think we can conclude, when we are dealing with a sufficiently long series to give small probable errors to the constants of the series, that:—

(i) If χ_δ^2 be so small as to warrant us in speaking of the distribution as a random variation on the frequency distribution determined from itself, then we may also speak of it as a random sample from a general population whose theoretical distribution differs only by quantities of the order of the probable errors of the constants, from the distribution deduced from the observed sample.

(ii) If χ_δ^2 be so large as to make it impossible for us to regard the observed distribution as a sample from a general population following the law of distribution deduced from the sample itself, it will be impossible to consider it as a sample from any general population following a distribution differing only by quantities of the order of the probable errors of the sample distribution constants from that sample distribution.

In other words, if a curve is a good fit to a sample, to the same fineness of grouping it may be used to describe other samples from the same general population. If it is a bad fit, then this curve cannot serve to the same fineness of grouping to describe other samples from the same population.

We thus seem in a position to determine whether a given form of frequency curve will effectively describe the samples drawn from a given population to a certain degree of fineness of grouping.

If it serves to this degree, it will serve for all rougher groupings, *but it does not follow* that it will suffice for still finer groupings. Nor again does it appear to follow that if the number in the sample be largely increased the same curve will still be a good fit. Roughly the χ^2's of two samples appear to vary for the same grouping as their total contents. Hence if a curve be a good fit for a large sample it will be good for a small one, but the converse is not true, and a larger sample may show that our theoretical frequency gives only an approximate law for samples of a certain size. In practice we must attempt to obtain a good fitting frequency for such groupings as are customary or utile. To ascertain the ultimate law of distribution of a population for any groupings, however small, seems a counsel of perfection.

6. Frequency Known or Supposed Known *A Priori*

Illustration I

The following data are due to Professor W.F.R. Weldon, F.R.S., and give the observed frequency of dice with 5 or 6 points when a cast of twelve dice was made 26,306 times:

No. of Dice in Cast with 5 or 6 Points	Observed Frequency, m'	Theoretical Frequency, m	Deviation, e
0	185	203	− 18
1	1149	1217	− 68
2	3265	3345	− 80
3	5375	5576	−101
4	6114	6273	−159
5	5194	5018	+176
6	3067	2927	+140
7	1331	1254	+ 77
8	403	392	+ 11
9	105	87	+ 18
10	14	13	+ 1
11	4	1	+ 3
12	0	0	+ 0
	26306	26306	

The results show a bias from the theoretical results, 5 and 6 points occurring more frequently than they should do. Are the deviations such as to forbid us to suppose the results due to random selection? Is there in apparently true dice a real bias towards those faces with the maximum number of points appearing uppermost?

We have:

Group	e^2	e^2/m
0	324	1.59606
1	4624	3.79951
2	6400	1.91330
3	10201	1.82945
4	25281	4.03013
5	30976	6.17298
6	19600	6.69628
7	5929	4.72807
8	121	0.30903
9	324	3.72414
10	1	0.07346
11	9	9.00000
12	0	.00000
Total	...	43.87241

Hence

$$\chi^2 = 43.87241 \text{ and } \chi = 6.623,625.$$

As there are 13 groups we have to find P from the formula:

$$P = e^{\chi^2/2}\left(1 + \frac{\chi^2}{2} + \frac{\chi^4}{2\cdot4} + \frac{\chi^6}{2\cdot4\cdot6} + \frac{\chi^8}{2\cdot4\cdot6\cdot8} + \frac{\chi^{10}}{2\cdot4\cdot6\cdot8\cdot10}\right),$$

which leads us to

$$P = .000016,$$

or the odds are 62,499 to 1 against such a system of deviations on a random selection. With such odds it would be reasonable to conclude that dice exhibit bias towards the higher points.

Illustration II

If we take the total number of fives and sixes thrown in the 26,306 casts of 12 dice, we find them to be 106,602 instead of the theoretical 105,224. Thus $106{,}602/12 \times 26{,}306 = .3377$ nearly, instead of 1/3.

Professor Weldon has suggested to me that we ought to take $26{,}306 \times (.3377 + .6623)^{12}$ instead of the binomial $26{,}306\,(1/3 + 2/3)^{12}$ to represent the theoretical distribution, the difference between .3377 and 1/3 representing the bias of the dice. If this be done we find:

Group	m'	m	e	e^2/m
0	185	187	− 2	.021,3904
1	1149	1146	+ 3	.007,8534
2	3265	3215	+ 50	.777,6050
3	5475	5465	+ 10	.018,2983
4	6114	6269	−155	3.991,8645
5	5194	5115	+ 79	1.220,1342
6	3067	3043	+ 24	.189,2869
7	1331	1330	+ 1	.000,7519
8	403	424	− 21	1.040,0948
9	105	96	+ 9	.841,8094
10	14	15	− 1	.666,6667
11	4	1	+ 3	9
12	0	0	0	0

Hence:

$$\chi^2 = 17.775{,}7555.$$

This gives us by the first formula in (ii) of art. 4:

$$P = .1227;$$

or the odds are now only 8 to 1 against a system of deviations as improbable as or more improbable than this one. It may be said accordingly that the dice experiments of Professor Weldon are consistent with the chance of five or six points being thrown by a single die being .3377, but they are excessively improbable, if the chance of all the faces is alike and equal to 1/6th.

Illustration III

In the case of runs of colour in the throws of the roulette-ball at Monte Carlo, I have shown* that the odds are at least 1000 millions to one against such a fortnight of runs as occurred in July 1892 being a random result of a true roulette. I now give χ^2 for the data printed in the paper referred to, *i.e.*:

4274 Sets at Roulette

Runs	1	2	3	4	5	6	7	8	9	10	11	12	Over 12
Actual	2462	945	333	220	135	81	43	30	12	7	5	1	0
Theory	2137	1068	534	267	134	67	33	17	8	4	2	1	0

From this we find $\chi^2 = 17243$, and the improbability of a series as bad as or worse than this is about $14.5/10^{30}$! From this it will be more than ever evident how little chance had to do with the results of the Monte Carlo roulette in July 1892.

7. Frequency of General Population Not Known *A Priori*

(*Editors' note*: Illustrations IV and V have been omitted.)

Illustration VI

In the current text-books of the theory of errors it is customary to give various series of actual errors of observation, to compare them with theory by means of a table of distribution based on the normal curve, or graphically by means of a plotted frequency diagram, and on the basis of these comparisons to assert that an experimental foundation has been established for the normal law of errors. Now this procedure is of peculiar interest. The works referred

* 'The Chances of Death,' vol. i.: The Scientific-Aspect of Monte Carlo Roulette, p. 54.

to generally give elaborate analytical proofs that the normal law of errors is the law of nature—proofs in which there is often a difficulty (owing to the complexity of the analysis and the nature of the approximations made) in seeing exactly what assumptions have been really made. The authors usually feel uneasy about this process of deducing a law of nature from Taylor's Theorem and a few more or less ill-defined assumptions; and having deduced the normal curve of errors, they give as a rule some meagre data of how it fits actual observation. But the comparison of observation and theory in general amounts to a remark—based on no quantitative criterion—of how well theory and practice really do fit! Perhaps the greatest defaulter in this respect is the late Sir George Biddell Airy in his text-book on the 'Theory of Errors of Observation.' In an Appendix he gives what he terms a "Practical Verification of the Theoretical Law for thc Frequency of Errors."

Now that Appendix really tells us *absolutely nothing* as to the goodness of fit of his 636 observations of the N.P.D. of Polaris to a normal curve. For, if we first take on faith what he says, namely, that positive and negative errors may be clubbed together, we still find that he has *thrice smoothed* his observation frequency distribution before he allows us to examine it. It is accordingly impossible to say whether it really does or does not represent a random set of deviations from a normal frequency curve. All we can deal with is the table he gives of observed and theoretical errors and his diagram of the two curves. These, of course, are not his proper data at all: it is impossible to estimate how far his three smoothings counter-balance or not his multiplication of errors by eight. But as I understand Sir George Airy, he would have considered such a system of errors as he gives on his p. 117 or in his diagram on p. 118 to be sufficiently represented by a normal curve. Now I have investigated his 37 groups of errors, observational and theoretical. In order to avoid so many different groups, I have tabulated his groups in .10″ units, and so reduced them to 21. From these 21 groups I have found χ^2 by the method of this paper. By this reduction of groups I have given Sir George Airy's curve even a better chance than it has, as it stands. Yet what do we find? Why,

$$\chi^2 = 36.2872.$$

Or, using the approximate equation,

$$P = .01423.$$

That is to say, only in one occasion out of 71 repetitions of such a set of observations on Polaris could we have expected to find a system of errors deviating as widely as this set (or more widely than this set) from the normal distribution. Yet Sir George Airy takes a set of observations, the odds against which being a random variation from the normal distribution are 70 to 1, to prove to us that the normal distribution applies to errors of observation. Nay, further, he cites this very improbable result as an experimental confirmation of the whole theory! "It is evident," he writes, "that the formula represents with all practicable accuracy the observed Frequency of Errors, upon which all the applications of the Theory of Probabilities are founded: and the validity of every investigation in this Treatise is thereby established."

Such a passage demonstrates how healthy is the spirit of scepticism in all inquiries concerning the accordance of theory and nature.

Illustration VII

It is desirable to illustrate such results a second time. Professor Merriman in his treatise on Least Squares* starts in the right manner, not with theory, but with actual experience, and then from his data deduces three axioms. From these axioms he obtains by analysis the normal curve as the theoretical result. But if these axioms be true, his data can only differ from the normal law of frequency by a system of deviations such as would reasonably arise if a random selection were made from material actually obeying the normal law. Now Professor Merriman puts in the place of honour 1000 shots fired at a line on a target in practice for the U.S. Government, the deviations being grouped according to the belts struck, the belts were drawn on the target of equal breadth and parallel to the line. The following table gives the distribution of hits and the theoretical frequency-distribution calculated from tables of the area of the normal curve†.

Belt	Observed Frequency	Normal Distribution	e	$\frac{e^2}{m}$
1	1	1	0	0
2	4	6	-2	.667
3	10	27	-17	10.704
4	89	67	$+22$	7.224
5	190	162	$+28$	4.839
6	212	242	-30	3.719
7	204	240	-36	5.400
8	193	157	$+36$	8.255
9	79	70	$+9$	1.157
10	16	26	-10	3.846
11	2	2	0	0
	1000	1000		$\chi^2 = 45.811$

Hence we deduce:

$$P = .000{,}00155.$$

* 'A Textbook on the Method of Least Squares,' 1891. p. 14.

† I owe the work of this illustration to the kindness of Mr. W. R. Macdonell, M.A., L.L.D.

Table of Values of P for Values of χ^2 and n'; χ^2 from 1 to 70, n' from 3 to 20*

χ^2 \ n'	3.	4.	5.	6.	7.	8.	9.	10.	11.
1	.606,531	.801,253	.909,796	.962,566	.985,612	.994,829	.998,249	.999,438	.999,828
2	.367,879	.572,407	.735,759	.849,146	.919,699	.959,839	.981,012	.991,446	.996,340
3	.223,130	.391,633	.557,825	.699,994	.808,847	.885,010	.934,357	.964,303	.981,424
4	.135,335	.261,470	.406,006	.549,422	.676,676	.779,783	.857,123	.911,418	.947,347
5	.082,085	.171,799	.287,298	.415,882	.543,813	.659,965	.757,576	.834,310	.891,178
6	.049,787	.111,611	.199,148	.306,220	.423,190	.539,750	.647,232	.739,919	.815,263
7	.303,197	.071,888	.135,888	.220,631	.320,847	.428,870	.536,632	.637,110	.725,544
8	.018,316	.046,012	.091,578	.156,236	.238,103	.332,594	.433,470	.534,146	.628,837
9	.011,109	.029,291	.061,099	.109,064	.173,578	.252,656	.342,296	.437,274	.532,104
10	.006,738	.018,567	.040,428	.075,236	.124,652	.188,574	.265,026	.350,486	.440,493
15	.000,553	.001,817	.004,701	.010,363	.020,256	.036,000	.059,145	.090,810	.132,061
20	.000,045	.000,170	.000,499	.001,250	.002,769	.005,570	.010,336	.017,913	.029,253
25	.000,004	.000,016	.000,050	.000,139	.000,341	.000,759	.001,554	.002,971	.005,345
30	.000,000	.000,001	.000,005	.000,015	.000,039	.000,095	.000,211	.000,439	.000,857
40	.000,000	.000,000	.000,000	.000,000	.000,000	.000,001	.000,003	.000,008	.000,017
50	.000,000	.000,000	.000,000	.000,000	.000,000	.000,000	.000,000	.000,000	.000,000
60	.000,000	.000,000	.000,000	.000,000	.000,000	.000,000	.000,000	.000,000	.000,000
70	.000,000	.000,000	.000,000	.000,000	.000,000	.000,000	.000,000	.000,000	.000,000

* I have to thank Miss Alice Lee, D.Sc., for help in the calculation of part of this table. The certainty, *i.e.* the 1. in columns 16 to 20, denotes, of course, something greater than .999,9995, *i.e.* unity to six figures.

* *Editor's note*: This table has been reproduced for n' from 3 to 11 only.

In other words, if shots are distributed on a target according to the normal law, then such a distribution as that cited by Mr. Merriman could only be expected to occur, on an average, some 15 or 16 times in 10,000,000 trials. Now surely it is very unfortunate to cite such an illustration as the foundation of those axioms from which the normal curve must flow! For if the normal curve flows from the axioms, then the data ought to be a probable system of deviations from the normal curve. But this they certainly are not. Now it appears to me that, if the earlier writers on probability had not proceeded so entirely from the mathematical standpoint, but had endeavoured first to classify experience in deviations from the average, and then to obtain some measure of the actual goodness of fit provided by the normal curve, that curve would never have obtained its present position in the theory of errors. Even today there are those who regard it as a sort of fetish; and while admitting it to be at fault as a means of generally describing the distribution of variation of a quantity x from its mean, assert that there must be some unknown quantity z of which x is an unknown function, and that z really obeys the normal law! This might be reasonable if there were but few exceptions to this universal law of error; but the difficulty is to find even the few variables which obey it, and these few are not those usually cited as illustrations by the writers on the subject!

Illustration VIII

The reader may ask: Is it not possible to find material which obeys within probable limits the normal law? I reply, yes; but this law is not a universal law of nature. We must hunt for cases. Out of three series of personal equations, I could only find one which approximated to the normal law. I took 500 lengths and bisected them with my pencil at sight. Without entering at length into experiments, destined for publication on another occasion, I merely give the observed and normal distribution of my own errors in 20 groups.

Group	Observation	Theory	Group	Observation	Theory
1	1	2.3	11	53	57.0
2	3	3.4	12	50.5	47.1
3	11	6.9	13	28.5	34.0
4	14.5	13.1	14	27	22.7
5	21.5	22.2	15	13.5	13.5
6	30	33.6	16	7.5	7.0
7	47	47.5	17	0	3.5
8	51.5	57.8	18	1	1.6
9	72	63.2	19	0	.6
10	65.5	62.7	20	2	.3

Calculating χ^2 in the manner already sufficiently indicated in this paper, we find

$$\chi^2 = 22.0422.$$

We must now use the more complex integral formula for P, and we find

$$P = .2817.$$

Or, in every three to four random selections, we should expect one with a system of deviations from the normal curve greater than that actually observed.

I think, then, we may conclude that my errors of judgment in bisecting straight lines may be fairly represented by a normal distribution. It is noteworthy, however, that I found other observers' errors in judgment of the same series of lines were distinctly skew.

(8) We can only conclude from the investigations here considered that the normal curve possesses no special fitness for describing errors or deviations such as arise either in observing practice or in nature. We want a more general theoretical frequency, and the fitness of any such to describe a given series can be investigated by aid of the criterion discussed in this paper. For the general appreciation of the probability of the occurrence of a system of deviations defined by χ^2 (or any greater value), the accompanying table has been calculated, which will serve to give that probability closely enough for many practical judgments, without the calculations required by using the formulae of art. 4.

Introduction to Student (1908) The Probable Error of a Mean

E.L. Lehmann
University of California at Berkeley

Testing a hypothesis about the mean ξ of a population on the basis of a sample $X_1, \ldots, X_n$ from that population was treated throughout the 19th century by a large-sample approach that goes back to Laplace. If the sample mean $\bar{X}$ is considered to be the natural estimate of ξ, the hypothesis $H: \xi = \xi_0$ should be rejected when $\bar{X}$ differs sufficiently from ξ_0. Furthermore, since for large n the distribution of $\sqrt{n}(\bar{X} - \xi_0)/\sigma$ is approximately standard normal under H (where $\sigma^2 < \infty$ is the variance of the X's), this suggests rejecting H when

$$\frac{\sqrt{n}|\bar{X} - \xi_0|}{\sigma} \tag{1}$$

exceeds the appropriate critical value calculated from that distribution.

Unfortunately, σ is usually unknown. However, if it is replaced by a consistent estimator, for example, by $\sqrt{S^2/(n-1)}$ where

$$S^2 = \sum (X_i - \bar{X})^2, \tag{2}$$

the limit distribution, as $n \to \infty$, of

$$|t| = \sqrt{n}|X - \xi_0| : \sqrt{\frac{S^2}{(n-1)}} \tag{3}$$

is the same as that of (1). Thus, (3) provides a test of H that for large n has about the right level.

This is essentially the approach described in the fourth paragraph of Student's paper* as "the usual method of determining that the mean of the population lies within a given distance of the mean of the sample."

* Student actually defines t with the denominator n instead of $n - 1$.

A crucial feature of the above approach is to act as if $S^2/(n-1)$ were a constant (the true value of σ^2) rather than a random variable. However, if the sample size n is small, S^2 will be subject to considerable variation. It was the effect of this variation that concerned Student, the pseudonym of W.S. Gosset, a chemist working for Guiness breweries.* He pointed out that if the form of the distribution of the X's is known, this variation can be taken into account since for any given n, the distribution of t is then determined exactly. He proposed to work out this distribution for the case in which the X's are normal.

Deriving the distribution of t was somewhat beyond Gosset's mathematical competence. He knew that the numerator of t was normal; to obtain the distribution of the denominator,† he derived the first four moments about 0 of S^2 and used these moments to fit a Pearson curve.‡ After recognizing that the higher moments of this distribution also agree with those of S^2, he came to the conclusion that "it is probable that the curve found represents the theoretical distribution of S^2, so that although we have no actual proof we shall assume it to do so in what follows."

Student next showed that the numerator and denominator are uncorrelated, believing this to mean that they were independent.§ On this basis, he then calculated the distribution of t. Of this derivation, Fisher (1915) writes, "This result, although derived by empirical methods, was established [by Student] almost beyond reasonable doubt." He goes on to say that "the form establishes itself instantly" by use of the geometric approach (which Fisher presents in this 1915 paper.)

To try out his distribution, Student performed a Monte Carlo study using real data (Sec. VI). He also provided a small table for sample sizes $n = 4(1)10$. In 1917, he extended the table and in 1925, refined it further using a method suggested by Fisher (1925). Later, much more detailed and complete tables were obtained by others; today the distribution is routinely available in statistical packages.

The importance of Student's new approach was not immediately realized by his contemporaries. Not too much blame should be attached to them for this lack of perspicuity since Student himself does not seem to have been aware of the significance of his contribution. As Fisher (1939) stated in his obituary of Student;

* For a brief account of Student's life and work, with references to earlier accounts, see C. Read, Gosset, William Sealy, *Encycl. Statis. Sci., Vol 3* (1983).

† This distribution was, in fact, also known; for a discussion of its history, see Heyde and Seneta (1977, p. 56).

‡ It has been pointed out to me by Professor Mosteller that Student might not have obtained this result if he had not lived in Great Britain, since it was only there that the system of Pearson curves was considered the natural family within which to fit.

§ Even the mathematically much more sophisticated Karl Pearson shared this belief as late as 1925; see Reid (1982, p.57).

> It is doubtful if "Student" ever realized the full importance of his contribution to the Theory of Errors. From correspondence with him before the War... I should form a confident judgment that at that time certainly he did not see how big a thing he had done... Probably he felt that, had the problem really been so important as it had once seemed, the leading authorities in English statistics would at least have given him the encouragement of recommending the use of his method; and better still, would have sought to gain similar advantages in more complex problems. Five years, however, passed without the writers in *Biometrika*, the journal in which he had published, showing any sign of appreciating the significance of his work.* This weighty apathy must greatly have chilled his enthusiasm.... It was sixteen years before, in 1924, the system of tests of which Student's was the prototype was logically complete. Only during the thirteen years which have since passed has "Student's" work found its proper place as an experiment resource.

It was, in fact, Fisher himself who in 1915, at age 25, put Student's test on a firm foundation when by means of his geometric method, in a few lines he obtained the joint distribution of the numerator and denominator of t and used the same approach to derive the distribution of the correlation coefficient. It was also Fisher who in 1922 pointed out that the t-distribution also applies in the two-sample case and to regression coefficients. It was finally Fisher who in 1923 and 1924 propagandized these tests, and followed this in 1925 with a survey paper "Applications of Student's distribution," in which he gives a unified exposition of this work. In the same year, there also appeared the first edition of Fisher's *Statistical Methods for Research Workers*, whose third chapter is given over nearly entirely to these tests and to illustrations of their application with many examples.

This work of Fisher's was a great achievement. With the inspiration only of Student's paper, to which he always gives full credit and which he often describes as revolutionary, Fisher single-handedly transformed both the theory and practice of hypothesis testing and made of it a powerful tool of nearly universal usefulness.

In the wake of Fisher's work, Student's t-test has had a fundamental impact on statistics in three different ways.

1. The t-test, if one includes its extension to two samples and to regression coefficients, has probably been used more widely (although not necessarily always wisely) than any other statistical test. It is applied extensively in all fields in which continuous variables are observed. Its usefulness has been greatly increased by the derivation and tabulation of the power of the test and development of the permutation t-test that extends its applicability (as an approximate test) to much more general circumstances.†
2. Student's paper proposed a new way of looking at statistics: a small-

* E.S. Pearson (1966, p. 6) writes about Karl Pearson, editor of *Biometrika*, concerning the shift from a large- to a small-sample point of view that "it was a shift which I think K.P. was never able or saw no need to make."

† Both these extensions were initiated by Fisher.

sample approach in which a specific family of models is assumed for the data and an exact inference procedure is developed within this framework.

3. Starting with Student's own paper, questions were raised regarding the effect of violation of the model assumptions. The resulting investigations of these effects on the t-test have spearheaded and motivated the subject of robustness that has played a central role in statistical practice and theory during the last 50 years.

The history of the t-test described in this introduction illustrates a surprisingly common pattern of discovery. A first scientist, dealing with a very specific situation, introduces a new idea either in the formulation of the problem or its solution. This idea strikes a spark in the imagination of a second scientist who sees possibilities of generalization and a multitude of applications far beyond those envisaged by the original discoverer. There is no point arguing which of the two deserves more credit. It took both Gosset and Fisher to make the t-test into the tool it has become and to begin the development of small-sample theory. But what remains true without question is that Student's paper is one of the seminal contributions to 20th century statistics.

References

Fisher, R.A. (1915). Frequency distribution of the values of the correlation coefficient in samples from an indefinitely large population, *Biometrika*, **10**, 507–521.

Fisher, R.A. (1922). The goodness of fit of regression formulae and the distribution of regression coefficients, *J.R. Statist. Soc.*, **85**, 597–612.

Fisher, R.A. (1923). Note on Dr. Burnside's recent paper on errors of observation, *Proc. Cambridge Philos. Soc.*, **21**, 655–658.

Fisher, R.A. (1924). On a distribution yielding the error functions of several well known statistics, *Proc. Int. Congress Math., Toronto*, **2**, 805–813.

Fisher, R.A. (1925a). Application of "Student's" distribution, *Metron*, **5**, 90–104.

Fisher, R.A. (1925b). Expansion of "Student's" integral in powers of n^{-1}, *Metron*, **5**, 109–120.

Fisher, R.A. (1925c). *Statistical Methods for Research Workers*. Oliver and Boyd, Edinburgh.

Fisher, R.A. (1939). "Student," *Ann. Eugen.*, **9**, 1–9.

Heyde, C.C., and Seneta, E. (1977). *I.J. Bienaymé* (*Statistical Theory Anticipated*). Springer-Verlag, New York.

Pearson, E.S. (1966). The Neyman-Pearson story, in *Research Papers in Statistics* (*Festschrift for J. Neyman*) (F.N. David, ed.). Wiley, London.

Reid, C. (1982). *Neyman–From Life*. Springer-Verlag, New York.

Student (1917). Tables for estimating the probability that the mean of a unique sample of observations lies beween $-\infty$ and any given distance of the mean of the population from which the sample is drawn, *Biometrika*, **11**, 414–417.

Student (1925). New tables for testing the significance of observations, *Metron*, **5**, 105–108.

The Probable Error of a Mean

Student

Introduction

Any experiment may be regarded as forming an individual of a "population" of experiments which might be performed under the same conditions. A series of experiments is a sample drawn from this population.

Now any series of experiments is only of value in so far as it enables us to form a judgment as to the statistical constants of the population to which the experiments belong. In a great number of cases the question finally turns on the value of a mean, either directly, or as the mean difference between the two quantities.

If the number of experiments be very large, we may have precise information as to the value of the mean, but if our sample be small, we have two sources of uncertainty:—(1) owing to the "error of random sampling" the mean of our series of experiments deviates more or less widely from the mean of the population, and (2) the sample is not sufficiently large to determine what is the law of distribution of individuals. It is usual, however, to assume a normal distribution, because, in a very large number of cases, this gives an approximation so close that a small sample will give no real information as to the manner in which the population deviates from normality: since some law of distribution must be assumed it is better to work with a curve whose area and ordinates are tabled, and whose properties are well known. This assumption is accordingly made in the present paper, so that its conclusions are not strictly applicable to populations known not to be normally distributed; yet it appears probable that the deviation from normality must be very extreme to lead to serious error. We are concerned here solely with the first of these two sources of uncertainty.

The usual method of determining the probability that the mean of the population lies within a given distance of the mean of the sample, is to assume a normal distribution about the mean of the sample with a standard deviation equal to $s/\sqrt{n}$, where s is the standard deviation of the sample, and to use the tables of the probability integral.

But, as we decrease the number of experiments, the value of the standard deviation found from the sample of experiments becomes itself subject to an increasing error, until judgments reached in this way may become altogether misleading.

In routine work there are two ways of dealing with this difficulty: (1) an experiment may be repeated many times, until such a long series is obtained that the standard deviation is determined once and for all with sufficient accuracy. This value can then be used for subsequent shorter series of similar experiments. (2) Where experiments are done in duplicate in the natural course of the work, the mean square of the difference between corresponding pairs is equal to the standard deviation of the population multiplied by $\sqrt{2}$. We can thus combine together several series of experiments for the purpose of determining the standard deviation. Owing however to secular change, the value obtained is nearly always too low, successive experiments being positively correlated.

There are other experiments, however, which cannot easily be repeated very often; in such cases it is sometimes necessary to judge of the certainty of the results from a very small sample, which itself affords the only indication of the variability. Some chemical, many biological, and most agricultural and large scale experiments belong to this class, which has hitherto been almost outside the range of statistical enquiry.

Again, although it is well known that the method of using the normal curve is only trustworthy when the sample is "large," no one has yet told us very clearly where the limit between "large" and "small" samples is to be drawn.

The aim of the present paper is to determine the point at which we may use the tables of the probability integral in judging of the significance of the mean of a series of experiments, and to furnish alternative tables for use when the number of experiments is too few.

The paper is divided into the following nine sections:

I. The equation is determined of the curve which represents the frequency distribution of standard deviations of samples drawn from a normal population.

II. There is shown to be no kind of correlation between the mean and the standard deviation of such a sample.

III. The equation is determined of the curve representing the frequency distribution of a quantity z, which is obtained by dividing the distance between the mean of a sample and the mean of the population by the standard deviation of the sample.

IV. The curve found in I. is discussed.

V. The curve found in III. is discussed.

VI. The two curves are compared with some actual distributions.
VII. Tables of the curves found in III. are given for samples of different size.
VIII and IX. The tables are explained and some instances are given of their use.
X. Conclusions.

Section I

Sample of n individuals are drawn out of a population distributed normally, to find an equation which shall represent the frequency of the standard deviations of these samples.

If s be the standard deviation found from a sample $x_1, x_2 \ldots x_n$ (all these being measured from the mean of the population), then

$$s^2 = \frac{S(x_1^2)}{n} - \left(\frac{S(x_1)}{n}\right)^2 = \frac{S(x_1^2)}{n} - \frac{S(x_1^2)}{n^2} - \frac{2S(x_1x_2)}{n^2}.$$

Summing for all samples and dividing by the number of samples we get the mean value of s^2 which we will write $\overline{s}^2$,

$$\overline{s}^2 = \frac{n\mu_2}{n} - \frac{n\mu_2}{n^2} = \frac{\mu_2(n-1)}{n},$$

where μ_2 is the second moment coefficient in the original normal distribution of x: since x_1, x_2, etc., not correlated and the distribution is normal, products involving odd powers of x_1 vanish on summing, so that $\frac{2S(x_1x_2)}{n^2}$ is equal to 0.

If M_R' represent the Rth moment coefficient of the distribution of s^2 about the end of the range where $s^2 = 0$,

$$M_1' = \mu_2\frac{(n-1)}{n}.$$

Again

$$\begin{aligned} s^4 &= \left\{\frac{S(x_1^2)}{n} - \left(\frac{S(x_1)}{n}\right)^2\right\}^2 \\ &= \left(\frac{S(x_1^2)}{n}\right)^2 - \frac{2S(x_1^2)}{n}\left(\frac{S(x_1)}{n}\right)^2 + \left(\frac{S(x_1)}{n}\right)^4 \\ &= \frac{S(x_1^4)}{n^2} + \frac{2S(x_1^2x_2^2)}{n^2} - \frac{2S(x_1^4)}{n^3} - \frac{4S(x_1^2x_2^2)}{n^3} + \frac{S(x_1^4)}{n^4} \\ &\quad + \frac{6S(x_1^2x_2^2)}{n^4} + \text{other terms involving odd powers of } x_1 \text{ etc., which will vanish on summation.} \end{aligned}$$

Now $S(x_1^4)$ has n terms but $S(x_1^2 x_2^2)$ has $\frac{1}{2}n(n-1)$, hence summing for all samples and dividing by the number of samples we get

$$M_2' = \frac{\mu_4}{n} + \mu_2^2\frac{(n-1)}{n} - \frac{2\mu_4}{n^2} - 2\mu_2^2\frac{(n-1)}{n^3} + \frac{\mu_4}{n^3} + 3\mu_2^2\frac{(n-1)}{n^3}$$

$$= \frac{\mu_4}{n}\{n^2 - 2n - 1\} + \frac{\mu_2^2}{n^3}(n-1)\{n^2 - 2n + 3\}.$$

Now since the distribution of x is normal, $\mu_4 = 3\mu_2^2$, hence

$$M_2' = \mu_2^2\frac{(n-1)}{n^3}\{3n - 3 + n^2 - 2n + 3\} = \mu_2^2\frac{(n-1)(n+1)}{n^2}.$$

In a similar tedious way I find:

$$M_3' = \mu_2^3\frac{(n-1)(n+1)(n+3)}{n^3},$$

and

$$M_4' = \mu_2^4\frac{(n-1)(n+1)(n+3)(n+5)}{n^4}.$$

The law of formation of these moment coefficients appears to be a simple one, but I have not seen my way to a general proof.

If now M_R be the Rth moment coefficient of s^2 about its mean, we have

$$M_2 = \mu_2^2\frac{(n-1)}{n^2}\{(n+1) - (n-1)\} = 2\mu_2^2\frac{(n-1)}{n^2},$$

$$M_3 = \mu_2^3\left\{\frac{(n-1)(n+1)(n+3)}{n^3} - \frac{3(n-1)}{n}\cdot\frac{2(n-1)}{n^2} - \frac{(n-1)^3}{n^3}\right\}$$

$$= \mu_2^3\frac{(n-1)}{n^3}\{n^2 + 4n + 3 - 6n + 6 - n^2 + 2n - 1\} = 8\mu_2^3\frac{(n-1)}{n^3},$$

$$M_4 = \frac{\mu_2^4}{n^4}\{(n-1)(n+1)(n+3)(n+5) - 32(n-1)^2 - 12(n-1)^3 - (n-1)^4\}$$

$$= \frac{\mu_2^4(n-1)}{n^4}\{n^3 + 9n^2 + 23n^n + 15 - 32n + 32 - 12n^2 + 24n - 12 - n^3 + 3n^2 - 3n + 1\}$$

$$= \frac{12\mu_2^4(n-1)(n+3)}{n^4}.$$

Hence

$$\beta_1 = \frac{M_3^2}{M_2^3} = \frac{8}{n-1}, \qquad \beta_2 = \frac{M_4}{M_2^2} = \frac{3(n+3)}{n-1},$$

$$\therefore\ 2\beta_2 - 3\beta_1 - 6 = \frac{1}{n-1}\{6(n+3) - 24 - 6(n-1)\} = 0.$$

Consequently a curve of Professor Pearson's type III. may be expected to fit the distribution of s^2.

The equation referred to an origin at the zero end of the curve will be

$$y = Cx^p e^{-\gamma x},$$

where

$$\gamma = 2\frac{M_2}{M_3} = \frac{4\mu_2^2(n-1)n^3}{8n^2\mu_2^3(n-1)} = \frac{n}{2\mu_2},$$

and

$$p = \frac{4}{\beta_1} - 1 = \frac{n-1}{2} - 1 = \frac{n-3}{2}.$$

Consequently the equation becomes

$$y = Cx^{(n-3)/2}e^{-nx/2\mu_2},$$

which will give the distribution of s^2.

The area of this curve is $C\int_0^\infty x^{(n-3)/2}e^{-nx/2\mu_2}\,dx = I$ (say).

The first moment coefficient about the end of the range will therefore be

$$\frac{C\int_0^\infty x^{(n-1)/2}e^{-nx/2\mu_2}\,dx}{I} = \frac{C\left[\dfrac{-2\mu_2}{n}x^{(n-1)/2}e^{-nx/2\mu_2}\right]_{x=0}^{x=\infty}}{I} + \frac{C\int_0^\infty \dfrac{n-1}{n}\mu_2 x^{(n-3)/2}e^{-nx/2\mu_2}\,dx}{I}.$$

The first part vanishes at each limit and the second is equal to

$$\frac{\dfrac{n-1}{n}\mu_2 I}{I} = \frac{n-1}{n}\mu_2,$$

and we see that the higher moment coefficients will bc formed by multiplying successively by $\frac{n+1}{n}\mu_2$, $\frac{n+3}{n}\mu_2$, etc., just as appeared to be the law of formation of M_2', M_3', M_4', etc.

Hence it is probable that the curve found represents the theoretical distribution of s^2; so that although we have no actual proof we shall assume it to do so in what follows.

The distribution of s may be found from this, since the frequency of s is equal to that of s^2 and all that we must do is to compress the base line suitably.

Now if

$$y_1 = \phi(s^2) \text{ be the frequency curve of } s^2$$

and

$$y_2 = \psi(s) \text{ be the frequency curve of } s,$$

then

$$y_1\, d(s^2) = y_2\, ds,$$

or

$$y_2\, ds = 2y_1 s\, ds,$$

$$\therefore\ y_2 = 2sy_1.$$

Hence

$$y_2 = 2Cs(s^2)^{(n-3)/2} e^{-ns^2/2\mu_2}$$

is the distribution of s.

This reduces to

$$y_2 = 2Cs^{n-2} e^{-ns^2/2\mu_2}.$$

Hence $y = Ax^{n-2}e^{-nx^2/2\sigma^2}$ will give the frequency distribution of standard deviations of samples of n, taken out of a population distributed normally with standard deviation σ. The constant A may be found by equating the area of the curve as follows:

$$\text{Area} = A \int_0^\infty x^{n-2} e^{-nx^2/2\sigma^2}\, dx. \qquad \left(\text{Let } I_p \text{ represent } \int_0^\infty x^p e^{-nx^2/2\sigma^2}\, dx.\right)$$

Then

$$I_p = \frac{\sigma^2}{n} \int_0^\infty x^{p-1} \frac{d}{dx}(-e^{-nx^2/2\sigma^2})\, dx$$

$$= \frac{\sigma^2}{n}[-x^{p-1}e^{-nx^2/2\sigma^2}]_{x=0}^{x=\infty} + \frac{\sigma^2}{n}(p-1)\int_0^\infty x^{p-2}e^{-nx^2/2\sigma^2}\, dx$$

$$= \frac{\sigma^2}{n}(p-1)I_{p-2},$$

since the first part vanishes at both limits.

By continuing this process we find

$$I_{n-2} = \left(\frac{\sigma^2}{n}\right)^{(n-2)/2} (n-3)(n-5)\ldots 3.1 I_0$$

or

$$= \left(\frac{\sigma^2}{n}\right)^{(n-3)/2} (n-3)(n-5)\ldots 4.2 I_1$$

according as n is even or odd.

But I_0 is

$$\int_0^\infty e^{-nx^2/2\sigma^2}\,dx = \sqrt{\frac{\pi}{2n}}\sigma,$$

and I_1 is

$$\int_0^\infty xe^{-nx^2/2\sigma^2}\,dx = \left[-\frac{\sigma^2}{n}e^{-nx^2/2\sigma^2}\right]_{x=0}^{x=\infty} = \frac{\sigma^2}{n}.$$

Hence if n be even,

$$A = \frac{\text{Area}}{(n-3)(n-5)\ldots 3.1\sqrt{\frac{\pi}{2}}\left(\frac{\sigma^2}{n}\right)^{(n-1)/2}},$$

and if n be odd

$$A = \frac{\text{Area}}{(n-3)(n-5)\ldots 4.2\left(\frac{\sigma^2}{n}\right)^{(n-1)/2}}.$$

Hence the equation may be written

$$y = \frac{N}{(n-3)(n-5)\ldots 3.1}\sqrt{\frac{2}{\pi}}\left(\frac{n}{\sigma^2}\right)^{(n-1)/2} x^{n-2}e^{-nx^2/2\sigma^2} \quad (n \text{ even})$$

or

$$y = \frac{N}{(n-3)(n-5)\ldots 4.2}\left(\frac{n}{\sigma^2}\right)^{(n-1)/2} x^{n-2}e^{-nx^2/2\sigma^2} \quad (n \text{ odd})$$

where N as usual represents the total frequency.

Section II

To show that there is no correlation between (*a*) the distance of the mean of a sample from the mean of the population and (*b*) the standard deviation of a sample with normal distribution.

(1) Clearly positive and negative positions of the mean of the sample are equally likely, and hence there cannot be correlation between the absolute value of the distance of the mean from the mean of the population and the

standard deviation, but (2) there might be correlation between the square of the distance and the square of the standard deviation.

Let

$$u^2 = \left(\frac{S(x_1)}{n}\right)^2 \text{ and } s^2 = \frac{S(x_1^2)}{n} - \left(\frac{S(x_1)}{n}\right)^2.$$

Then if m_1', M_1' be the mean values of u^2 and s^2, we have by the preceding part $M_1' = \mu_2\frac{(n-1)}{n}$ and $m_1' = \frac{\mu_2}{n}$.

$$\text{Now } u^2s^2 = \frac{S(x_1^2)}{n}\left(\frac{S(x_1)}{n}\right)^2 - \left(\frac{S(x_1)}{n}\right)^4$$

$$= \left(\frac{S(x_1^2)}{n}\right)^2 + 2\frac{S(x_1x_2)\cdot S(x_1^2)}{n^3} - \frac{S(x_1^4)}{n^4}$$

$$-\frac{6x_1^2x_2^2}{n^4} - \text{other terms of odd order which will vanish on summation.}$$

Summing for all values and dividing by the number of cases we get

$$R_{u^2s^2}\sigma_{u^2}\sigma_{s^2} + m_1M_1 = \frac{\mu_4}{n^2} + \mu_2^2\frac{(n-1)}{n^2} - \frac{\mu_4}{n^3} - 3\mu_2^2\frac{(n-1)}{n^3},$$

where $R_{u^2s^2}$ is the correlation between u^2 and s^2.

$$R_{u^2s^2}\sigma_{u^2}\sigma_{s^2} + \mu_2^2\frac{(n-1)}{n^2} = \mu_2^2\frac{(n-1)}{n^3}\{3+n-3\} = \mu_2^2\frac{(n-1)}{n^2}.$$

Hence $R_{u^2s^2}\sigma_{u^2}\sigma_{s^2} = 0$ or there is no correlation between u^2 and s^2.

Section III

To find the equation representing the frequency distribution of the means of samples of n drawn from a normal population, the mean being expressed in terms of the standard deviation of the sample.

We have $y = \frac{C}{\sigma^{n-1}}s^{n-2}e^{-ns^2/2\sigma^2}$ as the equation representing the distribution of s, the standard deviation of a sample of n, when the samples are drawn from a normal population with standard deviation σ.

Now the means of these samples of n are distributed according to the equation

$$y = \frac{\sqrt{n}\,N}{\sqrt{2\pi}\,\sigma}e^{-nx^2/2\sigma^2}*$$

* Airy, *Theory of Errors of Observations*, Part II. §6.

and we have shown that there is no correlation between x, the distance of the mean of the sample, and s, the standard deviation of the sample.

Now let us suppose x measured in terms of s, *i.e.* let us find the distribution of $z = \frac{x}{s}$.

If we have $y_1 = \phi(x)$ and $y_2 = \psi(z)$ as the equations representing the frequency of x and of z respectively, then

$$y_1\,dx = y_2\,dz = y_2\frac{dx}{s},$$

$$\therefore y_2 = sy_1.$$

Hence

$$y = \frac{N\sqrt{ns}}{\sqrt{2\pi}\,\sigma}e^{-ns^2/2\sigma^2}$$

is the equation representing the distribution of z for samples of n with standard deviation s.

Now the chance that s lies between s and $s + ds$ is:

$$\frac{\displaystyle\int_s^{s+ds}\frac{C}{\sigma^{n-1}}s^{n-2}e^{-ns^2/2\sigma^2}\,ds}{\displaystyle\int_0^{\infty}\frac{C}{\sigma^{n-1}}s^{n-2}e^{-ns^2/2\sigma^2}\,ds},$$

which represents the N in the above equation.

Hence the distribution of z due to values of s which lie between s and $s + ds$ is

$$y = \frac{\displaystyle\int_s^{s+ds}\frac{C}{\sigma^n}\sqrt{\frac{n}{2\pi}}s^{n-1}e^{-ns^2(1+z^2)/2\sigma^2}\,ds}{\displaystyle\int_0^{\infty}\frac{C}{\sigma^{n-1}}s^{n-2}e^{-ns^2/2\sigma^2}\,ds} = \frac{\displaystyle\sqrt{\frac{n}{2\pi}}\int_s^{s+ds}s^{n-1}e^{-ns^2(1+z^2)/2\sigma^2}\,ds}{\displaystyle\sigma\int_0^{\infty}s^{n-2}e^{-ns^2/2\sigma^2}\,ds},$$

and summing for all values of s we have as an equation giving the distribution of z

$$y = \frac{\sqrt{\dfrac{n}{2\pi}}}{\sigma}\frac{\displaystyle\int_0^{\infty}s^{n-1}e^{-ns^2(1+z^2)/2\sigma^2}\,ds}{\displaystyle\int_0^{\infty}s^{n-2}e^{-ns^2/2\sigma^2}\,ds}.$$

By what we have already proved this reduces to

$$y = \frac{1}{2}\frac{n-2}{n-3}\cdot\frac{n-4}{n-5}\cdots\frac{5}{4}\cdot\frac{3}{2}(1+z^2)^{-n/2} \quad \text{if } n \text{ be odd,}$$

and to

$$y = \frac{1}{\pi}\frac{n-2}{n-3}\cdot\frac{n-4}{n-5}\cdots\frac{4}{3}\cdot\frac{2}{1}(1+z^2)^{-n/2} \quad \text{if } n \text{ be even.}$$

Since this equation is independent of σ it will give the distribution of the distance of the mean of a sample from the mean of the population expressed in terms of the standard deviation of the sample for any normal population.

Section IV

Some Properties of the Standard Deviation Frequency Curve

By a similar method to that adopted for finding the constant we may find the mean and moments: thus the mean is at $\frac{I_{n-1}}{I_{n-2}}$, which is equal to

$$\frac{(n-2)}{(n-3)}\frac{(n-4)}{(n-5)}\cdots\frac{2}{1}\sqrt{\frac{2}{\pi}}\frac{\sigma}{\sqrt{n}} \quad \text{(if } n \text{ be even),}$$

or

$$\frac{(n-2)}{(n-3)}\frac{(n-4)}{(n-5)}\cdots\frac{3}{2}\sqrt{\frac{\pi}{2}}\frac{\sigma}{\sqrt{n}} \quad \text{(if } n \text{ be odd).}$$

The second moment about the end of the range is

$$\frac{I_n}{I_{n-2}} = \frac{(n-1)\sigma^2}{n}.$$

The third moment about the end of the range is equal to

$$\frac{I_{n+1}}{I_{n-2}} = \frac{I_{n+1}}{I_{n-1}}\cdot\frac{I_{n-1}}{I_{n-2}}$$
$$= \sigma^2 \times \text{the mean.}$$

The fourth moment about the end of the range is equal to

$$\frac{I_{n+2}}{I_{n-2}} = \frac{(n-1)(n+1)}{n^2}\sigma^4.$$

If we write the distance of the mean from the end of the range $\frac{D\sigma}{\sqrt{n}}$ and the moments about the end of the range v_1, v_2, etc. then

$$v_1 = \frac{D\sigma}{\sqrt{n}},\ v_2 = \frac{n-1}{n}\sigma^2,\ v_s = \frac{D\sigma^3}{\sqrt{n}},\ v_4 = \frac{n^2-1}{n^2}\sigma^4.$$

From this we get the moments about the mean

$$\mu_2 = \frac{\sigma^2}{n}(n - 1 - D^2),$$

$$\mu_3 = \frac{\sigma^3}{n\sqrt{n}}\{nD - 3(n-1)D + 2D^3\} = \frac{\sigma^3 D}{n\sqrt{n}}\{2D^2 - 2n + 3\},$$

$$\mu_4 = \frac{\sigma^2}{n^2}\{n^2 - 1 - 4D^2 n + 6(n-1)D^2 - 3D^4\}$$

$$= \frac{\sigma^4}{n^2}\{n^2 - 1 - D^2(3D^2 - 2n + 6)\}.$$

It is of interest to find out what these become when n is large.

In order to do this we must find out what is the value of D.

Now Wallis's expression for π derived from the infinite product value of $\sin x$ is

$$\frac{\pi}{2}(2n+1) = \frac{2^2 \cdot 4^2 \cdot 6^2 \cdots (2n)^2}{1^2 \cdot 3^2 \cdot 5^2 \cdots (2n-1)^2}.$$

If we assume a quantity $\theta\left(= a_0 + \frac{a_1}{n} + \text{etc.}\right)$ which we may add to the $2n + 1$ in order to make the expression approximate more rapidly to the truth, it is easy to show that $\theta = -\frac{1}{2} + \frac{1}{16n} - \text{etc.}$ and we get

$$\frac{\pi}{2}\left(2n + \frac{1}{2} + \frac{1}{16n}\right) = \frac{2^2 \cdot 4^2 \cdot 6^2 \cdots (2n)^2}{1^2 \cdot 3^2 \cdot 5^2 \cdots (2n-1)^2}.^*$$

From this we find that whether n be even or odd D^2 approximates to $n - \frac{3}{2} + \frac{1}{8n}$ when n is large.

Substituting this value of D we get

$$\mu_2 = \frac{\sigma^2}{2n}\left(1 - \frac{1}{4n}\right),\ \mu_3 = \frac{\sigma^3\sqrt{1 - \frac{3}{2n} + \frac{1}{16n^2}}}{4n^2},\ \mu_4 = \frac{3\sigma^4}{4n^2}\left(1 + \frac{1}{2n} - \frac{1}{16n^2}\right).$$

Consequently the value of the standard deviation of a standard deviation which we have found $\dfrac{\sigma}{\sqrt{2n}\sqrt{1 - \frac{1}{4n}}}$ becomes the same as that found for the normal curve by Professor Pearson $(\sigma/\sqrt{2n})$ when n is large enough to neglect the $1/4n$ in comparison with 1.

* This expression will be found to give a much closer approximation to π than Wallis's.

Neglecting terms of lower order than $\frac{1}{n}$ we find

$$\beta_1 = \frac{2n-3}{n(4n-3)}, \quad \beta_2 = 3\left(1 - \frac{1}{2n}\right)\left(1 + \frac{1}{2n}\right).$$

Consequently as n incrases β_2 very soon approaches the value 3 of the normal curve, but β_1 vanishes more slowly, so that the curve remains slightly skew.

Diagram I shows the theoreitcal distribution of the S.D. found from samples of 10.

$$y = \frac{N 10^{9/2}}{7 \cdot 5 \cdot 3} \sqrt{\frac{2}{\pi}} \frac{x^8}{\sigma^9} e^{-(10x^2/2\sigma^2)}.$$

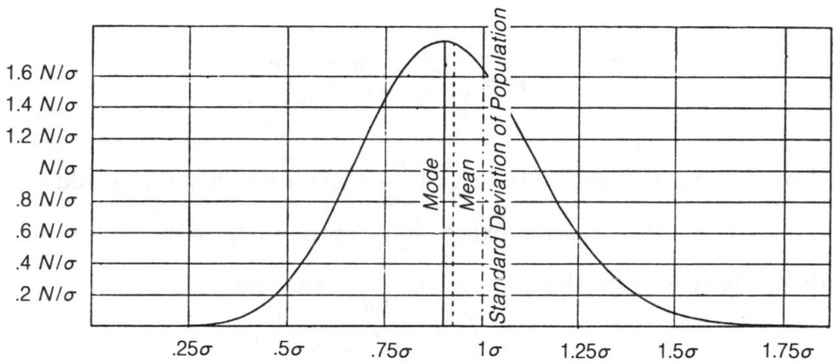

Diagram I. Frequency curve giving the distribution of Standard Deviations of samples of 10 taken from a normal population.

$$\text{Equation} \quad y = \frac{N}{7 \cdot 5 \cdot 3} \frac{10^{9/2}}{\sigma^9} \sqrt{\frac{2}{\pi}} x^8 e^{-(10x^2/2\sigma^2)}.$$

Section V

Some Properties of the Curve

$$y = \frac{n-2}{n-3} \cdot \frac{n-4}{n-5} \cdots \begin{pmatrix} \frac{4}{3} \cdot \frac{2}{\pi} \text{ if } n \text{ be even} \\ \frac{5}{4} \cdot \frac{3}{2} \text{ if } n \text{ be odd} \end{pmatrix} (1 + z^2)^{-n/2}$$

Writing $z = \tan\theta$ the equation becomes $y = \frac{n-2}{n-3} \cdot \frac{n-4}{n-5} \cdots \text{etc.} \times \cos^n\theta$, which affords an easy way of drawing the curve. Also $dz = d\theta/\cos^2\theta$.

Hence to find the area of the curve between any limits we must find

$$\frac{n-2}{n-3}\cdot\frac{n-4}{n-5}\cdots\text{etc.}\times\int\cos^{n-2}\theta\,d\theta$$

$$=\frac{n-2}{n-3}\cdot\frac{n-4}{n-5}\cdots\text{etc.}\left\{\frac{n-3}{n-2}\int\cos^{n-4}\theta\,d\theta+\left[\frac{\cos^{n-3}\theta\sin\theta}{n-2}\right]\right\}$$

$$=\frac{n-4}{n-5}\cdot\frac{n-6}{n-7}\cdots\text{etc.}\int\cos^{n-4}\theta\,d\theta+\frac{1}{n-3}\cdot\frac{n-4}{n-5}\cdots\text{etc.}[\cos^{n-3}\theta\sin\theta],$$

and by continuing the process the integral may be evaluated.

For example, if we wish to find the area between 0 and θ for $n = 8$ we have

$$\text{area}=\frac{6}{5}\cdot\frac{4}{3}\cdot\frac{2}{1}\cdot\frac{1}{\pi}\int_0^\theta\cos^6\theta\,d\theta$$

$$=\frac{4}{3}\cdot\frac{2}{\pi}\cdot\int_0^\theta\cos^4\theta\,d\theta+\frac{1}{5}\cdot\frac{4}{3}\cdot\frac{2}{\pi}\cos^5\theta\sin\theta$$

$$=\frac{\theta}{\pi}+\frac{1}{\pi}\cos\theta\sin\theta+\frac{1}{3}\cdot\frac{2}{\pi}\cos^2\theta\sin\theta+\frac{1}{5}\cdot\frac{4}{3}\cdot\frac{2}{\pi}\cos^5\theta\sin\theta,$$

and it will be noticed that for $n = 10$ we shall merely have to add to this same expression the term $\frac{1}{7}\cdot\frac{6}{5}\cdot\frac{4}{3}\cdot\frac{2}{\pi}\cos^7\theta\sin\theta$.

The tables at the end of the paper give the area between $-\infty$ and z

$$\left(\text{or } \theta=-\frac{\pi}{2}\text{ and }\theta=\tan^{-1}z\right).$$

This is the same as .5 + the area between $\theta = 0$, and $\theta = \tan^{-1} z$, and as the whole area of the curve is equal to 1, the tables give the probability that the mean of the sample does not differ by more than z times the standard deviation of the sample from the mean of the population.

The whole area of the curve is equal to

$$\frac{n-2}{n-3}\cdot\frac{n-4}{n-5}\cdots\text{etc.}\times\int_{-\pi/2}^{+\pi/2}\cos^{n-2}\theta\,d\theta,$$

and since all the parts between the limits vanish at both limits this reduces to 1.

Similarly the second moment coefficient is equal to

$$\frac{n-2}{n-3}\cdot\frac{n-4}{n-5}\cdots\text{etc.}\times\int_{-\pi/2}^{+\pi/2}\cos^{n-2}\theta\tan^2\theta\,d\theta$$

$$=\frac{n-2}{n-3}\cdot\frac{n-4}{n-5}\cdots\text{etc.}\times\int_{-\pi/2}^{+\pi/2}(\cos^{n-4}\theta-\cos^{n-2}\theta)\,d\theta$$

$$=\frac{n-2}{n-3}-1=\frac{1}{n-3}.$$

Hence the standard deviation of the curve is $1/\sqrt{n-3}$. The fourth moment coefficient is equal to

$$\frac{n-2}{n-3}\cdot\frac{n-4}{n-5}\cdots \text{etc.} \times \int_{-\pi/2}^{+\pi/2} \cos^{n-2}\theta \tan^4\theta\, d\theta$$

$$= \frac{n-2}{n-3}\cdot\frac{n-4}{n-5}\cdots \text{etc.} \times \int_{-\pi/2}^{+\pi/2} (\cos^{n-6}\theta - 2\cos^{n-4}\theta + \cos^{n-2}\theta)\, d\theta$$

$$= \frac{n-2}{n-3}\cdot\frac{n-4}{n-5} - \frac{2(n-2)}{n-3} + 1 = \frac{3}{(n-3)(n-5)}.$$

The odd moments are of course zero as the curve is symmetrical, so

$$\beta_1 = 0, \qquad \beta_2 = \frac{3(n-3)}{n-5} = 3 + \frac{2}{n-5}.$$

Hence as n increases the curve approaches the normal curve whose standard deviation is $1/\sqrt{n-3}$.

β_2 however is always greater than 3, indicating that large deviations are more common than in the normal curve.

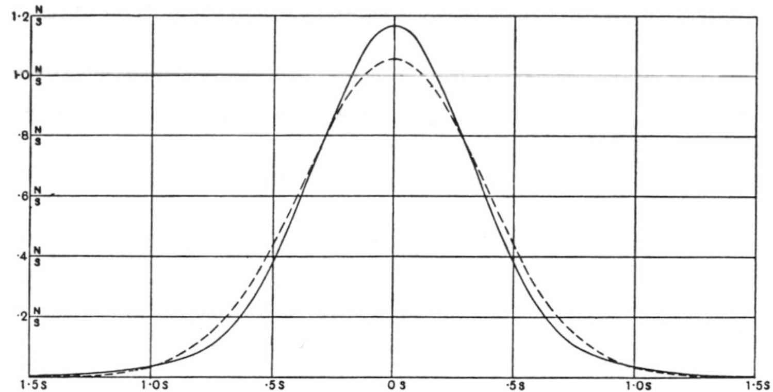

Diagram II. Solid curve $y = \frac{N}{S} \times \frac{8}{7}\cdot\frac{6}{5}\cdot\frac{4}{3}\cdot\frac{2}{\pi}\cos^{10}\theta$, $x/s = \tan\theta$. Broken line curve $y = \frac{\sqrt{7}\cdot N}{\sqrt{2\pi}\cdot s} e^{-(7x^2/2s^2)}$, the normal curve with the same S.D.

I have tabled the area for the normal curve with standard deviation $1/\sqrt{7}$ so as to compare with my curve for $n = 10$*. It will be seen that odds laid according to either table would not seriously differ till we reach $z = .8$, where the odds are about 50 to 1 that the mean is within that limit: beyond that the

* See table in Section VII.

normal curve gives a false feeling of security, for example, according to the normal curve it is 99,986 to 14 (say 7000 to 1) that the mean of the population lies between $-\infty$ and $+1.3s$ whereas the real odds are only 99,819 to 181 (about 550 to 1).

Now 50 to 1 corresponds to three times the probable error in the normal curve and for most purposes would be considered significant; for this reason I have only tabled my curves for values of n not greater than 10, but have given the $n = 9$ and $n = 10$ tables to one further place of decimals. They can be used as foundations for finding values for larger samples*.

The table for $n = 2$ can be readily constructed by looking out $\theta = \tan^{-1} z$ in Chambers' Tables and then $.5 + \theta/\pi$ gives the corresponding value.

Similarly $\frac{1}{2}\sin\theta + .5$ gives the values when $n = 3$.

There are two points of interest in the $n = 2$ curve. Here s is equal to half the distance between the two observations. $\tan^{-1}\frac{s}{s} = \frac{\pi}{4}$ so that between $+s$ and $-s$ lies $2 \times \frac{\pi}{4} \times \frac{1}{\pi}$ or half the probability, i.e. if two observations have been made and we have no other information, it is an even chance that the mean of the (normal) population will lie between them. On the other hand the second moment coefficient is

$$\frac{1}{\pi}\int_{-\pi/2}^{+\pi/2} \tan^2\theta\, d\theta = \frac{1}{\pi}\left[\tan\theta - \theta\right]_{-\pi/2}^{\pi/2} = \infty,$$

or the standard deviation is infinite while the probable error is finite.

Section VI. Practical Test of the Foregoing Equations

Before I had succeeded in solving my problem analyticallv, I had endeavoured to do so empirically. The material used was a correlation table containing the height and left middle finger measurements of 3000 criminals, from a paper by W.R. Macdonell (*Biometrika*, Vol. I. p. 219). The measurements were written out on 3000 pieces of cardboard, which were then very thoroughly shuffled and drawn at random. As each card was drawn its numbers were written down in a book which thus contains the measurements of 3000 criminals in a random order. Finally each consecutive set of 4 was taken as a sample—750 in all—and the mean, standard deviation, and correlation† of each sample determined. The difference between the mean of each sample and the mean of the population was then divided by the standard deviation of the sample, giving us the z of Section III.

* E.g. if $n = 11$, to the corresponding value for $n = 9$, we add $\frac{7}{8} \times \frac{5}{6} \times \frac{3}{4} \times \frac{1}{2} \times \frac{1}{2}\cos^8\theta\sin\theta$: if $n = 13$ we add as well $\frac{9}{10} \times \frac{7}{8} \times \frac{5}{6} \times \frac{3}{4} \times \frac{1}{2} \times \frac{1}{2}\cos^{10}\theta\sin\theta$ and so on.

† I hope to publish the results of the correlation work shortly. *Editor's note*: These results were published in the same volume of *Biometrika*, pp. 302–10.

This provides us with two sets of 750 standard deviations and two sets of 750 z's on which to test the theoretical results arrived at. The height and left middle finger correlation table was chosen because the distribution of both was approximately normal and the correlation was fairly high. Both frequency curves, however, deviate slightly from normality, the constants being for height $\beta_1 = .0026$, $\beta_2 = 3.175$, and for left middle finger lengths $\beta_1 = .0030$, $\beta_2 = 3.140$, and in consequence there is a tendency for a certain number of larger standard deviations to occur than if the distributions were normal. This, however, appears to make very little difference to the distribution of z.

Another thing which interferes with the comparison is the comparatively large groups in which the observations occur. The heights are arranged in 1 inch groups, the standard deviation being only 2.54 inches: while the finger lengths were originally grouped in millimetres, but unfortunately I did not at the time see the importance of having a smaller unit, and condensed them into two millimetre groups, in terms of which the standard deviation is 2.74.

Several curious results follow from taking samples of 4 from material disposed in such wide groups. The following points may be noticed:

(1) The means only occur as multiples of 25.

(2) The standard deviations occur as the square roots of the following types of numbers n, $n + .19$, $n + .25$, $n + .50$, $n + .69$, $2n + .75$.

(3) A standard deviation belonging to one of these groups can only be associated with a mean of a particular kind; thus a standard deviation of $\sqrt{2}$ can only occur if the mean differs by a whole number from the group we take as origin, while $\sqrt{1.69}$ will only occur when the mean is at $n \pm .25$.

(4) All the four individuals of the sample will occasionally come from the same group, giving a zero value for the standard deviation. Now this leads to an infinite value of z and is clearly due to too wide a grouping, for although two men may have the same height when measured by inches, yet the finer the measurements the more seldom will they be identical, till finally the chance that four men will have *exactly* the same height is infinitely small. If we had smaller grouping the zero values of the standard deviation might be expected to increase, and a similar consideration will show that the smaller values of the standard deviation would also be likely to increase, such as .436, when 3 fall in one group and 1 in an adjacent group, or .50 when 2 fall in two adjacent groups. On the other hand when the individuals of the sample lie far apart, the argument of Sheppard's correction will apply, the real value of the standard deviation being more likely to be smaller than that found owing to the frequency in any group being greater on the side nearer the mode.

These two effects of grouping will tend to neutralise each other in their effect on the mean value of the standard deviation, but both will increase the variability.

Accordingly we find that the mean value of the standard deviation is quite close to that calculated, while in each case the variability is sensibly greater. The fit of the curve is not good, both for this reason and because the frequency is not evenly distributed owing to effects (2) and (3) of grouping. On the other hand the fit of the curve giving the frequency of z is very good and as that is the only practical point the comparison may be considered satisfactory.

The following are the figures for height:

Mean value of standard deviations; calculated	$2.027 \pm .021$
Mean value of standard deviations; observed	2.026
Difference =	$-.001$

Standard deviation of standard deviations:

Calculated	$.8556 \pm .015$
Observed	$.9066$
Difference =	$+.0510$

In tabling the observed frequency, values between .0125 and .0875 were included in one group, while between .0875 and .0125 they were divided over the two groups. As an instance of the irregularity due to grouping I may mention that there were 31 cases of standard deviations 1.30 (in terms of the grouping) which is .5117 in terms of the standard deviation of the population, and they were therefore divided over the groups .4 to .5 and .5 to .6. Had they all been counted in groups .5 to .6 χ^2 would have fallen to 29.85 and P would have risen to .03. The χ^2 test presupposes *random* sampling from a frequency following the given law, but this we have not got owing to the interference of the grouping.

When, however, we test the z's where the grouping has not had so much effect we find a close correspondence between the theory and the actual result.

There were three cases of infinite values of z which, for the reasons given above, were given the next largest values which occurred, namely $+6$ or -6. The rest were divided into groups of .1; .04, .05 and .06, being divided between the two groups on either side.

The calculated value for the standard deviation of the frequency curve was $1 (\pm .017)$ while the observed was 1.039. The value of the standard deviation is really infinite, as the fourth moment coefficient is infinite, but as we have arbitrarily limited the infinite cases we may take as an approximation $\frac{1}{\sqrt{1500}}$ from which the value of the probable error given above is obtained. The fit of the curve is as follows:

This is very satisfactory, especially when we consider that as a rule observations are tested against curves fitted from the mean and one or more other moments of the observations, so that considerable correspondence is only to be expected; while this curve is exposed to the full errors of random sampling, its constants having been calculated quite apart from the observations.

The left middle finger samples show much the same features as those of the height, but as the grouping is not so large compared to the variability the

Comparison of Fit. Theoretical Equation: $y = \frac{16 \times 750}{\sqrt{2\pi}\,\sigma^3} x^2 e^{-2x^2/\sigma^2}$.

Scale in terms of standard deviation of population	0 to .1	.1 to .2	.2 to .3	.3 to .4	.4 to .5	.5 to .6	.6 to .7	.7 to .8	.8 to .9	.9 to 1.0	1.0 to 1.1	1.1 to 1.2	1.2 to 1.3	1.3 to 1.4	1.4 to 1.5	1.5 to 1.6	1.6 to 1.7	Greater than 1.7
Calculated frequency	$1\frac{1}{2}$	$10\frac{1}{2}$	27	$45\frac{1}{2}$	$64\frac{1}{2}$	$78\frac{1}{2}$	87	88	$81\frac{1}{2}$	71	58	45	33	23	15	$9\frac{1}{2}$	$5\frac{1}{2}$	7
Observed frequency	3	$14\frac{1}{2}$	$24\frac{1}{2}$	$37\frac{1}{2}$	107	67	73	77	$77\frac{1}{2}$	64	$52\frac{1}{2}$	$49\frac{1}{2}$	35	28	$12\frac{1}{2}$	9	$11\frac{1}{2}$	7
Difference	$+1\frac{1}{2}$	$+4$	$-2\frac{1}{2}$	-8	$+42\frac{1}{2}$	$-11\frac{1}{2}$	-14	-11	-4	-7	$-5\frac{1}{2}$	$+4\frac{1}{2}$	$+2$	$+5$	$-2\frac{1}{2}$	$-\frac{1}{2}$	$+6$	0

whence $\chi^2 = 48.06$, $P = .000,06$ (about).

Comparison of Fit. Theoretical Equation: $y = \frac{2}{\pi} \cos^4 \theta$, $z = \tan \theta$.

Scale of z	less than −3.05	−3.05 to −2.05	−2.05 to −1.55	−1.55 to −1.05	−1.05 to −.75	−.75 to −.45	−.45 to −.15	−.15 to +.15	+.15 to +.45	+.45 to +.75	+.75 to +1.05	+1.05 to +1.55	+1.55 to +2.05	+2.05 to +3.05	more than +3.05
Calculated frequency	5	$9\frac{1}{2}$	$13\frac{1}{2}$	$34\frac{1}{2}$	$44\frac{1}{2}$	$78\frac{1}{2}$	119	141	119	$78\frac{1}{2}$	$44\frac{1}{2}$	$34\frac{1}{2}$	$13\frac{1}{2}$	$9\frac{1}{2}$	5
Observed frequency	9	$14\frac{1}{2}$	$11\frac{1}{2}$	33	$43\frac{1}{2}$	$70\frac{1}{2}$	$119\frac{1}{2}$	$151\frac{1}{2}$	122	$67\frac{1}{2}$	49	$26\frac{1}{2}$	16	10	6
Difference	+4	+5	−2	$-1\frac{1}{2}$	−1	−8	$+\frac{1}{2}$	$+10\frac{1}{2}$	+3	−11	$+4\frac{1}{2}$	−8	$+2\frac{1}{2}$	$+\frac{1}{2}$	+1

whence $\chi^2 = 12.44$, $P = .56$

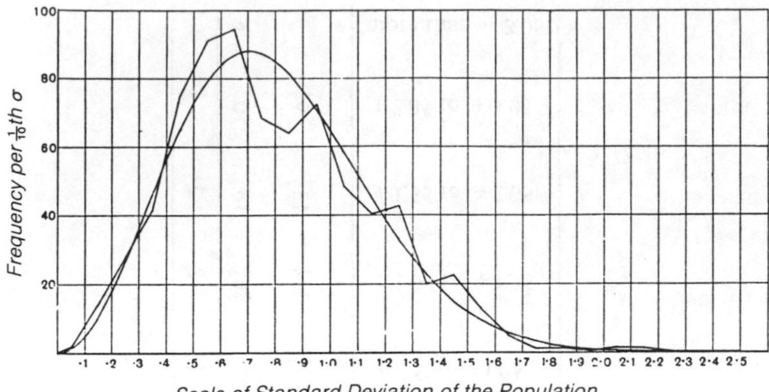

Diagram III. Comparison of Calculated Standard Deviation Frequency Curve with 750 actual Standard Deviations.

curves fit the observations more closely. Diagrams III.* and IV. give the standard deviations and the z's for this set of samples. The results are as follows:

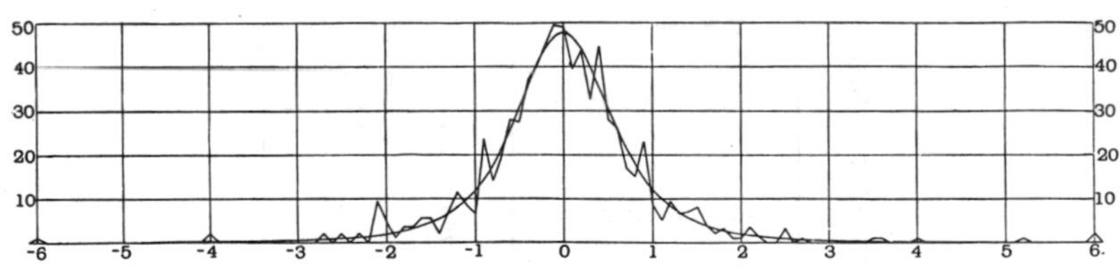

Diagram IV. Comparison of the theoretical frequency curve $y = \frac{1500}{\pi}\left(1 + \frac{x^2}{s^2}\right)^{-2}$, with an actual sample of 750 cases.

Mean value of standard deviations; calculated	$2.186 \pm .023$
Mean value of standard deviations; observed	2.179
Difference =	$-.007$

Standard deviaton of standard deviations:

Calculated	$.9224 \pm .016$
Observed	$.9802$
Difference =	$+.0578$

* There are three small mistakes in plotting the observed values in Diagram III., which make the fit appear worse than it really is.

Comparison of Fit. Theoretical Equation: $y = \frac{16 \times 750}{\sqrt{2\pi}\,\sigma^3} x^2 e^{-2x^2/\sigma^2}$.

Scale in terms of standard deviation of population	0 to .1	.1 to .2	.2 to .3	.3 to .4	.4 to .5	.5 to .6	.6 to .7	.7 to .8	.8 to .9	.9 to 1.0	1.0 to 1.1	1.1 to 1.2	1.2 to 1.3	1.3 to 1.4	1.4 to 1.5	1.5 to 1.6	1.6 to 1.7	Greater than 1.7
Calculated frequency	$1\frac{1}{2}$	$10\frac{1}{2}$	27	$45\frac{1}{2}$	$64\frac{1}{2}$	$78\frac{1}{2}$	87	88	$81\frac{1}{2}$	71	58	45	33	23	15	$9\frac{1}{2}$	$5\frac{1}{2}$	7
Observed frequency	2	14	$27\frac{1}{2}$	51	$64\frac{1}{2}$	91	$94\frac{1}{2}$	$68\frac{1}{2}$	$65\frac{1}{2}$	73	$48\frac{1}{2}$	$40\frac{1}{2}$	$42\frac{1}{2}$	20	$22\frac{1}{2}$	12	5	$7\frac{1}{2}$
Difference	$+\frac{1}{2}$	$+3\frac{1}{2}$	$+\frac{1}{2}$	$+5\frac{1}{2}$	—	$+12\frac{1}{2}$	$+7\frac{1}{2}$	$-19\frac{1}{2}$	-16	$+2$	$-9\frac{1}{2}$	$-4\frac{1}{2}$	$+9\frac{1}{2}$	-3	$+7\frac{1}{2}$	$+2\frac{1}{2}$	$-\frac{1}{2}$	$+\frac{1}{2}$

whence $\chi^2 = 21.80$, $P = .19$.

Comparison of Fit. Theoretical Equation: $y = \frac{2}{\pi} \cos^4 \theta,\ z = \tan \theta.$

Scale of z	less than −3.05	−3.05 to −2.05	−2.05 to −1.55	−1.55 to −1.05	−1.05 to −.75	−.75 to −.45	−.45 to −.15	−.15 to +.15	+.15 to +.45	+.45 to +.75	+.75 to +1.05	+1.05 to +1.55	+1.55 to +2.05	+2.05 to +3.05	more than +3.05
Calculated frequency	5	$9\frac{1}{2}$	$13\frac{1}{2}$	$34\frac{1}{2}$	$44\frac{1}{2}$	$78\frac{1}{2}$	119	141	119	$78\frac{1}{2}$	$44\frac{1}{2}$	$34\frac{1}{2}$	$13\frac{1}{2}$	$9\frac{1}{2}$	5
Observed frequency	4	$15\frac{1}{2}$	18	$33\frac{1}{2}$	44	75	122	138	$120\frac{1}{2}$	71	$46\frac{1}{2}$	36	11	9	6
Difference	−1	+6	$+4\frac{1}{2}$	−1	$-\frac{1}{2}$	$-3\frac{1}{2}$	+3	−3	$+1\frac{1}{2}$	$-7\frac{1}{2}$	+2	$+1\frac{1}{2}$	$-2\frac{1}{2}$	$-\frac{1}{2}$	+1

whence $\chi^2 = 7.39$, $P = .92$

Calculated value of standard deviation 1(±.017)

Observed value of standard deviation .982

Difference = −.018

A very close fit.

We see then that if the distribution is approximately normal our theory gives us a satisfactory measure of the certainty to be derived from a small sample in both the cases we have tested; but we have an indication that a fine grouping is of advantage. If the distribution is not normal, the mean and the standard deviation of a sample will be positively correlated, so that although both will have greater variability, yet they will tend to counteract each other, a mean deviating largely from the general mean tending to be divided by a larger standard deviation. Consequently I believe that the tables at the end of the present paper may be used in estimating the degree of certainty arrived at by the mean of a few experiments, in the case of most laboratory or biological work where the distributions are as a rule of a 'cocked hat' type and so sufficiently nearly normal.

Section VII

Tables of $\frac{n-2}{n-3}\frac{n-4}{n-5}\cdots\left(\begin{matrix}\frac{3}{2}\cdot\frac{1}{2}\,n\text{ odd}\\ \frac{2}{1}\cdot\frac{1}{\pi}\,n\text{ even}\end{matrix}\right)\int_{-(\pi/2)}^{\tan^{-1}z}\cos^{n-2}\theta\,d\theta$ for values of n from 4 to 10 Inclusive. Together with $\frac{\sqrt{7}}{\sqrt{2\pi}}\int_{-\infty}^{x}e^{-(7x^2/2)}\,dx$ for comparison when $n = 10$.

$z\left(=\frac{x}{s}\right)$	$n=4$	$n=5$	$n=6$	$n=7$	$n=8$	$n=9$	$n=10$	For comparison $\left(\frac{\sqrt{7}}{\sqrt{2\pi}}\int_{-\infty}^{x}e^{-7x^2/2}\,dx\right)$
.1	.5633	.5745	.5841	.5928	.6006	.60787	.61462	.60411
.2	.6241	.6458	.6634	.6798	.6936	.70705	.71846	.70159
.3	.6804	.7096	.7340	.7549	.7733	.78961	.80423	.78641
.4	.7309	.7657	.7939	.8175	.8376	.85465	.86970	.85520
.5	.7749	.8131	.8428	.8667	.8863	.90251	.91609	.90691
.6	.8125	.8518	.8813	.9040	.9218	.93600	.94732	.94375
.7	.8440	.8830	.9109	.9314	.9468	.95851	.96747	.96799
.8	.8701	.9076	.9332	.9512	.9640	.97328	.98007	.98253
.9	.8915	.9269	.9498	.9652	.9756	.98279	.98780	.99137
1.0	.9092	.9419	.9622	.9751	.9834	.98890	.99252	.99820

1.1	.9236	.9537	.9714	.9821	.9887	.99280	.99539	.99926
1.2	.9354	.9628	.9782	.9870	.9922	.99528	.99713	.99971
1.3	.9451	.9700	.9832	.9905	.9946	.99688	.99819	.99986
1.4	.9531	.9756	.9870	.9930	.9962	.99791	.99885	.99989
1.5	.9598	.9800	.9899	.9948	.9973	.99859	.99926	.99999
1.6	.9653	.9836	.9920	.9961	.9981	.99903	.99951	
1.7	.9699	.9864	.9937	.9970	.9986	.99933	.99968	
1.8	.9737	.9886	.9950	.9977	.9990	.99953	.99978	
1.9	.9770	.9904	.9959	.9983	.9992	.99967	.99985	
2.0	.9797	.9919	.9967	.9986	.9994	.99976	.99990	
2.1	.9821	.9931	.9973	.9989	.9996	.99983	.99993	
2.2	.9841	.9941	.9978	.9992	.9997	.99987	.99995	
2.3	.9858	.9950	.9982	.9993	.9998	.99991	.99996	
2.4	.9873	.9957	.9985	.9995	.9998	.99993	.99997	
2.5	.9886	.9963	.9987	.9996	.9998	.99995	.99998	
2.6	.9898	.9967	.9989	.9996	.9999	.99996	.99999	
2.7	.9908	.9972	.9991	.9997	.9999	.99997	.99999	
2.8	.9916	.9975	.9992	.9998	.9999	.99998	.99999	
2.9	.9924	.9978	.9993	.9998	.9999	.99998	.99999	
3.0	.9931	.9981	.9994	.9998	—	.99999	—	—

Section VIII. Explanation of Tables

The tables give the probability that the value of the mean, measured from the mean of the population, in terms of the standard deviation of the sample, will lie between $-\infty$ and z. Thus, to take the table for samples of six, the probability of the mean of the population lying between $-\infty$ and once the standard deviation of the sample is .9622 or the odds are about 24 to 1 that the mean of the population lies between these limits.

The probability is therefore .0378 that it is greater than once the standard deviation and .0756 that it lies outside ± 1.0 times the standard deviation.

(*Editors' note*: Section IX. Illustrations of Method has been omitted.)

Section X

Conclusions

I. A curve has been found representing the frequency distribution of standard deviations of samples drawn from a normal population.

II. A curve has been found representing the frequency distribution of values of the means of such samples, when these values are measured from the mean of the population in terms of the standard deviation of the sample.

III. It has been shown that this curve represents the facts fairly well even when the distribution of the population is not strictly normal.

IV. Tables are given by which it can be judged whether a series of experiments, however short, have given a result which conforms to any required standard of accuracy or whether it is necessary to continue the investigation.

Finally I should like to express my thanks to Professor Karl Pearson, without whose constant advice and criticism this paper could not have been written.

Introduction to
Fisher (1925) Statistical Methods for Research Workers

S.C. Pearce

R.A. Fisher had an abiding interest in inference, which came out in many of his writings. The inaugural meeting of the British Region of the Biometric Society was notable for his far-reaching presidential address (Fisher, 1948) in which he saw statistical science as doing for inference what mathematics had done for deduction. Among his contributions to the subject, the most influential has surely been the concept of significance, widely used from the time of its appearance. Although immensely valuable as an intellectual tool, its widespread and uncritical use has arguably been harmful as well as beneficial.

Like many other good ideas, it is simple and, what is more, it systematizes a way in which people often think. If we believe something, we shall not convince others if they can explain events as readily from their point of view as we can from ours. If the data could well have arisen even assuming the null hypothesis, the hypothesis may be correct. Such arguing from the opposition viewpoint is the basis of the Socratic method. An echo is found in Shakespeare, where Mark Antony confounds Caesar's critics by pretending to agree with them and then showing the inconsistencies of their position. Significance testing is reduction to absurdity in a probabilistic sense.

Historical Background

Fisher's early employment as a statistician was in agricultural research at the Rothamsted Experimental Station, where he was appointed in 1919. There he encountered the problems of interpreting data from comparative experiments. Some time previously, Mercer and Hall (1911) had revealed the complexity of fertility patterns in fields that were apparently uniform, a conclusion that had encouraged the practice of replication, i.e., the use of several small plots for each treatment so as to better sample the area of the experiment. It was usual to allocate treatments to plots systematically in order to balance out any trends. This provided further evidence that local variation could be considerable but suggested no way of measuring it. Fisher's innovation was to show that the error variance could be estimated—and hence, the significance level quantified—by allocating the treatments at random instead of systematically. He also introduced the idea of blocking to remove gross differences in fertility and devised the analysis of variance to simplify the calculations that his method required, but they were incidental. At the heart of his approach were the twin ideas of randomization and significance.

The two met with very different responses. Significance was received with delight because it met a psychological need. Without it, the interpreter of experimental results had been in a cleft stick. Was it safe to build much on a small difference? Would not the skeptics retort that it was all a matter of chance? On the other hand, too much caution might lead other critics to object that important conclusions were being missed. It became much safer when some differences could be declared significant and the others not so.

Randomization, on the other hand, was resisted from the first. It led to many difficulties in the field and was therefore rejected as unpractical. More than that, to people accustomed to considering their land carefully so as to balance out differences, it seemed absurd to leave matters to chance and perhaps arrive at an allocation with obvious disadvantages. Further, it raised awkward questions. Suppose one did randomize but obtained a systematic result? Also, one might accept the randomizers' argument that systematic designs biased the error and led to extreme values, but if that was the difficulty, should the treatments not be allocated in some way that secured a median value? The discussions at the Royal Statistical Society were prolonged and sometimes acrimonious [Neyman (1935) and Gosset (1936)], but they were dialogues of the deaf. The real difference between the two sides did not concern method but objective. One party wanted to minimize the error mean square; the other to estimate it without bias and the two were incompatible. Eventually, Fisher and his associates won, not because experimenters came to like randomization, but because the experts insisted that it was essential if significance was to be quantified. A method that gave apparent certainty, where before there was doubt, had powerful appeal and even randomization could be accepted as the price.

The Influence of Fisher's Book

It is difficult to overstate the influence of Fisher's book, *Statistical Methods for Research Workers*, on agricultural studies. As has been said, it met a psychological need. In fact, it touched on many topics and Fisher's mind illuminated them all, but its prime object, as Sec. 4 declared, was to teach the use of significant tests and that was what most of its readers wanted. For some time after its publication, writers on agricultural experimentation assumed without question that the aim was to discern which differences were significant, e.g., Hoblyn (1931), Yates (1933), Wishart and Sanders (1936). Later writers, e.g., Bailey (1959), found it necessary to explain with some care what a significant test could and could not do, thereby implying that there were circumstances in which it might not be relevant. Also, a growing number of writers [e.g., Finney 1955)] mention estimation, i.e., the setting of confidence limits around an estimated mean or difference of means, in parallel with testing, a point that will be considered below. Later, Dyke (1974) was to write, "*The object of almost all field experiments is the estimation of effects.... The testing of significance is important but secondary. The above sentences, executed in poker-work, ought to hang over the desk of every field experimenter.*" Today that view is held by most applied statisticians who have to listen to the questions posed by experimenters, although not all express themselves so well.

Nevertheless, the old idea has not died out. There are respected modern texts in which the testing of hypotheses is seen as the main statistical task, if not the only one. Journals of agricultural research provide many instances of irrelevant testing. Indeed, some still require that all differences be tested, a condition that has led to wry comment [Little (1978); Bryan-Jones and Finney (1983)], which the journal in question has had the grace to publish. (There are one or two other papers that look as if the author, tongue in cheek, was duping an unwary editor into condemning himself.) Worst of all, however, are journals that publish tables giving the results, mostly unintelligible, of multiple range tests, the said results receiving no mention in the text. The last fault arises possibly from the misconceived idea that the property of significance resides in the data themselves, not in the contrasts they estimate. Accordingly, if the data are "significant," the author is free to comment on any feature however trivial; if they are not, interpretation is deemed impermissible.

Difficulties with Significance

Nevertheless, a good idea is not invalidated by misuse or misunderstanding. Significance is a concept of the highest value, but like other ideas, it is used to best advantage by those who have faced any difficulties it raises. The first difficulty concerns the probability level required to dispel skepticism. Professor G.A. Barnard, who knew Fisher well, relates [Barnard (1990)] that the 0.05 level was derived from the long established practice of astronomers in checking observations that lie more than three probable errors from their expected value. It may fairly be commented that accepted theories are not discredited or new ones established by evidence just sufficient to cast doubt on an observation. It is quite wrong for editors of journals to publish a result that is "significant," as if the word were synonymous with "proven," especially at a level like one in 20. In fact, initial opinions about an effect are commonly diverse and there is no single level that will carry conviction with everyone. Fisher was well aware of that and chose the word significant as indicating the level at which the possibility of the effect should receive serious consideration, not at which its existence was established.

Also, there should be no suggestion that statistical analyses exist to find significant differences. Experiments are conducted in order to find answers to the questions being asked. Possibly no one doubts that a difference exists. If so, the task is to estimate its size and not to test for its existence. Further, a difference of means may be significant but not important or vice versa. The most insidious misuse of significance, however, comes with composite tests in which several contrasts are studied together and here Fisher set a bad example with his z-test. If a substance is tested at three equally spaced levels and the middle dose gives a response about midway between the extremes, when the two degrees of freedom are tested together, the negligible quadratic effect may disguise the linear, the whole being nonsignificant. Even worse are the multiple range tests, which examine together all the contrasts that involve only two treatments. As a result, more complicated contrasts, like those that indicate interactions or curvature, are neglected. Further, the analysis is the same whatever the purpose of the experiment and the possible need for estimation is disregarded.

Some have sought to avoid the difficulty by reporting only treatment means and standard errors, leaving interpretation to the reader, but the solution is not ideal. First, there are technical difficulties. Many experiments, e.g., those in blocks, do not estimate means anyway but only contrasts between them. (Also, without full knowledge of the design, it is not possible to derive the standard errors of contrasts from those of the component means.) Second, while accepting that interpretation can be very personal, the procedure does nothing to resolve the problem of the two-stage argument, when a writer has to decide after the first stage what form the second should take. To continue with the example of three equally spaced levels, a significant quadratic effect may dispel belief in a straight-line relationship. In that case, it makes no sense to estimate the slope of the nonexistent line or to inquire about its intercepts. On the other hand, if the response does appear to be linear, those questions may arise and require answers. In presenting results, a writer has to justify the preliminary decision about the quadratic effect; otherwise, there may be accusations of arbitrary and prejudiced interpretation. It helps to quote a significance level.

An example

In Sec. 42, Fisher gives examples of his methods. He had examined the data of Example 41 before (Fisher and Mackenzie, 1923) and now does so in more detail. They come from an experiment that could be modern. To use present terminology, there is a completely randomized design with three replicates of 12 cultivars of potato. Each plot is divided into three subplots, one of which receives no additional potassium, one is given the element in the form of sulphate, and the last in the form of chloride. There are two analyses of variance, one for the stratum of plots within the whole and the second for that of subplots with plots.

Considering the purpose for which the z- (or F-)test was devised, one is left wondering why it is used in the first analysis. The 12 cultivars were chosen as being different. What purpose is served by showing that they are so? (What interpretation would have been placed on the data if their means had *not* differed significantly?) This automatic testing of everything that can be tested is part of Fisher's legacy to the science of experimentation and it has done great harm. The objection can be taken further. The cultivars cannot be there on their own account because their characteristics are known. They are needed only to provoke an interaction with the fertilizer treatments. It would be enough to have a mean yield for each at that particular location, so why is the first analysis needed at all?

In the second analysis, Fisher distinguishes the two questions implied by the fertilizer treatments. Namely (1) do potatoes gain from being given potassium? (2) Does it matter which salt is used? (Actually, the second test should come first. If the sulphate does experience a different effect from the chloride, each needs separate comparison with an untreated control. That will serve as an example of a two-stage argument.) In doing so, he shows a regard for the scientific purpose of the experiment that many others have missed.

Rather mysteriously, however, Fisher does not partition the interaction similarly. Here again, the argument has two stages. If there is no interaction, an effect of the fertilizer may be assessed over all cultivars taken together; if there is, it must be studied for each separately. The first step therefore is to take the two contrasts for fertilizers and to ask if there is an interaction of either or both with the cultivars. According to the outcome, the rest of the analysis follows. There is a point to be noted here. It is true that if A interacts with B, then B must interact with A, but in practice, it usually happens that only one factor is under study, the other having been introduced just to see if there is an interaction that qualifies conclusions about the first. If that is the position, as it is here, the interaction is all-important, but many have followed Fisher's example of regarding the main effects as the object of study and the interactions as no more than interesting adornments. In fact, to the practitioner (i.e., the person for whose benefit the experiment exists), it may be the interaction that matters. A farmer does not want to apply fertilizer to a cultivar that does not need it or to use chloride, which is cheaper, on a cultivar that may suffer leaf scorch as a result. The main effects can often be foreseen. In agriculture and biological science, generally the problems mostly relate to the interactions, but no one would think so from reading the statistical literature.

Further, because of the importance that should be attached to interactions and the possibility that the study of the data may take various forms depending on their magnitude, a number of difficult problems arise at the design stage [Pearce (1989)], which also have received too little attention.

Assessment

All this sounds very critical. It is not meant to be. Fisher was a genius and in these passages he was pointing the way as no one had pointed it before. Like many geniuses, he has suffered from uncritical disciples, but reading the words he actually wrote shows that his claims were modest and completely justified. He called his book *Statistical Methods for Research Workers* and that is what he set out to provide. He did not claim to present all methods that might be useful. Indeed, Sec. 4 implies a very limited scope and he much exceeded his promise. Further, he did not claim to have invented the idea of significance but recognized freely that it was implicit in Karl Pearson's use of the χ^2-test, introduced some decades earlier. He thought the concept important and, giving credit where credit was due, he greatly extended the range of tests. If any blame is to be ascribed to him, it is for not using the later editions, of which there were many, to correct the faults of over-zealous disciples. His own use of the approach may be open to minor criticisms, such as those presented here, but he systematized a wide range of thinking and did so in an effective and comprehensible way.

T.H. Huxley, the 19th century biologist, is attributed with the remark that new truths start as heresies and end as superstitions. Significance was never a heresy, but it certainly became a superstition. Fisher's text was accorded an uncritical reverence that for many people precluded thinking about its meaning and left room only for dogmatism. Indeed, several writers [e.g., Preece 1984)] have likened much conventional statistical practice to a religious rite, understood and valued by some but required of everybody as a mark of respectability. It should not be like that. The analysis of variance is wonderfully flexible and significance testing is a task it performs well, but it can do more. It can be applied to a wide range of operations according to the needs of inquiry.

It is interesting to look again at Fisher's own example of his method, because it exemplifies the very faults that many complain about in the uncritical use of the analysis of variance today. First, that of testing everything that can be tested, regardless of relevance. Second, that of failing to isolate the questions, so as to ask them one by one in logical order, finding answers to each in turn. It would be unfair to blame Fisher for not seeing further when he was introducing the method for the first time. He was laying the foundations, not topping off the tower. The blame belongs rather to those disciples who followed him slavishly even in nonessentials, e.g., in making significance testing the supreme goal and in missing the importance of interactions, while failing to see the potential for development in the approach that he had made available to them. Despite it all, Fisher's vision of a rational system of inference remains.

References

Bailey, N.J.T. (1959). *Statistical Methods in Biology*. English Universities Press, London.

Barnard, G.A. (1990). Must clinical trials be large? The interpretation of *P*-values and the combination of test results, *Statistics in Med.*, **9**, 601–614.

Bryan-Jones, J., and Finney, D.J. (1983). An error in "Instructions to authors," *HortSci.*, **18**, 279–282.

Dyke, G.V. (1974). *Comparative Experiments with Field Crops*. Butterworth, London.

Finney, D.J. (1955). *Experimental Design and Its Statistical Basis*. Cambridge University Press, London.

Fisher, R.A. (1948). Biometry, *Biometrics*, **4**, 217–219.

Fisher, R.A., and Mackenzie, W.A. (1923). Studies in crop variation. II. The manurial response of different potato varieties. *J. Agric. Sci.* (*Cambridge*), **13**, 311–320.

Gosset, W.S. (1936). Cooperation in large-scale experiments (with discussion), *J. Roy. Statist. Soc., Suppl.*, **3**, 114–136.

Hoblyn, T.N. (1931). *Field Experiments in Horticulture*. Tech. Comm. 2., Commonwealth of Bur. Fruit Production, East Malling, Kent, England.

Little, T.M. (1978) If Galileo published in *HortScience.*, *HortSci.*, **13**, 504–506.

Mercer, W.B., and Hall, A.D. (1911). The experimental error of field trials, *J. Agric. Sci.* (*Cambridge*), **4**, 107–132.

Neyman, J. (1935). Statistical problems in agricultural experimentation. (with discussion), *J. Roy. Statist. Soc., Suppl.*, **2**, 107–180.

Pearce, S.C. (1989) The size of a comparative experiment, *J. Appl. Statist.*, **16**, 3–6.

Preece, D.A. (1984). Biometry in the Third World: Science, not ritual. *Biometrics*, **40**, 519–523.

Wishart, J., and Sanders, H.G. (1936). *Principles and Practice of Field Experiments*. Empire Cotton Growing Corp., London.

Yates, F. (1933). The principles of orthogonality and confounding in replicated experiments, *J. Agric. Sci.* (*Cambridge*), **23**, 108–145.

Statistical Methods for Research Workers

R.A. Fisher
Rothamsted Experimental Station

4. Scope of this Book

The prime object of this book is to put into the hands of research workers, and especially of biologists, the means of applying statistical tests accurately to numerical data accumulated in their own laboratories or available in the literature.

Ex. 41. Analysis of Variation in Experimental Field Trials

The table on the following page gives the yield in 1b. per plant in an experiment with potatoes (Rothamsted data). A plot of land, the whole of which had received a dressing of dung, was divided into 36 patches, on which 12 varieties were grown, each variety having 3 patches scattered over the area. Each patch was divided into three lines, one of which received, in addition to dung, a basal dressing only, containing no potash, while the other two received additional dressings of sulphate and chloride of potash respectively.

From data of this sort a variety of information may be derived. The total yields of the 36 patches give us 35 degrees of freedom, of which 11 represent differences among the 12 varieties, and 24 represent the differences between different patches growing the same variety. By comparing the variance in these two classes we may test the significance of the varietal differences in yield for the soil and climate of the experiment. The 72 additional degrees of freedom given by the yields of the separate rows consist of 2 due to manurial treatment, which we can subdivide into one representing the differences due

Editor's note: This example could be read in conjunction with Fisher's Paper (1926) reproduced in this volume.

Table 46

Variety	Sulphate Row	Sulphate Row	Sulphate Row	Chloride Row	Chloride Row	Chloride Row	Basal Row	Basal Row	Basal Row
Ajax	3.20	4.00	3.86	2.55	3.04	4.13	2.82	1.75	4.71
Arran Comrade	2.25	2.56	2.58	1.96	2.15	2.10	2.42	2.17	2.17
British Queen	3.21	2.82	3.82	2.71	2.68	4.17	2.75	2.75	3.32
Duke of York	1.11	1.25	2.25	1.57	2.00	1.75	1.61	2.00	2.46
Epicure	2.36	1.64	2.29	2.11	1.93	2.64	1.43	2.25	2.79
Great Scot	3.38	3.07	3.89	2.79	3.54	4.14	3.07	3.25	3.50
Iron Duke	3.43	3.00	3.96	3.33	3.08	3.32	3.50	2.32	3.29
K. of K.	3.71	4.07	4.21	3.39	4.63	4.21	2.89	4.20	4.32
Kerr's Pink	3.04	3.57	3.82	2.96	3.18	4.32	2.00	3.00	3.88
Nithsdale	2.57	2.21	3.58	2.04	2.93	3.71	1.96	2.86	3.56
Tinwald Perfection	3.46	3.11	2.50	2.83	2.96	3.21	2.55	3.39	3.36
Up-to-Date	4.29	2.93	4.25	3.39	3.68	4.07	4.21	3.64	4.11

to a potash dressing as against the basal dressing, and a second representing the manurial difference between the sulphate and the chloride; and 70 more representing the differences observed in manurial response in the different patches. These latter may in turn be divided into 22 representing the difference in manurial response of the different varieties, and 48 representing the differences in manurial response in different patches growing the same variety. To test the significance of the manurial effects, we may compare the variance in each of the two manurial degrees of freedom with that in the remaining 48; to test the significance of the differences in varietal response to manure, we compare the variance in the 22 degrees of freedom with that in the 48; while to test the significance of the difference in yield of the same variety in different patches, we compare the 24 degrees of freedom representing the differences in the yields of different patches growing the same variety with the 48 degrees representing the differences of manurial response on different patches growing the same variety.

For each variety we shall require the total yield for the whole of each patch, the total yield for the 3 patches and the total yield for each manure; we shall also need the total yield for each manure for the aggregate of the 12 varieties; these values are given in Table 47.

Table 47

		Manuring				Plot	
Variety	Sulphate	Chloride	Basal	Total	I	II	III
Ajax	11.06	9.72	9.28	30.06	8.57	8.79	12.70
Arran Comrade	7.39	6.21	6.76	20.36	6.63	6.88	6.85
British Queen	9.85	9.56	8.82	28.23	8.67	8.25	11.31
Duke of York	4.61	5.32	6.07	16.00	4.29	5.25	6.46
Epicure	6.29	6.68	6.47	19.44	5.90	5.82	7.72
Great Scot	10.34	10.47	9.82	30.63	9.24	9.86	11.53
Iron Duke	10.39	9.73	9.11	29.23	10.26	8.40	10.57
K. of K.	11.99	12.23	11.41	35.63	9.99	12.90	12.74
Kerr's Pink	10.43	10.46	8.88	29.77	8.00	9.75	12.02
Nithsdale	8.36	8.68	8.38	25.42	6.57	8.00	10.85
Tinwald Perfection	9.07	9.00	9.30	27.37	8.84	9.46	9.07
Up-to-Date	11.47	11.14	11.96	34.57	11.89	10.25	12.43
Total	111.25	109.20	106.26	326.71	...	...	...

The sum of the squares of the deviations of all the 108 values from their mean is 71.699; divided, according to patches, in 36 classes of 3, the value for the 36 patches is 61.078; dividing this again according to varieties into 12 classes of 3, the value for the 12 varieties is 43.638. We may express the facts so far as follows:

Table 48

Variance	Degrees of Freedom	Sum of Squares	Mean Square	Log (S.D.)
Between varieties	11	43.6384	3.967	.6890
Between patches for same variety	24	17.4401	.727	−.1594
Within patches	72	10.6204	...	...
Total	107	71.6989		

The value of z, found as the difference of the logarithms in the last column, is .8484, the corresponding 1 per cent. value being about .564; the effect of variety is therefore very significant.

Of the variation within the patches the portion ascribable to the two differences of manurial treatment may be derived from the totals for the three manurial treatments. The sum of the squares of the three deviations, divided by 36, is .3495; of this the square of the difference of the totals for the two potash dressings, divided by 72, contributes .0584, while the square of the difference between their mean and the total for the basal dressing, divided by 54, gives the remainder, .2911. It is possible, however, that the whole effect of the dressings may not appear in these figures, for if the different varieties had responded in different ways, or to different extents, to the dressings, the whole effect would not appear in the totals. The 70 remaining degrees of freedom would not be homogeneous. The 36 values, giving the totals for each manuring and for each variety, give us 35 degrees of freedom, of which 11 represent the differences of variety, 2 the differences of manuring, and the remaining 22 show the differences in manurial response of the different varieties. The analysis of this group is shown below:

Table 49

Variance due to	Degrees of Freedom	Sum of Squares	Mean Square
Potash dressing	1	.2911	.2911
Sulphate *v.* chloride	1	.0584	.0584
Differential response of varieties	22	2.1911	.0996
Differential response in patches with same variety	48	8.0798	.1683
Total	72	10.6204	

To test the significance of the variation observed in the yield of patches bearing the same variety, we may compare the value .727 found above from 24 degrees of freedom, with .1683 just found from 48 degrees. The value of z, half the difference of the logarithms, is .7316, while the 1 per cent. point is about .394. The evidence for unequal fertility of the different patches is therefore unmistakable. As is always found in careful field trials, local irregularities in the nature or depth of the soil materially affect the yields. In this case the soil irregularity was perhaps combined with unequal quality or quantity of the dung supplied.

There is no sign of differential response among the varieties; indeed, the difference between patches with different varieties is less than that found for patches with the same variety. The difference between the values is not significant; $z = .2623$, while the 5 per cent. point is about .33.

Finally, the effect of the manurial dressings tested is small; the difference due to potash is indeed greater than the value for the differential effects, which we may now call random fluctuations, but z is only .3427, and would require to be about .7 to be significant. With no total response, it is of course to be expected, though not as a necessary consequence, that the differential effects should be insignificant. Evidently the plants with the basal dressing had all the potash necessary, and in addition no apparent effect on the yield was produced by the difference between chloride and sulphate ions.

Introduction to Fisher (1926) The Arrangement of Field Experiments

T.P. Speed
University of California at Berkeley

Background and Significance

In 1919, the Director of Rothamsted Experimental Station, Sir John Russell, invited Ronald Aylmer Fisher, a young mathematician with interests in evolution and genetics, to join the small group of scientists at Rothamsted in order that [see Russell (1966, p. 327)] "after studying our records he should tell me whether they were suitable for proper statistical examination and might be expected to yield more information than we had extracted." Fisher accepted the invitation and in a very short time Russell realized (loc. cit.) "that he was more than a man of great ability; he was in fact a genius who must be retained." In the few years that followed, Fisher introduced the subdivision of sums of squares now known as an analysis of variance (anova) table (1923), derived the exact distribution of the (log of the) ratio of two independent chi-squared variates (1924), introduced the principles of blocking and randomization, as well as the randomized block, Latin square, and split-plot experiments, the latter with two anova tables (1925), promoted factorial experiments, and foreshadowed the notion of confounding (1926). Of course Fisher made many contributions to theoretical statistics over this same period [see Fisher (1922)], but the above relate directly to the design and analysis of field experiments, the topic of the paper that follows. It was an incredibly productive period for Fisher, with his ideas quickly transforming agricultural experimentation in Great Britain and more widely, and in major respects these ideas have remained the statistical basis of agricultural experimentation to this day. Other fields of science and technology such as horticulture, manufacturing industry, and later psychology and education also adopted Fisher's statistical principles of experimentation, and again they continue to be regarded as fundamental today.

The story of Fisher and the design and analysis of experiments has been told many times before and at greater length than is possible here; see Mahalanobis (1938), Hotelling (1951), Youden (1951), Yates (1964, 1975), Yates and Mather (1965), Russell (1966, pp. 325–332), Neyman (1967), Cochran (1976), and Box (1978). In this short introduction to Fisher's delightful 1926 essay, we will briefly review the background to the paper and comment on its immediate and longer-term impact.

The main topics discussed in Fisher's essay are significance testing, replication, randomization, local control (the elimination of heterogeneity), and factorial (there called complex) experimentation. Replication, randomization, and local control are all linked together in Fisher's discussion of the dual tasks of reducing the error in comparisons of interest and in obtaining a valid estimate of that error. Most of these ideas had been published in the previous two or three years, either in articles on particular field experiments or in the book, *Statistical Methods for Research Workers*; see Pearce (in this volume). What distinguishes this essay from these earlier works is its clarity, its comprehensiveness, its compactness (the material in the book was somewhat scattered), and the total absence of numerical or algebraic calculations. It is truly a discussion of principles. This form was undoubtedly dictated by the journal in which the essay appeared; indeed, it is probable that the paper by Russell listed as Ref. (3) was the direct cause of this paper being written. Russell's essay was aimed at farmers interested in the results of field experiments and others interested in methods of carrying them out, and no doubt, Fisher saw this as an excellent opportunity to explain his new ideas to their natural audience.

Fisher does not appear to have had any direct contact with field experiments prior to his joining Rothamsted in 1919, but his Cambridge teacher, F.J.M. Stratton, had written on the application of the theory of errors to agriculture [Wood and Stratton (1910)], and it seems likely that Fisher would have been familiar with this paper and a later one [see Mercer and Hall (1911)] on the same topic. Sir A. Daniel Hall was Russell's predecessor as Director of Rothamsted and published an expository paper [Hall (1925)] on the principles of agricultural experiments in the same volume in which Russell (1926) appeared.

Fisher's interest in the correlation coefficient and in evolutionary theory and genetics came together in his earlier paper [Fisher (1918)] on the correlation between relatives, and this provided the foundation for his work on the design and analysis of field experiments. In that paper, the modern term "variance" was introduced as the square of the standard deviation in a normal population. Fisher was considering measurements such as stature in humans, and his interest was in ascribing to various causes fractions of the total variance that they combine to produce. After explaining the role of Mendelian inheritance in his analysis, Fisher derived a partition of the variance of such a measurement that, in its simplest form, was into two parts: an additive part and a second part, described as the effect of dominance (within locus interaction). His definitions involved what we would now describe as the weighted least-squares fitting of a linear model, with the first part being that "explained" by the model, and the second the residual. Later Fisher introduced a further term due to epistacy (between locus interaction) and he also considered multiple alleles at each locus. Thus by 1918, he already had a rather more complex than usual form of linear model for a number of factors, and a clear formulation of the idea of partitioning variation into ascribable causes.

On March 20, 1923, the paper by Fisher and Mackenzie (1923) was received by the *Journal of Agricultural Science* in Cambridge. This paper contained the analysis of a series of field experiments begun two years earlier at Rothamsted and was probably the first in which Fisher was involved. The innovations in this paper have been discussed elsewhere [Seal (1967) and Yates (1975)], and its main interest to us lies in the fact that we find there the first published analysis of variance table, including terms due to the main effects and interactions of varieties and the manuring factor. This experiment was later reanalyzed as a split-plot design, with two analysis of variance tables, in Fisher (1925); see Pearce (this volume) for a detailed analysis. Nine days after the Fisher and Mackenzie paper was received, Gosset ("Student") wrote to Fisher, [Gosset (1970, letter 20)], requesting assistance on a problem concerning the appropriate error when comparing a number of varieties, each replicated an equal number of times in a field experiment. Fisher's solution, one of the few cases in which he presents an explicit linear model, appeared as a footnote in Gosset (1923) and was identical to that embodied in his paper with Mackenzie.

At this point, it is clear that Fisher knew how to analyze field experiments: by a combination of (usually implicit) linear models for the estimation of effects, together with an analysis of variance table for the estimation of error, although it should be noted that he also tried a multiplicative model in Fisher and Mackenzie (1923). It is not clear when Fisher hit on the idea of eliminating heterogeneity through blocking or the use of Latin squares. Interestingly, it appears that he did not get an opportunity to design and conduct an experiment at Rothamsted along his own lines until after the publication of his 1925 book and the paper that follows. It is stated in Box (1978, pp. 156–158) that Fisher was working on Latin squares in 1924, and that in the same year he had used a Latin square design for a forestry experiment. The results of such an experiment would naturally not become available for some time afterwards, and as a consequence, the numerical illustrations of the increase in precision achievable through blocking and the use of Latin squares offered in his book refer to layouts randomly imposed on the mangold uniformity data of Mercer and Hall (1911).

It is equally unclear when (and how) Fisher arrived at the idea of randomization. The role which randomization plays in validating the comparison of treatment and error mean squares was touched upon, but not explained very fully, in the 1925 book, and so the present paper is the first careful discussion of this issue. On its own, the argument is clear and compelling, but the whole topic continues to puzzle students of the subject, even now. A possible explanation of this lies in the absence of any direct connection between the linear models, according to which such experiments are usually analyzed, and the randomization argument. The connection was made in the work of Kempthorne (1952) and his students, but in most treatments of the design and analysis of experiments, the problem remains.

Summarizing the discussion so far, we see that the principles of field experimentation outlined by Fisher in the paper below were based, in part, on the result of his analyzing experiments designed by others and, in part, by numerical experiments he carried out on the uniformity trials of Mercer and Hall (1911). At the time of the publication of this paper, he could not have pointed to a single experiment successfully designed and analyzed according to the principles expounded. It is clear that this in no way inhibited him from vigorously expounding his ideas, but it is interesting to note that, at the time, they were just that: ideas.

The impact of Fisher's principles for designing and analyzing field experiments was dramatic. His methods spread quickly in Great Britain and were also rapidly taken up by research workers in other countries and in other fields. In a sense, the 1935 publication of *Design of Experiments* signaled the end, not the beginning of the Fisherian revolution in field experimentation, for by 1935, many national agricultural research organizations were using Fisher's ideas. A major factor in this rapid transformation was undoubtedly the central role played by the Rothamsted Experimental Station in agricultural research in Great Britain [see Russell (1966)], and in the British Empire more generally.

Box (1978, p. 157) describes the inquiries and requests from agricultural workers running experiments, the world travels of Sir John Russell, the establishment of the Commonwealth Agricultural Bureaux, and the rapidly increasing number of voluntary workers who joined the statistics department at Rothamsted during this period. One such report can be found in Goulden (1931). Later in 1931, Fisher also visited the United States. For further remarks and details on early experiments run according to Fisher's principles, we refer to Kerr (1988, sugar in Australia in 1928), Haines (1929, 1930a, 1930b, rubber in Malaya), Kirk (1929, potatoes in Canada), Richey (1930, agronomy in the United States), Gregory et al. (1932, cotton in the Sudan), Arnold (1985) for references to New Zealand work, Harrison et al. (1935, tea in India), Tippett (1935, textiles in Great Britain), Wishart (1935, silk in China). We also draw attention to the photograph given in Box (1978, Plate 6) of the Latin square experiment Fisher designed in 1929 for the Forestry Commission of Great Britain. Shortly before the publication of *Design of Experiments*, Snedecor (1934) appeared, and this work contributed greatly to the spread of Fisher's ideas in the United States and more widely, not only in the sphere of agriculture, but in other fields as well.

Randomized block experiments, both complete and incomplete, have continued to be popular in field experimentation, because of their flexibility, simplicity, and robustness; see Patterson and Silvey (1980). In field experimentation, at least, the many special designs (Latin and Graeco-Latin squares, lattice designs, etc.) [see, e.g., Pearce (1983) and Mead (1988)], are being replaced by the more flexible although less symmetric α-designs [Patterson et al. (1978)] and related row and column designs [Williams and John (1989); see also Seeger et al. (1987)]. Essential use is made of computers to generate and test such designs (for efficiency), as well as to analyze data from the resulting experiment.

Factorial experiments have become widely used in most areas of applied science and technology. In these fields, at least, Fisher's view that Nature should be asked more than one question at a time has clearly won the day. The idea of confounding foreshadowed at the end of the paper was quickly developed by Yates, Barnard, Bose, and others in the 1930s and 1940s and remains an important component of field and industrial experimentation today.

Although Fisher's approach to the design and analysis of field experiments was quickly and widely adopted, it was not accepted in every detail. Discussion concerning the relative merits of systematic and randomized designs began with Gosset (1931), and this debate continued until after Gosset's death in 1937 [see, e.g., Gosset (1936a, 1936b, 1937–38), Arnold (1985)], with Fisher, Yates, Neyman, Pearson, Jeffreys, and others participating [see, e.g., Barbacki and Fisher (1936), Yates (1939), and references therein].

Two questions concerning randomization that Fisher did not adequately address were (1) what, if anything, should be done if an obviously undesirable arrangement arises as a result of randomization, and (2) whether the analysis of a field experiment should be conditional upon the actual arrangement obtained through randomization, as that is an obvious ancillary. The relationship between model-based and randomization-based inference continues to be debated in the context of field experiments, as it does in other areas of statistics. The whole subject of randomization continues to puzzle some Bayesians [see Savage (1976)] and others, and it seems safe to say that the recent resurgence of interest in neighbor methods sparked by the publication of Wilkinson et al. (1983) raises the topic once more in the context of field experiments.

For a quite different approach to field experiments and references to a Bayesian view of randomization, see Neyman (1990) and the remarks of Rubin which follow that translation.

The paper that follows, far better than the scattered outline in Fisher (1925), the more detailed manual of Fisher and Wishart (1930), or the book by Fisher (1935) that followed, can be viewed as outlining in simple terms Fisher's statistical principles of field experimentation. The fact that every one of the ideas so lucidly expounded here, for essentially the first time, remains central to the subject even today, is clear evidence that it deserves to be included in any collection entitled *Breakthroughs in Statistics*.

Notes on the Paper

Fisher asks, in the section entitled, "*When Is a Result Significant*?," how one should interpret the result of an acre plot with manure yielding 10% more than a similarly located plot that was not treated with manure, but that was treated similarly in all other respects. He describes the classical interpretation of a significance level and then goes on to say that about 500 years' experience would be required to estimate the upper 5% point of the distribution of such ratios, giving a simple argument using the order statistics for so doing. How did Fisher come up with the number 500, and why did he focus on the 5% point? His answer to the second question is clearly spelled out in the last paragraph on p. 83; see also Hall and Selinger (1986) for further discussion of this point. As for the number 500, we can be sure that Fisher was well aware of the fact that the (asymptotic) variance of the [Np]th order statistic in a random sample of size N is approximately

$$\frac{p(1-p)}{N}\cdot\frac{1}{f(x_p)^2},$$

where f is the common density and x_p its upper 100% point. With $p = .05$ and $N = 500$, the square root of the first factor is .01 and, of course, the second factor is unknown when f (not to mention x_p) is unknown. It seems reasonable to surmise that Fisher fixed upon $N = 500$ in order to make this standard deviation on the order of 1%, well aware that he could not know the other factor. (For the normal density, this term is about 10.)

Having obtained this unrealistically large number, Fisher goes on to explain how the experimenter can use the t-distribution if the previous 10 years' records are available. He argues as follows: The difference d in the year of an experiment should be divided by the estimated standard error s based on the previous 10 years of trials with a uniform treatment and compared with the upper 5% point of the t-distribution on 10 degrees of freedom, i.e., with 2.238. Fisher measures the difference of yields and the standard error as percents of the mean yield and concludes that if the standard error based on 10 years' data was 3%, an observed difference of 10% exceeds $3 \times 2.238 = 6.684$ and so is significant at the 5% level. He concludes by explaining how to calculate s.

In effect, Fisher has argued that we do not need 500 years of uniformity data as he originally suggested; some number on the order of 10 will suffice "if we put our trust in the theory of errors." His next and boldest step is to argue that no uniformity trials are necessary at all, provided that the experiment is properly randomized. In so doing, he is implicitly assuming that the magnitude of the year-to-year variation he had just been discussing coincides with that of the error provided by the experiment itself, i.e., of the plot-to-plot variation, within a year. Of course, this is not generally true, and a valid criticism of Fisher's emphasis on randomization, replication, and local control in the context of agricultural trials is that he focuses on the component of variation of least practical importance. The within-experiment error with which he is so concerned in this paper is frequently far smaller than the variation observed from year to year, at one location, from location to location, in one year, and sizable location-by-year interaction components of variation are not uncommon [see Patterson and Silvey (1980)].

In the paragraph entitled "Errors Wrongly Estimated," Fisher (footnote, p. 84) disclaims responsibility for the design of an experiment discussed by Russell (1926). There must have been a misunderstanding here, because in the article Russell (pp. 996–7) introduces and displays a balanced (i.e., systematic) plan, with the following "An instance is afforded by the experiment designed by Mr. R.A. Fisher for a detailed study of the effect of phosphates and of nitrogen on crop yield."

The paragraph beginning near the bottom of p. 85 finds Fisher explaining the sense in which the estimate of error obtained by randomization is *valid.* Much effort has been devoted to the justification of these and similar remarks; and so it seems worthwhile to note some of the details. Suppose that $2r$ plots are labeled $i = 1, \ldots, 2r$, and that a subset $\mathbf{T}$ is of r plots assigned to a treatment, and the remainder $\mathbf{C}$ are controls, the assignment being at random, so that all $\binom{2r}{r}$ possible assignments are equally likely. In what follows, $\mathbf{E}_R$ denotes the expectation taken over the set of all such assignments, equally weighted, and under the additional assumption (the *null hypothesis*) of no differences between treated and control plots. If the yields are $y_1, \ldots, y_{2r}$, and $\bar{y}^T, \bar{y}^C$, and $\bar{y}$ denote the averages of the treated, control, and all plots, respectively, then the treatment and error sums of squares are given by

$$\mathrm{SST} = r\{(\bar{y}^T - \bar{y})^2 + (\bar{y}^C - \bar{y})^2\},$$

$$\mathrm{SSE} = \sum_{i \in T} (y_i - \bar{y}^T)^2 + \sum_{i \in C} (y_i - \bar{y}^C)^2.$$

Write $\sigma^2 = (2r)^{-1} \sum_{i=1}^{2r} (y_i - \bar{y})^2$. Then it is an easy calculation based on the symmetry of the covariance matrix induced by the random assignment to prove that

$$\mathbf{E}\{\mathrm{MST}\} = \mathbf{E}\{\mathrm{MSE}\} = \sigma^2,$$

where $\mathrm{MSE} = \mathrm{SSE}/2(r-1)$ is Fisher's "estimated error," and σ^2 his "real" error in this case.

Fisher asserts (p. 85) that "... the ratio of the real to the estimated error, calculated afresh for each of these arrangements, will be actually distributed in the theoretical distribution by which the significance of the result is tested." This will be trivially true if the actual randomization distribution is used, which can be done these days, but Fisher undoubtedly had in mind a chi-squared approximation to the distribution of SSE/σ^2, and a corresponding F approximation to the distribution of MST/MSE. Eden and Yates (1933) carried out a simulation study that supported this conclusion. Wald and Wolfowitz (1944) proved a theorem that justifies the chi-squared approximation in the simple case discussed above, whereas Welch (1937) and Pitman (1938) presented results that supported, to some extent, the use of the F

distribution to approximate the randomization distribution of MST/MSE in the case of randomized complete block designs and Latin squares. The matter is not a simple one [see Davis and Speed (1988) for some related calculations], and it seems fair to say that the assertion of Fisher quoted above has been shown to be a reasonable approximation to the truth, under certain conditions, for certain designs. The matter is really academic now, for we can actually calculate (or sample) the exact randomization distribution if we wish.

The discussion on p. 86 what is known as Beavan's half-drill strip method [see "Student" (Ref. 2) cited by Fisher, for more complete details] concerns the dispute between Fisher and "Student" mentioned on page 75 above. Fisher felt that the estimate of error used by "Student" in this type of experiment was not valid, whereas "Student" maintained to his death that the problem was more theoretical than practical, that the gain in precision which resulted was more than offset by any theoretical lack of validity. There seems to be little doubt that "Student" was correct on this point.

In his discussion of the Latin square, Fisher notes that the two types of squares he illustrates, the diagonal and Knight's move (also called Knut Vik) squares, had been used previously for variety trials in Ireland and Denmark. Both of these countries were carrying out co-operative agricultural research before the turn of the century [see Gossett (1936) for some historical remarks on their early activities, and Cochran (1976) and references therein]. It is not hard to see that the diagonal square generally underestimates the error, whereas the Knight's move square generally overestimates it [see the discussion in Sec. 34 of Fisher (1935)]. An interesting historical sidelight is the following remark concluding the section just cited: "It is a curious fact that the bias of the Knut Vik square, which was unexpected, appears to be actually larger than that of the diagonal square, which all experienced experimenters would confidently recognise." As Yates (1965–66) points out, this assertion was based on a small arithmetical error committed by Tedin in some numerical experiments; in fact, the "bias" or loss of precision of the diagonal square is exactly equal to the gain in precision of the Knut Vik square.

So that the reader can better appreciate Fisher's remark about asking Nature few questions, it is worth quoting from Russell (1926, p. 989, our italics)

> A committee or an investigator considering a scheme of experiments should first ... ask whether each experiment or question is framed in such a way that a definite answer can be given. The chief requirement is simplicity: *only one question should be asked at a time.*

Biographical Notes

Ronald Aylmer Fisher was born on February 17, 1890 in a suburb of London, England. He showed a special ability in mathematics at an early age and received a good mathematical education at school prior to entering Gonville and Caius College, Cambridge in 1909. At Cambridge, he studied mathematics and physics, and he also developed an interest in genetics and evolutionary theory, an interest that was to remain with him throughout his life. He graduated with first-class honors in 1912 and spent a further year at Cambridge on a studentship in physics.

Upon leaving Cambridge, Fisher successively worked in a finance office, on a farm in Canada, and as a public school mathematics and physics teacher. His first paper, on the fitting of frequency curves by the method we now know as maximum likelihood, was published in 1912, and not long afterwards, he obtained the exact (null) distribution of the normal correlation coefficient. This paper also drew attention to the 1908 paper of "Student" (see this volume), which had been neglected up to that time. Over this period, Fisher was developing his ideas on the statistical aspects of genetics, and in 1918, he published his path-breaking synthesis of ideas from the Mendelian and biometric schools: "The correlation between relatives on the supposition of Mendelian inheritance."

Shortly after the publication of this paper, Fisher joined Rothamsted Experimental Station and went on to make a number of fundamental contributions to statistics and genetics. Further details of his life and work can be found in Yates and Mather (1965) and Box (1978).

References

Arnold, G.C. (1985). The Hudson–Gosset correspondence, *The New Zealand Statistician*, **20**, 20–25.

Barbacki, S., and Fisher, R.A. (1936). A test of the supposed precision of systematic arrangements, *Ann. Eugen.*, **7**, 189–193.

Box, J. F. (1978). *R.A. Fisher. The Life of a Scientist.* Wiley, New York.

Cochran, W.G., (1976). Early development of techniques in comparative experimentation, in *On the History of Statistics and Probability* (D.B. Owen, ed.). New York: Marcel Dekker, pp. 3–25.

Davis, A.W., and Speed, T.P. (1988). An Edgewoth expansion for the distribution of the F-ratio under a randomization model for the randomized block design, in *Statistical Decision Theory and Related Topics IV*, vol. **2**, (S.S. Gupta, and J.O. Berger, eds.). Springer-Verlag, New York, pp. 119–130.

Eden, T., and Yates, F. (1933). On the validity of Fisher's z test when applied to an actual sample of non-normal data, *J. Agric. Sci*, **23**, 6–17.

Fisher, R.A. (1918). The correlation between relatives on the supposition of Mendelian inheritance, *Trans. Roy. Soc. Edinburgh*, **52**, 399–433. (reprinted with a commentary by P.A.P. Moran and C.A.B. Smith as Eugenics Laboratory Memoir XLI, Cambridge University Press, London, 1966).

Fisher, R.A. (1922). On the mathematical foundations of theoretical statistics, *Philos. Trans. Roy. Soc., Lon. Ser. A*, **222**, 309–368.

Fisher, R.A. (1923). Appendix to Gosset (1923).

Fisher, R.A. (1924). On a distribution yielding the error functions of several well known statistics, *Proc. Internat. Congress. Math. Toronto*, **2**, 805–813.

Fisher, R.A. (1925). *Statistical Methods for Research Workers*, Oliver and Boyd, Edinburgh.

Fisher, R.A. (1935). *The Design of Experiments.* Oliver and Boyd, Edinburgh.
Fisher, R.A., and Mackenzie, W.A. (1923). Studies in crop variation. II. The manurial response of different potato varieties, *J. Agric. Sci.*, **13**, 311–320.
Fisher, R.A., and Wishart, J. (1930). The arrangement of field experiments and the statistical reduction of the results, *Imperial Bur. Soil Sci., Tech. Comm.*, **10**, 23.
Gosset, W.S. ("Student") (1923). On testing varieties of cereals, *Biometrika*, **15**, 271–293.
Gosset, W.S. ("Student") (1931). Yield trials, in *Baillière's Encyclopedia of Scientific Agriculture*, London. [reprinted as paper 15 in Gosset (1942)].
Gosset, W.S. ("Student") (1936a). Cooperation in large-scale experiments, *J. Roy. Statist. Soc., Suppl.*, **3**, 115–122.
Gosset, W.S. ("Student") (1936b). The half-drill strip system, Agricultural experiments, letter to *Nature*, **138**, 971.
Gosset, W.S. ("Student") (1937–8). Comparison between balanced and random arrangements of field plots, *Biometrika*, **29**, 191–208.
Gosset, W.S. (1942). *"Student's" Collected Papers* (E.S. Pearson and J. Wishart, eds., with a foreword by L. McMullen). Biometrika Office, University College, London.
Gosset, W.S. (1970). *Letters from W.S. Gosset to R.A. Fisher* (1915—1936) (Summaries by R.A. Fisher with a foreword by L . McMullen). Private circulation.
Goulden, C.H. (1931). Modern methods of field experimentation, *Scientif. Agri.*, **11**, 681.
Gregory, F.G., Crowther, F. and Lambert, A.R. (1932). The interrelation of factors controlling the production of cotton under irrigation in the Sudan, *J. Agric. Sci.*, **22**, 617.
Haines, W.B. (1929). Block 6. Manuring experiment, RRI Experiment Station. First report, *Quar. J. Rubber Res. Inst. Malaya*, **1**, 241–244.
Haines, W.B. (1930a). Block 6. Manuring experiment, RRI Experiment Station. Second report, *Quar. J. Rubber Res. Inst. Malaya*, **2**, 31–35.
Haines, W.B. (1930b). Manuring of Rubber—II. Technique of plot experimentation, *Quar. J. Rubber Res. Inst. Malaya*, **2**, 51–60.
Hall, A.D. (1925). The principles of agricultural experiments, *J. Min. Agric. Great Britain*, **32**, 202–210.
Hall, P., and Selinger, B. (1986). Statistical significance: Balancing evidence against doubt, *Austral. J. Statist.*, **28**, 354–370.
Harrison, C.J., Bose, S.S., and Mahalanobis, P.C. (1935). The effect manurial dressings, weather conditions and manufacturing processes on the quality of tea at Tocklai experimental station, Assam, *Sankhyā*, **2**, 33–42.
Hotelling, H. (1951). The impact of R.A. Fisher on statistics, *J. Amer. Statist. Assoc.*, **46**, 35–46.
Kempthorne, O. (1952). *The Design and Analysis of Experiments.* Wiley, New York.
Kerr, J.D. (1988). Introduction of statistical design and analysis by the Queensland Board of Sugar Experiment Stations, *Austral. J. Statist.*, **30B**, 44–53.
Kirk, L.E. (1929). Field plot technique with potatoes with special reference to the Latin square, *Scientif. Agri.*, **9**, 719–729.
Mahalanobis, P.C. (1938). Professor Ronald Aylmer Fisher, *Sankhyā*, **4**, 265–272.
Mead, R. (1988). *The Design of Experiments: Statistical Principles for Practical Applications.* Cambridge University Press, New York.
Mercer, W.B., and Hall, A.D. (1911). The experimental error of field trials, *J. Agric. Sci.*, **4**, 107–132.
Neyman, J. (1967). R.A. Fisher (1890–1962): An appreciation. *Sci.*, **156**, 1456–1460.
Neyman, J. (1990). On the application of probability theory to agricultural experiments. Essay on principles, *Statist. Sci.* **5**, 465–480. (translation from the original Polish, *Rocz. Nauk. Roln X* (1923) pp. 27–42 by D. Dabrowska, T.P. Speed, ed.).

Patterson, H.D., and Silvey. V. (1980). Statutory and recommended list trials of crop varieties in the United Kingdom, *J. Roy. Statist. Soc., Ser. A.*, **143**, 219–252.
Patterson, H.D., Williams, E.R., and Hunter, E.A. (1978). Block designs for variety trials, *J. Agric. Sci. Cambridge*, **90**, 395–400.
Pearce, S.C. (1983). *The Agricultural Field Experiment. A Statistical Examination of Theory and Practice.* Wiley, New York.
Pitman, E.J.G. (1938). Significance tests which may be applied to samples from any populations: III. The analysis of variance tests, *Biometrika*, **29**, 322–335.
Richey, F.D. (1930). Some applications of statistical methods to agronomic experiments, *J. Amer. Statist. Assoc.*, **25**, 269–283.
Russell, E.J. (1926). Field experiments: How they are made and what they are, *J. Min. Agric. Great Britain*, **32**, 989–1001.
Russell, E.J. (1966). *A History of Agricultural Science in Great Britain.* George Allen & Unwin, London.
Savage, L.J. (1976). On rereading R.A. Fisher (J.W. Pratt, ed.). *Ann. Statist.*, **4**, 441–500.
Seal, H. (1967). The historical development of the Gauss linear model, *Biometrika*, **54**, 1–24.
Seeger, P., Kristensen, K., and Norell, L. (1987). Experimental designs in two dimensions for official sugar beet variety trials in Scandinavia. Report 286, Department of Economics and Statistics, Svenges Lantbruksuniversitet (Swedish University of Agricultural Sciences), Uppsala.
Snedecor, G.W. (1934). *Computation and Interpretation of Analysis of Variance and Covariance.* Collegiate Press, Ames, Iowa.
Tippett, L.H.C. (1935). Some applications of statistical methods to the study of variation of quality in the production of cotton yarn, *J. Roy. Statist. Soc. Suppl.*, **2**, 27–62.
Wald, A., and Wolfowitz J. (1944). Statistical tests based on permutations of the observations. *Ann. Math. Statist.* **15**, 358–372.
Welch, B.L. (1937). On the *z*-test in randomized blocks and Latin squares, *Biometrika*, **29**, 21–51.
Wilkinson, G.N., Eckhart, S.R., Hancock, T.W., and Mayo, O. (1983). Nearest neighbour (NN) analysis of field experiments (with discussion), *J. Roy. Statist. Soc., Ser. B*, **45**, 151–211.
Williams, E.R., and John, J.A. (1989). Construction of row and column designs with contiguous replicates, *Appl. Statist.*, **38**, 149–154.
Wishart, J. (1935). Contribution to the discussion of Tippett (1935), pp. 58–59.
Wood, T.B., and Stratton, F.J.M. (1910). The interpretation of experimental results, *J. Agric. Sci.*, **3**, 417–440.
Yates, F. (1939). The comparative advantages of systematic and randomized arrangements in the design of agricultural and biological experiments, *Biometrika*, **30**, 440–466.
Yates, F. (1964). Sir Ronald Fisher and the design of experiments, *Biometrics*, **20**, 307–321.
Yates, F. (1965–66). A fresh look at the basic principles of the design and analysis of experiments, in *Proceedings of the 5th Berkeley Symposium on Mathematical Statistics and Probability*, Vol. 4. Berkeley: University of California Press. (J. Neyman and E.L. Scott, eds.) pp. 777–790.
Yates, F. (1975). The early history of experimental design, *in A Survey of Statistical Design and Linear Models* (J.N. Srivastava, ed.) North-Holland, Amsterdam: pp. 581–592.
Yates, F., and Mather, K. (1965). Ronald Aylmer Fisher, *Biographical Memoirs of Fellows of the Roy. Soc.*, **9**, 91–120.
Younden, W.J. (1951). The Fisherian revolution in methods of experimentation, *J. Amer. Statist. Assoc.*, **45**, 47–50.

The Arrangement of Field Experiments

R.A. Fisher
Rothamsted Experimental Station

The Present Position

The present position of the art of field experimentation is one of rather special interest. For more than fifteen years the attention of agriculturalists has been turned to the *errors* of field experiments. During this period, experiments of the uniformity trial type have demonstrated the magnitude and ubiquity of that class of error which cannot be ascribed to carelessness in measuring the land or weighing the produce, and which is consequently described as due to "soil heterogeneity"; much ingenuity has been expended in devising plans for the proper arrangement of the plots; and not without result, for there can be little doubt that the standard of accuracy has been materially, though very irregularly, raised. What makes the present position interesting is that it is now possible to demonstrate (*a*) that the actual position of the problem is very much more intricate than was till recently imagined, but that realising this (*b*) the problem itself becomes much more definite and (*c*) its solution correspondingly more rigorous.

The conception which has made it possible to develop a new and critical technique of plot arrangement is that an estimate of field errors derived from any particular experiment may or may not be a valid estimate, and in actual field practice is usually not a valid estimate, of the actual errors affecting the averages or differences of averages of which it is required to estimate the error.

When Is a Result Significant?

What is meant by a valid estimate of error? The answer must be sought in the use to which an estimate of error is to be put. Let us imagine in the broadest outline the process by which a field trial, such as the testing of a material of real or supposed manurial value, is conducted. To an acre of ground the manure is applied; a second acre, sown with similar seed and treated in all other ways like the first, receives none of the manure. When the produce is weighed it is found that the acre which received the manure has yielded a crop larger indeed by, say, 10 per cent. The manure has scored a success, but the confidence with which such a result should be received by the purchasing public depends wholly upon the manner in which the experiment was carried out.

The first criticism to be answered is—"What reason is there to think that, even if no manure had been applied, the acre which actually received it would not still have given the higher yield?" The early experimenter would have had to reply merely that he had chosen the land fairly, that he had no reason to expect one acre to be better than the other, and (possibly) that he had weighed the produce from these two acres in previous years and had never known them to differ by 10 per cent. The last argument alone carries any weight. It will illustrate the meaning of tests of significance if we consider for how many years the produce should have been recorded in order to make the evidence convincing.

First, if the experimenter could say that in twenty years experience with uniform treatment the difference in favour of the acre treated with manure had never before touched 10 per cent., the evidence would have reached a point which may be called the verge of significance; for it is convenient to draw the line at about the level at which we can say: "Either there is something in the treatment, or a coincidence has occurred such as does not occur more than once in twenty trials." This level, which we may call the 5 per cent. point, would be indicated, though very roughly, by the greatest chance deviation observed in twenty successive trials. To locate the 5 per cent. point with any accuracy we should need about 500 years' experience, for we could then, supposing no progressive changes in fertility were in progress, count out the twenty-five largest deviations and draw the line between the twenty-fifth and the twenty-sixth largest deviation. If the difference between the two acres in our experimental year exceeded this value, we should have reasonable grounds for calling the result significant.

If one in twenty does not seem high enough odds, we may, if we prefer it, draw the line at one in fifty (the 2 per cent. point), or one in a hundred (the 1 per cent. point). Personally, the writer prefers to set a low standard of significance at the 5 per cent. point, and ignore entirely all results which fail to reach this level. A scientific fact should be regarded as experimentally established only if a properly designed experiment *rarely fails* to give this level of significance. The very high odds sometimes claimed for experimental results should usually be discounted, for inaccurate methods of estimating error have far more influence than has the particular standard of significance chosen.

Since the early experimenter certainly could not have produced a record of 500 years' yields, the direct test of significance fails; nevertheless if he had only ten previous years' records he might still make out a case, if he could claim that under uniform treatment, the difference had never come *near* to 10 per cent. His argument is now much less direct; he wishes to convince us that such an error as 10 per cent. would occur by chance in less than 5 per cent. of fair trials, and he can only appeal to ten trials. On the other hand, for those ten years he knows the actual value of the error. From these he can calculate a standard error, or rather an estimate of the standard error, to which the experiment is subject; and, if the observed difference is many times greater than this standard error, he claims that it is significant. At how many times greater should he draw the line? This factor depends on the amount of experience upon which the standard error is based. If on ten values, we look in the appropriate published table for "the 5 per cent. value of t, when $n = 10$" and find (1 p. 137) the value 2.228. If, then, the standard error is only 3 per cent., the 5 per cent. point is at 6.684 per cent., and we can admit significance for a difference of 10 per cent.

If we thus put our trust in the theory of errors, all the calculation necessary is to find the standard error. In the simple case chosen above (in which, for simplicity, it is assumed that each of the two acres beats the other equally often) all that is necessary is to multiply each of the ten errors by itself, thus forming its square, to find the average of the ten squares and to find the square root of the average. The average of the ten squares is called the variance, and its square root is called the standard error. The procedure outlined above, relying upon the theory of errors, involves some assumptions about the nature of field errors; but these assumptions arc not in fact disputed, and have been extensively verified in the examination of the results of uniformity trials.

Measurement of Accuracy by Replication

It would be exceedingly inconvenient if every field trial had to be preceded by a succession of even ten uniformity trials; consequently, since the only purpose of these trials is to provide an estimate of the standard error, means have been devised for obtaining such an estimate from the actual yields of the trial year.

The method adopted is that of replication. If we had challenged, as before, the result of an experiment performed, say, ten years ago, we should not probably have been referred to the experience of previous years, but should have learnt that each trial acre was divided into, say, four separate quarters; and that the two acres were systematically intermingled in eight strips arranged ABBAABBA, where A is the manured portion, and B the unmanured.*

* This principle was employed in an experiment on the influence of weather on the effectiveness of phosphates and nitrogen alluded to by Sir John Russell (3). The author must disclaim all responsibility for the design of this experiment, which is, however, a good example of its class.

Besides affording an estimate of error such intermingling of experimental plots is of value in diminishing the actual error representing the difference in actual fertility between the two acres. For it is obvious that such differences in fertility will generally be greater in whole blocks of land widely separated, than in narrow adjacent strips. This important advantage of reducing the standard error of the experiment has often been confused with the main purpose of replication in providing an estimate of error; and, in this confusion, types of systematic arrangement have been introduced and widely employed which provide altogether false estimates of error, because the conditions, upon which a replicated experiment provides a valid estimate of error, have not been adhered to.

Errors Wrongly Estimated

The error of which an estimate is required is that in the difference in yield between the area marked A and the area marked B, *i.e.*, it is an error in the difference between plots treated differently in respect of the manure tested. The *estimate* of error afforded by the replicated trial depends upon differences between plots treated alike. An estimate of error so derived will only be valid for its purpose if we make sure that, in the plot arrangement, pairs of plots treated alike are not nearer together, or further apart than, or in any other relevant way, distinguishable from pairs of plots treated differently. Now in nearly all systematic arrangements of replicated plots care is taken to put the unlike plots as close together as possible, and the like plots consequently as far apart as possible, thus introducing a flagrant violation of the conditions upon which a valid estimate is possible.

One way of making sure that a valid estimate of error will be obtained is to arrange the plots deliberately at random, so that no distinction can creep in between pairs of plots treated alike and pairs treated differently; in such a case an estimate of error, derived in the usual way from the variations of sets of plots treated alike, may be applied to test the significance of the observed difference between the averages of plots treated differently.

The estimate of error is valid, because, if we imagine a large number of different results obtained by different random arrangements, the ratio of the real to the estimated error, calculated afresh for each of these arrangements, will be actually distributed in the theoretical distribution by which the significance of the result is tested. Whereas if a group of arrangements is chosen such that the real errors in this group are on the whole less than those appropriate to random arrangements, it has now been demonstrated that the errors, as estimated, will, in such a group, be higher than is usual in random arrangements, and that, in consequence, within such a group, the test of significance is vitiated. It is particularly to be noted that those methods of arrangement, at which experimenters have consciously aimed, and which reduce the real errors, will appear from their (falsely) estimated standard errors to be not more but less accurate than if a random arrangement had been applied; whereas, if the experimenter is sufficiently unlucky, as must often be the case, to *increase* by his systematic arrangement the real errors, then the (falsely) estimated standard error will now be smaller, and will indicate that the experiment is not less, but more accurate. Opinions will differ as to which event is, in the long run, the more unfortunate; it is evident that in both cases quite misleading conclusions will be drawn from the experiment.

A Necessary Distinction

The important question will be asked at this point as to whether it is necessary, in order to obtain a valid estimate of error, to give up all the advantage in accuracy to be obtained from growing plots, which it is desired to compare, as closely adjacent as possible. The answer is that it is not necessary to give up any such advantage. Two things are necessary, however: (a) that a sharp distinction should be drawn between those components of error which are to be eliminated in the field, and those which are not to be eliminated; and that while the elimination of the one class shall be complete, no attempt shall be made to eliminate the other; (b) that the statistical process of the estimation of error shall be modified so as to take account of the field arrangement, and so that the components of error actually eliminated in the field shall equally be eliminated in the statistical laboratory.

In reconciling thus the two *desiderata* of the *reduction of error* and of the *valid estimation* of error, it should be emphasised that no principle is in the smallest degree compromised. An experiment either admits of a valid estimate of error, or it does not; whether it does so, or not, depends not on the actual arrangement of plots, but only on the way in which that arrangement was arrived at: If the arrangement ABBAABBA was arrived at by writing down a succession of "sandwiches" ABBA, it does not admit of any estimate of certain validity, although "Student" (2) has shown reasons to think that by treating each "sandwich" as a unit, the uncertainties of the situation are much reduced. If, however, the same arrangement happened to occur subject to the conditions that each pair of strips shall contain an A and a B, but that which came first shall be decided by the toss of a coin, then a valid estimate may be obtained from the four differences in yield in the four pairs of strips. It is not now the "sandwiches" but the pairs of strips which provide independent units of information, and these units are double the number of the "sandwiches."

Moreover, if the experiment is repeated, either by replication on the same field, or at different farms scattered over the country, the arrangement must be obtained afresh by chance for each replication, so that in only a small and calculable proportion of cases will the sandwich arrangement be reproduced.

Thus validity of estimation can be guaranteed by appropriate methods of arrangement, and on the other hand there is reason to think that well-designed experiments, yielding a valid estimate of error, and therefore capable of genuine significance tests, will give actual errors as small as even the most ingenious of systematic arrangements. It is difficult to prove this assertion save by experimenting on the data provided by uniformity trials, because, in the absence of any satisfactory estimate of error, it is impossible to tell for certain how accurate, or inaccurate, such systematic arrangements really are; while the aggregate of the uniformity trial data, hitherto available, is scarcely adequate for any such test. What can be said for certain is, that experiments capable of genuine tests of significance can easily be designed to be very much more accurate than any experiments ordinarily conducted.

A Useful Method

The distinction between errors eliminated in the field, and the errors which are to be carefully randomized in order to provide a valid estimate of the errors which cannot be eliminated, may be made most clear by one of the most useful and flexible types of arrangement, namely, the arrangement in "randomized blocks." Let us suppose that five different varieties are to be tested, and that it is decided to give each variety seven plots, making thirty-five in all. It would be a perfectly valid experiment to divide the land into thirty-five equal portions, *in any way one pleased*, and then to assign seven portions chosen wholly at random to each treatment. In such a case, as has been stated above, no modification is introduced in the process of estimating the standard error from the results, for no portion of the field heterogeneity has been eliminated. On most land, however, we shall obtain a smaller standard error, and consequently a more valuable experiment, if we proceed otherwise. The land is divided first into seven blocks, which, for the present purpose, should be as compact as possible; each of these blocks is divided into five plots, and these are assigned in each case to the five varieties, independently, and wholly at random. If this is done, those components of soil heterogeneity which produce differences in fertility *between plots of the same block* will be completely randomized, while those components which produce differences in fertility between different blocks will be completely eliminated. In calculating an estimate of error from such an experiment, care must of course be taken to eliminate the variance due to differences between blocks, and for this purpose exact methods have been developed (1. pp. 176–232).

Most experimenters on carrying out a random assignment of plots will be shocked to find how far from equally the plots distribute themselves; three or four plots of the same variety, for instance, may fall together at the corner where four blocks meet. This feeling affords some measure of the extent to which estimates of error are vitiated by systematic regular arrangements, for, as we have seen, if the experimenter rejects the arrangement arrived at by chance as altogether "too bad," or in other ways "cooks" the arrangement to suit his preconceived ideas, he will either (and most probably) increase the standard error as estimated from the yields; or, if his luck or his judgment is bad, he will increase the real errors while diminishing his estimate of them.

The Latin Square

For the purpose of variety trials, and of those simple types of manurial trial in which every possible comparison is of equal importance, the problem of designing economical and effective field experiments, reduces to two main principles (*i*) the division of the experimental area into the plots as small as possible subject to the type of farm machinery used, and to adequate precautions against edge effect; (*ii*) the use of arrangements which eliminate a maximum fraction of the soil heterogeneity, and yet provide a valid estimate of the residual errors. Of these arrangements, by far the most efficient, as judged by experiments upon uniformity trial data, is that which the writer has named the Latin Square.

Systematic arrangements in a square, in which the number of rows and of columns is equal to the number of varieties, such as

A	B	C	D	E		A	B	C	D	E
E	A	B	C	D		D	E	A	B	C
D	E	A	B	C		B	C	D	E	A
C	D	E	A	B		E	A	B	C	D
B	C	D	E	A		C	D	E	A	B

have been used previously for variety trials in, for example, Ireland and Denmark, but the term "Latin Square" should not be applied to any such systematic arrangements. The problem of the Latin Square, from which the name was borrowed, as formulated by Euler, consists in the enumeration of *every possible* arrangement, subject to the conditions that each row and each column shall contain one plot of each variety. Consequently, the term Latin Square should only be applied to a process of randomization by which one is selected at random out of the total number of Latin Squares possible; or, at least, to specify the agricultural requirement more strictly, out of a number of Latin Squares in the aggregate, of which every pair of plots, not in the same row or column, belongs equally frequently to the same treatment.

The actual laboratory technique for obtaining a Latin Square of this random type, will not be of very general interest, since it differs for 5×5 and 6×6 squares, these being by far the most useful sizes. They may be obtained quite rapidly, and the Statistical Laboratory at Rothamsted is prepared to supply these, or other types of randomized arrangements, to intending experimenters; this procedure is considered the more desirable since it is only too probable that new principles will, at their inception, be, in some detail or other, misunderstood and misapplied; a consequence for which their originator, who has made himself responsible for explaining them, cannot be held entirely free from blame.

Complex Experimentation

Only a minority of field experiments are of the simple type, typified by variety trials, in which all possible comparisons are of equal importance. In most experiments involving manuring or cultural treatment, the comparisons involving single factors, *e.g.*, with or without phosphate, are of far higher interest and practical importance than the much more numerous possible comparisons involving several factors. This circumstance, through a process of reasoning, which can best be illustrated by a practical example, leads to the remarkable consequence that large and complex experiments have a much higher efficiency than simple ones. No aphorism is more frequently repeated in connection with field trials, than that we must ask Nature few questions, or, ideally, one question, at a time. The writer is convinced that this view is wholly mistaken. Nature, he suggests, will best respond to a logical and carefully thought out questionnaire; indeed, if we ask her a single question, she will often refuse to answer until some other topic has been discussed.

A good example of a complex experiment with winter oats is being carried out by Mr. Eden at Rothamsted this year, and is shown in the diagram.

Nitrogenous manure in the form of Sulphate (S), or Muriate (M) of ammonia, is applied as a top dressing *early*, or *late* in the season, in quantities represented by 0, 1, 2. When no manure is applied, we cannot, of course, distinguish between sulphate and chloride, or between early and late applications; nevertheless, since the general comparison 0 *versus* 1 dose is one of the important comparisons to be made, the number of plots receiving no nitrogenous manure (corresponding roughly to the so-called "control" plots of the older experiments) are made to be equal in number to those plots receiving one or two doses. This makes twelve treatments, and these are replicated in the above sketch in eight randomized blocks. Note what a "bad" distribution chance often supplies; the chloride plots are all bunched together in the middle of the first block, while they form a solid band across the top block on the right; in the bottom block on the right, too, all the early plots are on one side, and all the late plots on the other.

The value of such large and complex experiments is that all the necessary comparisons can be made with known and with, probably, high accuracy; any general difference between sulphate and chloride, between early and late

	2 M EARLY	2 S LATE		2 S LATE			1 S EARLY
1 S EARLY	1 M EARLY	1 M LATE	1 S LATE	2 M EARLY	2 M LATE	1 M EARLY	1 M LATE
	2 M LATE		2 S EARLY		1 S LATE		2 S EARLY
2 S EARLY	2 M EARLY		1 M LATE		2 S EARLY	2 S LATE	2 M LATE
	1 S LATE	1 S EARLY	1 M EARLY	1 M LATE			1 S LATE
2 M LATE		2 S LATE		2 M EARLY		1 M EARLY	1 S EARLY
2 S EARLY	2 M LATE	1 S EARLY	2 M EARLY	2 S LATE	2 S EARLY	2 M EARLY	
		1 M LATE		1 M EARLY	2 M LATE		1 M LATE
2 S LATE	1 M EARLY		1 S LATE			1 S EARLY	1 S LATE
2 M EARLY	1 M EARLY	2 M LATE	2 S LATE	1 S EARLY			1 S LATE
1 S LATE			1 M LATE	1 M EARLY	2 S EARLY	2 M LATE	
1 S EARLY		2 S EARLY			2 M EARLY	2 S LATE	1 M LATE

Figure 1. A complex experiment with winter oats.

application, or ascribable to quantity of nitrogenous manure, can be based on thirty-two comparisons, each of which is affected only by such soil heterogeneity as exists between plots in the same block. To make these three sets of comparisons only, with the same accuracy, by single question methods, would require 224 plots, against our 96; but in addition many other comparisons can also be made with equal accuracy, for all combinations of the factors concerned have been explored. Most important of all, the conclusions drawn from the single-factor comparisons will be given, by the variation of non-essential conditions, a very much wider inductive basis than could be obtained, by single question methods, without extensive repetitions of the experiment.

In the above instance no possible interaction of the factors is disregarded; in other cases it will sometimes be advantageous deliberately to sacrifice all possibility of obtaining information on some points, these being believed confidently to be unimportant, and thus to increase the accuracy attainable on questions of greater moment. The comparisons to be sacrificed will be deliberately confounded with certain elements of the soil heterogeneity, and with them eliminated. Some additional care should, however, be taken in reporting and explaining the result of such experiments.

References

(1) R.A. Fisher: *Statistical Methods for Research Workers.* (Oliver & Boyd, Edinburgh, 1925).
(2) "Student": *On Testing Varities of Cereals.* (*Biometrika*, XV, pp. 271–293, 1923.)
(3) Sir John Russell: Field Experiments: How They are Made and What They are. (*Jour. Min. Agric.*, XXXII, 1926, pp. 989–1001.)

Introduction to
Kolmogorov (1933) On the Empirical Determination of a Distribution

M.A. Stephens
Simon Fraser University

1. Introduction

In 1933, A.N. Kolmogorov (1933a) published a short but landmark paper in the Italian Giornale dell'Istituto Italiano degli Attuari. He formally defined the empirical distribution function (EDF) and then enquired how close this would be to the true distribution $F(x)$ when this is continuous. This leads naturally to the definition of what has come to be known as the Kolmogorov statistic (or sometimes the Kolmogorov–Smirnov statistic) D, and Kolmogorov not only then demonstrates that the difference between the EDF and $F(x)$ can be made as small as we please as the sample size n becomes larger, but also gives a method for calculating the distribution of D at specified points, for finite n, and uses this to give the asymptotic distribution of D. The ideas in this paper have formed a platform for a vast literature, both of interesting and important probability problems, and also concerning methods of using the Kolmogorov statistic (and other statistics) for testing fit to a distribution. This literature continues with great strength today, after over 50 years, showing no signs of diminishing. It is evident that the ideas set in motion by Kolmogorov are of paramount importance in statistical analysis, and variations on the probabilistic problems, including modern methods of treating them, continue to hold attention.

2. A.N. Kolmogorov: Early Years and Position in 1933

Andrei Nikolaevich Kolmogorov was born on April 25, 1903. His father was an agronomist who later died in the aftermath of the Revolution; his mother died shortly after his birth and he was brought up by his mother's sister. He was taught by his aunts until he was seven and then went to a gymnasium in Moscow, to which he later gave much credit for his early training. He was interested early on in mathematics, but also in biology and Russian history: these interests widened even more in later life to include, for example, methods of education and poetry. He entered Moscow University in 1920 to study physics and mathematics, but continued his studies in history. He was a student during very difficult times in Russia and in 1922, to augment his income, he became a schoolteacher while still a student, a position he held for three years. Nevertheless, he quickly came to the attention of the Professors at Moscow University and, as quickly, began to produce original results in various areas of mathematics, especially in set theory and Fourier series. In 1924, his lifetime interest in probability theory began and in 1925, he published his first paper in this field with A.Y. Khinchin. Also in 1925, Kolmogorov graduated from Moscow University and became a postgraduate student.

In the years that followed, he published fundamental work on the laws of large numbers; he regarded such laws, thc study of which began with Bernoulli, as the true beginnings of probability. By the time he finished as a postgraduate (as in many European countries at the time, a thesis degree was not deemed necessary), Kolmogorov had written nearly twenty mathematical papers, and in June 1929, he joined the Institute of Mathematics and Mechanics at Moscow University as a faculty member. Two years later, he became professor and two years more saw him appointed Director of the Scientific and Research Institute of Mathematics at the University. Earlier, he had begun his fundamental work in measure theory applied to probability, arising from his concern to provide a rigorous axiomatic foundation for the subject. This was first addressed in his writings in 1929, and then in 1933, the same year as the paper introduced here, he produced his classical monograph on the foundations of probability theory, which was to prove so influential to the development of this subject. Between these two works appeared, in 1933, "On Methods of Analysis in Probability Theory", in which he exhibited the relationships between the theory of probability and the classical analytic methods of theoretical physics. This too was to become a seminal work in the theory of random processes.

The paper considered here was thus written when Kolmogorov was thirty years old, at the height of his mathematical powers, already recognized in the Soviet Union, and increasingly well known outside its borders. It is a brilliant combination of his skill with classical probability arguments combined, as we shall see, with his abilities in mathematical analysis.

For the above summary, the author is greatly indebted to the review of Kolmogorov's life by Shiryaev (1989); a biography of Kolmogorov is also given in Kotz, Johnson, and Read (1989).

3. Summary of the Paper

In this section, the contents of the paper will be outlined in more detail than that given earlier; in subsequent sections, we show some of the ways in which this short article led to advances across the broad fields of probability and statistics.

Suppose a random sample is given of values of X; these are ordered and labeled so that $X_1 \leq X_2 \leq \cdots \leq X_n$. In more modern notation, this would be written $X_{(1)} \leq X_{(2)} \leq \cdots \leq X_{(n)}$, but Kolmogorov's original notation will be used here.

(A) The function $F_n(x)$, called the empirical distribution function (EDF), is defined as

$$F_n(x) = 0, \qquad x < X_1;$$

$$F_n(x) = \frac{k}{n}, \qquad X_k \leq x < X_{k+1} \quad k = 1, 2, \ldots n-1;$$

$$\vdots$$

$$F_n(x) = 1, \qquad X_n \leq x$$

(B) Kolmogorov states that we are "almost naturally" led to ask if $F_n(x)$ is approximately equal to $F(x)$ when n assumes a very large value, and he refers to von Mises' (1931) book where, only two years earlier, he had introduced another statistic to measure how close $F_n(x)$ is to $F(x)$. Kolmogorov defines

$$D = \sup_x |F_n(x) - F(x)|$$

and points out the importance of answering whether $Pr(D < \varepsilon)$ tends to 1 as $n \to \infty$, however small the ε.

(C) He answers the question by proving the following asymptotic result, expressed as Theorem I. Let $\Phi(\lambda) = Pr(D < \lambda/\sqrt{n})$; then $\Phi(\lambda)$, as $n \to \infty$ uniformly in λ, tends to

$$\Phi(\lambda) = \sum_{k=-\infty}^{\infty} (-1)^k e^{-2k^2\lambda^2}$$

for any continuous distribution function $F(x)$. Some values of $\Phi(\lambda)$ are given for various λ; it is pointed out that, for small λ, $\Phi(\lambda)$ converges slowly, and the first term of the equivalent formula

$$\Phi(\lambda) = \frac{\sqrt{2\pi}}{\lambda} \sum_{k=1}^{\infty} \exp\left[-\frac{(2k-1)^2\pi^2}{(8\lambda^2)}\right]$$

then gives excellent results for $\lambda < 0.6$.

(D) The proof of the theorem first involves the probability integral transformation $Y = F(X)$, showing that the distribution of Y is $F(y) = y$, $0 \leq y \leq 1$, namely, the uniform distribution. Also, if D_y is calculated from the EDF of the Y values given by $Y_i = F(X_i)$, $i = 1, 2, \ldots, n$, then D_y will equal D. Thus, the result required may be deduced by assuming that the original values have a uniform distribution between 0 and 1, which we shall write $U(0, 1)$.
(E) The calculations are based on the following argument. Suppose lines U ($y = x + d$) and L ($y = x - d$) are drawn parallel to $y = F(x) = x$. For $D < d$, all the "corners" of $F_n(x)$ must lie between U and L. Suppose P_{ik} is the probability that E_{ik} occurs: E_{ik} is the event that $F_n(x)$ lies between U and L at the values $x = j/n$, for all $j \leq k$, while also, at $x = k/n$, $|F_n(k/n) - (k/n)| = i/n$. Clearly, $P(D < d)$ is then P_{0n}. Kolmogorov gives a formula for P_{ik^*}, where $k^* = k + 1$, as a linear combination of the P_{ij} for $j \leq k$; the coefficients in the expression are conditional probabilities $Q_{ji}(k)$ that E_{ik^*} occurs given that E_{jk} has occurred. These linear equations can be solved for P_{ik} and, hence, for the required P_{0n}.

For practical calculations, Kolmogorov defines new quantities R_{ik} as functions of the P_{ik}; these enable R_{ik^*} to be expressed as linear combinations of R_{jk}, similar to the equations for P_{ik^*}, but with easier coefficients.
(F) At this point, Kolmogorov's analytic skills are brought to bear. Theorem II is given, describing the behavior of a random walk with steps Y_j which are integral multiples of a constant ε. Suppose

$$S_k = \sum_{j=1}^{k} Y_j,$$

and let $S_n = i\varepsilon$ for some i. Kolmogorov gives a result for R_{in}^-, the probability that S_k always lies between certain bounds, in terms of the Green's function of classical mathematical physics. The theorem gives the solution to a much more general problem than that discussed here; it is not proven in detail, but reference is made to an existing note and to one forthcoming [see Kolmogorov (1933b)]. For the particular problem concerning D, the Y_i are made to be Poisson variables, and R_{in}^- is shown to be the same as R_{in} in (E) above.

The steps ε now approach zero, and the random walk becomes "tied down" to zero at the *nth* step, thus becoming the Brownian bridge of modern notation; the application of Theorem II with appropriate boundaries leads to the asymptotic result given in Theorem I.

4. Contemporary Work and the Impact of the Paper

It seems fair to say that Kolmogorov regarded his paper as the solution of an interesting problem in probability, following his interests of the time, rather than a paper on statistical methodology. Apart from the casual remark that $F_n(x)$ should closely estimate $F(x)$ in some sense, no suggestion is made that

$F_n(x)$ should be used for *testing* that $F(x)$ is the distribution of x. This was, nevertheless, to become one of the major outgrowths of the article. Suggestions that $F_n(x)$ should be used for such a test were in the air at the time. Cramér (1928) had proposed expanding $F(x)$ in a type of Gram–Charlier series and then using as test statistics integrals of the type

$$I_j = \int \{\Delta_j(x)\}^2\, dx,$$

where $\Delta_j(x) = F_n(x) - \hat{F}_j(x)$, and $\hat{F}_j(x)$ is the expansion of $F(x)$ up to the *jth* term. The integral is over the support of x. The term $\Delta_j(x)$ can be thought of as the *jth* component of the difference $F_n(x) - F(x)$, and the approach is reminiscent of Neyman's work on smooth tests, which appeared a few years later. In 1931, von Mises suggested that a test could be based on the statistic

$$\omega^2 = n \int \lambda(x)[F_n(x) - F(x)]^2\, dx,$$

where $\lambda(x)$ is a suitably chosen weight function. von Mises suggested that $\lambda(x)$ should be constant, chosen so that $E(\omega^2) = 1$, and with this $\lambda(x)$ von Mises gave a computing formula for ω^2. The distribution of the criterion will vary with $F(x)$ under test [and also, of course, with $\lambda(x)$] even when this is completely specified; von Mises gave no distribution theory, but evaluated some variances of the criterion when the true distribution is uniform or normal.

Several years later, the Soviet mathematician and statistician Smirnov (1936, 1937) made a significant change in the definition of ω^2. This was to write

$$\omega^2 = n \int \lambda(F(x))[F_n(x) - F(x)]^2\, dF(x),$$

so that the integral is with respect to $F(x)$ rather than to x. The criterion now becomes based on the values of $Z_i = F(X_i)$, which, as was seen above in paragraph 3(*D*), will be $U(0, 1)$; it will now be distribution-free, that is, not dependent on the true $F(x)$. This version of the statistic, with $\lambda(F(x)) = 1$, has come to be known as W^2, the Cramér–von Mises statistic. A notable achievement of Smirnov was to find the asymptotic distribution of W^2, in the form of a sum of weighted χ_1^2 variables.

Smirnov (1939a, 1939b) was also interested in Kolmogorov's work; he extended it to encompass one-sided tests and also two-sample tests. Let

$$D^+ = \sup_x\{F_n(x) - F(x)\} \qquad \text{and} \qquad D^- = \sup_x\{F(x) - F_n(x)\}.$$

These will have the same asymptotic distribution, which was found by Smirnov

$$\lim_{n\to\infty} P(\sqrt{n}\, D^+ < \lambda) = 1 - e^{-2\lambda^2}.$$

For two samples, suppose $F_n(x)$ and $G_m(x)$ are the EDF's of two independent random samples of sizes n and m respectively; define $N = mn/(m + n)$, and let

$$D^+_{n,m} = \sup_x\{F_n(x) - G_m(x)\}, \qquad D^-_{n,m} = \sup_x\{G_m(x) - F_n(x)\}, \qquad \text{and}$$

$$D_{n,m} = \sup_x|F_n(x) - G_m(x)|.$$

Smirnov shows that the asymptotic distribution of $\sqrt{N}D_{m,n}$ is the same as that of $\sqrt{n}D$ given in Kolmogorov's Theorem I.

Smirnov (1939a) also examined $V_n(\lambda)$, the number of crossings of $F_n(x)$ with the lines $F(x) \pm \lambda\sqrt{n}$, and shows that as $n \to \infty$, $P(V_n(\lambda) \leq t\sqrt{n})$ converges to

$$\Theta(t, \lambda) = 1 - 2 \sum_{m=0}^{\infty} \frac{(-1)^m}{m!} \frac{d^m}{dt^m}\left[t^m \exp\left\{-\frac{(t + 2\lambda m + 2\lambda)^2}{2}\right\}\right].$$

He also gave a new proof of Kolmogorov's Theorem I and tabulated the asymptotic distribution $\Phi(\lambda)$ in Smirnov (1939b); in Smirnov (1944), he found the distribution of $\sqrt{n}D^+$. The table of $\Phi(\lambda)$ was later reproduced in English in Smirnov (1948). Statistics of the D^+, D^-, and D type are often referred to as Kolmogorov–Smirnov statistics.

5. The War and Afterwards

Thus, over a period of about 10 years, the foundations were laid by a number of distinguished mathematicians of methods of testing fit to a distribution based on the EDF. To test the null hypothesis H_0 that $F(x)$, completely specified, is the true distribution of X, the statistics above may be calculated and referred to the appropriate distribution.

At this point, the war intervened and much momentum in this field was certainly lost. Kolmogorov himself became involved in the war effort (he worked, for example, on artillery problems), which certainly brought him into greater contact with statistical analysis, and may account for an increasing interest in statistics itself. In 1948, he edited and wrote a preface to the Russian edition of Cramér's *Mathematical Methods of Statistics*; he protested the overly theoretical basis of the training of Soviet statisticians, a lament familiar enough outside the Russian borders. Perhaps also, Kolmogorov was impressed by Cramér's opening, which gives great credit to British and American statisticians for advances in statistics, while admiring France and Russia for their excellence in probability; at any rate, in that year he spoke at the Tashkent Conference on Mathematical Statistics, on "Basic Problems of Theoretical Statistics" and also enlightened the assembled statisticians on "The Real Meaning of the Analysis of Variance." This was to be followed, as time passed, by many more contributions to the mainstream of statistics, while, of course, his other wide interests were maintained. These came to include, with the years, an increasing interest in the teaching of both mathematics and statistics.

In the 1950s, there was a surge of interest in Russia in Kolmogorov–Smirnov statistics, particularly in combinatoric problems associated with crossings and with two-sample statistics. Gnedenko and Korolyuk (1951) found the exact distributions of $D^+_{n,n}$, and of $D_{n,n}$, to compare two empirical distributions from independent samples both of size n: later Korolyuk (1955) found exact distribution theory when m is an integral multiple of n, $m = np$. By allowing $p \to \infty$, he deduced the exact distribution of D^+ and also the more difficult distribution of D itself. Gnedenko and Rvaceva (1952) obtained the joint distribution of $D^+_{n,n}$ and $D^-_{n,n}$ and verified the asymptotic joint distribution already found by Smirnov in 1939; further results were given by Gnedenko (1952). Gnedenko and Mihalevic (1952a, 1952b) discussed the number of crossings, when one distribution function $F_n(x)$ crosses the other $G_m(x)$. The interest spread to Hungary and across Asia to China: Renyi (1953) proposed several variations of Kolmogorov's statistic, such as $\sup_x |\{F_n(x) - F(x)\}/F(x)|$; Chang (1955) examined the ratio of $F_n(x)/F(x)$, closely related to Renyi's statistics; and Cheng (1958) gave further results on crossings.

Meantime, in the western world also, EDF statistics were attracting attention. An elegant paper by Feller (1948) appeared, giving more accessible proofs of the results of both Kolmogorov and Smirnov: where Kolmogorov used Green's function to find the asymptotics from the equations for $P(D < c/n)$, Feller introduced generating functions for the component probabilities and then examined their limiting forms. He also gave a theorem on the asymptotic expectation of the number of crossings $V_n(\lambda)$ of $F_n(x)$ with the boundaries $F(x) \pm \lambda/\sqrt{n}$. At about the same time, there were significant advances in methodology. Doob, in 1949, suggested that the asymptotic behaviour of EDF statistics based on $\varepsilon_n(x) = F_n(x) - F(x)$ could be found by examining the limiting behaviour of $\sqrt{n}\,\varepsilon_n(x)$, a Gaussian process, and calculating the statistics from this limiting process. According to Khmaladze (1986), in an article presenting the 1933 paper in Kolmogorov's collected works, Kolmogorov himself put forward similar ideas in a Moscow seminar toward the end of 1948, and Smirnov (1949) wrote a brief paper on the asymptotics of the Cramér–von Mises statistic. These ideas, those of Doob made rigorous by Donsker (1952), laid the foundation for a great deal of later work on the asymptotics of EDF statistics. Anderson and Darling (1952) used them to examine such statistics and introduced the statistic A^2, for which the weight function in Smirnov's version of ω^2 is $1/[F(x)\{1 - F(x)\}]$. This compensates for the fact that $\varepsilon_n(x)$ must necessarily become small in the tails, by essentially dividing by the variance of $\varepsilon_n(x)$, and gives due weight to tail observations.

These developments demonstrated elegant techniques of combinatorics and analysis in the field of probability, but apart from some asymptotic tables, the practical statistician was largely neglected. However, in the 1950s, other authors were filling the gap. Massey (1950, 1951a), and Birnbaum and Tingey (1951), using new formulas and difference equations, gave tables of percentage points and of probabilities for finite sample size n, for D and D^+; these were later augmented by Miller (1956). Birnbaum (1952), using the original techniques of Kolmogorov himself, gave complete tables of the distribution of D, and a table of percentage points for n up to 100. Thus, at last—nearly twenty years after the statistic was suggested!—practical formulas and tables were available to make D available to test that $F(x)$ is a completely specified continuous distribution.

Many years later again, Stephens (1970) used these tables to derive a modification of D. This is an expression in D and n that gives D^*; this is to be compared, for testing purposes, with the asymptotic points for $\sqrt{n}D$ given by Kolmogorov's Theorem I. The test is thus made easy to use without extensive tables of points for every n. Stephens (1970) also found similar modifications for D^+ and D^-, for $V = D^+ + D^-$ (see Sec. 7 below), and for the Cramér–von Mises W^2.

For two samples, Massey (1951b) and Drion (1952) gave tables for $D_{n,n}$ and Massey (1952) for $D_{m,n}$, mostly for $n = mp$, where p is an integer; practical formulas for the calculation of these statistics also began to appear in the literature. In addition, it was pointed out [see Wald and Wolfowitz (1939), Massey (1950), Birnbaum and Tingey (1951)] that D can be used to give a confidence interval for $F(x)$, and D^+ a one-sided interval.

At this point, all attempt will be abandoned to survey exhaustively the enormous literature which has developed on Kolmogorov–Smirnov statistics and on other EDF statistics; many more properties of D, D^+, and D^- have been discovered, new methods of computing distributions proposed, and variants of the basic statistics suggested. Durbin (1973) provides a comprehensive and unifying account of developments up to that time, with many references; Niederhausen (1981b) also offers references and brings together many of the computational procedures. A survey of goodness-of-fit tests may be found in Kendall and Stuart (1979, Vol. 2, Chap. 30) and another was given by Sahler (1968). EDF tests are surveyed by Stephens (1986).

6. The Problem of Unknown Parameters

Despite the interest of mathematical statisticians and the availability of tables, it has taken many years for the Kolmogorov–Smirnov statistics, and other EDF statistics, to become part of the regular arsenal of applied statisticians. No doubt, this is because major new problems are presented if tests are to be made on $F(x)$, which we now call $F(x; \theta)$, when $F(x; \theta)$ is a continuous distribution containing parameters that are components of the vector θ, and when one or more of these components must be estimated from the given data set. For the well-established Pearson X^2 test, provided the estimation of parameters is done correctly—but how often it is not!—the asymptotic χ^2 distribution on H_0 merely changes its degrees of freedom, but for D^+, D^-, and D (and for other EDF statistics), the distribution theory will depend on the particular $F(x; \theta)$ being tested. This is so even when the unknown components of θ are estimated by maximum likelihood or another efficient method; the distributions, even asymptotic, are now stochastically much smaller than for the case when $F(x; \theta)$ is completely known. For Kolmogorov–Smirnov statistics, they depend asymptotically on the distribution of the maximum of a Gaussian process with mean zero, tied down at 0 and 1; even though the process covariance can be found, this distribution remains unknown and the early techniques of Kolmogorov will not find it. The discovery of the asymptotics of D^+, D^-, and D, when parameters must be estimated, thus remains a major theoretical problem in the area of Kolmogorov–Smirnov statistics.

If the unknown components of θ are only location or scale parameters, however, the distribution theory of all EDF statistics, even for finite n, will depend only on the family tested, and not on the true values of these parameters, a fact early recognized by David and Johnson (1948). In these circumstances, Durbin (1973, 1975) has shown how exact distributions of D^+ and D can be calculated for the exponential distribution $F(x; \theta) = 1 - \exp(-x/\theta)$, $x \leq 0$, with unknown scale θ, and has provided points for test purposes; for other distributions, including the normal, extreme-value, Weibull, and logistic distributions, several authors have produced Monte Carlo tables. For Cramér–von Mises statistics, the situation is different; asymptotic distributions can be found (see, for example, Darling, 1955, Durbin, 1973, and Stephens, 1976) and percentage points for finite n converge rapidly to the asymptotic points. Also, for some important distributions with shape parameters, for example, the von Mises and gamma, the asymptotic points for Cramér–von Mises statistics do not depend strongly on the true value of the shape, and a test using the estimated shape can be used [Lockhart and Stephens (1985a, 1985b)]. The tests described above, for parameters known or unknown, have been collected in Stephens (1986).

7. Further Developments

We conclude this introduction by giving only a brief summary of some of the more important developments of Kolmogorov–Smirnov tests, with references either to basic introductory sources or to articles that themselves survey the particular area and give references.

Kolmogorov-Smirnov tests have been developed for use with right- or left-censored data (or both): these mostly use D, but some variations of the Renyi type, such as taking the supremum of $F_n(x) - F(x)$ over a restricted range of $F(x)$ or of $F_n(x)$, have also been suggested. Randomly censored data is an important problem, for example, with survival data: tests with such data often use the Kaplan–Meier estimate of $F(x)$. Hall and Wellner (1980) give a review and show how confidence bounds for the distribution can be found. A recent technique for censored data was given by Guilbaud (1988).

The statistic $V = D^+ + D^-$ has been proposed [Kuiper (1960)] for use with data on a circle, because the value of V, in contrast to those of D^+, D^-, or D, does not depend on the choice of origin. Of course, V can also be used for data on a line. Pettitt and Stephens (1977) produced tables for D for the uniform distribution for discrete data, and Niederhausen (1981a) for a variance-weighted D, similar to A^2. A test for the symmetry of a distribution was proposed by Smirnov (1947) and has since been extended; Gibbons (1983) gives a review of such tests. Tables for some of the above tests, and further discussion and references, can be found in Stephens (1983, 1986). An interesting area for future work is to provide tests for multivariate distributions.

Statistics closely related to D^+, D^-, and D were proposed by Pyke (1969). Suppose x_i, $i = 1, \ldots, n$ are the order statistics for a sample from the uniform distribution; C^+ is $\max_i [x_i - i/(n+1)]$, $C^- = \max_i [i/(n+1) - x_i]$, and $C = \max(C^+, C^-)$. These arise naturally in examining the Poisson process or the periodogram in time series analysis: they are discussed by Durbin (1973).

8. Power

In terms of power, Kolmogorov–Smirnov tests tend to fall between the Pearson X^2 and Cramér–von Mises tests. On the one hand, this might be expected, since X^2 loses information in a test for a continuous distribution by grouping the data into cells. Kac, Kiefer, and Wolfowitz (1955) showed that if equi-probable cells are used for X^2 and if $\Delta = \sup_x |F_1(x) - F(x)|$, where $F_1(x)$ is the true distribution and $F(x)$ the tested distribution, D requires $n^{4/5}$ observations compared with n observations for X^2 to attain the same power for a given Δ, for large n. Thus, in these circumstances, X^2 will have asymptotic relative efficiency equal to zero compared with D. Many Monte Carlo studies have confirmed this superiority of D over X^2 in most situations, especially with small samples.

On the other hand, Cramér–von Mises statistics might well be expected to be superior to D, since they make a comparison of $F_n(x)$ with $F(x)$ all along the range of x, rather than looking for a marked difference at one point. If the alternative is directional, that is, if $F_1(x) - F(x)$ is mostly positive or mostly negative, the one-sided D^+ or D^- can be very powerful. Of all the different families of goodness-of-fit statistics, Cramér–von Mises statistics provide overall powerful tests. [Stephens (1974); also Kendall and Stuart (1979) and Stephens (1986) for further discussion.]

9. Concluding Remarks.

If these remarks on power appear to weaken the appeal of D and its related statistics, it should nonetheless be emphasized that they are preferable to the much used X^2 statistic. They also have the value that they can be used, by simply adding a constant to $F_n(x)$, and subtracting it from $F_n(x)$, to give a confidence interval for $F(x)$, an attraction in today's world where graphical display is increasingly available.

The final assessment of the article by Kolmogorov must be based not only on the elegance and power of the paper itself, but also on the pioneering role it has played in the development of statistics in the succeeding 50 years and more. It launched the use of the EDF $F_n(x)$ as an estimator of $F(x)$, to be followed by its use in testing a given $F(x)$; it was the first article to give a statistic that would not depend (when the null hypothesis was true) on the distribution $F(x)$ tested; it was also the first to introduce a statistic whose asymptotic distribution could be found and easily tabulated. Kolmogorov also gave the essential technique for finding the distribution for finite samples. More than 50 years later, interest in Kolmogorov's and other EDF statistics continues unabated. It is fitting, in conclusion, to note the resurgence of $F_n(x)$ in the wide use of the bootstrap: this technique, making use of the power that modern computers provide, is based on the use of $F_n(x)$ to estimate $F(x)$, just as was proposed by Kolmogorov in 1933.

References

Anderson, T.W., and Darling, D.A. (1952). Asymptotic theory of certain goodness of fit criteria based on stochastic processes, *Ann. Math. Statist.*, **23**, 193–212.

Birnbaum, Z.W. and Tingey, F.H. (1951). One-sided confidence contours for distribution functions, *Ann. Math. Statist.*, **22**, 592–596.

Birnbaum, Z.W. (1952). Numerical tabulation of the distribution of Kolmogorov's statistic for finite sample size, *J. Amer. Statist. Assoc.*, **47**, 425–441.

Cramér, H. (1928). On the composition of elementary errors. Second paper: Statistical Applications. *Skand. Aktuarietidskrift*, **11**, 171–180.

Chang, L.C. (1955). On the ratio of an empirical distribution function to the theoretical distribution function, *Acta Math. Sinica*, **5**, 347–368 [also *Selected Translations in Math. Statist. Prob.*, **4**, (1963), 17–38].

Cheng, P. (1958). Non-negative jump points of an empirical distribution function relative to a theoretical distribution function, *Acta Math. Sinica*, **8**, 333–347 [also *Selected Translations in Math. Statist. Prob.*, **3**, (1962), 205–224].

Darling, D.A. (1955). The Cramér–Smirnov test in the parametric case, *Ann. Math. Statist.*, **26**, 1–20.

David, F.N., and Johnson N.L. (1948). The probability integral transformation when parameters are estimated from the sample. *Biometrika*, **35**, 182–192.

Donsker, M.D. (1952). Justification and extension of Doob's heuristic approach to the Kolmogorov–Smirnov theorems. *Ann. Math Statist.* **23**, 277–281.

Doob, J.L. (1949). Heuristic approach to the Kolmogorov–Smirnov theorems, *Ann. Math. Statist.*, **20**, 393–403.

Drion, E.F. (1952). Some distribution-free tests for the difference between two empirical cumulative distribution functions, *Ann. Math. Statist.*, **23**, 563–574.

Durbin, J. (1973). *Distribution Theory for Tests Based on the Sample Distribution Function.* CBMS-NSF Regional Conference Series in Applied Mathematics, Society for Industrial and Applied Mathematics, Philadelphia, Pa.

Durbin, J. (1975). Kolmogorov-Smirnov tests when parameters are estimated with applications to tests of exponentiality and tests on spacings, *Biometrika*, **62**, (1), 5–22.

Feller, W. (1948). On the Kolmogorov-Smirnov limit theorems for empirical distributions, *Ann. Math. Statist.*, **19**, 177–189.

Gibbons, J.D. (1983). Kolmogorov–Smirnov symmetry test, in *Encyclopedia of Statistical Sciences* (S. Kotz, N.L. Johnson, and C.B. Read, eds.) vol. 4. Wiley, New York, pp. 396–398.

Gnedenko, B.V. (1952). Some results on the maximum discrepancy between two empirical distributions, *Dokl. Akad. Nauk SSSR*, **82**, 661–663 [also *Selected Translations in Math. Statist. Prob.*, **1**, (1961), 73–76].

Gnedenko, B.V., and Korolyuk, V.S. (1951): On the maximum discrepancy between two empirical distributions, *Dokl. Akad. Nauk SSSR*, **80**, 525–528 [also *Selected Translations in Math. Statist. Prob.*, **1** (1961), 13–22].

Gnedenko, B.V., and Mihalevic, V.S. (1952a). On the distribution of the number of excesses of one empirical distribution function over another, *Dokl. Akad. Nauk SSSR*, **82**, 841–843 [also *Selected Translations in Math. Statist. Prob.*, **1** (1961), 83–85].

Gnedenko, B.V., and Mihalevic, V.S. (1952b). Two theorems on behaviour of empirical distribution functions, *Dokl. Akad. Nauk SSSR*, **85**, 25–27 [also *Selected Translations in Math. Statist. Prob.*, **1** (1961), 55–58].

Gnedenko, B.V., and Rvaceva, E.L. (1952). On a problem of the comparison of two empirical distributions, *Dokl. Akad. Nauk SSSR*, **82**, 513–516 [also *Selected Translations in Math. Statist. Prob.*, **1** (1961), 69–72].

Guilbaud, O. (1988). Exact Kolmogorov-type tests for left truncated and/or right-censored data, *J. Amer. Statist. Assoc.*, **83**, 213–221.

Hall, W.J., and Wellner, J.A. (1980). Confidence band for a survival curve from censored data, *Biometrika*, **67**, 133–143.

Kac, M., Kiefer, J., and Wolfowitz, J. (1955). On tests of normality and other tests of goodness of fit based on distance methods, *Ann. Math. Statist.*, **26**, 189–211.

Khmaladze, E.V. (1986). Introduction to Kolmogorov (1933). In *Teoriia Veroiatnostei I Matematicheskcia Statistika.* Moskva Nautka: Moscow.

Kendall, M.G., and Stuart, A. (1979). *The Advanced Theory of Statistics*, 4th ed. McMillan, New York.

Kolmogorov, A. (1931). Über die analytischen Methoden in der Wahrscheinlichkeitsrechnung, *Math. Ann.*, **104**, 415–458.

Kolmogorov, A. (1933a). Sulla determinazione empirica di una legge di distribuzione, *Ist. Ital. Attuari*, G., **4**, 1–11.

Kolmogorov, A. (1933b). Über die Grenzwertsätze der Wahrscheinlichkeitsrechnung, *Bull. (Izvestija) Acad. Sci. URSS*, 363–372.

Koroljuk, V.S. (1955). On the discrepancy of empiric distributions for the case of two independent samples, *Izv. Akad. Nauk SSSR Ser. Mat.*, **19**, 91–96 [also *Selected Translations in Math. Statist. Prob.*, **4** (1963), 105–122].

Kotz, S., Johnson, N.L., and Read C.B. (eds.) (1989). Kolmogorov, Andrei Nikoleyevich, in *Encyclopedia of Statistical Sciences, Suppl. Volume.* Wiley, New York, 78–80.

Kuiper, N.H. (1960): Tests concerning random points on a circle. *Proc. Koninkl. Neder. Akad. van. Wetenschappen*, **A**, 63, 38–47.

Lockhart, R.A., and Stephens, M.A. (1985a). Goodness-of-fit tests for the gamma distribution. Technical report, Department of Mathematics and Statistics, Simon Fraser University.

Lockhart, R.A., and Stephens, M.A. (1985b). Goodness-of-fit tests for the von Mises distribution. *Biometrika*, **72**, 647–652.

Massey, F.J. (1950). A note on the estimation of a distribution function by confidence limits, *Ann. Math. Statist.*, **21**, 116–119.

Massey, F.J. (1951a). The Kolmogorov–Smirnov tests for goodness of fit, *J. Amer. Statist. Assoc.*, **46**, 68–78.

Massey, F.J. (1951b). The distribution of the maximum deviation between two sample cumulative step functions, *Ann. Math. Statist.*, **22**, 125–128.

Massey, F.J. (1952). Distribution table for the deviation between sample cumulatives. *Ann. Math. Statist.*, **23**, 435–441.

Miller, L.H. (1956). Table of percentage points of Kolmogorov statistics, *J. Amer. Statist. Assoc.*, **51**, 111–121.

Niederhausen, H. (1981a). Tables of significant points for the variance-weighted Kolmogorov–Smirnov statistics. Technical report, Department of Statistics, Stanford University.

Niederhausen, H. (1981b). Sheffer polynomials for computing exact Kolmogorov–Smirnov and Renyi type distributions, *Ann. Statist.* **5**, 923–944.

Pettitt, A.N., and Stephens, M.A. (1977): The Kolmogorov–Smirnov goodness-of-fit statistics with discrete and grouped data, *Technometrics*, **19**, 205–210.

Pyke, R. (1959). The supremum and infimum of the Poisson process, *Ann. Math. Statist.*, **30**, 568–576.

Renyi, A. (1953). On the theory of order statistics, *Acta Math. Acad. Sci. Hungary*, **4**, 191–231.

Sahler, W. (1968). A survey of distribution-free statistics based on distances between distribution functions. *Metrika*, **13**, 149–169.

Shiryaev, A.N. (1989). Kolmogorov's life and creative activities, *Ann. Prob.* **17**, 866–944.

Smirnov, N.V. (1936, 1937). Sur la distribution de ω^2 (criterium de M.R. von Mises), *C. R. Acad. Sci. (Paris)*, **202**, (1936), 449–452, [paper with the same title in Russian, *Recueil Math.*, **2** (1937), 973–993].

Smirnov, N.V. (1939a). Ob uklonenijah empiriceskoi krivoi raspredelenija, *Recueil Math. Mat. Sbornik, N.S.*, **6** (48), 13–26.

Smirnov, N.V. (1939b). On the estimation of the discrepancy between empirical curves of distributions for two independent samples, *Bull. Math. Univ. Moscou*, **2**, 2.

Smirnov, N.V. (1944). Approximate laws of distribution of random variables from empirical data, *Uspehi Mat. Nauk*, **10**, 179–206.

Smirnov, N.V. (1947). Sur un critére de symetrie de la loi de distribution d'une variable aléatoire, *Akad. Nauk SSSR, C.R. (Dokladi)*, **56**, 11–14.

Smirnov, N.V. (1948). Table for estimating the goodness of fit of empirical distributions, *Ann. Math. Statist.*, **19**, 279–281.

Smirnov, N.V. (1949). On the Cramér–von Mises criterion (in Russian), Uspehi. Mat. Nauk, **14**, 196–197.

Stephens, MA. (1970). Use of the Kolmogorov–Smirnov, Cramér–von Mises and related statistics without extensive tables, *J. Roy. Statist. Soc., Ser. B*, **32**, 115–122.

Stephens, M.A. (1974). EDF statistics for goodness-of-fit and some comparisons, *J. Amer. Statist. Assoc.*, **69**, 730–737.

Stephens, M.A. (1976). Asymptotic results for goodness-of-fit statistics with unknown parameters. *Ann. Statist.*, **4**, 357–369.

Stephens, M.A. (1983). Kolmogorov–Smirnov statistics; Kolmogorov–Smirnov tests of fit, in *Encyclopedia of Statistical Sciences* (S. Kotz, N.L. Johnson, and C.B. Read, eds.) vol. 4. Wiley, New York, 393–396; 398–402.

Stephens, M.A. (1986). Tests based on EDF statistics, in *Goodness-of-Fit Techniques* (R.B. D'Agostino, and M.A. Stephens, eds.). Marcel Dekker, New York, Chap. 4.

von Mises, R. (1931). *Vorlesungen aus dem Gebiete der Angewandten Mathematik*, **1**, *Wahrscheinlichkeitsrechnung und ihre Anwendung in der Statistik und theoretischen Physik*. Springer: Wien.

Wald, A. and Wolfowitz, J. (1939). Confidence limits for continuous distribution functions, *Ann. Math. Statist.*, **10**, 105–118.

On the Empirical Determination of a Distribution Function

A. Kolmogorov
[translated from the Italian by Dr. Quirino Meneghini (1990)]

Summary. Contribution to the study of the possibility of determining a distribution function, knowing the results of a finite number of trials.

1. Let $X_1, X_2, \ldots, X_n$ be the results of n observations, mutually independent, ordered in an increasing manner $X_1 \leq X_2 \leq \cdots \leq X_n$ and let

$$F(x) = P\{X \leq x\}$$

be a distribution function that corresponds to that succession of values.

The function $F_n(x)$ is known as the *empirical distribution function* and is defined by the relations,

$$\begin{aligned} F_n(x) &= 0, && x < X_1; \\ F_n(x) &= k/n, && X_k \leq x < X_{k+1}, k = 1, 2, \ldots, n-1; \\ F_n(x) &= 1, && X_n \leq x. \end{aligned}$$

Consequently, $nF_n(x)$ represents the number of values X_k that do not exceed the value x. We are almost naturally led to ask ourselves if $F_n(x)$ is approximately equal to $F(x)$ when n assumes a very large value. A theorem that addresses this question has been given by von Mises[1] under the name of ω^2 method. However, I do not believe the fundamental proposition that the probability of the inequality

$$D = \sup_x |F_n(x) - F(x)| < \varepsilon$$

tends towards unity for $n \to \infty$, whatever the number ε, has been, to this day,

[1] Wahrscheinlichkeitsrechnung (1931), pp. 316–335.

Editor's note: Corrections supplied by M.A. Stephens are incorporated in this translation.

expressly formulated, although the proof of this can be shown in many simple ways.

I will deduce it as an immediate consequence of the following theorem. I must add that I was led to the formulation of the problem, here treated, by recent research by Glivenko.[2]

Theorem I. *The probability* $\Phi_n(\lambda)$ *of the inequality*

$$D = \sup_x |F_n(x) - F(x)| < \lambda\sqrt{n}$$

tends, for $n \to \infty$ *uniformly in* λ, *towards*

$$\Phi(\lambda) = \sum_{-\infty}^{+\infty} (-1)^k e^{-2k^2\lambda^2} \tag{1}$$

for any continuous distribution function $F(x)$.

Here are, in the meantime, some values of $\Phi(\lambda)$ calculated by N. Kogeknikov:

λ	$\Phi(\lambda)$	λ	$\Phi(\lambda)$	λ	$\Phi(\lambda)$
0.0	0.0000	1.0	0.7300	2.0	0.99932
0.2	0.0000	1.2	0.8877	2.2	0.99986
0.4	0.0028	1.4	0.9603	2.4	0.999973
0.6	0.1357	1.6	0.9880	2.6	0.9999964
0.8	0.4558	1.8	0.9969	2.8	0.99999966

From these values, one notes that the inequality $D \leq 2.4/\sqrt{n}$ can be considered as practically certain. Also note that

$$\Phi(0.83) \cong 0.5.$$

When λ is very small, the series (1) converges very slowly; in that case, one can use the asymptotic formula

$$\Phi(\lambda) \cong \frac{\sqrt{2\pi}}{\lambda} \exp\left(\frac{-\pi^2}{8\lambda^2}\right).$$

This formula, for $\lambda = 0.6$, gives us

$$\Phi(0.6) \cong 0.1327,$$

instead of

$$\Phi(0.6) = 0.1357,$$

calculated from the exact formula (1).

[2] Refer to the treatment, using the same argument, in this issue of the "Journal".

2. Lemma. *The probability function $\Phi_n(\lambda)$ is independent of the distribution function $F(x)$ on condition that the latter be continuous.*

Proof. Let X be a random variable with a continuous distribution function $F(x)$; to the random variable $Y = F(X)$ corresponds evidently a distribution function $F^0(x)$ such that

$$F^{(0)}(x) = 0, \qquad x \le 0;$$

$$F^{(0)}(x) = x, \qquad 0 \le x \le 1;$$

$$F^{(0)}(x) = 0, \qquad x \ge 1.$$

Given that $F_n(x)$ and $F_n^{(0)}(x)$ represent the empirical distribution functions for X and Y after n observations, the following equalities hold:

$$F_n(x) - F(x) = F_n^{(0)}[F(x)] - F^{(0)}[F(x)] = F_n^{(0)}(y) - F^{(0)}(y),$$

$$\sup_x |F_n(x) - F(x)| = \sup_x |F_n^{(0)}(y) - F^{(0)}(y)|.$$

It then follows that the probability function $\Phi_n(\lambda)$, corresponding to an arbitrary continuous distribution function $F(x)$, is identical to that which corresponds to the function $F^{(0)}(x)$. □

One can therefore limit oneself, in proving our theorem, to the consideration of the case $F(x) = F^{(0)}(x)$.

We will denote henceforth $F_n^{(0)}(x)$ by $F_n(x)$ and we will refer to the segment $0 \le x \le 1$ in which $F^{(0)}(x) = x$. Our problem then reduces to that of the determination of the probability $\Phi_n(\lambda)$, that is

$$D = \sup_x |F_n(x) - x| < \lambda/\sqrt{n}, \qquad 0 \le x \le 1. \tag{2}$$

Suppose furthermore that λ is of the form

$$\lambda = \mu/\sqrt{n}$$

with μ an integer. Substituting in (2) one will obtain

$$\Phi_n(\lambda) = P\{\sup_x |F_n(x) - x| < \mu/n\}, \qquad \lambda = \mu/\sqrt{n}.$$

The function $F_n(x)$ assumes only values equal to various multiples of $1/n$; say, for example, $F_n(x) = i/n$ and $x = j/n + \varepsilon$, $(0 \le \varepsilon \le 1/n)$.

Keeping in mind that $F_n(x)$, being a distribution function, is monotone, it follows immediately that

$$F_n(x) - x = \frac{i-j}{n} - \varepsilon,$$

$$F_n\left(\frac{j}{n}\right) - \frac{j}{n} \le F_n(x) - (x - \varepsilon) = \frac{i-j}{n},$$

$$F_n\left(\frac{j+1}{n}\right) - \frac{j+1}{n} \ge F_n(x) - \left(x + \frac{1}{n} - \varepsilon\right) = \frac{i-j-1}{n}.$$

For the inequality

$$|F_n(x) - x| = \left|\frac{i-j}{n} - \varepsilon\right| \geq \mu/n$$

to hold, it is therefore necessary that at at least one of the two following inequalities be satisfied:

$$F_n\left(\frac{j}{n}\right) - \frac{j}{n} \leq \frac{i-j}{n} \leq -\frac{\mu}{n},$$

$$F_n\left(\frac{j+1}{n}\right) - \frac{j+1}{n} > \frac{i-j-1}{n} \geq \frac{\mu}{n}.$$

One can conclude that formula (3) can be replaced by the following:

$$\Phi_n(\lambda) = P\{\max|F_n(i/n) - i/n| < \mu/n\}, i = 0, 1, \ldots, n. \tag{4}$$

Now, let P_{ik} be the probability of occurence of the event E_{ik}, under which the following relations are simultaneously valid:

$$\left|F_n\left(\frac{j}{n}\right) - \frac{j}{n}\right| < \frac{\mu}{n}, \qquad j = 0, 1, 2, \ldots, k;$$

$$\left|F_n\left(\frac{k}{n}\right) - \frac{k}{n}\right| = \frac{i}{n}. \tag{5}$$

One notes that

$$\Phi_n(\lambda) = P_{0n}. \tag{6}$$

For $k = 0$ one has, evidently,

$$P_{00} = 1, \qquad P_{i0} = 0, \qquad (i \neq 0). \tag{7}$$

In general

$$P_{ik} = 0 \tag{8}$$

for

$$|i| \geq \mu$$

because in this case, the conditions (5) are contradictory. One then has

$$P_{i,k+1} = \sum_j P_{jk} Q_{ji}^{(k)}, \qquad |i| < \mu, \tag{9}$$

where $Q_{ji}^{(k)}$ indicates the probability that the event $E_{i,k+1}$ occurs, given the hypothesis that the event E_{jk} has occurred, that is, the probability that the relation

$$F_n\left(\frac{k+1}{n}\right) - F_n\left(\frac{k}{n}\right) = \frac{i-j+1}{n} \tag{10}$$

occurs, under the condition

$$F_n\left(\frac{k}{n}\right) = \frac{k+j}{n}. \tag{11}$$

Because of (11) one has that, amongst the n results of the observations X_1, $X_2, \ldots, X_n$, exactly $n - k - j$ find themselves on the segment $k/n < x \leq 1$; Equation (10) can be satisfied only when $i - j + 1$ of the $n - k - j$ above indicated observations are in the interval $k/n < x \leq (k+1)/n$.

The probability that this circumstance should occur, given that the distribution of the X_m is uniform, is therefore equal to

$$Q_{ji}^{(k)} = \binom{n-k-j}{i-j+1} \cdot \left(1 - \frac{1}{n-k}\right)^{n-k-i-1} \cdot \left(\frac{1}{n-k}\right)^{i-j+1}. \tag{12}$$

Formulas (7), (8), (9), (12) and (6) permit us to calculate the probability $\Phi_n(\lambda)$, with $\lambda = \mu/\sqrt{n}$.

It is now possible to replace these formulas by others that are more practical. With this goal in mind, let us suppose

$$R_{ik} = \frac{(n-k-i)!\, n^n}{(n-k)^{n-k-i} n!} e^{-k} P_{ik}. \tag{13}$$

Conditions (7) and (8) are transformed, by R_{ik}, into the following:

$$R_{00} = 1, \quad R_{i0} = 0, \qquad i \neq 0; \tag{14}$$

$$R_{ik} = 0, \qquad |i| \geq \mu. \tag{15}$$

Relation (9) leads us, after some elementary calculations, to the formula

$$R_{i,k+1} = \sum_j R_{jk} \frac{1}{(i-j+1)!} e^{-1}. \tag{16}$$

Finally, for the formulas (6) and (13), one obtains

$$\Phi_n(\lambda) = \frac{n!}{n^n} e^n R_{0n}. \tag{17}$$

Formulas (14), (15), (16) and (17) are equally sufficient for the calculation of $\Phi_n(\lambda)$ in the case $\lambda = \mu/\sqrt{n}$.

3. Consider now $Y_1, Y_2, \ldots, Y_n$ a sequence of random variables, mutually independent, with a distribution function characterized by the formulas

$$P\left\{Y_k = \frac{i-1}{\mu}\right\} = \frac{1}{i!} e^{-1}, \qquad i = 0, 1, 2, \ldots. \tag{18}$$

If one supposes

$$S_k = Y_1 + Y_2 + \cdots + Y_k$$

it is easy to verify that the probability $\bar{R}_{ik}$, that the relations

$$|S_j| < \mu, \qquad j = 0, 1, 2, \ldots, k;$$

$$S_k = \frac{i}{\mu},$$

hold simultaneously, satisfies the very same conditions (14), (15) and (16) that R_{ik} satisfies; that is to say, one has $\bar{R}_{ik} = R_{ik}$.

It is moreover possible to give an asymptotic expression of R_{ik}, as $n \to \infty$. To reach this goal, we can make use of the following general theorem.

Theorem II. *Let $Y_1, Y_2, \ldots, Y_n$ be a succession of random variables, mutually independent, that only assume values that are multiples of a constant ε. Moreover, let*

$$E(Y_k) = 0, \quad E(Y_k^2) = 2b_k, \quad E(|Y_k^3|) = d_k,$$

$$S_k = Y_1 + Y_2 + \cdots + Y_k,$$

$$t_k = b_1 + b_2 + \cdots + b_k$$

and $a(t)$ and $b(t)$ two continuous and differentiable functions that satisfy the inequalities

$$a(t) < b(t),$$

$$a(0) < 0 < b(t).$$

Denoting by R_{in} the probability that the relations

$$a(t_k) < S_k < b(t_k), \qquad k = 1, 2, \ldots, n,$$

$$S_n = i\varepsilon$$

hold true simultaneously, and by $u(\sigma, \tau, s, t)$, Green's function related to the heat equation

$$\frac{\partial f}{\partial t} = \frac{\partial^2 f}{\partial s^2}$$

in a domain G defined by the inequalities

$$a(t) < s < b(t),$$

one will have

$$R_{in} = \varepsilon \cdot \{u(0, 0, i\varepsilon, t_n) + \Delta\}$$

with Δ tending to zero uniformly with ε when the following conditions are statisfied:

(i) *$a(t)$ and $b(t)$ remain constant for n contained between certain fixed limits*

$$0 < T_1 < t_n < T_2;$$

(ii) *there exists a constant $C > 0$ such that $d_k/b_k \leq C\varepsilon$;*
(iii) *there exists a constant $K > 0$ such that for each k and one i_k, appropriately chosen, the following inequalities are satisfied:*

$$P\{Y_k = i_k \cdot \varepsilon\} > K,$$
$$P\{Y_k = (i_k + 1) \cdot \varepsilon\} > K;$$

(iv) *there exists a constant A such that*

$$a(t_n) + A < i\varepsilon < b(t_n) - A.$$

In addition to the nature of the variables Y_k, their number n and the number i, all of these quantities can vary arbitrarily with ε.

This theorem belongs to the same order of ideas as the theorem enunciated in one of my notes.[3] The conclusion of the theorem here enunciated is however more precise: the theorem in the note cited above would permit us only to affirm that

$$\sum_{i=p}^{i=q} R_{in} = \int_{p\varepsilon}^{q\varepsilon} u(0, 0, z, t_n)\, dz + \Delta',$$

given Δ' an infinitesimal with ε, if conditions (i) and (ii) are satisfied. Condition (iii), which is an essential part of our new theorem, has already been used by von Mises for entirely analogous considerations.[4]

In our case we have

$$\begin{aligned}
\varepsilon &= 1/\mu, \\
E(Y) = 0, \quad E(Y_k^2) &= 2b_k = 1/\mu^2, \quad E(|Y_k^3|) = d_k = C/\mu^3, \\
d_k/b_k &= C/\mu = C\varepsilon, \quad t_n = n/(2\mu^2) = 1/(2\lambda^2), \\
a(t) &= -1, \quad b(t) = +1, \\
R_{0n} &= \varepsilon \cdot \{u(0, 0, 0, 1/(2\lambda^2)) + \Delta\}, \\
u(0, 0, s, t) &= \frac{1}{2\sqrt{\pi t}} \sum_{-\infty}^{+\infty} (-1)^k e^{-(s-2k)^2/(4t)}, \\
\Phi_n(\lambda) &= P_{0n} = \frac{n!}{n^n} e^n R_{0n} \\
&= \{\sqrt{2\pi n} + \delta\} \cdot \frac{1}{\mu} \cdot \left\{ \frac{\lambda}{\sqrt{2\pi}} \sum_{-\infty}^{+\infty} (-1)^k e^{-2k^2\lambda^2} + \Delta \right\} \\
&= \sum_{-\infty}^{+\infty} (-1)^k e^{-2k^2\lambda^2} + R = \Phi(\lambda) + R.
\end{aligned}$$

[3] A. Kolmogorov, *Eine Verallgemeinerung des Laplace-Liapounoffschen Satzes.* "Bulletin de l'Académie des Sciences de l'U.S.S.R." (1931), pp. 959–962. The complete proof of this theorem will soon appear in this same Bulletin (1932).

[4] Loc. cit. [1], p. 212, *Erste Fundamentalsatz für arithmetische Verteilungen.*

The remainder R in this formula tends uniformly to zero for $n \to \infty$ in the case where λ is greater then a predetermined $\lambda_0 > 0$. Indeed, $\varepsilon = 1/\mu = 1/(\lambda\sqrt{n})$, in which case, it tends to zero as $n \to \infty$.

The statement in Theorem I is therefore proved for values of λ of the form $\mu/\sqrt{n}$ and under the condition $\lambda > \lambda_0$. Because the limiting distribution function $\Phi(\lambda)$ is continuous and has as the limiting value $\Phi(0) = 0$, it is easily seen that the indicated restrictions are not in fact necessary. In reference to the proof of Theorem II, we propose to return to the subject in a subsequent note.

Introduction to Neyman (1934) On the Two Different Aspects of the Representative Method: The Method of Stratified Sampling and the Method of Purposive Selection

T. Dalenius

Neyman was born in 1894 in Russia to Polish parents. An extensive account of his private and professional life is found in Reid (1982), which divides Neyman's life into three periods.

The first period, 1894–1921 was "the Russian period". In 1912, Neyman entered the University of Kharkov, where he studied mathematics (especially measure theory and integration). During his university years, he became acquainted with K. Pearson's *The Grammar of Science*, which made a deep and lasting impression on him.

The second period, 1921–38, was the Polish–English period. In 1921, Neyman moved to Poland, where he first worked at the National Agricultural Institute and later at the State Meteorological Institute as an applied statistician. In 1924, Neyman obtained his Ph.D. in mathematics. In the early 1930s, Neyman served at the Institute for Social Problems as a consultant in survey sampling. Finally, in 1934, he moved to England, more specifically to University College in London, on the faculty of which were R.A. Fisher, K. Pearson, and E.S. Pearson. While there, he wrote his 1934 paper, the subject of this introduction. In 1937, he visited the United States and gave his "Lectures and Conferences on Mathematical Statistics and Probability," a section of which presented his views about the representative method.

The third period, 1938–81, was "the American period." He took up an appointment as professor of statistics and director of the statistical laboratory at the department of statistics in Berkeley, California. He was active there until his death in 1981. The following account of Neyman's paper has three purposes:

1. To place his paper in its historical context.
2. To identify its major contributions to the theory, methods, and application of survey sampling.
3. To review its impact on the consequent developments.

1. The Historical Context

Up to the end of the 1800s, there were a multitude of sample surveys with the objective of providing inferences about some finite population. Typically, little if any attention was paid to the development and use of *theory* to guide the design and analysis of these sample surveys.

By the end of the 1800s, significant changes took place. Many of these changes reflect the endeavors of A.N. Kiaer, head of the Norwegian Central Statistical Bureau. In Kiaer (1897), he presented his views about "the representative method." Kiaer advocated the use of that method at several sessions of the International Statistical Institute (ISI). With the forceful support of the British academician A.L. Bowley, the ISI endorsed Kiaer's ideas in a resolution passed in 1903.

In 1924, the ISI appointed a commission to study the representative method; Bowley was one of its members. The commission presented in 1926 its "Report on the Representative Method in Statistics." In Bowley (1926), two basic methods of selecting a sample of individuals were considered. One method called for using the individuals as sampling units using simple random sampling or proportional stratified sampling. The other method called for selecting a sample of groups of individuals (i.e., clusters), which in respect of certain controls yield a "miniature of the population, a representative sample."

Mention should also be made of the development in the 1920s of the theory and methods for the design and analysis of experiments by R.A. Fisher at the Rothamsted Experimental Station. The theory was based on certain principles, including randomization, replication, and local control, which stimulated Neyman's work.

2. The Major Contributions of Neyman (1934)

We will review these contributions under three headings:

1. the inferential basis of the representative method,
2. sampling designs, and
3. Neyman's assessment of purposive selection.

1. The Inferential Basis of the Representative Method

Neyman's view of the representative method was concisely summarized as follows: "Obviously the problem of the representative method is *par excellence* the problem of statistical estimation." More specifically, while not trying to define the "generally representative sample," Neyman specified two components of the representative method: a "representative method of sampling" and a "consistent method of estimation." Using both of these methods would make it possible to estimate the accuracy of the sample results, "*irrespective of the unknown properties of the population studied.*"

The method of sampling called for "random sampling," what is nowadays also known as probability sampling. The probability of selecting for the sample an individual in the population had to be known, but it did not have to be the same for all individuals.

The method of estimation called for using a consistent estimator. If the methods of sampling and estimation allow ascribing to every possible sample a confidence interval (X_1, X_2) such that the frequency of errors in the statements

$$X_1 \leq \bar{X} \leq X_2,$$

where X is the population value to be estimated, does not exceed a limit α, say, prescribed in advance, Neyman would call the method of sampling "representative" and the method of estimation "consistent." Today, Neyman's specification would be referred to in terms of a "measurable sample design."

2. Sample Designs

We will consider three powerful contributions.

I. *Stratified Sampling*

Neyman pointed out that the most favorable allocation of a sample is not proportional to the corresponding strata sizes (M_i, $i = 1, \ldots, k$). in which case the sample would be a miniature of the population. He derived the allocation that makes the variance of a linear estimator a minimum, i.e., minimum variance allocation. When Neyman prepared his paper, he was not aware of the fact that Tschuprow (1923) contained the same result, something that Neyman later acknowledged.*

As applied to a skewed population, minimum variance allocation amounts to using selection probabilities that are to be larger in strata with large variances and smaller in strata with small variances.

* Incidentally, the same result had been derived by others, independently of Tschuprow.

We add here an observation about the concept of proportional allocation yielding a miniature of the population. This fact per se is not a guarantee that the sample yields an estimate of measurable quality and especially not a guarantee that the estimate will be close to the population value. This circumstance was long overlooked by proponents of polls based on "quota sampling," the approach made famous by the American pollster George Gallup in 1934!

II. *Cluster Sampling*

Neyman realized that in many cases it would not be possible to efficiently use the individuals of the population as the sampling units. This being the case, groups of individuals—in today's terminology, "clusters"—had to be used as sampling units. He criticized sampling a *small* number of *large* groups. Instead, he advocated sampling a large number of small groups.

III. *Use of Supplementary Information*

Neyman considered the case in which there is available supplementary information about the population at the time of the sample design. He had no doubt faced this situation when he was advising the Institute for Social Problems about its use of sampling, as discussed earlier in this introduction.

Neyman pointed to two ways of using supplementary information in the sampling design. The first method called for using such information for dividing the population into strata and then selecting a sample from each stratum; he did not elaborate on how to demarcate the strata.

The second approach called for using the supplementary information in the estimation procedure by way of ratio estimation.

3. Neyman's Assessment of Purposive Selection

Neyman compared "purposive selection" and "random sampling." He did not on principle reject purposive selection, but emphasized that in order for purposive selection to yield satisfactory results (i.e., estimates close to the values in the population), certain basic assumptions about the relation between the estimation variable and the controls must be fulfilled. Especially, if this relation was expressed in terms of a regression line, it would be rather dangerous to assume any definitive hypothesis concerning that regression line: If the hypothesis is wrong, the resulting estimate may be correspondingly wrong.

3. The Consequent Developments

Almost from the first year of its appearance, Neyman (1934) left its stamp on the volume and scope of applications of sample surveys and the developments of methods and theory, and it did so on a global scale.

As to applications, the impact was especially marked in the United States, India, and Western Europe. A related consequence was the appearance, around 1950, of several textbooks, one of which was written at the request of the United Nations Sub-Commission on Statistical Sampling, a body created to promote the sound use of sample surveys and standard terminology.

Neyman's contributions to the methods and theory of sample surveys has been described as "pioneering" (by M.H. Hansen and W.G. Madow), "the Neyman watershed" (by W.H. Kruskal and F. Mosteller), and the start of the "Neyman revolution" (by T.M.F. Smith).

As for the development of new methods and theory, it is worth noting that in response to a question from the audience at the lectures and conferences in Washington, D.C. in 1937, Neyman himself developed a theory for double-sampling (also called "two-phase sampling") [see Neyman (1938)].

In the balance of this section, we will review developments of methods and theory by statisticians other than Neyman. The emphasis will be on topics explicitly discussed in Neyman (1934).

Those topics will be reviewed in Sec. 3.1 and 3.2. In Sec. 3.3, we will briefly comment on some developments concerning topics not explicitly discussed in Neyman (1934). For a more comprehensive account, the reader is referred to Hansen et al. (1985) and Sukhatme (1966). Incidentally, this second-mentioned reference was written by one of Neyman's students from University College, London.

1. The Inferential Basis of the Representative Method

Neyman's view about the inferential basis of the representative method as summarized in Sect. 2 was soon accepted by the statistical community and especially by survey statisticians working in official statistics agencies, but not necessarily everywhere.*

While Neyman's *specification* of the representative method was accepted by most survey statisticians, his definition of a consistent method of estimation was challenged. Thus, in Cochran (1953), an estimate was termed "consistent" if it becomes equal to the population value for $n = N$. And in Hansen et al. (1953), an estimate was said to be "consistent" if the proportion of sample estimates that differ from the value being estimated by less than any

* As an example, the original Gallup poll, while using schemes for "local control", advocated the use of nonrandom sampling (especially quota sampling).

specified small amount approaches 100% as the size of the sample is increased (p. 20).

In the last few decades, two important lines of theoretical development have occurred. One line extends Neyman's view that it is not necessary that the probabilities of selection be equal for all individuals. Such an extension is represented by Horvitz and Thompson (1952), who consider a design that associates with each individual an individual probability of being selected for the sample.

The second line focuses on the role of random sampling and especially the role of models of the population in the selection procedure. A comprehensive account of this line of development is given in Cassel et al. (1977). Today, we have two schools of thought. Some authors have advocated the use of *model-dependent* designs: Which sample to observe should be solely determined by a model of the population concerned. Other authors, echoing Neyman's criticism of purposive selection, have argued in favor of *model-based* designs: Although a model may be used in the design, insurance against any inaccuracy of that model should be provided by using random sampling, as discussed in Hansen et al. (1983).

2. Sampling Designs

We will consider developments related to Neyman's contributions as discussed in Sec. 2.

I. *Stratified Sampling*

Neyman's basic notion of stratified sampling has been much extended as illustrated by the following topics:

1. dividing a population into strata,
2. choosing the number of strata, and
3. allocation of a sample to estimate several parameters.

For a discussion of these topics, see Dalenius (1957). Additional related topics are poststratification and such extensions of the basic idea of stratified sampling as Latin square sampling, deep stratification, and controlled selection.

II. *Cluster Sampling*

Cluster sampling as discussed in Neyman (1934) was tailored to the following situation. For a survey to be conducted by the Institute for Social Problems, data were to be collected from the records of the 1930 population census. These records were kept by "statistical districts"; altogether there were 123,383 such districts. The districts were internally relatively homogeneous. They were divided into 113 strata, and from each stratum a sample of districts (and hence records) was selected; in all 1621 districts were selected. Collecting the data needed for the survey called for processing the records for each selected district; there was no field-work (such as interviewing) involved.

This type of design is, however, not efficient in the context of a nationwide survey in a large country, for which the data are to be collected by interviewers. The interviewing will be expensive, and so will the supervision of their work.

In order to place such applications on the basis of measurable designs (in the sense of Neyman), a new theory was needed. Such a theory was presented in Hansen and Hurvitz (1943) and some subsequent documents for use in such surveys as the Current Population Survey carried out by the U.S. Bureau of the Census. The theory is for multistage sampling. At the first stage, large internally heterogenous sampling units (referred to as primary sampling units) are selected from each one of a set of internally homogeneous strata. The selection is carried out by means of probabilities proportional to some measure of size ("PPS sampling"), followed by random subsampling from each selected primary unit; this subsampling may be carried out in one or more stages.

III. *The Use of Supplementary Information*

As demonstrated by the developments in the last half-century, supplementary information may be exploited for all aspects of the sample design, the definition of sampling units, the selection design, and the estimation method.

We will focus here on the use of supplementary information in the estimation. The supplementary information is X, with *known mean* $\bar{X}$. The population value $\bar{Y}$ is to be estimated. Assume that $\bar{y}$ is an unbiased estimate of $\bar{Y}$ and $\bar{x}$ an unbiased estimate of $\bar{X}$. Consider the following expression:

$$\hat{y}_\theta = \bar{y} + \theta(\bar{X} - \bar{x}).$$

For $\theta = \bar{y}/\bar{x}$, $\hat{y}_\theta$ becomes the ratio estimate. For $\theta = b$, the sample estimate of the regression coefficient, $\bar{y}_\theta$ becomes the regression estimate. And for $\theta = k$, a constant, $\hat{y}_\theta$ becomes the difference estimate. Clearly, the difference estimate is unbiased. More important, however, it is linear; consequently, an *exact* expression can be derived for its variance. This is not true of the ratio and regression estimates. For a more comprehensive discussion of this topic and pertinent reference, see Hansen et al. (1985).

Another development is multivariate ratio estimation as discussed in Olkin (1958). Obviously, the same idea may be exploited to define multivariate difference estimation.

3. Some Additional Developments

The impact of Neyman (1934) has not been limited to the developments reviewed above. We will be satisfied here by presenting a *list* of some selected additional developments, especially developments due to statisticians adhering to Neyman's statistical philosophy:

1. systematic sampling
2. estimation of values for "domains of study"
3. estimation of "change over time"
4. variance estimation
5. use of classical inference tools (such as regression analysis) on survey data
6. coping with specific sources of nonsampling errors (such as nonresponse)
7. development of theory for measuring and controlling the total error of an estimate, i.e., "survey models"

References

Bowley, A.L. (1926). Measurement of the precision attained in sampling, *Proc. Internat. Statist. Inst.*, **12** (1), 6–62.

Cassel, C.-M., Särndal, C.E., and Wretman, J.H. (1977). *Foundations of Inference in Survey Sampling.* Wiley, New York.

Cochran, W.G. (1942). Sampling theory when the sampling units are of unequal sizes, *J. Amer. Statist. Assoc.*, **37**, 199–212.

Cochran, W.G. (1953). *Sampling Techniques.* Wiley, New York.

Dalenius, T (1957). *Sampling in Sweden.* Almqvist and Wiksell, Stockholm.

Hansen, M.H., Dalenius, T., and Tepping, B.J. (1985). The development of sample surveys of finite populations, in *A Celebration of Statistics* (A.C. Atkinson, and S.E. Feinberg, eds.). Springer-Verlag, New York.

Hansen, M.H., and Hurwitz, W.N. (1943). On the theory of sampling from finite populations, *Ann. Math. Statist.*, **14**, 333–362.

Hansen, M.H., Hurwitz, W.N., and Madow, W.G. (1953). *Sample Survey Methods and Theory*, Vol. I. Wiley, New York.

Hansen, M.H., Madow, W.G., and Tepping, B.J. (1983). An evaluation of model-dependent and probability-sampling inference in sample surveys, *J. Amer. Statist. Assoc.*, **78**, 776–793.

Horvitz, D.G., and Thompson, D.J. (1952). A generalization of sampling without replacement from a finite universe, *J. Amer. Statist. Assoc.*, **47**, 663–685.

Kiaer, A.N. (1897). *The Representative Method of Statistical Surveys* (original in Norwegian; translated and reprinted by Statistisk Sentralbyrå, Oslo).

Neyman, J. (1938). Contribution to the theory of sampling human populations, *J. Amer. Statist. Assoc.*, **33**, 101–116.

Neyman, J. (1952). *Lectures and Conferences on Mathematical Statistics and Probability.* Graduate School, U.S. Department of Agriculture, Washington, D.C.

Olkin, I. (1958). Multivariate ratio estimation for finite populations, *Biometrika*, **45**, 154–165.

Reid, C. (1982). *Neyman from Life.* Springer-Verlag, New York.

Sukhatme, P.V. (1956). Major developments in sampling theory and practice, in (F.N. David, ed.). *Research Papers in Statistics*: *Festschrift for J. Neyman.* Wiley, New York.

Tschuprow, A.A. (1923). On the mathematical expectation of the moments of frequency distributions in the case of correlated observation, *Metron*, **2**, 461–493, 646–680.

On the Two Different Aspects of the Representative Method: the Method of Stratified Sampling and the Method of Purposive Selection

Jerzy Neyman
Biometric Laboratory, Nencki Institute
Varsoviensis, Warsaw

I. Introductory

Owing to the work of the International Statistical Institute,* and perhaps still more to personal achievements of Professor A.L. Bowley, the theory and the possibility of practical applications of the representative method has attracted the attention of many statisticians in different countries. Very probably this popularity of the representative method is also partly due to the general crisis, to the scarcity of money and to the necessity of carrying out statistical investigations connected with social life in a somewhat hasty way. The results are wanted in some few months, sometimes in a few weeks after the beginning of the work, and there is neither time nor money for an exhaustive research.

But I think that if practical statistics has acquired something valuable in the representative method, this is due primarily to Professor A.L. Bowley, who not only was one of the first to apply this method in practice,† but also wrote a very fundamental memoir‡ giving the theory of the method. Since then the representative method has been often applied in different countries and for different purposes.

My chief topic being the theory of the representative method, I shall not go into its history and shall not quote the examples of its practical application

* See "The Report on the Representative Method in Statistics" by A. Jensen, *Bull. Inst. Intern. Stat.*, XXII. 1ère Livr.

† A.L. Bowley "Working Class Households in Reading." *J.R.S.S.*, June, 1913.

‡ A.L. Bowley: "Measurement of the Precision Attained in Sampling." Memorandum published by the Int. Stat. Inst., *Bull. Int. Stat. Inst.*, Vol. XXII. 1ère Livr.

however important—unless I find that their consideration might be useful as an illustration of some points of the theory.

There are two different aspects of the representative method. One of them is called the method of random sampling and the other the method of purposive selection. This is a division into two very broad groups and each of these may be further subdivided. The two kinds of method were discussed by A.L. Bowley in his book, in which they are treated as it were on equal terms, as being equally to be recommended. Much the same attitude has been expressed in the Report of the Commission appointed by the International Statistical Institute for the purpose of studying the application of the Representative Method in Statistics.* The Report says: "In the selection of that part of the material which is to be the object of direct investigation, one or the other of the following two principles can be adopted: in certain instances it will be possible to make use of a combination of both principles. The one principle is characterized by the fact that the units which are to be included in the sample are selected at random. This method is only applicable where the circumstances make it possible to give every single unit an equal chance of inclusion in the sample. The other principle consists in the samples being made up by purposive selection of groups of units which it is presumed will give the sample the same characteristics as the whole. There will be especial reason for preferring this method, where the material differs with respect to composition from the kind of material which is the basis of the experience of games of chance, and where it is therefore difficult or even impossible to comply with the aforesaid condition for the application of selection at random. Each of these two methods has certain advantages and certain defects...."

This was published in 1926. In November of the same year the Italian statisticians C. Gini and L. Galvani were faced with the problem of the choice between the two principles of sampling, when they undertook to select a sample from the data of the Italian General Census of 1921. All the data were already worked out and published and the original sheets containing information about individual families were to be destroyed. In order to make possible any further research, the need for which might be felt in the future, it was decided to keep for a longer time a fairly large sample of the census data, amounting to about 15 per cent. of the same.)

The chief purpose of the work is stated by the authors as follows:† "To obtain a sample which would be representative of the whole country with respect to its chief demographic, social, economic and geographic characteristics."

At the beginning of the work the original data were already sorted by provinces, districts (circondarî) and communes, and the authors state that the

* Bull. Int. Stat. Inst., XXII. 1ère Livr. p. 376.

† *Annali di Statistica*, Ser. VI. Vol. IV. p. 1. 1929.

easiest method of obtaining the sample was to select data in accordance with the division of the country in administrative units. As the purpose of the sample was among others to allow local comparisons to be made in the future, the authors expressed the view that the selection of the sample, taking administrative units as elements, was the only possible one.

For various reasons, which, however, the authors do not describe, it was impossible to take as an element of sampling an administrative unit smaller than a commune. They did not, however, think it satisfactory to use communes as units of selection because (p. 3 *loc. cit.*) their large number (8,354) would make it difficult to apply the method of purposive selection. So finally the authors fixed districts (circondarî) to serve as units of sampling. The total number of the districts in which Italy is divided amounts to 214. The number of the districts to be included in the sample was 29, that is to say, about 13.5 per cent. of the total number of districts.

Having thus fixed the units of selection, the authors proceed to the choice of the principle of sampling: should it be random sampling or purposive selection? To solve this dilemma they calculate the probability, π, that the mean income of persons included in a random sample of $k = 29$ districts drawn from their universe of $K = 214$ districts will differ from its universe-value by not more than 1.5 per cent. The approximate value of this probability being very small, about $\pi = .08$, the authors decided that the principle of sampling to choose was that of purposive selection.*

The quotation from the Report of the Commission of the International Statistical Institute and the choice of the principle of sampling adopted by the Italian statisticians, suggest that the idea of a certain equivalency of both principles of random sampling and purposive selection is a rather common one. As the theory of purposive selection seems to have been extensively presented only in the two papers mentioned, while that of random sampling has been discussed probably by more than a hundred authors, it seems justifiable to consider carefully the basic assumptions underlying the former. This is what I intend to do in the present paper. The theoretical considerations will then be illustrated on practical results obtained by Gini and Galvani, and also on results of another recent investigation, carried out in Warsaw, in which the representative method was used. As a result of this discussion it may be that the general confidence which has been placed in the method of purposive selection will be somewhat diminished.

* It may be noted, however, that the choice of the principle seems to have been predetermined by the previous choice of the unit of sampling.

II. Mathematical Theories Underlying the Representative Method

1. The Theory of Probabilities *a posteriori* and the Work of R.A. Fisher

Obviously the problem of the representative method is *par excellence* the problem of statistical estimation. We are interested in characteristics of a certain population, say π, which it is either impossible or at least very difficult to study in detail, and we try to estimate these characteristics basing our judgment on the sample. Until recently it has been usually assumed that the accurate solution of such a problem requires the knowledge of probabilities *a priori* attached to different admissible hypotheses concerning the values of the collective characters* of the population π. Accordingly, the memoir of A.L. Bowley may be regarded as divided into two parts. Each question is treated from two points of view: (*a*) The population π is supposed to be known; the question to be answered is: what could be the samples from this population? (*b*) We know the sample and are concerned with the probabilities *a posteriori* to be ascribed to different hypotheses concerning the population.

In sections which I classify as (*a*) we are on the safe ground of classical theory of probability, reducible to the theory of combinations.†

In sections (*b*), however, we are met with conclusions based, *inter alia*, on some quite arbitrary hypotheses concerning the probabilities *a priori*, and Professor Bowley accompanies his results with the following remark: "It is to be emphasized that the inference thus formulated is based on assumptions that are difficult to verify and which are not applicable in all cases."

However, since Bowley's book was written, an approach to problems of this type has been suggested by Professor R.A. Fisher which removes the difficulties involved in the lack of knowledge of the *a priori* probability law.‡ Unfortunately the papers referred to have been misunderstood and the validity of statements they contain formally questioned. This I think is due largely to the very condensed form of explaining ideas used by R.A. Fisher, and perhaps also to a somewhat difficult method of attacking the problem. Avoiding the necessity of appeals to the somewhat vague statements based on

* This is a translation of the terminology used by Bruns and Orzecki. Any characteristics of the population or sample is a collective character.

† In this respect I should like to call attention to the remarkable paper of the late L. March published in *Metron*, Vol. VI. There is practically no question of probabilities and many classical theorems of this theory are reduced to the theory of combinations.

‡ R.A. Fisher: *Proc. Camb. Phil. Soc.*, Vol. XXVI, Part 4, Vol. XXVIII, Part 3, and *Proc. Roy. Soc.*, A. Vol. CXXXIX.

probabilities *a posteriori*, Fisher's theory becomes, I think, the very basis of the theory of representative method. In Note I in the Appendix I have described its main lines in a way somewhat different from that followed by Fisher.

The possibility of solving the problems of statistical estimation independently from any knowledge of the *a priori* probability laws, discovered by R.A. Fisher makes it superfluous to make any appeals to the Bayes' theorem.

The whole procedure consists really in solving the problems which Professor Bowley termed direct problems: given a hypothetical population, to find the distribution of certain characters in repeated samples. If this problem is solved, then the solution of the other problem, which takes the place of the problem of inverse probability, can be shown to follow.

The form of this solution consists in determining certain intervals, which I propose to call the confidence intervals (see Note I), in which we may assume are contained the values of the estimated characters of the population, the probability of an error in a statement of this sort being equal to or less than $1 - \varepsilon$, where ε is any number $0 < \varepsilon < 1$, chosen in advance. The number ε I call the confidence coefficient. It is important to note that the methods of estimating, particularly in the case of large samples, resulting from the work of Fisher, are often precisely the same as those which are already in common use. Thus the new solution of the problems of estimation consists mainly in a rigorous justification of what has been generally considered correct more or less on intuitive grounds.*

Here I should like to quote the words of Laplace, that the theory of probability is in fact but the good common sense which is reduced to formulæ.

* I regret that the necessarily limited size of the paper does not allow me to go into the details of this important question. It has been largely studied by R.A. Fisher. His results in this respect form a theory which he calls the Theory of Estimation. The above-mentioned problems of confidence intervals are considered by R.A. Fisher as something like an additional chapter to the Theory of Estimation, being perhaps of minor importance. However, I do not agree in this respect with Professor Fisher. I am inclined to think that the importance of his achievements in the two fields is in a relation which is inverse to what he thinks himself. The solution of the problem which I described as the problem of confidence intervals has been sought by the greatest minds since the work of Bayes 150 years ago. Any recent book on the theory of probability includes large sections concerning this problem. These sections are crowded with all sorts of "paradoxes," etc. The present solution means, I think, not less than a revolution in *the theory* of statistics. On the other hand, the problem of the choice of estimates has—as far as I can see—mainly *a practical* importance. If this is not properly solved (granting that the problem of confidence intervals has been solved correctly) the resulting confidence intervals will be unnecessarily broad, but our statements about the values of estimated collective characters will still remain correct. Thus I think that the problems of the choice of the estimates are rather the technical problems, which, of course, are extremely important from the point of view of practical work, but the importance of which cannot be compared with the importance of the other results of R.A. Fisher, concerning the very basis of the modern statistical theory. These are, of course, "qualifying judgments," which may be defended and may be attacked, but which anyone may accept or reject, according to his personal point of view and the perspective on the theory of statistics.

It is able to express in exact terms what the sound minds feel by a sort of instinct, sometimes without being able to give good reasons for their beliefs.

2. The Choice of the Estimates

However, it may be observed that there remains the question of the choice of the collective characters of the samples which would be most suitable for the purpose of the construction of confidence intervals and thus for the purposes of estimation. The requirements with regard to these characters in practical statistics could be formulated as follows:

1. They must follow a frequency distribution which is already tabled or may be easily calculated.
2. The resulting confidence intervals should be as narrow as possible.

The first of these requirements is somewhat opportunistic, but I believe as far as the practical work is concerned this condition should be borne in mind.*

Collective characters of the samples which satisfy both conditions quoted above and which may be used in the most common cases, are supplied by the elegant method of A.A. Markoff,† used by him when dealing with the theory of least squares. The method is not a new one, but as it was published in Russian it is not generally known.‡ This method, combined with some results of R.A. Fisher and of E. S. Pearson concerning the extension of "Student's" distribution allows us to build up the theory of different aspects of representative method to the last details.

Suppose θ is a certain collective character of a population π and

$$x_1, x_2, \ldots x_n \tag{1}$$

is the sample from this population. We shall say that a function of these x's, say

$$\theta' = \theta'(x_1, x_2, \ldots x_n) \tag{2}$$

is a "mathematical expectation estimate"§ of θ, if the mean value of θ' in repeated samples is equal to θ. Further, we shall say that the estimate θ' is the best linear estimate of θ if it is linear with regard to x's, *i.e.*

$$\theta' = \lambda_1 x_1 + \lambda_2 x_2 + \cdots + \lambda_n x_n + \lambda_0 \tag{3}$$

* The position is a different one if we consider the question from the point of view of the theory. Here I have to mention the important papers of R.A. Fisher on the theory of likelihood.

† A.A. Markoff: *Calculus of Probabilities*. Russian. Edition IV, Moscow 1923. There was a German edition of this book, Leipzig 1912, actually out of print.

‡ I doubt, for example, whether it was known to Bowley and to Gini and Galvani when they wrote their papers.

§ Only the estimates of this kind will we consider below.

and if its standard error is less than the standard error of any other linear estimate of θ.

Of course, in using the words "best estimate" I do not mean that the estimate defined has unequivocable advantages over all others. This is only a convention and, as long as the definition is borne in mind, will not cause any misunderstanding. Still, the best linear estimates have some important advantages:

1. If n be large, their distribution practically always follows closely the normal law of frequency. This is important, as in applying the representative method in social and economic statistics we are commonly dealing with very large samples.
2. In most cases they are easily found by applying Markoff's method.
3. The same method provides us with the estimate of their standard errors.
4. If the estimate θ' of θ is a linear estimate, and if μ is the estimate of its standard error, then, in cases when the sampled population is normally distributed, the ratio

$$t = \frac{\theta' - \theta}{\mu} \tag{4}$$

follows the "Student's" distribution, which is dependent only upon the size of the sample. This is the result due to R.A. Fisher. Moreover, R. A. Fisher has provided tables giving the values of t such that the probability of their being exceeded by $|\theta' - \theta|/\mu$ has definite values such as .01, .02, ... etc. This table* was published long before any paper dealing with the solution of the problem of estimation independent of the probabilities *a priori*. However, this solution is already contained in the table. In fact it leads directly to the construction of the confidence intervals. Suppose the confidence coefficient chosen is $\varepsilon = .99$. Obtain from Fisher's table the value of t, say t_ε, corresponding to the size of the sample we deal with and to a probability of its being exceeded by $|\theta' - \theta|/\mu$ equal to $1 - \varepsilon = .01$. It may then be easily shown that the confidence interval, corresponding to the coefficient $\varepsilon = .99$ and to the observed values of θ' and μ, will be given by the inequality

$$\theta' - \mu t_\varepsilon \leq \theta \leq \theta' + \mu t_\varepsilon. \tag{5}$$

5. The previous statement is rigorously true if the distribution of the x's is normal. But, as it has been experimentally shown by E.S. Pearson,† the above result is very approximately true for various linear estimates by fairly skew distributions, provided the sample dealt with is not exceedingly small, say not smaller than of 15 individuals. Obviously, when applying the representative method to social problems this is a limitation of no importance. In fact, if the samples are very large, the best linear estimates follow the normal law of

* R.A. Fisher: *Statistical Methods for Research Workers*, London, 1932, Edition IV.

† This *Journal*, Vol. XCVI, Part I.

frequency, and the multiplier t_ε in the formula giving the confidence interval may be found from any table of the normal integral.*

The above properties of the linear estimates make them exceedingly valuable from the point of view of their use in applying the representative method. I proceed now to the Markoff method of finding the best linear estimates.

This may be applied under the following conditions, which are frequently satisfied in practical work.

Suppose we are dealing with k populations,

$$\pi_1, \pi_2, \ldots \pi_k \tag{6}$$

from which we may draw random samples. Let

$$x_{i1}, x_{i2}, \ldots x_{in_i} \tag{7}$$

be a sample, Σ_i, of n_i, individuals randomly drawn (with replacement or not) from the population π_i. Let A_i be the mean of the population π_i. We have now to make some assumption about the variances, σ_i^2, of the populations π_i. The actual knowledge of these variances is not required. But we must know numbers which are proportional to σ_i^2. Thus we shall assume that

$$\sigma_i^2 = \frac{\sigma_0^2}{P_i} \tag{8}$$

σ_0^2 being an unknown factor, and P_i a known number.† It would be a special case of the above conditions if it were known that

$$\sigma_1 = \sigma_2 = \cdots = \sigma_k \tag{9}$$

the common value of the σ's being unknown.

Suppose now we are interested in the values of one or several collective characters of the populations, π_i, each of them being a linear function of the means of these populations, say

$$\theta_j = a_{j1}A_1 + a_{j2}A_2 + \cdots + a_{jk}A_k \tag{10}$$

where the a's are some known coefficients. Markoff gives now the method of finding linear functions of the x's determined by samples from all the populations, namely,

$$\begin{aligned}\theta_j' = {} & \lambda_{11}x_{11} + \lambda_{12}x_{12} + \cdots + \lambda_{1n_1}x_{1n_1} \\ & + \cdots\cdots\cdots\cdots\cdots\cdots \\ & \cdots\cdots\cdots\cdots\cdots\cdots \\ & + \lambda_{k1}x_{k1} + \lambda_{k2}x_{k2} + \cdots + \lambda_{kn_k}x_{kn_k}\end{aligned} \tag{11}$$

such, *that whatever the value of unknown* θ_j:

* For example, Table I of the Pearson's *Tables for Statisticians and Biometricians*, Part I, may be used.

† Sometimes, in special problems, even this knowledge is not required.

(*a*) Mean θ_j' in repeated samples $= \theta_j$.

(*b*) Standard error of θ_j' is less than that of any other linear function, satisfying (*a*).

(The details concerning this method are given in Note II of the Appendix.)

It is worth considering the statistical meaning of the two conditions (*a*), (*b*), when combined with the fact that if the number of observations is large, the distribution of θ' in repeated sampling tends to be, and for practical purposes *is* actually normal. The condition (*a*) means that the most frequent values of θ' will be those close to θ. Therefore, if ψ is some linear function of the x's, which does not satisfy the condition (*a*), but instead the condition.

Mean ψ in repeated samples $= \ + \Delta$, (say), then, using ψ as an estimate of θ, we should commit systematic errors, which most frequently would be near Δ. Such estimates as ψ are called biased.

The condition (*b*) assures us that when using θ''s as estimates of θ's, we shall get confidence intervals corresponding to a definite confidence coefficient, narrower than those obtained using any other linear estimate. In other words, using linear estimates satisfying the conditions (*a*) and (*b*) we may be sure that we shall not commit systematic errors, and that the accuracy of the estimate will be the greatest.

III. Different Aspects of the Representative Method

We may now proceed to consider the two aspects of the representative method.

1. The Method of Random Sampling

The method of random sampling consists, as it is known, in taking at random elements from the population which it is intended to study. The elements compose a sample which is then studied. The results form the basis for conclusions concerning the population. The nature of the population is arbitrary. But we shall be concerned with populations of inhabitants of some country, town, etc. Let us denote this population by Π. Its elements will be single individuals, of which we shall consider a certain character x, which may be measurable or not (*i.e.* an attribute). Suppose we want to estimate the average value of the character x, say X, in all individuals forming the population Π. It is obvious that in the case where x is an attribute, which may be possessed or not by the individuals of the population, its numerical value in these individuals will be 0 or 1, and its mean value X will be the proportion of the individuals having actually the attribute x.

The method of random sampling may be of several types:

(*a*) The sample, Σ, which we draw to estimate X is obtained by taking at random single individuals from the population Π. The method of sampling may be either that with replacement or not. This type has been called by Professor Bowley that of unrestricted sampling.

(*b*) Before drawing the random sample from the population Π this is divided into several "strata," say

$$\Pi_1, \Pi_2, \ldots \Pi_k \tag{12}$$

and the sample Σ is composed of k partial samples, say

$$\Sigma_1, \Sigma_2, \ldots \Sigma_k \tag{13}$$

each being drawn (with replacement or not) from one or other of the strata. This method has been called by Professor Bowley the method of stratified sampling. Professor Bowley considered only the case when the sizes, say, m_i', of the partial samples are proportionate to the sizes of corresponding strata. I do not think that this restriction is necessary and shall consider the case when the sizes of the strata, say

$$M_1', M_2', \ldots M_k' \tag{14}$$

and the sizes of partial samples, say

$$m_1', m_2', \ldots m_k' \tag{15}$$

are arbitrary.

In many practical cases the types of sampling described above cannot be applied. Random sampling means the method of including in the sample single elements of the population with equal chances for each element. Human populations are rarely spread in single individuals. Mostly they are grouped. There are certainly exceptions. For instance, when we consider the population of insured persons, they may appear in books of the insurance offices as single units. This circumstance has been used among others by A.B. Hill,* who studied sickness of textile workers, using a random sample of persons insured in certain Approved Societies. But these cases are rather the exceptions. The process of sampling is easier when the population from which we want a sample to be drawn is not a population of persons who are living miles apart, but some population of cards or sheets of paper on which are recorded the data concerning the persons. But even in this simplified position we rarely find ungrouped data. Mostly, for instance when we have to take a sample from the general census data, these are grouped in some way or other, and it is exceedingly difficult to secure an equal chance for each individual to be included in the sample. The grouping of the general census data—for the sake of definiteness we shall bear this example in mind—has generally several grades. The lowest grade consists perhaps in groupings according to lodgings: the inhabitants of one apartment are given a single sheet. The next grouping may include sheets corresponding to apartments in several neighbouring

* A.B. Hill: *Sickness amongst Operatives in Lancashire Cotton Spinning Mills*, London 1930.

houses* visited by the same officer collecting the data for the Census. These groups are then grouped again and again. Obviously it would be practically impossible to sample at random single individuals from data subject to such complex groupings. Therefore it is useful to consider some further types of the random sampling method.

(c) Suppose that the population Π of M' individuals is grouped into M_0 groups. Instead of considering the population Π we may now consider another population, say π, having for its elements the M_0 groups of individuals, into which the population Π is divided. Turning to the example of the Polish Census, in which the material has been kept in bundles, containing data from single statistical districts, it was possible to substitute the study of the population π of $M_0 = 123{,}383$ statistical districts, for the study of the population Π of $M' = 32$ million individuals. If there are enormous difficulties in sampling individuals at random, these difficulties may be greatly diminished when we adopt groups as the elements of sampling. This being so, it is necessary to consider, whether and how our original problem of estimating X, the average value of the character x of individuals forming the population Π, may be transformed into a problem concerning the population π of groups of individuals.

The number we wish to estimate is

$$X = \frac{1}{M'} \sum_{i=1}^{M'} (x_i) \tag{16}$$

where x_i means the value of the character x of the i-th individual. Obviously there is no difficulty in grouping the terms of the sum on the right-hand side of the above equation so that each group of terms refers to a certain group of individuals, forming the population π. Suppose that these groups contain respectively

$$v_1, v_2, \ldots v_M \tag{17}$$

individuals and that the sums of the x's corresponding to these individuals are

$$u_1, u_2, \ldots u_M \tag{18}$$

With this notation we shall have

$$M' = v_1 + v_2 + \cdots + v_M = \Sigma(v) \tag{19}$$

$$\sum_{i=1}^{M'} (x_i) = u_1 + u_2 + \cdots u_M = \Sigma(u) \text{ (say)} \tag{20}$$

The problem of estimating X is now identical with the problem of estimating the character of the population π, namely,

* This was the grouping used in the Polish General Census in 1931. The corresponding groups will be called "statistical districts." The number of persons in one statistical district varied from 30 to about 500.

$$X = \frac{\Sigma(u)}{\Sigma(v)} \tag{21}$$

We have now to distinguish two different cases: (*a*) the number M' of individuals forming the population Π is known, and (*b*) this number is not known.

In the first case the problem of estimating X reduces itself to that of estimating the sum of the u's in the numerator of (21). In the other case we have also to estimate the sum of the v's in the denominator and, what is more, the ratio of the two sums. Owing to the results of S. Bernstein and of R.C. Geary this may be easily done if the estimates of both the numerator and the denominator in the formula giving X are the best linear estimates. The theorem of S. Bernstein* applies to such estimates, and states that under ordinary conditions of practical work their simultaneous distribution is representable by a normal surface with constants easy to calculate. Of course there is the limiting condition that the size of the sample must be large. The result of Geary† then makes it possible to determine the accuracy of estimation of X by means of the ratio of the separate estimates of the numerator and the denominator.

Thus we see that if it is impossible or difficult to organize a random sampling of the individuals forming the population to be studied, the difficulty may be overcome by sampling groups of individuals. Here again we may distinguish the two methods of unrestricted and of stratified sampling. It is indisputable that the latter has definite advantages both from the point of view of the accuracy of results and of the ease in performing the sampling. Therefore we shall further consider only the method of stratified sampling from the population π, the elements of which are groups of individuals forming the population Π. It is worth noting that this form of the problem is very general. It includes the problem of unrestricted sampling, as this is the special case when the number of strata $k = 1$. It includes also the problem of sampling individuals from the population Π, as an individual may be considered as a group, the size of which is $v = 1$. We shall see further on that the method of stratified sampling by groups includes as a special case the method of purposive selection.

2. The Method of Purposive Selection

Professor Bowley did not consider in his book the above type (*c*) of the method of random sampling by groups.‡ Then, therefore, he speaks about the

* S. Bernstein: "Sur l'extension du théorème limite du calcul des probabilites." *Math. Ann.*, Bd. 97.

† R.C. Geary: "The Frequency Distribution of the Quotient of Two Normal Variates." *J.R.S.S.*, Vol. XCIII, Part III.

‡ Though he applied it in practical work.

principle of random sampling he is referring to the sampling of individuals. According to Bowley, the method of purposive selection differs from that of random sampling mainly in the circumstances that "in purposive selection the unit is an aggregate, such as a whole district, and the sample is an aggregate of these aggregates, while in random selection the unit is a person or thing, which may or may not possess an attribute, or with which some measurable quantity is associated.... Further, the fact that the selection is purposive very generally involves intentional dependence on correlation, the correlation between the quantity sought and one or more known quantities. Consequently the most important additional investigation in this section relates to the question how far the precision of the measurements is increased by correlation, and how best an inquiry can be arranged to maximize the precision."

It is clear from this quotation that the terminology of Professor Bowley and that which I am using do not quite fit together. In fact the circumstance that the elements of sampling are not human individuals, but groups of these individuals, does not necessarily involve a negation of the randomness of the sampling. Therefore I have thought it useful to consider the special type of random sampling by groups, and the nature of the elements of sampling will not be further considered as constituting any essential difference between random sampling and purposive selection.

The words purposive selection will be used to define the method of procedure described by Bowley, Gini and Galvani. This may be divided into two parts: (*a*) the method of obtaining the sample, and (*b*) the method of estimation of such an average as X, described above.

The method of obtaining the sample assumes that the population Π of individuals is divided into several, M, districts forming the population π, that the number of individuals in each district, say v_i, is known and, moreover, that there is known for each district the value of one or more numerical characters, which Professor Bowley calls "controls." There is no essential difference between cases where the number of controls is one or more, so we shall consider only the case where there is one control, which we shall denote by y_i for the i-th district. We shall retain our previous notation and denote by u_i the sum of values of x, corresponding to the i-th district or group. Consider next, say, $\bar{x}_i = u_i/v_i$ or the mean value of the character x in the i-th district. The basic hypothesis of the method of purposive selection is that the numbers $\bar{x}_i$ are correlated with the control y_i and that the regression of $\bar{x}_i$ on y_i is linear. As we shall have to refer again to this hypothesis, it will be convenient to describe it as the hypothesis H.

Assuming that the hypothesis H is true, the method of forming the sample consists in "purposive selection" of such districts for which the weighted mean

$$Y' = \frac{\Sigma(vy)}{\Sigma(v)} \tag{22}$$

has the same value, or at least as nearly the same as it is possible, as it has for the whole population, say Y. It is assumed that the above method of selection may supply a fairly representative sample, at least with regard to the character x. As it follows from the quotation from the work of Gini and Galvani, it was also believed that by multiplying the controls it would be possible to obtain what could be termed a generally representative sample with regard to many characters. Otherwise the method of purposive selection could not be applied to supply a sample which could be used in the future for purposes not originally anticipated.

This is the method of obtaining the sample. As we shall easily see, it is a special case of stratified random sampling by groups. In fact, though the three authors think of districts as of rather large groups with populations attaining sometimes one million persons, they assume that the number M of these districts is not very small. In the Italian investigation it was over 200. If we consider the values of the control, y, calculated for each district, we shall certainly find such districts for which the value of y is practically the same. Thus the districts may be grouped in strata, say of the first order

$$\pi_{y_1}, \pi_{y_2}, \dots \pi_{y_k} \tag{23}$$

each corresponding to a given value of y. Now each of the first order strata of districts may be subdivided into several second order strata, according to the values of v in the districts. Denote by π_{yv} a stratum containing, say M_{yv} districts, all of which have practically the same values of the control, y, and the same number of individuals v. Denote further by m_{yv} the number of the districts belonging to π_{yv} to be included in the sample. If the principle directing the selection consists only in the fulfilment of the condition that the weighted mean of the control with v's as weights should be the same in the sample and in the population, then it means nothing but a random sampling of some m_{yv} districts from each second order stratum, the numbers m_{yv} being fixed in advance, some of them being probably zero. This is obvious, since for purposes of keeping the weighted mean $Y' = Y =$ constant, two different districts belonging to the same second order stratum are of equal value. Hence we select one of them at random.*

Thus we see that the method of purposive selection consists, (*a*) in dividing the population of districts into second order strata according to values of y and v, and (*b*) in selecting randomly from each stratum a definite number of districts. The numbers of samplings are determined by the condition of maintenance of the weighted average of the y. Comparing the method of purposive

* It must be emphasized that the above interpretation of the method of purposive selection is a necessary one if we intend to treat it from the point of view of the theory of probability. There is no room for probabilities, for standard errors, etc., where there is no random variation or random sampling. Now if the districts are *selected* according to the corresponding values of the control y and also of the number of individuals, v, they contain, the only possible variate which is left to chance is $\bar{x}_i$. If the districts are very large and therefore only very few, then the majority of second order strata will contain no or only one district. In this case, of course, the process of random sampling from such a stratum is an imaginary one.

selection with that of stratified sampling by groups we have to bear in mind these two special features of the former.

IV. Comparison of the Two Methods of Sampling

1. Estimates of Bowley and of Gini and Galvani

Suppose now the sample is drawn and consider the methods of estimation of the average X. In this respect the Italian statisticians do not agree with Bowley, so we shall have to consider two slightly different procedures. I could not exactly follow the method proposed by Professor Bowley. It is more clearly explained by the Italian writers, but I am not certain whether they properly understood the idea of Bowley. It consists in the following:

Denote by X_Σ the weighted mean of values $\bar{x}_i$ deduced from the sample, Σ; by $\bar{x}$, the unweighted mean of the same numbers, also deduced from the sample. Y will denote the weighted mean of the control y, having *ex hypothesi* equal values for the sample and for the population. $\bar{y}$ will denote the unweighted mean of the control y, calculated for the population, and finally g the coefficient of regression of $\bar{x}_i$ on y_i, calculated partly from the sample and partly from the population.

As a first approximation to the unknown X, X_Σ may be used. But it is possible to calculate a correction, K, to be subtracted from X_Σ so that the difference $X_\Sigma - K$ should be considered as the second approximation to X. The correction K is given by the formula

$$K = -(X - \bar{x}) + g(Y - \bar{y}). \tag{24}$$

As the value of X is unknown, its first approximation X_Σ may be substituted in its place. In this way we get as a second approximation to X the expression, say,

$$X' = X_\Sigma + (X_\Sigma - \bar{x}) - g(Y - \bar{y}). \tag{25}$$

I do not know whether this is the method by which Bowley has calculated the very accurate estimates in the examples he considers in his paper. At any rate the method as described above is inconsistent: even if applied to a sample including the whole population and even if the fundamental hypothesis H about the linearity of regression of $\bar{x}_i$ on y_i is exactly satisfied, it may give wrong results:

$$X' \neq X \tag{26}$$

This may be shown on the following simple example. Suppose that the population π consists only of four districts characterized by the values of $\bar{x}_i$, y_i and v_i as shown in the following Table I.

Table I.

Districts	x_i	y_i	v_i	$u_i = \bar{x}_i v_i$	$y_i v_i$
I.	.07	.09	100	7	9
II.	.09	.09	400	36	36
III.	.11	.12	100	11	12
IV.	.13	.12	900	117	108
Totals	.40	.42	1500	171	165
Means	$\bar{x} = .100$	$\bar{y} = .105$	—	$X = .114$	$Y = .110$

Owing to the fact that the control y has only two different values, .09 and .12, there is no question about the hypothesis H concerning the linearity of regression, which is certainly satisfied. The regression line passes through the points with co-ordinates ($y = .09$, $x = .08$) and ($y = .12$, $x = .12$). Thus the coefficient of regression $g = \frac{4}{3}$. Assume now we have a sample from the above population, which includes the whole of it and calculate the estimate X' of $X = 114$. We shall have

$$\begin{aligned} X_\Sigma = & \quad = .114 \\ (X_\Sigma - \bar{x}) = & \quad = .014 \\ -g(Y - \bar{y}) = & -\frac{.02}{3} = -.007 \\ & \overline{X' = .121,} \end{aligned} \tag{27}$$

which is not equal to $X_\Sigma = .114$.

Gini and Galvani applied Bowley's method to estimate the average rate of natural increase of the population of Italy, using a sample of 29 out of 214 circondarî. They obtained results which they judged to be unsatisfactory, and they proposed another method of estimation. This consists in the following:

They start by finding what could be called the weighted regression equation. If there are several controls, say $y^{(1)}, y^{(2)}, \ldots y^{(s)}$, the weighted regression equation

$$x = b_0 + b_1 y^{(1)} + b_2 y^{(2)} + \cdots + b_s y^{(s)} \tag{28}$$

is found by minimizing the sum of squares

$$\Sigma v_i (x_i - b_0 - b_1 y_i^{(1)} - \cdots - b_s y_i^{(s)})^2 \tag{29}$$

with regard to the coefficients $b_0, b_1, b_2, \ldots b_s$. This process would follow from the ordinary formulæ if we assumed that one district with the number of individuals v_i and the mean character $\bar{x}_i$ is equivalent to v_i individuals, each having the same value of the character $x = \bar{x}_i$. Having noticed this, it is not necessary to go any further into the calculations. If there is only one control, y, then the weighted regression equation will be different from the ordinary one in that it will contain weighted sample means of both $\bar{x}_i$ and y_i instead of the unweighted ones, and that in the formula of the regression coefficient we should get weighted instead of unweighted sums. The weighted regression equation is then used by Gini and Galvani to estimate the value of $\bar{x}_i$ for each district, whether included in the sample or not. This is done by substituting into the equation the values of the control y_i corresponding to each district and in calculating the value of the dependent variable. The estimates of the means $\bar{x}_i$ thus obtained, say $\bar{x}_i'$, are then used to calculate their weighted mean

$$X' = \frac{\Sigma(v_i \bar{x}_i')}{\Sigma(v_i)}, \tag{30}$$

which is considered as an estimate of the unknown mean X.

Simple mathematical analysis of the situation proved (see Note III) that this estimate is consistent when a special hypothesis, H', about the linearity of regression of $\bar{x}_i$ on y_i holds good, and even that it is the best linear estimate under an additional condition, H_1, concerning the variation of the $\bar{x}_i$ in strata corresponding to different fixed values of y and v.

The hypothesis H' consists in the assumption that the regression of $\bar{x}$ on y is linear not only if we consider the whole population π of the districts, but also if we consider only districts composed of a fixed number of individuals. It is seen that the hypothesis H' is a still more limiting than the hypothesis H.

The other condition, H_1, is as follows. Consider a stratum, π', defined by the values $y = y'$ and $v = v'$ and consider the districts belonging to this stratum. Let

$$\bar{x}_1, \bar{x}_2, \ldots \bar{x}_p \tag{31}$$

be the values of the means $\bar{x}$ corresponding to these districts. The hypothesis, say H_1, under which the estimate of X proposed by Gini and Galvani is the best linear estimate, consists in the assumption that the standard deviation, say σ' of the $\bar{x}_i$ corresponding to the stratum π' may be presented by the formula

$$\sigma' = \frac{\sigma}{\sqrt{v'}} \tag{32}$$

σ being a constant, independent of the fixed value of $v = v'$. This hypothesis would be justifiable if the population of each district could be considered as a random sample of the whole population Π. In fact, then the standard deviation of means, $\bar{x}_i$ corresponding to districts having their population equal to v would be proportional to $v^{-\frac{1}{2}}$. The population of a single district is certainly not a random sample from the population of the country, so the estimate of Gini and Galvani is not the best linear estimate—at least in most cases.

Having got so far we may consider whether and to what extent there is justification for the principle of choosing the sample so that the weighted mean of the control in the sample should be equal to the weighted mean of the population. The proper criterion to use in judging seems to be the standard error of the estimate of X'. This is given by a function (see Note III) which, *cæteris paribus*, has smaller values when the weighted sample mean of the control is equal to its population value, and when the sum of weights $\Sigma(v)$, calculated for the sample, has the greatest possible value. Thus the principle of purposive selection is justified. The analysis carried out in Note III suggests also that if the number of districts to be included in the sample is fixed we should get greater accuracy by choosing larger districts rather than smaller ones. This conclusion, however, depends largely upon the assumptions made concerning the standard deviations within the districts and the linearity of regression.

2. The Hypotheses Underlying both Methods and the Conditions of Practical Work

We may now consider the questions: (1) Are we likely to find in practice instances where the hypotheses underlying the method of purposive selection are satisfied, namely, the hypothesis H' concerning the linearity of regression and the hypothesis H_1 concerning the variation of the character sought within the strata of second order? (2) If we find instances where these hypotheses are not satisfied exactly, then what would be the result of our ignoring this fact and applying the method of purposive selection? (3) Is it possible to get any better method than that of purposive selection?*

With regard to (1), I have no doubt that it is possible to find instances, when the regression of a certain character $\bar{x}_i$ on the control y_i is fairly nearly linear. This may be the case especially when one of the characters $\bar{x}$ and y is some linear function of the other, say if $\bar{x}$ is the rate of natural increase of the population and y the birth-rate. This is the example considered by Gini and Galvani. I think, however, that this example is rather artificial. When y is known for any district, in most cases we shall probably have all the necessary data to enable us to compute the $\bar{x}$ without any appeal to the representative method. In other cases, however, when the connection between the character sought and the possible control is not so straightforward, I think it is rather dangerous to assume any definite hypothesis concerning the shape of the regression line. I have worked out the regression of the mean income $\bar{x}_i$ of people inhabiting different circondarî on the first of the controls used by Gini and Galvani, i.e. the birth-rate, y_i. The figures I and II give respectively the approximate spot diagram of the correlation table of those characters, and the graph of the weighted regression line of $\bar{x}_i$ and y_i. It is to be remembered that the data concern the whole population, and thus the graph represents the "true" regression line. This is far from being straight. It is difficult, of course,

* *I.e.* a method which would not lose its property of being consistent when the hypothesis H' is not satisfied.

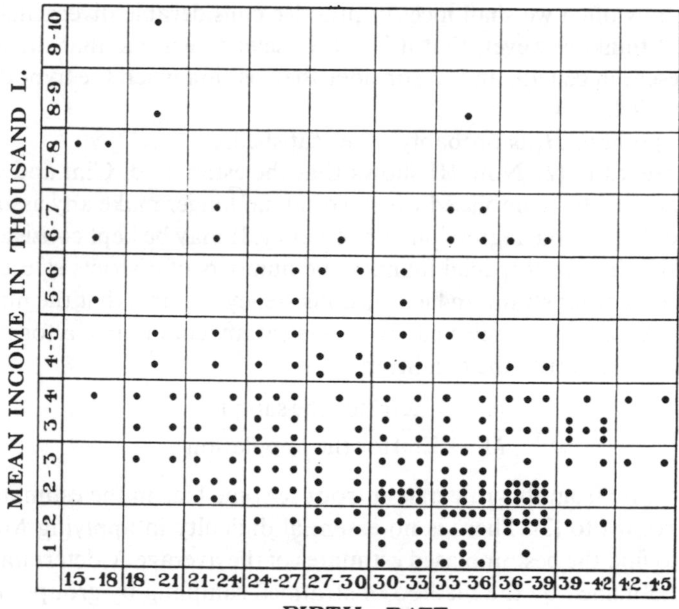

REGRESSION OF x ON y.

Figure I.

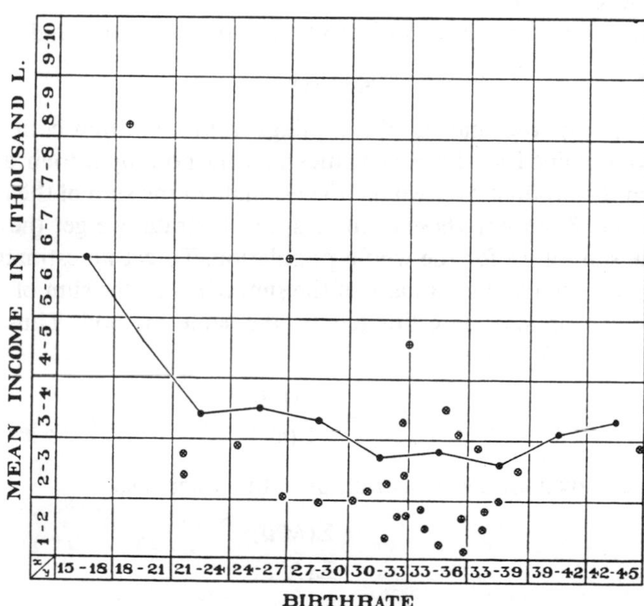

Figure II.

to judge how often we shall meet in practice considerable divergencies from linearity. I think, however, that it is rather safer to assume that the linearity is not present in general and to consider the position when the hypothesis H' is not satisfied.

The hypothesis H_1 is probably never satisfied.

With regard to (2): Note III shows that the estimate of Gini and Galvani generally ceases to be unbiased when we can no longer make any assumption about the shape of the regression line of $\bar{x}$ on y. It may be kept consistent only by adjusting in a very special manner the numbers of districts selected from single second order strata. In fact the consistency requires that the number of districts, say m' to be selected from a stratum containing altogether M' districts, should satisfy the condition

$$\frac{m'}{M'} = \frac{\Sigma(v) \text{ for the sample}}{\Sigma(v) \text{ for the population}}. \tag{33}$$

Any departure from this rule may introduce some bias in the estimate.

With regard to (3): There is no essential difficulty in applying Markoff's method to find the best unbiased estimates of the average X determined from a sample obtained by the method of stratified sampling by groups. This has been done in full detail in my Polish publication (there is an English summary)* concerning the theory of the representative method. The principle of stratifying, *i.e.* of the division of the original population of districts into strata, does not affect the method of obtaining the estimate. In any case, and whatever the variances of the $\bar{x}_i$ within the strata, the best linear estimate of X is always the same.

I shall return here to variables introduced previously and shall use

$$u_i = v_i \bar{x}_i \tag{34}$$

instead of $\bar{x}_i$. Suppose that in m' samplings from a stratum containing M' districts, we obtained m' different values of u. Denote by $\bar{u}$ their arithmetic mean. Then the product $M'\bar{u}$ will be the estimate of the sum of the u's for the whole stratum. Summing these estimates for all strata, we get the best estimate of the sum of u's for the whole population. To get an estimate of X it remains only to divide the estimate of the sum of u's by the sum of v's, which may be known or may be estimated by the same method. Thus the final estimate of X say X'' is either

$$X'' = \frac{\Sigma(M'\bar{u})}{\Sigma(v)} \tag{35}$$

if the v's are known for every district, or in the other case

$$X'' = \frac{\Sigma(M'\bar{u})}{\Sigma(M'\bar{v})} \tag{36}$$

* J. Neyman: *An Outline of the Theory and Practice of Representative Method, Applied in Social Research.* Institute for Social Problems, Warsaw, 1933.

where $\overline{v}$ means the arithmetic mean of v's, calculated from the sample separately for each stratum.

The consistency of the estimates $\Sigma(M'\overline{u})$ and $\Sigma(M'\overline{v})$ does not depend upon any arbitrary hypothesis concerning the sampled population. The only condition, which must be satisfied is that the sample should contain districts from every stratum. So we may safely apply these estimates, whatever the properties of single strata and irrespective of variations of u's and v's within the strata. But the standard errors of the two estimates do depend both upon the variability of the characters of districts within the strata and upon the relationship of numbers m' and M'. It is known that the formula giving the variance, say σ^2, of the estimate $\Sigma(M'\overline{u})$ is as follows:

$$\sigma^2 = \sum \left\{ \frac{M_i^2}{m_i} \frac{M_i - m_i}{M_i - 1} \sigma_i^2 \right\} \tag{37}$$

where m_i and M_i refer to the i-th stratum, σ_i^2 is the variance of the u's in the i-th stratum and the summation Σ extends over all strata. The dependence of σ^2 upon the σ_i^2 is obvious. If we succeed in dividing the population π into strata which would be very homogeneous with regard to the character u of the districts, σ_i^2 will be small and so will be σ^2. It is also obvious that by increasing the numbers, m_i, of districts to be selected from the strata we shall also improve the accuracy of the estimate. By taking $m_i = M_i$ the accuracy will be absolute, but then we shall have an exhaustive enquiry. It will probably be necessary to assume that the actual conditions of the research fix a certain number, say

$$m_0 = \Sigma(m_i) \tag{38}$$

of districts to be selected from the population. Our problem will then consist in distributing the total number of samplings among single strata so as to have the minimum possible value of σ^2.

Simple calculations show that the variance (37) may be written in the form

$$\sigma^2 = \frac{M_0 - m_0}{m_0} \Sigma(M_i S_i^2) + \Sigma m_i \left(\frac{M_i S_i}{m_i} - \frac{\Sigma(M_i S_i)}{m_0} \right)^2 - \frac{M_0}{m_0} \Sigma M_i \left(S_i - \frac{\Sigma(M_i S_i)}{M_0} \right)^2 \tag{39}$$

where S^2 stands for $M_i\sigma_i^2/(M_i - 1)$. We see that only the middle term of the right-hand side depends upon the values of the m's. The other terms remain constant whatever the system of m's provided their sum, m_0, remains unchanged. Thus the method of diminishing the value of σ^2 consists in diminishing the middle term of the right hand side of (39). This has its minimum value, zero, when the numbers m_i are proportional to the products $M_i S_i$. Thus if it is possible to estimate the variances σ_i^2 of the u's within any given stratum, the most favourable system of m_i's is not that for which the m_i are proportional to the M_i. Denote the three terms of the right-hand side of (39) respectively by A, B and $-C$. If we assume that the m_i's are proportional to M_i, then we shall find that the term $B = C$ and the variance σ^2 is reduced to

$$\sigma^2 = \frac{M_0 - m_0}{m_0} \Sigma(M_i S_i^2) = A* \tag{40}$$

If, however, m_i are proportional to $M_i S_i$, then the positive term B in (39) vanishes and we get

$$\sigma^2 = A - C \tag{41}$$

which is the optimum value of σ^2.

If the research is carried out with regard to several highly correlated characters of groups forming the elements of sampling, then by means of a preliminary enquiry it is possible to estimate the numbers S_i, which, if calculated for the different characters sought, would be also correlated. Hence we could then by a proper choice of the numbers m_i if not reduce the middle term of the righthand side of (39) to zero, then at least diminish it sensibly.

Such was the case in the Warsaw enquiry already referred to, carried out by the Institute for Social Problems. The purpose of this enquiry was to describe the structure of the working class in Poland, according to different characters, such as the age distribution of males and females, whether married or single, the distribution of the number of children in families, etc., and this separately for three different categories of workers. Obviously all characters of the elements of sampling sought are highly correlated with the number of workers in each element. As there are in Poland large districts where the percentage of workers is negligible and others where they are numerous, the numbers S_i calculated for the different characters sought varied from stratum to stratum in broad limits. Accordingly, an adjustment of numbers m_i was made in order to diminish variances of the estimates.

The necessity of these adjustments is not difficult to appreciate. One feels intuitively that it would be unreasonable to include in the sample, equal percentages of statistical districts from two strata A and B in one of which, A, the percentage of workers, amounts to say 60 per cent. and in the other, B, to 5 per cent. It may even be assumed that in such cases it would be advisable to omit totally the stratum B. However, I do not think it is really always advisable, since the total number of workers in the stratum B may be sometimes equal to or even larger than those in stratum A, and the structure of family conditions in both strata may be very different.

Of course this sort of research is a rather special one. In many cases the characters sought are not likely to be highly correlated. In other cases—as in the work of Gini and Galvani—it is impossible to state at the time of sampling which characters of the elements of sampling will be the matter of research. Any adjustments of the numbers, m_i, are then impossible, since a

* *Editor's note*: Author's misprints have been corrected in this reproduction.

wrong adjustment may give to σ^2 a value larger than that corresponding to the system of proportional sampling. The best we can do is to sample proportionately to the sizes of strata.*

Thus the principle that the numbers m_i should be proportional to M_i, suggested by Professor Bowley, is just the best that one could advise in the most general case.

Up to this point I have considered the possibility of reducing the value of σ^2 by adjusting properly the numbers m_i of samplings from different strata. I assumed, in fact, that the districts forming the elements of sampling and their total number m_0 to be included in the sample are fixed. Now I shall suppose that the districts are not fixed except that their size will not be very different, and that all that is known is that the sample should include a certain percentage of districts, whatever be their kind.

In other words, I intend to consider the situation in which we decide to include in the sample some, *e.g.* 10 per cent. of the population, and are considering the question what should be our "districts," forming the elements of sampling: whether they should include about, say 200 or about 20,000 persons, etc.

I wish to call attention to the fact, that the ratios m_i/M_i being fixed in some way or other, the value of σ^2 (see (37)) depends upon the products $M_i S_i^2 = M_i^2 \sigma_i^2/(M_i - 1)$, or practically upon the products $M_i \sigma_i^2$, and may be influenced by a proper choice of the element of sampling. In fact, if we consider two different systems of division of a stratum into larger and smaller districts, then the values of u's corresponding to several smaller districts forming a larger one, will be very generally positively correlated. As the result of this the value of $M_i \sigma_i^2$, corresponding to a subdivision of strata into smaller districts, will be less than that corresponding to a subdivision into larger districts. This point may be illustrated on an extreme case. Suppose, for instance, that X represents the proportion of agricultural workers aged 20 to 21. Then for every individual of the population x will have the value $x = 1$ if this individual is an agricultural worker aged 20 to 21, and $x = 0$ in all other cases. If now we consider as elements of sampling the statistical districts including 50 inhabitants, then in a stratum we may have (in the most unfavourable case) one half of the districts composed only of agricultural workers at the fixed age, thus having $u = 50$, while in the other half of the district $u = 0$. The standard deviation σ_i would be 25. On the other hand, if the districts were to include not 50 persons, but, say, 500, the maximum possible value of σ_i would be tenfold, 250. The term $M_i \sigma_i^2$ in this second case would be ten times larger than in the former. Of course it may be argued that taking larger districts we decrease the chance of their being extremely differentiated. This is certainly so, but on the other hand I think it extremely probable that the products $M_i S_i^2$ calculated for districts including tens of thousands or hundreds of thousands of people must be expected to be incomparably larger than those calculated

* It is to be remembered that "the size of the stratum" is the number, M_i, of its elements, not the number of individuals.

for the districts including on the average two or three hundred people. And this for the majority of imaginable characters which could be the matter of statistical research.*

The effect of choosing smaller units of sampling may be roughly illustrated on another example of a game of chance, in which the probability of a gain is equal to $\frac{1}{2}$. Suppose we dispose of a sum of £100 for the game, which we may either bet at once or divide in a hundred separate bettings. In the first case it is obviously impossible to predict the result. In the other case, however, we may be pretty certain that the gain or loss will not exceed some £15 or £20.

Similarly, if we want to obtain a representative sample, say amounting to 15 per cent. of the population, it is much safer to make, say, 3,000 samplings of small units rather than 30 of larger ones, and this is probably true, whatever the stratification.

(*Editors' note*: Subsection 3 has been omitted.)

V. Conclusions

Let us now turn to the question, which I raised at the beginning of the paper, whether the idea of a certain equivalency of the two aspects of the representative method is really justified. We shall have to consider both the theory and the practical results obtained by both methods. Professor Bowley, who was first to give the theory of the method of purposive selection, has not, I believe, used it in practice. The most important research, known to me, by which the representative method was used, is the *New Survey of London Life and Labour*. It has been directed by Bowley, who chose the method of random sampling by groups. This is, I think, an example of the intuition to which Laplace referred.

The Italian statisticians, who applied the method of purposive selection of very few (29) and very large districts with populations from about 30,000 to about 1 million persons, did not find their results to be satisfactory. The comparison between the sample and the whole country showed, in fact, that though the average values of seven controls used are in a satisfactory agreement, the agreement of average values of other characters, which were not used as controls, is often poor. The agreement of other statistics besides the means, such as the frequency distributions, etc., is still worse. This applies also to the characters used as controls. The statement of the above facts is followed

* I do not know whether these were the reasons for which Gini and Galvani expressed the view that the results of their sampling would have been much better if the method of selection adopted were that of stratified sampling, and if the element of sampling were a commune. The reasons for not applying this method seems to be that "nobody could under-appreciate the difficulty in a stratification of the communes simultaneously with regard to different characters." (Page 6, *loc. cit.*). I think, however, that a stratification assuming the 214 circondarî as strata, each containing about 40 communes, which might be considered as elements of sampling, would be quite sufficient. Of course the results would be probably still better if the elements of sampling were smaller than a commune.

in the paper by Gini and Galvani by general considerations concerning the concept of a representative sample. They question whether it is possible to give any precise sense to the words "a generally representative sample." I think it is, and I agree also that an exhaustive enquiry is the only method which can give absolutely true results. However, the need for a representative method is an urgent one and many enquiries would be impossible if we were not able to use this method. In fact we are often forced to apply sampling for general purposes, so as to get a "generally representative" sample, which might be used for a variety of different purposes.

If there are difficulties in defining the "generally representative sample," I think it is possible to define what should be termed a *representative method of sampling* and a *consistent method of estimation*. These I think may be defined accurately as follows. I should use these words with regard to the method of sampling and to the method of estimation, if they make possible an estimate of the accuracy of the results obtained in the sense of the new form of the problem of estimation, *irrespectively of the unknown properties of the population* studied. Thus, if we are interested in a collective character X of a population π and use methods of sampling and of estimation, allowing us to ascribe to every possible sample, Σ, a confidence interval $X_1(\Sigma)$, $X_2(\Sigma)$ such that the frequency of errors in the statements

$$X_1(\Sigma) \leq X \leq X_2(\Sigma) \tag{43}$$

does not exceed the limit $1 - \varepsilon$ prescribed in advance, *whatever the unknown properties of the population*, I should call the method of sampling representative and the method of estimation consistent. We have seen that the method of random sampling allows a consistent estimate of the average X whatever the properties of the population. Choosing properly the elements of sampling we may deal with large samples, for which the frequency distribution of the best linear estimates is practically normal, and there are no difficulties in calculating the confidence intervals. Thus the method of random stratified sampling may be called a representative method in the sense of the word I am using. This, of course, does not mean that we shall always get correct results when using this method. On the contrary, erroneous judgments of the form (43) must happen, but it is known how often they will happen in the long run: their probability is equal to ε.

On the other hand, the consistency of the estimate suggested by Gini and Galvani, based upon a purposely selected sample, depends upon hypotheses which it is impossible to test except by an extensive enquiry.

If these hypotheses are not satisfied, which I think is a rather general case, we are not able to appreciate the accuracy of the results obtained. Thus this is not what I should call a representative method. Of course it may give sometimes perfect results, but these will be due rather to the uncontrollable intuition of the investigator and good luck than to the method itself. Even if the underlying hypotheses are satisfied, we have to remember that the elements of sampling which it is possible to use when applying the purposive

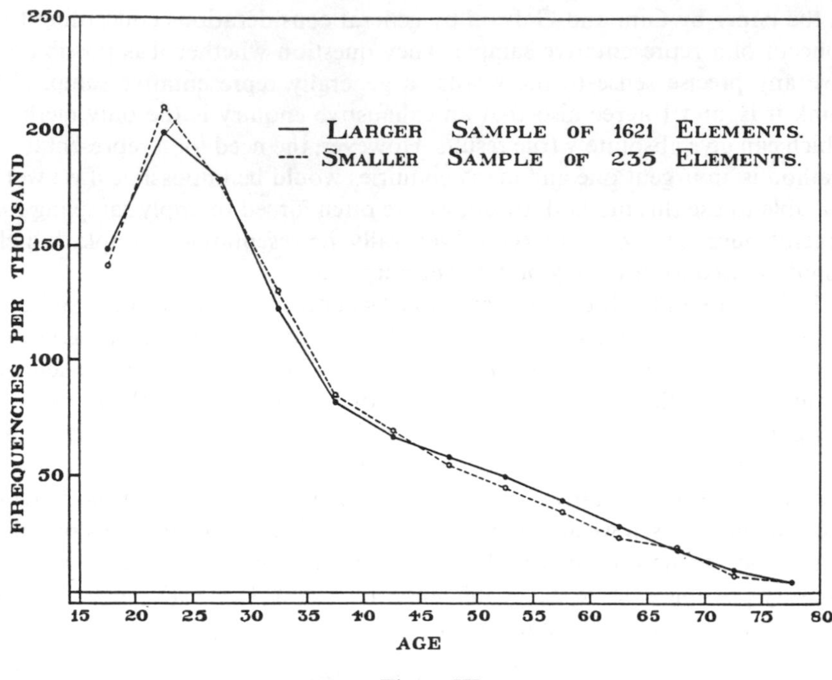

Figure III

selective method, must be very few in number and very large in size. Consequently I think that when using this method we are very much in the position of a gambler, betting at one time £100.

For the above reasons I have advised the Polish Institute for Social Problems to use the method of random stratified sampling by groups when carrying out the enquiry on the structure of Polish workers.*

Poland was divided into 113 strata, containing 123,383 elements of sampling (statistical districts). The average number of persons within an element of sampling was about 250 persons. There were considerable variations from stratum to stratum. The random stratified sample contained altogether 1,621 elements, thus about 1.24 per cent. of the whole population. I am not yet able to state how accurate are the results obtained, as the respective data of the General Census are not yet published. All that was possible in testing their accuracy was to compare the age distribution of workers found in the whole sample with the age distribution computed from a minor sample of 235 elements selected for an introductory enquiry which aimed at testing the variability within the strata. The results are presented in Figure III and in

* The results of the enquiry are to be found in the publication of J. Piekalkiewicz: *Rapport sur les recherches concernant la structure de la population ouvrière en Pologne selon la méthode répresentative.* Institute for Social Problems, Warsaw, 1934.

Table IV, and seem to be satisfactory. However, even if through the chances of sampling they had been bad, I think I was justified in advising the method of stratified sampling by groups, because I was able to calculate that (with the probability of an error equal to .01) the error of actuarial calculations, based upon the tables which were computed as the result of enquiry, could not exceed 4.5 per cent.

The method of stratified sampling by groups has been recently used by

Table IV. Age Distribution of Polish Workers. Males.

Age	Larger Sample	Smaller Sample
15–19 ...	148	141
20–24 ...	199	213
25–29 ...	178	176
30–34 ...	122	130
35–39 ...	82	85
40–44 ...	67	69
45–49 ...	58	54
50–54 ...	49	44
55–59 ...	39	34
60–64 ...	28	23
65–69 ...	18	19
70–74 ...	9	7
75–79 ...	4	4
Totals ...	1001	999

Professor O. Anderson,* who directed an enquiry into the farming conditions in Bulgaria. The process of getting the sample with which he was faced was a more difficult one, as this was not a sample of sheets of paper containing the necessary information, but a sample of villages from which it was necessary to collect the original data. In fact the enquiry in question was a substitute for a general agricultural census. The element of the sampling was a village. The total number of about 5,000 villages was divided into 28 strata. Out of each stratum 2 per cent. of the villages were selected to form the sample. There is only one detail in this enquiry which I am not certain is justifiable. When selecting the villages from single strata special attention was paid to selecting villages which according to the last General Census in 1926 showed a distribution of different characters of farms, similar to that in the whole stratum. I think that the variability of farms and villages is also a character of their population which may be of interest. This character, however, if the efforts of Bulgarian investigators were successful, would be biased in the sample.

The final conclusion which both the theoretical considerations and the above examples suggest is that the only method which can be advised for

* *Bull. de Statistique*, publ. Direction Gen. de Statistique de Bulgarie, No. 8, 1934.

general use is the method of stratified random sampling. If the conditions of the practical work allow, then the elements of the sampling should be individuals. Otherwise we may sample groups, which, however, should be as small as possible. The examples of enquiries in London, in Bulgaria, and in Poland show that random sampling by groups does not present unsurmountable difficulties.

There are instances when we may select individuals purposely with great success. Such is, for instance, the case when we are interested in regression of some variate y on x, in which case the selection of individuals with values of x varying within broad limits would give us more precision. But these cases are rather exceptional.*

(*Editors' note*: Section VI, Appendix, has been omitted.)

* Interesting remarks in this respect are to be found in the excellent book of M. Ezekiel: *Methods of Correlation Analysis* (1930).

Introduction to
Hotelling (1936) Relations Between Two Sets of Variates

T.W. Anderson
Stanford University

1. Canonical Correlations

Let the random vector* x with mean 0 be partitioned into subvectors of s and t components, respectively, $x = (x^{(1)\prime}, x^{(2)\prime})'$, and let the covariance matrix matrix $\mathscr{E}xx' = \Sigma$ be partitioned conformally

$$\mathscr{E}xx' = \begin{bmatrix} \mathscr{E}x^{(1)}x^{(1)\prime} & \mathscr{E}x^{(1)}x^{(2)\prime} \\ \mathscr{E}x^{(2)}x^{(1)\prime} & \mathscr{E}x^{(1)}x^{(1)\prime} \end{bmatrix} = \begin{bmatrix} \Sigma_{11} & \Sigma_{12} \\ \Sigma_{21} & \Sigma_{22} \end{bmatrix} = \Sigma. \tag{1}$$

Assume that Σ is nonsingular. This paper treats many questions concerning the relationship between $x^{(1)}$ and $x^{(2)}$ in terms of the covariances, actually in terms of the correlations. The most notable innovation is the definition of canonical variates and canonical correlations.

Hotelling asks first for the linear function of $x^{(1)}$, say, $u = a'x^{(1)}$, that has maximum correlation with a linear function of $x^{(2)}$, say, $v = b'x^{(2)}$. The correlation between these two linear functions is

$$\frac{\mathscr{E}uv}{\sqrt{\mathscr{E}u^2}\sqrt{\mathscr{E}v^2}} = \frac{a'\Sigma_{12}b}{\sqrt{a'\Sigma_{11}a}\sqrt{b'\Sigma_{22}b}}; \tag{2}$$

This is to be maximized with respect to a and b. The use of Lagrange multipliers to impose the condition $a'\Sigma_{11}a = 1$ and $b'\Sigma_{22}b = 1$ gives the function

$$a'\Sigma_{12}b - \frac{1}{2}\lambda a'\Sigma_{11}a - \frac{1}{2}\mu b'\Sigma b. \tag{3}$$

* Hotelling carried out his algebra in terms of components; it is much easier to express the algebra in terms of vectors and matrices, as done here.

The derivatives of (2) with respect to the elements of a and b set equal to 0 are

$$\Sigma_{12}b - \lambda\Sigma_{11}a = 0, \tag{4}$$

$$\Sigma_{21}a - \mu\Sigma_{22}b = 0. \tag{5}$$

The multiplication of (4) on the left by a' and of (5) on the left by b' shows

$$\lambda = \mu = a'\Sigma_{12}b \tag{6}$$

for solutions a, b of (4) and (5). The equations (4) and (5) can then be written

$$\begin{pmatrix} -\lambda\Sigma_{11} & \Sigma_{12} \\ \Sigma_{21} & -\lambda\Sigma_{22} \end{pmatrix}\begin{pmatrix} a \\ b \end{pmatrix} = 0. \tag{7}$$

For nontrivial solutions, it is necessary that

$$\begin{vmatrix} -\lambda\Sigma_{11} & \Sigma_{12} \\ \Sigma_{21} & -\lambda\Sigma_{22} \end{vmatrix} = 0. \tag{8}$$

The determinant on the left-hand side of (8) is a polynomial of degree $s + t$ (since Σ_{11} and Σ_{22} are nonsingular). Let the roots be $\rho_1 \geq \cdots \geq \rho_{s+t}$. Then the maximum correlation is ρ_1 and linear combinations with this correlation are $u_1 = a^{(1)\prime}x^{(1)}$ and $v_1 = b^{(1)\prime}x^{(2)}$, where $a^{(1)}$ and $b^{(1)}$ are a solution to (7) for $\lambda = \rho_1$. Note that $\mathscr{E}u_1^2 = a^{(1)\prime}\Sigma_{11}a^{(1)} = 1$, $\mathscr{E}v_1^2 = b^{(1)\prime}\Sigma_{22}b^{(1)} = 1$, and $\mathscr{E}u_1v_1 = a^{(1)\prime}\Sigma_{12}b^{(1)} = \rho_1$.

Next consider another linear combination $u = a'x^{(1)}$ uncorrelated with u_1 and another linear combination $v = b'x^{(2)}$ uncorrelated with v_1 that have maximum correlation. It follows from the definition of u_1 and v_1 that u is uncorrelated with v_1 and v uncorrelated with u_1. Proceeding in a manner similar to that outlined above, Hotelling found that the maximum correlation is ρ_2, and it is the correlation between $u_2 = a^{(2)\prime}x^{(1)}$ and $v_2 = b^{(2)\prime}x^{(2)}$, where $a^{(2)}$, $b^{(2)}$ is a solution to (7) for $\lambda = \rho_2$. Similarly, the linear combinations $u_3 = a^{(3)\prime}x^{(1)}, \ldots, u_s = a^{(s)\prime}x^{(1)}$ and $v_3 = b^{(3)\prime}x^{(2)}, \ldots, v_s = b^{(s)\prime}x^{(2)}$ are found, having correlation $\rho_3, \ldots, \rho_s$, respectively. For convenience, assume that $s \leq t$. If $s < t$, there are $t - s$ more linear combinations $v_{s+1} = b^{(s+1)\prime}x^{(2)}, \ldots, v_t = b^{(t)\prime}x^{(2)}$ that are mutually uncorrelated and uncorrelated with earlier u's and v's.

Let

$$A = \begin{pmatrix} a^{(1)\prime} \\ \vdots \\ a^{(s)\prime} \end{pmatrix}, \quad B = \begin{pmatrix} b^{(1)\prime} \\ \vdots \\ b^{(t)\prime} \end{pmatrix} \quad u = \begin{pmatrix} u_1 \\ \vdots \\ u_s \end{pmatrix}, \quad b = \begin{pmatrix} v_1 \\ \vdots \\ v_s \end{pmatrix}. \tag{9}$$

Then

$$\begin{bmatrix} u \\ v \end{bmatrix} = \begin{bmatrix} Ax^{(1)} \\ Bx^{(2)} \end{bmatrix} = \begin{bmatrix} A & 0 \\ 0 & B \end{bmatrix}\begin{bmatrix} x^{(1)} \\ x^{(2)} \end{bmatrix}. \tag{10}$$

The covariance matrix of (u', v') is

$$\mathscr{E}\begin{bmatrix} u \\ v \end{bmatrix}[u', v'] = \begin{bmatrix} A & 0 \\ 0 & B \end{bmatrix}\begin{bmatrix} \Sigma_{11} & \Sigma_{12} \\ \Sigma_{21} & \Sigma_{22} \end{bmatrix}\begin{bmatrix} A' & 0 \\ 0 & B' \end{bmatrix}$$

$$= \begin{bmatrix} A\Sigma_{11}A' & A\Sigma_{12}B' \\ B\Sigma_{21}A' & B\Sigma_{22}B' \end{bmatrix}. \tag{11}$$

Hotelling has found A and B so that this covariance matrix is ($s \le t$)

$$\begin{bmatrix} A\Sigma_{11}A' & A\Sigma_{12}B' \\ B\Sigma_{21}A' & B\Sigma_{22}B' \end{bmatrix} = \begin{bmatrix} I_s & R & 0 \\ R & I_s & 0 \\ 0 & 0 & I_{t-s} \end{bmatrix}, \tag{12}$$

where R is a diagonal matrix with diagonal elements $\rho_1 \ge \rho_2 \ge \cdots \ge \rho_s \ge 0$. The other roots of (7) are $-\rho_1, \ldots, -\rho_s$, and 0 of multiplicity $t - s$. These roots are the invariants of Σ under the group of transformations of the form of (10), and (12) is a canonical form of the covariances (or correlations) between the two sets of variates.

The transformation of x to $(u', v')'$ can be considered a transformation to a new coordinate system that describes in simplest terms the structure of association between the two sets of variates. The new variables are uncorrelated except in pairs $(u_1, v_1), \ldots, (u_s, v_s)$. The pairs are ordered so that u_1, v_1 has the maximum correlation. Earlier Hotelling (1935) termed u_1 the most predictable criterion and v_1 the best predictor. If $s = 1$, the first (and only) canonical correlation is the multiple correlation between the scalar $x^{(1)}$ and the vector $x^{(2)}$.

The algebra of canonical correlations and canonical variates can be developed in another way. Let C_1 and C_2 be nonsingular matrices such that

$$C_1\Sigma_{11}C_1' = I_s, \qquad C_2\Sigma_{22}C_2' = I_t. \tag{13}$$

Define $G = C_1\Sigma_{12}C_2'$, and let the singular value decomposition of G be

$$G = P\Lambda Q_1, \tag{14}$$

where Λ is diagonal, P is an orthogonal matrix of order s, and $Q_1Q_1' = I_s$. Then

$$GG' = P'\Lambda^2 P', \tag{15}$$

$$G'G = Q_1'\Lambda^2 Q_1. \tag{16}$$

If $\lambda_1 \ge \cdots \ge \lambda_s$ are the diagonal elements of Λ, they satisfy

$$|GG' - \lambda^2 I_s| = 0, \tag{17}$$

$$|G'G - \lambda^2 I_t| = 0. \tag{18}$$

Since $\Sigma_{22}^{-1} = C_2'C_2$, (17) is

$$|C_1\Sigma_{12}\Sigma_{22}^{-1}\Sigma_{21}C_1' - \lambda^2 I_s| = 0. \tag{19}$$

The multiplication of (19) on the left by $|(C_1)^{-1}|$ and on the right by $|C_1'^{-1}|$

yields

$$|\Sigma_{12}\Sigma_{22}^{-1}\Sigma_{21} - \lambda^2\Sigma_{11}| = 0 \tag{20}$$

since $\Sigma_{11}^{-1} = C_1' C_1$. However, from (8) we obtain

$$\begin{aligned} 0 &= \begin{vmatrix} I_s & \frac{1}{\lambda}\Sigma_{12}\Sigma_{22}^{-1} \\ 0 & I_t \end{vmatrix} \begin{vmatrix} -\lambda\Sigma_{11} & \Sigma_{12} \\ \Sigma_{21} & -\lambda\Sigma_{22} \end{vmatrix} \begin{vmatrix} I_s & 0 \\ \frac{1}{\lambda}\Sigma_{12}^{-1}\Sigma_{21} & I_t \end{vmatrix} \\ &= \begin{vmatrix} \frac{1}{\lambda}\Sigma_{12}\Sigma_{22}^{-1}\Sigma_{21} - \lambda\Sigma_{11} & 0 \\ 0 & -\lambda\Sigma_{22} \end{vmatrix} \\ &= (-\lambda)^{t-s}|-\Sigma_{22}||\Sigma_{12}\Sigma_{22}^{-1}\Sigma_{21} - \lambda^2\Sigma_{11}|. \end{aligned} \tag{21}$$

Thus, the roots of (19) are roots of (7); that is, $\lambda_1^2 = \rho_1^2, \ldots, \lambda_s^2 = \rho_s^2$.

Let

$$Q = \begin{pmatrix} Q_1 \\ Q_2 \end{pmatrix} \tag{22}$$

be orthogonal of order t. Then

$$P'GQ' = (\Lambda, 0). \tag{23}$$

Let $A = P'C_1$, $B = QC_2$. Then A and B satisfy (12).

In Sec. 5, Hotelling considers the corresponding sample problem. If $x_f = (x_f^{(1)\prime}, x_f^{(2)\prime})'$, $f = 1, \ldots, N$, is a sample and the expected value, $\mathscr{E}x = \mu$, is unknown, an unbiased estimate of Σ is

$$\frac{1}{n}\sum_{f=1}^{N}(x_f - \bar{x})(x_f - \bar{x})' = S = \begin{bmatrix} S_{11} & S_{12} \\ S_{21} & S_{22} \end{bmatrix}, \tag{24}$$

where $n = N - 1$. The algebra for Σ can be applied to S to obtain estimates of A, B, and $\rho_1, \ldots, \rho_s$. We denote the latter as $r_1 > r_2 > \cdots > r_s$.

The matrix with $(x_f^{(1)} - \bar{x}^{(1)})'$ as the fth row defines an s-dimensional linear space in an N-dimensional space, and the matrix with $(x_f^{(2)} - \bar{x}^{(2)})'$ as the fth row defines a t-dimensional space in this N-dimensional space. (A column of either matrix is a vector in the N space.) Then y_1 is the cosine of the smallest angle between a line in the s space and a line in the t space. In turn, y_2 is the cosine of the smallest angle between a line in the first subspace orthogonal to the line determined by the first maximization and a line in the second subspace orthogonal to the line in that space determined by the first maximization. The other canonical correlation can be interpreted similarly.

Hotelling showed that if $\rho_1 > \rho_2 > \cdots > \rho_p$, $p \leq s$, and x has the multivariate normal distribution, then $\sqrt{n}(r_1 - \rho_1), \ldots, \sqrt{n}(r_p - \rho_p)$ has a limiting normal distribution with a diagonal covariance matrix; the variance of the limiting distribution of $\sqrt{n}(r_\gamma - \rho_\gamma)$ is $(1 - \rho_\gamma^2)^2$.

In Sec. 6, Hotelling gives a numerical example in which $s = t = 2$ and $N = 140$. In this example, the asymptotic theory is used to test the null hypothesis that $\rho_2 = 0$; the hypothesis is accepted; $r_2 = .0688$ is not significantly different from 0. Hotelling notes that the null hypothesis $\rho_1 = 0$ implies that $\rho_1 = \rho_2 = 0$. The asymptotic theory of Sec. 5 does not apply because the roots then are not distinct. A test of $\rho_1 = \rho_2 = 0$ is a test that $\Sigma_{12} = 0$, which (in the case of normality) is independence of $x^{(1)}$ and $x^{(2)}$. Methods of computation are discussed in detail; during the next decade, Hotelling was to become very involved with matrix calculation.

In Sec. 4, Hotelling proposed two summary measures of association between $x^{(1)}$ and $x^{(2)}$. One he calls the vector correlation coefficient, which is the (positive) square root of

$$Q^2 = \begin{vmatrix} 0 & \Sigma_{12} \\ \Sigma_{21} & \Sigma_{22} \end{vmatrix} = \prod_{j=1}^{s} \rho_j^2. \tag{25}$$

The other, called the vector alienation coefficient, is

$$Z = \frac{|\Sigma|}{|\Sigma_{11}| \cdot |\Sigma_{22}|} = \prod_{j=1}^{s} (1 - \rho_j^2). \tag{26}$$

When x is normally distributed, the property $\Sigma_{12} = 0$, that is, $\rho_1 = \cdots = \rho_s = 0$, is equivalent to $x^{(1)}$ and $x^{(2)}$ being independent. Note that $Z = 1$ if and only if $\Sigma_{12} = 0$, that is, if $\rho_1 = \cdots = \rho_s = 0$. The likelihood ratio criterion for testing the hypothesis $\Sigma_{12} = 0$ is the $N/2$ power of $z = \prod_{j=1}^{s} (1 - r_j^2)$, derived by Wilks (1935).

In Sec. 7, Hotelling notes that in the sample $q = \prod_{j=1}^{s} r_j$. He finds the exact sampling distribution of q in Sec. 8 in terms of a hypergeometric function and its moments in Sec. 9, with further discussion in Sec. 10. The titles of the other sections are 11, "Tests for Complete Independence; 12, "Alternants of a Plane and of a Sample; 13, "The Bivariate Distribution for Complete Independence ($s = t = 2$, $n = 4$)" 14, "Theorem on Circularly Distributed Variates;" 15, Generalization of Section 13 for Samples of Any Size and 16, Further Problems. Sections 7–16 are not reprinted here as they are not now of great interest.

2. The Life and Works of Harold Hotelling

A review of the life and works of Harold Hotelling was given in the introduction to Harold Hotelling, "The Generalization of Student's Ratio. "For more information, see *The Collected Economics Articles by Harold Hotelling*, edited by Adrian C. Darnell.

3. This Paper in the Context of Current Statistical Research

Three years before this paper appeared, Hotelling (1933) wrote a long paper (in two parts) on principal components. In that study, Hotelling raised the question of what linear function of a random vector x has maximum variance. More precisely, suppose $\mathscr{E}x = 0$ and $\mathscr{E}xx' = \Sigma$. For $\alpha'\alpha = 1$, maximize

$$\mathscr{E}(\alpha'x)^2 = \mathscr{E}\alpha'xx'\alpha = \alpha'\Sigma\alpha$$

with respect to α. The maximum is the largest characteristic root of Σ. Let $\lambda_1 \geq \cdots \geq \lambda_p$ be the roots of

$$|\Sigma - \lambda I| = 0. \tag{27}$$

Then $\max_{\alpha'\alpha=1}\alpha'\Sigma\alpha = \lambda_1$. The next root λ_2 is the maximum of the variance of $\alpha'x$ for $\alpha'\alpha = 1$ among linear functions uncorrelation with the first. The matrix of the linear function defines a transformation $y = Ax$ such that

$$\mathscr{E}yy' = A\Sigma A' = \Lambda, \tag{28}$$

where Λ is a diagonal matrix and A an orthogonal matrix. This can be considered a canonical form for the covariance matrix. Thus, the paper being introduced is in the same genre.

Hotelling's development of canonical correlations was entirely within the framework of x being random. However, the analysis is relevant to one vector random and one nonstochastic. If x is normally distributed with mean $\mu = (\mu^{(1)'}, \mu^{(2)'})'$, the conditional distribution of $x^{(1)}$ given $x^{(2)}$ is normal with mean vector $\mu^{(1)} + \Gamma(x^{(2)} - \mu^{(2)})$ and covariance matrix Ψ, where $\Gamma = \Sigma_{12}\Sigma_{22}^{-1}$ and $\Psi = \Sigma_{11} - \Sigma_{12}\Sigma_{22}^{-1}\Sigma_{21}$. Now consider a set of observations $(x_f^{(1)'}, x_f^{(2)'})'$, $f = 1, \ldots, N$, where $x_f^{(1)}$ is an observation from $N[\tau + \Gamma(x_f^{(2)} - \bar{x}^{(2)}), \Psi]$, $x_1^{(2)}, \ldots, x_N^{(2)}$ are nonstochastic, and $\bar{x}^{(2)} = (1/N)\sum_{f=1}^N x_f^{(2)}$. For an arbitrary vector a, the linear combination $a'x_f^{(1)}$ has variance $a'\Psi a$ and expected value $a'\tau + a'\Gamma(x_f^{(2)} - \bar{x}^{(2)})$; this last is the best predictor of $a'x_f^{(1)}$. The sample average of these expected values is $\mathscr{E}a'\bar{x}^{(1)} = a'\tau$, and the mean sum of squares due to $x^{(2)}$ is

$$\sum_{f=1}^{N} (\mathscr{E}a'x_f^{(1)} - a'\tau)^2 = a'\Gamma S_{22}\Gamma'a. \tag{29}$$

When Γ is estimated by $S_{12}S_{22}^{-1}$ and Ψ is estimated by $S_{11} - S_{12}S_{22}^{-1}S_{21}$, the ratio of the mean sum of squares due to $x^{(2)}$ to the variance is

$$\frac{a'S_{12}S_{22}^{-1}S_{21}a}{a'(S_{11} - S_{12}S_{22}^{-1}S_{21})a}. \tag{30}$$

The maximum of this ratio is $r_1^2/(1 - r_1^2)$, and the maximizing linear combination $a'x^{(1)}$ is the first sample canonical variate. The algebra for $x^{(1)}$ and $x^{(2)}$ random carries over to the case of $x^{(1)}$ random and $x^{(2)}$ fixed. [See Sec. 12.6

of Anderson (1984) for more details.] This model is sometimes known as "reduced rank regression."

The sample equivalent of (20) is

$$|S_{12}S_{22}^{-1}S_{21} - r^2 S_{11}| = 0. \tag{31}$$

The joint distribution of $r_1^2, \ldots, r_s^2$ when $x_1, \ldots, x_N$ are normally distributed and $\rho_1 = \cdots = \rho_s = 0$ was found independently and almost simultaneously by Fisher (1939), Girshick (1939), Hsu (1939), Mood (1951), and Roy (1939). The distribution when there is no condition on $\rho_1^2, \ldots, \rho_s^2$ was obtained by Bartlett (1947) and Constantine (1963) in terms of zonal polynomials. Hsu derived the asymptotic distribution of $r_1^2, \ldots, r_s^2$ as $N \to \infty$ when there is no restriction on $\rho_1^2, \ldots, \rho_s^2$ [that is, multiple roots of (1) are permitted] for $x^{(2)}$ nonrandom (1941a) and $x^{(2)}$ random (1941b). Anderson (1951a) found the joint asymptotic distribution of $r_1^2, \ldots, r_s^2$ and the estimators of A and B in the case of $x^{(2)}$ nonrandom. Glynn and Muirhead (1978) showed that if $\theta_i = \operatorname{arctanh} \rho_i$ and $z_i = \operatorname{arctanh} r_i$, then

$$\hat{\theta} = z_i - \frac{1}{2Nr_i}\left[p - 2 + r_i^2 + 2(1 - r_i^2)\sum_{j \neq i} \frac{r_j^2}{r_i^2 - r_j^2}\right] \tag{32}$$

is unbiased for θ_i up to order $O(N^{-2})$ and the variance of $\hat{\theta}$ is $1/N + O(N^{-2})$. Dempster (1966) showed the effect of using the jackknife to remove the bias of r_i as an estimate of ρ_i.

As mentioned earlier, under normality the hypothesis that $x^{(1)}$ and $x^{(2)}$ are independent is equivalent to the hypothesis that $\Sigma_{12} = 0$, which in turn is equivalent to the hypothesis that $\rho_1 = \cdots = \rho_s = 0$. The criterion for testing this hypothesis, which Hotelling called the vector alienation coefficient, is equivalent to the likelihood ratio criterion. It was studied by Wilks (1935) and generalized to an arbitrary number of sets of variates. As noted above, the hypothesis that $\Sigma_{12} = 0$ is equivalent to the hypothesis that the regression of $x^{(1)}$ on $x^{(2)}$ is 0 and the likelihood ratio test for the regression matrix $\Gamma = 0$ is identical to that for $\Sigma_{12} = 0$. Thus, under the null hypothesis, the distributions of the criteria are the same. Tables of factors relating significance points of -2 times the logarithm of the likelihood ratio criterion are given in Pearson and Hartley (1972) and reprinted in Anderson (1984). Kiefer and Schwartz (1965) showed the admissibility of the likelihood ratio test.

Another statistic for testing independence, proposed by Nagao (1973), is

$$\frac{1}{2}\operatorname{tr}\left[\begin{pmatrix} S_{11}^{-1/2} & 0 \\ 0 & S_{22}^{-1/2} \end{pmatrix}\begin{pmatrix} S_{11} & S_{12} \\ S_{21} & S_{22} \end{pmatrix}\begin{pmatrix} S_{11}^{-1/2} & 0 \\ 0 & S_{22}^{-1/2} \end{pmatrix} - I\right]^2$$

$$= \operatorname{tr}(S_{11}^{-1}S_{12}S_{22}^{-1}S_{21}) = \sum_{j=1}^{s} r_j^2. \tag{33}$$

This criterion times N has a limiting χ^2 distribution (under the null hypothesis) with st degrees of freedom, the same as the limiting distribution of -2

times the logarithm of the likelihood ratio criterion. The number of degrees of freedom is the number of components in Σ_{12}.

The number of ρ_j's different from 0 is the rank of Σ_{12} or the rank of Γ. That number represents the complexity of the relationship between $x^{(1)}$ and $x^{(2)}$. Anderson (1951b) showed that the likelihood ratio criterion for testing the hypothesis that the rank of Γ is k is

$$\prod_{j=k+1}^{s} (1 - r_j^2)^{N/2}. \tag{34}$$

Independence of $x^{(1)}$ and $x^{(2)}$ (that is, $\Gamma = 0$) corresponds to $k = 0$. Under the null hypothesis of rank k, the criterion $-N\sum_{j=k+1}^{s} \log(1 - r_j^2)$ has a limiting χ^2 distribution with $(s - k)(t - k)$ degrees of freedom. Since $A\Sigma_{12} = (R, 0)B^{1^{-1}}$ or $A\Gamma = (R, 0)B^{1^{-1}}\Sigma_{22}^{-1}$, if the rank of Σ_{12} or of Γ is k, the last $s - k$ rows of R are 0 and $a_j'\Sigma_{12} = 0$ or $a_j'\Gamma = 0$, $j = k + 1, \ldots, s$. Then $u_j = a_j'x^{(1)}$ is uncorrelated with $x^{(1)}$ or the regression of u_j on $x^{(2)}$ is 0, $j = k + 1, \ldots, s$. Anderson (1951b) showed that a set of estimates of $a_{k+1}, \ldots, a_s$ when the rank restriction is applied are solutions to the sample equivalent of (4). Of course, the likelihood ratio criterion for $x^{(2)}$ normal with unrestricted covariance matrix Σ_{22} is the same since the joint density of $x^{(1)}$ and $x^{(2)}$ is the product of the conditional density of $x^{(1)}$ given $x^{(2)}$ and the marginal density $n(x^{(2)}|\mu^{(2)}, \Sigma_{22})$ of $x^{(2)}$. The first term of a Taylor series expansion of $-2\log$ times (34) is $N\sum_{j=k+1}^{s} r_j^2$, which yields another criterion.

In the form of the conditional distribution of $x^{(1)}$ given $x^{(2)}$, this analysis has been popularized as *reduced-rank regression* (see Izenman (1975) and (1980).) The analysis has been applied to autoregressive time series with $x_t^{(2)\prime} = (x_{t-1}^{(1)\prime}, \ldots, x_{t-q}^{(1)})$; see Robinson (1973) and Velu et al. (1986).

Canonical correlation has been used more generally in time series analysis. Let $\{y_t\}$ be a stationary process, let

$$x_t^{(1)} = \begin{bmatrix} y_{t-(q-1)} \\ \vdots \\ y_t \end{bmatrix} \tag{35}$$

denote a finite number of past and present vector observations; and let

$$x_t^{(2)} = \begin{bmatrix} y_{t+1} \\ \vdots \\ y_{t+r} \end{bmatrix} \tag{36}$$

denote r future observations. Suppose $\mathscr{E}y_t = 0$ and

$$\mathscr{E}y_t y_{t-s}' = \Omega(t - s) = \Omega'(s - t). \tag{37}$$

Then

$$\Sigma_{11} = \begin{bmatrix} \Omega(0) & \Omega'(1) & \cdots & \Omega'(q-1) \\ \Omega(1) & \Omega(0) & \cdots & \Omega'(q-2) \\ \vdots & \vdots & & \vdots \\ \Omega(q-1) & \Omega(q-2) & \cdots & \Omega(0) \end{bmatrix}, \tag{38}$$

$$\Sigma_{22} = \begin{bmatrix} \Omega(0) & \Omega'(1) & \dots & \Omega'(r-1) \\ \Omega(1) & \Omega(0) & \dots & \Omega'(r-2) \\ \vdots & \vdots & & \vdots \\ \Omega(r-1) & \Omega(r-2) & \dots & \Omega(0) \end{bmatrix}, \tag{39}$$

$$\Sigma'_{12} = \Sigma_{21} = \begin{bmatrix} \Omega(q) & \Omega(q-1) & \dots & \Omega(1) \\ \Omega(q+1) & \Omega(q) & \dots & \Omega(2) \\ \vdots & \vdots & & \vdots \\ \Omega(q+r-1) & \Omega(q+r-2) & \dots & \Omega(q) \end{bmatrix}. \tag{40}$$

Then the canonical correlations indicate the kind of dependence of the future on the present and the past. See Quenouille (1957), Box and Tiao (1977), Jewell and Bloomfield (1983), Jewell et al. (1983), and Tsay and Tiao (1955). Brillinger (1969) applied canonical analysis to the frequency domain.

Hotelling's analysis actually yields the singular value decomposition. If $\Sigma_{11} = I_s$ and $\Sigma_{22} = I_t$, the matrices A and B are orthogonal. Then

$$\begin{bmatrix} A & 0 \\ 0 & B \end{bmatrix}\begin{bmatrix} I_s & \Sigma_{12} \\ \Sigma_{21} & I_t \end{bmatrix}\begin{bmatrix} A' & 0 \\ 0 & B' \end{bmatrix} = \begin{bmatrix} I_s & R & 0 \\ R & I_s & 0 \\ 0 & 0 & I_{t-s} \end{bmatrix}, \tag{41}$$

$$\begin{aligned} \begin{bmatrix} I_s & \Sigma_{12} \\ \Sigma_{21} & I_t \end{bmatrix} &= \begin{bmatrix} A' & 0 \\ 0 & B' \end{bmatrix}\begin{bmatrix} I_s & R & 0 \\ R & I_s & 0 \\ 0 & 0 & I_{t-s} \end{bmatrix}\begin{bmatrix} A & 0 \\ 0 & B \end{bmatrix} \\ &= \begin{bmatrix} I & A'(R, 0)B \\ B'\begin{pmatrix} R \\ 0 \end{pmatrix} & I_t \end{bmatrix}. \end{aligned} \tag{42}$$

Let $B' = (B'_1, B'_2)$. Then

$$\begin{aligned} \Sigma_{12} &= A'(R, 0)\begin{pmatrix} B_1 \\ B_2 \end{pmatrix} \\ &= A'RB_1. \end{aligned} \tag{43}$$

The singular value decomposition, however, has a long history. The first derivation of it seems to have seen obtained by Beltrami (1873).

References

Anderson, T.W. (1951a). The asymptotic distribution of certain characteristic roots and vectors, *Proc. Sec. Berkeley Symp. Math. Statist. Prob.* (Jerzy Neyman, ed.). University of California Press, Berkeley, 1951, pp. 103–130.

Anderson, T.W. (1951b). Estimating linear restrictions on regression coefficients for multivariate normal distributions, *Ann. Math. Statist.*, **22**, 327–351 (correction, *Ann. Statist.*, **8** (1980), 1400).

Anderson, T.W. (1984). *An Introduction to Multivariate Statistical Analysis Sec.* ed. Wiley, New York.
Bartlett, M.S. (1947). The general canonical correlation distribution, *Ann. Math. Statist.*, **18**, 1–17.
Beltrami, E. (1873). On bilinear functions (Italian), *J. Math.* (*Naples*), **11**, 98–106.
Box, G.E.P., and Tiao, G.C. (1977). A canonical analysis of multiple time series, *Biometrika*, **64**, 355–365.
Brillinger, D.R. (1969). The canonical analysis of stationary times series, in *Multivariate Analysis* (P.R. Krishnaiah, ed.). Academic Press, New York, pp. 331–350.
Constantine, A.G. (1963). Some noncentral distribution problems in multivariate analysis, *Ann. Math. Statist.*, **34**, 1270–1285.
Darnell, A.C. (1990). *The Collected Economics Articles of Harold Hotelling*. Springer-Verlag, New York.
Dempster, A.P. (1966). Estimation in multivariate analysis, in *Multivariate Analysis* (P.R. Krishnaiah, ed.). Academic Press, New York pp. 315–334.
Fisher, R.A. (1939). The sampling distribution of some statistics obtained from nonlinear equations, *Ann. Eugen.*, **9**, 238–249.
Girshick, M.A. (1939). On the sampling theory of roots of determinantal equations, *Ann. Math. Statist.*, **10**, 203–224.
Glynn, W.J., and Muirhead, R.J. (1978). Inference in canonical correlation analysis, *J. Multiv. Anal.*, **8**, 468–478.
Hotelling, H. (1933). Analysis of a complex of statistical variables into principal components, *J. Educ. Psychol.*, **24**, 417–441, 498–520.
Hotelling, H. (1935). The most predictable criterion, *J. Educ. Psychol.*, **26**, 139–142.
Hsu, P.L. (1939). On the distribution of the roots of certain determinantal equations, *Ann. Eugen.*, **9**, 250–258.
Hsu, P.L. (1941a). On the limiting distribution of roots of a determinantal equation, *Proc. London Math. Soc.*, **16**, 183–194.
Hsu, P.L. (1941b). On the limiting distribution of the canonical correlations, *Biometrika*, **32**, 38–45.
Izenman, A.J. (1975). Reduced-rank regression for the multivariate linear model, *J. Multiv. Anal.*, **5**, 248–264.
Izenman, A.J. (1980). Assessing dimensionality in multivariate regression, in *Analysis of Variance, Handbook of Statistics*, Vol. 1 (P.R. Krishnaiah, ed.). North-Holland, Amsterdam, pp. 571–591.
Jewell, N.P., and Bloomfield, P. (1983). Canonical correlations of past and future for time series: Definitions and theory, *Ann. Statist.*, **11**, 837–847.
Jewell, N.P., Bloomfield, P., and Bartmann, F.C. (1983). Canonical correlations of past and future for time series: Bounds and computation, *Ann. Statist.*, **11**, 848–855.
Kiefer, J., and Schwartz, R. (1965). Admissible Bayes character of T^2-, R^2-, and other fully invariant tests for classical multivariate normal problems, *Ann. Math. Statist.*, **36**, 747–770.
Mood, A.M. (1951). On the distribution of the characteristic roots of normal second-moment matrices, *Ann. Math. Statist.*, **22**, 266–273.
Nagao, H. (1973). On some test criteria for covariance matrix, *Ann. Statist.*, **1**, 700–709.
Pearson, E.S., and Hartley, H.O. (1972). *Biometrika Tables for Statisticians*, Vol. 2., Published for the Biometrika Trustees at University Press, Cambridge, England.
Quenouille, M.H. (1957). *The Analysis of Multiple Time Series*. Griffin, London.
Robinson, P.M. (1973). Generalized canonical analysis for time series, *J. Multiv. Anal.*, **3**, 141–160.
Roy, S.N. (1939). *p*-statistics or some generalizations in analysis of variance appropriate to multivariate problems, *Sankhyā*, **4**, 381–396.

Tsay, R.S., and Tiao, G.C. (1985). Use of canonical analysis in time series model identification, *Biometrika*, **72**, 299–315.
Velu, R.P., Reinsel, G.C., and Wichern, D.W. (1986). Reduced rank models for multiple time series, *Biometrika*, **73**, 105–118.
Wilks, S.S. (1935). On the independence of *k* sets of normally distributed statistical variables, *Econometrica*, **3**, 309–326.

Relations Between Two Sets of Variates*

Harold Hotelling
Columbia University

1. The Correlation of Vectors. The Most Predictable Criterion and the Tetrad Difference

Concepts of correlation and regression may be applied not only to ordinary one-dimensional variates but also to variates of two or more dimensions. Marksmen side by side firing simultaneous shots at targets, so that the deviations are in part due to independent individual errors and in part to common causes such as wind, provide a familiar introduction to the theory of correlation; but only the correlation of the horizontal components is ordinarily discussed, whereas the complex consisting of horizontal and vertical deviations may be even more interesting. The wind at two places may be compared, using both components of the velocity in each place. A fluctuating vector is thus matched at each moment with another fluctuating vector. The study of individual differences in mental and physical traits calls for a detailed study of the relations between sets of correlated variates. For example the scores on a number of mental tests may be compared with physical measurements on the same persons. The questions then arise of determining the number and nature of the independent relations of mind and body shown by these data to exist, and of extracting from the multiplicity of correlations in the system suitable characterizations of these independent relations. As another example, the inheritance of intelligence in rats might be studied by applying not one but s different mental tests to N mothers and to a daughter

* Presented before the American Mathematical Society and the Institute of Mathematical Statisticians at Ann Arbor, September 12, 1935.

of each. Then $\frac{s(s-1)}{2}$ correlation coefficients could be determined, taking each of the mother-daughter pairs as one of the N cases. From these it would be possible to obtain a clearer knowledge as to just what components of mental ability are inherited than could be obtained from any single test.

Much attention has been given to the effects of the crops of various agricultural commodities on their respective prices, with a view to obtaining demand curves. The standard errors associated with such attempts, when calculated, have usually been found quite excessive. One reason for this unfortunate outcome has been the large portion of the variance of each commodity price attributable to crops of other commodities. Thus the consumption of wheat may be related as much to the prices of potatoes, rye, and barley as to that of wheat. The like is true of supply functions. It therefore seems appropriate that studies of demand and supply should be made by groups rather than by single commodities*. This is all the more important in view of the discovery that demand and supply curves provide altogether inadequate foundation for the discussion of related commodities, the ignoring of the effects of which has led to the acceptance as part of classical theory of results which are wrong not only quantitatively but qualitatively. It is logically as well as empirically necessary to replace the classical one-commodity type of analysis, relating for example to the incidence of taxation, utility, and consumers' surplus, by a simultaneous treatment of a multiplicity of commodities†.

The relations between two sets of variates with which we shall be concerned are those that remain invariant under internal linear transformations of each set separately. Such invariants are not affected by rotations of axes in the study of wind or of hits on a target, or by replacing mental test scores by an equal number of independently weighted sums of them for comparison with physical measurements. If measurements such as height to shoulder and difference in height of shoulder and top of head are replaced by shoulder height and stature, the invariant relations with other sets of variates will not be affected. In economics there are important linear transformations corresponding for example to the mixing of different grades of wheat in varying proportions‡. Both prices and quantities are then transformed linearly.

In this case, besides the invariants to be discussed in this paper, there will

* The only published study known to the writer of groups of commodities for which standard errors were calculated is the paper of Henry Schultz, "Interrelations of Demand," in *Journal of Political Economy*, Vol. XLI. pp. 468–512, August, 1933. Some at least of the coefficients obtained are significant.

† Harold Hotelling, "Edgeworth's Taxation Paradox and the Nature of Demand and Supply Functions" in *Journal of Political Economy*, Vol. XL, pp. 577–616, October, 1932, and "Demand Functions with Limited Budgets" in *Econometrica*, Vol. III, pp. 66–78, January, 1935.

‡ Harold Hotelling, "Spaces of Statistics and their Metrization" in *Science*, Vol. LXVII, pp. 149–150, February 10, 1928.

be other resulting from the fact that the transformation of quantities is not independent of that of the prices, but is contragredient to it. (Cf. Section 16 below.)

Sets of variates, which may also be regarded as many-dimensional variates, or as vector possessed of frequency distributions, have been investigated from several different standpoints. The work of Gauss on least squares and that of Bravais, Galton, Pearson, Yule and others on multivariate distributions and multiple correlation are early examples. In "The Generalization of Student's Ratio*," the writer has given a method of testing in a manner invariant under linear transformations, and with full statistical efficiency, the significance of sets of means, of regression coefficients, and of differences of means or regression coefficients. A procedure generalizing the analysis of variance to vectors has been applied to the study of the internal structure of cell by means of Brownian movements, for which the vectors representing displacements in consecutive fifteen-second intervals were compared with their resultants to demonstrate the presence of invisible obstructions restricting the movement†. Finally, S.S. Wilks has published important work on relations among two or more sets of variates which are invariant under internal linear transformations‡. Denoting by A, B and D respectively the determinants of sample correlations within a set of s variates, within a set of t variates, and in the set consisting of all these $s + t$ variates, the distribution of the statistic,

$$z = \frac{D}{AB} \tag{1.1}$$

was determined exactly by Wilks under the hyothesis that the distribution is normal, with no population correlation between any variate in one set and any in the other. Wilks also found distributions of analogous functions of three or more sets, and of other related statistics.

The statistic (1.1) is invariant under internal linear transformations of either set, as will be proved in Section 4. Another example of such a statistic is provided by the maximum multiple correlation with either set of a linear function of the other set, which has been the subject of a brief study§. This problem of finding, not only a best predictor among the linear functions of one set, but at the same time the function of the other set which is predicts most accurately, will be solved in Section 3 in a more symmetrical manner. When the influence of these two linear functions is eliminated by partial

* *Annals of Mathematical Statistics*, Vol. II. pp. 360–378, August, 1931.

† L.G.M. Baas-Becking, Henriette van de Sande Bakhuyzen, and Harold Hotelling, "The Physical State of Protoplasm" in *Verhandelingen der Koninklijke Akademie van Wetenschappen te Amsterdam*, Second Section, Vol. v. (1928).

‡ "Certain Generalizations in the Analysis of Variance" in *Biometrika*, Vol. XXIV. pp. 471–494, November, 1932.

§ Harold Hotelling, "The Most Predictable Criterion" in *Journal of Educational Psychology*, Vol. XXVI. pp. 139–142, February, 1935.

correlation, the process may be repeated with the residuals. In this way we may obtain a sequence of pairs of variates, and of correlations between them, which in the aggregate will fully characterize the invariant relations between the sets, in so far as these can be represented by correlation coefficients. They will be called *canonical variates* and *canonical correlations*. Every invariant under general linear internal transformations, such for example as z, will be seen to be a function of the canonical correlations.

Observations of the values taken in N cases by the components of two vectors constitute two matrices, each of N columns. If each vector has s components, then each matrix has s rows. In this case we may consider the correlation coefficient between the C_s^N-rowed determinants in one matrix and the corresponding determinants in the other. Since a linear transformation of the variates in either set effects a linear transformation of the rows of the matrix of observations, which merely multiplies all these determinants by the same constant, it is evident that the correlation coefficient thus calculated is invariant in absolute value. We shall call it the *vector correlation* or *vector correlation coefficient*, and denote it by q. When $s = 2$, if we call the variates of one set x_1, x_2, and those of the other x_3, x_4, and r_{ij} the correlation of x_i with x_j, then it is easy to deduce with the help of the theorems stated in Section 2 below that

$$q = \frac{r_{13}r_{24} - r_{14}r_{23}}{\sqrt{(1 - r_{12}^2)(1 - r_{34}^2)}} \tag{1.2}$$

For larger values of s, q will have as its numerator the determinant of correlations of variates in one set with variates in the other, and as its denominator the geometric mean of the two determinants of internal correlations. A generalization of q for sets with unequal numbers of components will be given in Section 4.

Corresponding to the correlation coefficient r between two simple variates, T.L. Kelley has defined the *alienation coefficient* as $\sqrt{1 - r^2}$. The square of the correlation coefficient between x and y is the fraction of the variance of y attributable to x, while the square of the alienation coefficient is the fraction independent of x. If we adopt this apportionment of variance as a basis of generalization, we shall be consistent in calling $\sqrt{z}$ the *vector alienation coefficient*.

If there exists a linear function of one set equal to a linear function of the other—if for example x_1 is identically equal to x_3—the expression (1.2) for q reduces to a partial correlation coefficient. If one set consists of a single variate and the other of two or more, the vector correlation coincides with the multiple correlation. If each set contains only one variate, q is the simple correlation between the two. Thus the vector correlation coefficient provides a generalization of several familiar concepts.

The numerator of (1.2), known as the tetrad difference or tetrad, has been of much concern to psychologists. The vanishing in the population of all the tetrads among a set of tests is a necessary condition for the theory, propounded by Spearman, that scores on the tests are made up of a component common in varying degrees to all of them, and of independent components specific to each. The vanishing of some but not all of the tetrads is a condition for certain variants of the situation*. The sampling errors of the tetrad have therefore received much attention. In dealing with them it has been thought necessary to ignore at least three types of error:

(1) The standard error formulae used are only asymptotically valid for very large samples, with no means of determining how large a sample is necessary.
(2) The assumption is made implicitly that the distribution of the tetrad is normal, though this cannot possibly be the case, since the range is finit†.
(3) Since the standard error formulae involve unknown population values, these are in practice replaced by sample values. No limit is known for the errors committed in this way.

Now it is evident that to test whether the population value of the tetrad is zero—the only value of interest—is the same thing as to test the vanishing of any multiple of the tetrad by a finite non-vanishing quantity. Wishart‡ considered the tetrad of covariances, which is simply the product of the tetrad of correlations by the four standard deviations. For this function he found exact values of the mean and standard error, thus eliminating the first source of error mentioned above.

The exact distribution of q found in Section 8 below may be used to test the vanishing of the tetrad, eliminating the first and second sources of error. Unfortunately even this distribution involves a parameter of the population, one of the canonical correlations, which must usually be estimated from the sample, introducing again an error of the third type. However there may be cases in which this one parameter will be known from theory or from a larger sample.

Now it will be shown that q is the product of the canonical correlations. Hence at least one of these correlations is zero if the tetrad is. Thus still another test of the same hypothesis may be made in this way. Now we shall obtain for a canonical correlation vanishing in the population the extremely

* Truman L. Kelley, *Crossroads in the Mind of Man*, Stanford University Press, 1928. This book, in addition to relevant test data and discussion, contains references to the extensive literature, a standard error formula for the tetrad, and other mathematical developments.

† The first proof that the distribution of the tetrad approaches normality for large samples was given by J.L. Doob in an article, "The Limiting Distributions of Certain Statistics," in the *Annals of Mathematical Statistics*, Vol. VI. pp. 160–169 (September, 1935). The proof is applicable only if the population value of z is different from unity, i.e. if the sets x_1, x_2 and x_3, x_4 are not completely independent. If they are completely independent, the limiting distribution is of the form $\frac{1}{2}ce^{-c|t|}\,dt$, as Doob showed. What the distribution of the tetrad is for any finite number of cases no one knows.

‡ "Sampling Errors in the Theory of Two Factors" in *British Journal of Psychology*, Vol. XIX. pp. 180–187 (1928).

simple standard error formula $\frac{1}{\sqrt{n}}$, involving no unknown parameter. Thus this test evades errors of the third kind, but is subject to those of the first two, although the second is somewhat mitigated by an ultimate approach to normality, since the canonical correlations satisfy the criterion for approach to normality derived by Doob in the article cited. Further research is needed to find an exact test involving no unknown parameter. The question of whether this is possible raises a very fundamental problem in the theory of statistical inference. We shall, however, find exact distributions appropriate for testing a variety of hypotheses.

2. Theorems on Determinants and Matrices

We shall have frequent occasion to refer to the following well-known theorem, the proofs of which parallel those of the multiplication theorem for determinants, and which might advantageously be used in expounding many parts of the theory of statistics:

Given two arrays, each composed of m rows and n columns ($m \leqslant n$). The determinant formed by multiplying the rows of one array by those of the other equals the sum of the products of the m-rowed determinants in the first array by the corresponding m-rowed determinants in the second.

When the two arrays are identical, we have the corollary that the symmetrical determinant of the products of rows by rows of an array of m rows and n columns ($m \leqslant n$) equals the sum of the squares of the m-rowed determinants in the array, and is therefore not negative.

3. Canonical Variates and Canonical Correlations. Applications to Algebra and Geometry

If x_1, x_2, ... are variates having zero expectations and finite covariances, we denote these covariances by

$$\sigma_{\alpha\beta} = Ex_\alpha x_\beta,$$

where E stands for the mathematical expectation of the quantity following. If new variates $x'_1, x'_2, \ldots$ are introduced as linear functions of the old, such that

$$x'_\gamma \sum_\alpha c_{\gamma\alpha} x_a,$$

then the covariances of the new variates are expressed in terms of those of the old by the equations

$$\sigma'_{\gamma\delta} = \sum_{\alpha\beta}\sum c_{\gamma\alpha} c_{\delta\beta} \sigma_{\alpha\beta} \tag{3.1}$$

obtained by substituting the equations above directly in the definition

$$\sigma'_{\gamma\delta} = Ex'_{\gamma} x'_{\delta},$$

and taking the expectation term by term.

Now (3.1) gives also the formula for the transformation of the coefficients of a quadratic form $\sum\sum \sigma_{\alpha\beta} x_{\alpha} x_{\beta}$ when the variables are subjected to a linear transformation. Hence the problem of standardizing the covariances among a set of variates by linear transformations is algebraically equivalent to the canonical reduction of a quadratic form. The transformation of the quadratic form into a sum of squares corresponds to replacing a set of variates by uncorrelated components. It is to be observed that the fundamental nature of covariances implies that $\sum\sum \sigma_{\alpha\beta} x_{\alpha} x_{\beta}$ is a positive definite quadratic form, and that only real transformations are relevant to statistical problems.

Considering two sets of variates $x_1, \ldots, x_s$ and $x_{s+1}, \ldots, x_{s+t}$, we shall denote the covariances, in the sense of expectations of products, by $\sigma_{\alpha\beta}$, $\sigma_{\alpha i}$, and σ_{ij}, using Greek subscripts for the indices $1, 2, \ldots, s$ and Latin subscripts for $s + 1, \ldots, s + t$. Determination of invariant relations between the two sets by means of the correlations or covariances among the $s + t$ variates is associated with the algebraic problem, which appears to be new, of determining the invariants of the system consisting of two positive definite quadratic forms

$$\sum_{\alpha\beta}\sum \sigma_{\alpha\beta} x_{\alpha} x_{\beta}, \qquad \sum_{ij}\sum \sigma_{ij} x_i x_j,$$

in two separate sets of variables, and of a bilinear form

$$\sum_{\alpha i}\sum \sigma_{\alpha i} x_{\alpha} x_i$$

in both sets, under real linear non-singular transformations of the two sets separately.

Sample covariances are also transformed by the formula (3.1). The ensuing analysis might therefore equally well be carried out for a sample instead of for the population. Correlations might be used instead of covariances, either for the sample or for the population, by introducing appropriate factors, or by assuming the standard deviations to be unity.

We shall assume that there is no fixed linear relation among the variates, so that the determinant of their covariances or correlations is not zero. This implies that there is no fixed linear relation among any subset of them; consequently every *principal* minor of the determinant of $s + t$ rows is different from zero.

If we consider a function u of the variates in the first set and a function v of those in the second, such that

$$u = \sum_{\alpha} a_{\alpha} x_{\alpha}, \qquad v = \sum_{i} b_i x_i,$$

the conditions

$$\sum\sum \sigma_{\alpha\beta} a_\alpha a_\beta = 1, \qquad \sum\sum \sigma_{ij} b_i b_j = 1 \tag{3.2}$$

are equivalent to requiring the standard deviations of u and v to be unity. The correlation of u with v is then

$$R = \sum_{\alpha i}\sum \sigma_{\alpha i} a_\alpha b_i. \tag{3.3}$$

If u and v are chosen so that this correlation is a maximum, the coefficients a_α and b_i will satisfy the equations obtained by differentiating

$$\sum\sum \sigma_{\alpha i} a_\alpha b_i - \tfrac{1}{2}\lambda \sum\sum \sigma_{\alpha\beta} a_\alpha a_\beta - \tfrac{1}{2}\mu \sum\sum \sigma_{ij} b_i b_j,$$

namely

$$\sum_i \sigma_{\alpha i} b_i - \lambda \sum_\beta \sigma_{\alpha\beta} a_\beta = 0 \tag{3.4}$$

$$\sum_\alpha \sigma_{\alpha i} a_\alpha - \mu \sum_j \sigma_{ij} b_j = 0. \tag{3.5}$$

Here λ and μ are Lagrange multipliers. Their interpretation will be evident upon multiplying (3.4) by a_α and summing with respect to α, then multiplying (3.5) by b_i and summing with respect to i. With (3.2) and (3.3), this process gives

$$\lambda = \mu = R.$$

The $s + t$ homogeneous linear equations (3.4) and (3.5) in the $s + t$ unknowns a_α and b_i will determine variates u and v making R a maximum, a minimum, or otherwise stationary, if their determinant vanishes. Since $\lambda = \mu$, this condition is

$$\begin{vmatrix} -\lambda\sigma_{11} \cdots -\lambda\sigma_{1s} & \sigma_{1,s+1} \cdots\cdots\cdots \sigma_{1,s+t} \\ \cdots\cdots\cdots & \cdots\cdots\cdots \\ -\lambda\sigma_{s1} \cdots -\lambda\sigma_{ss} & \sigma_{s,s+1} \cdots\cdots\cdots \sigma_{s,s+t} \\ \sigma_{s+1,1} \cdots \sigma_{s+1,s} & -\lambda\sigma_{s+1,s+1} \cdots -\lambda\sigma_{s+1,s+t} \\ \cdots\cdots\cdots & \cdots\cdots\cdots \\ \sigma_{s+t,1} \cdots \sigma_{s+t,s} & -\lambda\sigma_{s+t,s+1} \cdots -\lambda\sigma_{s+t,s+t} \end{vmatrix} = 0. \tag{3.6}$$

This symmetrical determinant is the discriminant of a quadratic form $\phi - \lambda\psi$, where

$$\phi = 2\sum_{\alpha i}\sum \sigma_{\alpha i} z_\alpha z_i, \qquad \psi = \sum_{\alpha\beta}\sum \sigma_{\alpha\beta} z_\alpha z_\beta + \sum_{ij}\sum \sigma_{ij} z_i z_j.$$

Here ψ is positive definite because it is the sum of two positive definite quadratic forms. Consequently* all the roots of (3.6) are real. Moreover the elementary divisors are all of the first degree†. This means that the matrix of

* Maxime Bôcher, *Introduction to Higher Algebra*, New York, 1931, p. 170, Theorem 1.

† Bôcher, p. 305, Theorem 4; p. 267, Theorem 2; p. 271, Definition 3.

the determinant in (3.6) is reducible, by transformations which do not affect either its rank or its linear factors, to a matrix having zeros everywhere except in the principal diagonal, while the elements in this diagonal are polynomials

$$E_1(\lambda), \qquad E_2(\lambda), \ldots, E_{s+t}(\lambda),$$

none of which contains any linear factor of the form $\lambda - \rho$ raised to a degree higher than the first. Therefore, if a simple root of (3.6) is substituted for λ, the rank is $s + t - 1$; but substitution of a root of multiplicity m for λ makes the rank $s + t - m$. Consequently if a simple root is substituted for λ and μ in (3.4) and (3.5) these equations will determine values of $a_1, a_2, \ldots, a_s, b_{s+1}, \ldots, b_{s+t}$, uniquely except for constant factors whose absolute values are determinable from (3.2). Not all these quantities are zero; from this fact, and the form of (3.4) and (3.5), it is evident that at least one a_α and at least one b_i differ from zero, provided the value put for λ is not zero. The variates u and v will then be fully determinate except that they may be replaced by the pair $-u$, $-v$. But for a root of multiplicity m there will be m linearly independent solutions instead of one in a complete set of solutions. From these may be obtained m different pairs of variates u and v.

The coefficient of the highest power of λ in (3.6) is the product of two principal minors, both of which differ from zero because the variates have been assumed algebraically independent. The equation is therefore of degree $s + t$. We assume as a mere matter of notation, if $s \neq t$, that $s < t$. Then of the $s + t$ roots at least $t - s$ vanish; for the coefficients of λ^{t-s-1} and lower powers of λ are sums of principal minors of $2s + 1$ or more rows, in which λ is replaced by zero, and every such minor vanishes, as can be seen by a Laplace expansion. Also, the sign of λ may be changed in (3.6) without changing the equation, for this may be accomplished by multiplying each of the first s rows and last t columns by -1. Therefore the negative of every root is also a root. The $s + t - (t - s) = 2s$ roots that do not necessarily vanish consist therefore of s positive or zero roots $\rho_1, \rho_2, \ldots, \rho_s$, and of the negatives of these roots. These s roots which are positive or zero we shall call the *canonical correlations* between the sets of variates; the corresponding linear functions u, v whose coefficients satisfy (3.2), (3.4) and (3.5) we call *canonical variates**. It is clear that every canonical correlation is the correlation coefficient between a pair of canonical variates. Hence no canonical correlation can exceed unity. The greatest canonical correlation is the maximum multiple correlation with either set of a disposable linear function of the other set. If u, v are canonical variates corresponding to ρ_γ, then the pair $u, -v$ or $v, -u$ is associated with the root $-\rho_\gamma$.

If a pair of canonical variates corresponding to a root ρ_γ is

$$u_\gamma = \sum_\alpha a_{\alpha\gamma} x_\alpha, \qquad v_\gamma = \sum_i b_{i\gamma} x_i, \tag{3.7}$$

* The word "canonical" is used in the algebraic theory of invariants with a meaning consistent with that of this paper.

the coefficients must satisfy (3.4) and (3.5), so that

$$\sum_i \sigma_{\alpha i} b_{i\gamma} = \rho_\gamma \sum_\beta \sigma_{\alpha\beta} a_{\beta\gamma}, \tag{3.8}$$

$$\sum_\alpha \sigma_{\alpha i} a_{\alpha\gamma} = \rho_\gamma \sum_j \sigma_{ij} b_{j\gamma}. \tag{3.9}$$

Also let

$$u_\delta = \sum_\beta a_{\beta\delta} x_\beta, \qquad v_\delta = \sum_i b_{i\delta} x_i \tag{3.10}$$

be canonical variates associated with a canonical correlation ρ_δ. Among the four variates (3.7) and (3.10) there are six correlations. Apart from ρ_γ and ρ_δ these are obviously

$$\left.\begin{aligned} Eu_\gamma u_\delta &= \sum\sum \sigma_{\alpha\beta} a_{\beta\gamma} a_{\alpha\delta}, & Eu_\gamma v_\delta &= \sum\sum \sigma_{\alpha i} a_{\alpha\gamma} b_{i\delta} \\ Ev_\gamma u_\delta &= \sum\sum \sigma_{\alpha i} b_{i\gamma} a_{\alpha\delta}, & Ev_\gamma v_\delta &= \sum\sum \sigma_{ij} b_{i\gamma} b_{j\delta} \end{aligned}\right\}. \tag{3.11}$$

We shall prove that the last four are all zero. Multiply (3.8) by $a_{\alpha\delta}$ and sum with respect to α. The result, with the help of (3.11), may be written

$$Ev_\gamma u_\delta = \rho_\gamma Eu_\gamma u_\delta. \tag{3.12}$$

Multiplying (3.9) by $b_{i\delta}$ and summing with respect to i, we get

$$Eu_\gamma v_\delta = \rho_\gamma Ev_\gamma u_\delta. \tag{3.13}$$

Interchanging γ and δ in this and then using (3.12), we obtain

$$\rho_\gamma Eu_\gamma u_\delta = \rho_\delta Ev_\gamma v_\delta. \tag{3.14}$$

Again interchanging γ and δ, we have

$$\rho_\delta Eu_\gamma u_\delta = \rho_\gamma Ev_\gamma v_\delta.$$

If $\rho_\gamma^2 \neq \rho_\delta^2$, the last two equations show that $Eu_\gamma u_\delta = Ev_\gamma v_\delta = 0$. Hence, by (3.12) and (3.13), $Ev_\gamma u_\delta$ and $Eu_\gamma v_\delta$ vanish. Thus all the correlations among canonical variates are zero except those between the canonical variates associated with the same canonical correlation.

If ρ_α is a root of multiplicity m, it is possible by well-known processes to obtain m solutions of the linear equations such that, if

$$a_{1\gamma}, \ldots, a_{s\gamma}, \quad b_{s+1,\gamma}, \ldots, b_{s+t,\gamma},$$

$$a_{1\delta}, \ldots, a_{s\delta}, \quad b_{s+1,\delta}, \ldots, b_{s+t,\delta},$$

are any two of these solutions, they will satisfy the orthogonality condition

$$\sum_\alpha a_{\alpha\gamma} a_{\alpha\delta} + \sum_i b_{i\gamma} b_{i\delta} = 0. \tag{3.15}$$

There is no loss of generality in supposing that each of the original variates was uncorrelated with the others in the same set and had unit variance. In this case (3.15) is equivalent to

$$Eu_\gamma u_\delta + Ev_\gamma v_\delta = 0,$$

where u_γ, v_γ, u_δ are given by (3.7) and (3.10). For this case of equal roots we have also from (3.14),

$$\rho_\alpha(Eu_\gamma u_\delta - Ev_\gamma v_\delta) = 0.$$

If $\rho_\alpha \neq 0$, the last two equations show that $Eu_\gamma u_\delta = Ev_\gamma v_\delta = 0$, and then from (3.12) and (3.13) we have that $Ev_\gamma u_\delta = Eu_\gamma v_\delta = 0$. These correlations also vanish if $\rho_\alpha = 0$, for then the right-hand members of (3.8) and (3.9) vanish, leaving two distinct sets of equations in disjunct sets of unknowns. The solutions may therefore be chosen so that the two sums in (3.15) vanish separately.

A double zero root determines uniquely, if $s = t$, a pair of canonical variates. If $s < t$, such a root determines a canonical variate for the less numerous set, and leaves $t - s$ degrees of freedom for the choice of the other.

The reduction of our sets of variates to canonical form may be completed by the choice of new variates $v_{s+1}, v_{s+2}, \ldots, v_t$ as linear functions of the second and more numerous set (unless the numbers in the two sets are equal), uncorrelated with each other and with the canonical variates v_γ previously determined, and having unit variance. This may be done in infinitely many ways, as is well known. These variates will also be uncorrelated with the canonical variates u_γ. Indeed, if

$$v_k = \sum b_{jk} x_j$$

is one of them, its correlation with u_γ is, by (3.7) and (3.9),

$$Eu_\gamma v_k = \sum\sum \sigma_{\alpha j} a_\gamma b_{jk} = \rho_\gamma \sum\sum \sigma_{ij} b_{i\gamma} b_{jk} = \rho_\gamma Ev_\gamma v_k,$$

which vanishes because v_k was defined to be uncorrelated with v_γ.

The normal form of two sets of variates under internal linear transformations is thus found to consist of linear functions $u_1, u_2, \ldots, u_s$ of one set, and $v_1, v_2, \ldots, v_t$ of the other, such that all the correlations among these linear functions are zero, except that the correlation of u_γ with v_γ is a positive number ρ_γ ($\gamma = 1, 2, \ldots, s$). Therefore *the only invariants of the system under internal linear transformations are* $\rho_1, \rho_2, \ldots, \rho_s$, *and functions of these quantities.*

The solution of the algebraic problem mentioned at the beginning of this section, by steps exactly parallel to those just taken with the statistical problem, is the following:

The positive definite quadratic forms $\sum\sum \sigma_{\alpha\beta} x_\alpha x_\beta$, *and* $\sum\sum \sigma_{ij} x_i x_j$, *and the bilinear form* $\sum\sum \sigma_{\alpha i} x_\alpha x_i$ *with real coefficients, where the Latin subscripts are summed from* 1 *to* s *and the Greek subscripts from* $s + 1$ *to* $s + t$, *and* $s \leqslant t$, *may be reduced by a real linear transformation of* $x_1, \ldots, x_s$ *and a real linear transformation of* $x_{s+1}, \ldots, x_{s+t}$ *simultaneously to the respective forms* $x_1^2 + \cdots + x_s^2$, $x_{s+1}^2 + \cdots + x_{s+t}^2$, *and* $\rho_1 x_1 x_{s+1} + \rho_2 x_2 x_{s+2} + \cdots + \rho_s x_s x_{2s}$. *A fundamental system of invariants under such transformations consists of* $\rho_1, \ldots, \rho_s$.

This algebraic theorem holds also if the quadratic forms are not restricted to be positive definite, provided (3.6) has no multiple roots and the forms are non-singular.

The normalization process we have defined may also be carried out for a sample, yielding canonical correlations $r_1, r_2, \ldots, r_s$, which may be regarded as estimates of $\rho_1, \rho_2, \ldots, \rho_s$, and associated canonical variates. With sampling problems raised in this way we shall largely be concerned in the remainder of this paper.

A further application is to geometry. In a space of N dimensions a sample of N values of a variate may be represented by a point whose coordinates are the observed values. The sample correlation between two variates is the cosine of the angle between lines drawn from the origin to the representing points, with the proviso, since deviations from means are used in the expression for a correlation, that the sum of all the coordinates of each point be zero. A sample of $s + t$ variates determines a flat space of s and one of t dimensions, intersecting at the origin, and containing the points representing the two sets of variates. In typical cases these two flat spaces do not intersect except at this one point. A complete set of metrical invariants of a pair of flat spaces is esaily seen from the foregoing analysis to consist of s angles whose cosines are r_1, $\ldots, r_s$. Indeed, like all correlations, they are invariant under rotations of the N-space about the origin, and they do not depend on the particular points used to define the two flat spaces. Each of these invariants is the angle between a line in one flat space and a line in the other. One of the invariants is the *minimum* angle of this kind, and the others are in a sense stationary. The condition that the two flat spaces intersect in a line is that one of the invariant quantities $r_1, \ldots, r_s$ be unity. They intersect in a plane if two of these quantities equal unity. For two planes through a point in space of four or more dimensions, there will be two invariants r_1, r_2, of which one is the cosine of the minimum angle. If $r_1 = r_2$, the planes are *isocline.* Every line in each plane then makes the minimum angle with some line in the other. If $r_1 = r_2 = 0$, the planes are completely perpendicular; every line in one plane is then perpendicular to every line in the other. If one of these invariants is zero and the other is not, the planes are semi-perpendicular; every line in each plane is perpendicular to a certain line in the other.

The determinant of the correlations among canonical variates is

$$\Delta = \begin{vmatrix} 1 & 0 & \ldots & 0 & \rho_1 & 0 & \ldots & 0 & \ldots & 0 \\ 0 & 1 & \ldots & 0 & 0 & \rho_2 & \ldots & 0 & \ldots & 0 \\ \multicolumn{4}{c}{\cdots\cdots\cdots\cdots} & \multicolumn{6}{c}{\cdots\cdots\cdots\cdots\cdots\cdots} \\ 0 & 0 & \ldots & 1 & 0 & 0 & \ldots & \rho_s & \ldots & 0 \\ \rho_1 & 0 & \ldots & 0 & 1 & 0 & \ldots & 0 & \ldots & 0 \\ 0 & \rho_2 & \cdots & 0 & 0 & 1 & \ldots & 0 & \ldots & 0 \\ \multicolumn{4}{c}{\cdots\cdots\cdots\cdots} & \multicolumn{6}{c}{\cdots\cdots\cdots\cdots\cdots\cdots} \\ 0 & 0 & \ldots & \rho_s & 0 & 0 & \ldots & 1 & \ldots & 0 \\ \multicolumn{4}{c}{\cdots\cdots\cdots\cdots} & \multicolumn{6}{c}{\cdots\cdots\cdots\cdots\cdots\cdots} \\ 0 & 0 & \ldots & 0 & 0 & 0 & \ldots & 0 & \ldots & 1 \end{vmatrix}$$

$$= (1 - \rho_1^2)(1 - \rho_2^2)\ldots(1 - \rho_s^2). \tag{3.16}$$

The rank of the matrix

$$\left\| \begin{matrix} \rho_{1,s+1} \cdots\cdots\cdots \rho_{1,s+t} \\ \cdots\cdots\cdots\cdots\cdots\cdots \\ \rho_{s,s+1} \cdots\cdots\cdots \rho_{s,s+t} \end{matrix} \right\|.$$

of correlations between the two sets is invariant under non-singular linear transformations of either set. Transformation to canonical variate reduces this matrix to

$$\left\| \begin{matrix} \rho_1 & 0 & \ldots & 0 & \cdots\cdots & 0 \\ 0 & \rho_2 & \ldots & 0 & \cdots\cdots & 0 \\ \cdots & \cdots & \cdots & \cdots & \cdots & \cdots \\ 0 & 0 & \ldots & \rho_s & \cdots\cdots & 0 \end{matrix} \right\|.$$

The rank is therefore the number of canonical correlations that do not vanish. This is the number of independent components common to the two sets. In the parlance of mental testing, the number of "common factors" of two sets of tests (e.g. mental and physical, or mathematical and linguistic tests) is the number of non-vanishing canonical correlations.

4. Vector Correlation and Alienation Coefficients

In terms of the covariances among the variates in the two sets $x_1, \ldots, x_s$ and $x_{s+1}, \ldots, x_{s+t}$, we define the following determinants, maintaining the convention that Greek subscripts take values from 1 to s, and Latin subscripts take values from $s+1$ to $s+t$. It will be assumed throughout that $s \leqslant t$. A is the determinant of the covariances among the variates in the first set, arranged in order: that is, the element in the αth row and βth column of A is $\sigma_{\alpha\beta}$. B is the determinant of the covariances among variates in the second set, likewise ordered. D is the determinant of $s+t$ rows containing in order all the covariances among all the variates of both sets. C is obtained from D by replacing the covariances among the variates of the first set, including their variances, by zeros. Symbolically,

$$A = |\sigma_{\alpha\beta}|, \quad B = |\sigma_{ij}|, \quad C = \left| \begin{array}{c|c} 0 & \sigma_{i\alpha} \\ \hline \sigma_{\alpha i} & \sigma_{ij} \end{array} \right|, \quad D = \left| \begin{array}{c|c} \sigma_{\alpha\beta} & \sigma_{i\alpha} \\ \hline \sigma_{\alpha i} & \sigma_{ij} \end{array} \right|.$$

Suppose now that new variates $x_1', \ldots, x_s'$ are defined in terms of the old variates in the first set by the s equations

$$x_\gamma' = \sum c_{\gamma\alpha} x_\alpha.$$

The new covariances are then expressed in terms of the old by (3.1). The determinant of these new covariances, which we shall denote by A', may by

(3.1) and the multiplication theorem of determinants be expressed as the product of three determinants, of which two equal the determinant $c = |c_{\gamma\alpha}|$ of the coefficients of the transformation, while the third is A. If the variates of the second set are subjected to a transformation of determinant d, the determinants of covariances among the new variates analogous to those defined above are readily seen in this way to equal

$$A' = c^2 A, \quad B' = d^2 B, \quad C' = c^2 d^2 C, \quad D' = c^2 d^2 D. \tag{4.1}$$

Thus A, B, C, D are *relative invariants* under internal transformations of the two sets of variates.

The ratios

$$Q^2 = \frac{(-1)^s C}{AB} \quad \text{and} \quad Z = \frac{D}{AB} \tag{4.2}$$

we shall call respectively the squares of the *vector correlation coefficient* or *vector correlation*, and of the *vector alienation coefficient*. It is evident that both are absolute invariants under internal transformations of the two sets, since their values computed from transformed variates have numerators and denominators multiplied by the same factor $c^2 d^2$, in accordance with (4.1).

The notation just used is appropriate to a population, but the same definitions and reasoning may be applied to a sample. We denote by q^2 and z the same functions of the sample covariances that Q^2 and Z, respectively, have been defined to be of the population covariances.

A particularly simple linear transformation consists of dividing each variate by its standard deviation. The covariances among the new variates are then the same as their correlations, which are also the correlations among the old variates. Hence, in the definitions of the vector correlation and alienation coefficients, the covariances may be replaced by the correlations. For example, if $s = t = 2$, the squared vector correlation in a sample may be written

$$q^2 = \frac{\begin{vmatrix} 0 & 0 & r_{13} & r_{14} \\ 0 & 0 & r_{23} & r_{24} \\ r_{31} & r_{32} & 1 & r_{34} \\ r_{41} & r_{42} & r_{43} & 1 \end{vmatrix}}{\begin{vmatrix} 1 & r_{12} \\ r_{12} & 1 \end{vmatrix} \cdot \begin{vmatrix} 1 & r_{34} \\ r_{34} & 1 \end{vmatrix}} = \frac{(r_{13}r_{24} - r_{14}r_{23})^2}{(1 - r_{12}^2)(1 - r_{34}^2)}. \tag{4.3}$$

The vector correlation coefficient will always be taken as the positive square root of q^2 or of Q^2 (which are seen below to be positive) when $s < t$, and usually also when $s = t$. However, if in accordance with (4.3) we write

$$q = \frac{r_{13}r_{24} - r_{14}r_{23}}{\sqrt{(1 - r_{12}^2)(1 - r_{34}^2)}}, \tag{4.4}$$

it is evident that q may be positive for some samples of a particular set of variates, and negative for other samples. It may sometimes be advantageous, as in testing whether two samples arose from the same population, to retain the sign of q for each sample, since this provides evidence in addition to that given by the absolute value of q. But unless otherwise stated we shall always regard q as the positive root of q^2. Likewise, Q, $\sqrt{z}$ and $\sqrt{Z}$ will denote the positive roots unless otherwise specifically indicated in each case. A transformation of either set will reverse the sign of the algebraic expression (4.4) if the determinant of the transformation is negative. This will be true of a simple interchange of two variates; for example, $x_1' = x_2$, $x_2' = x_1$ has the determinant -1. On the other hand, the sign is conserved if the determinant of the transformation is positive. Such considerations apply whenever $s = t$.

Since the vector correlation and alienation coefficients are invariants, they may be computed on the assumption that the variates are canonical. In this case $A = B = 1$, and D is given by (3.16). To obtain C we replace the first s 1's in the principal diagonal of (3.16) by 0's. It then follows that

$$C = (-1)^s \rho_1^2 \rho_2^2 \ldots \rho_s^2.$$

This confirms that the value of Q^2 given in (4.2) is positive. In this way the vector correlation and alienation coefficients are expressible in terms of the canonical correlations by the equations

$$Q = \pm \rho_1 \rho_2 \ldots \rho_s, \qquad Z = (1 - \rho_1^2)(1 - \rho_2^2)\ldots(1 - \rho_s^2), \tag{4.5}$$

$$q = \pm r_1 r_2 \ldots r_s, \qquad z = (1 - r_1^2)(1 - r_2^2)\ldots(1 - r_s^2). \tag{4.6}$$

From these results it is obvious that both the vector correlation and vector alienation coefficients are confined to values not exceeding unity. Also Z and z are necessarily positive, except that they vanish if, and only if, all the variates in one set are linear functions of those in the other.

Since the denominator of (4.4) is obviously less than unity, and since we have just shown that $q \leqslant 1$, the tetrad must be still less. This simple proof that the tetrad is between -1 and $+1$ shows the falsity of the idea that the range of the tetrad is from -2 to $+2$, which has gained some currency. An equivalent proof in vector notation was communicated to the writer by E.B. Wilson.

The only case in which Z can attain its maximum value unity is that in which all the canonical correlations vanish. In this case no variate in either set is correlated with any variate in the other, so that the two sets are completely independent, at least if the distribution is normal. Moreover, $Q = 0$. On the other hand, the only case in which Q can be unity is that in which all the canonical correlations are unity. In this event, $Z = 0$; also, the variates in the first set are linear functions of those in the second. Thus either z, $1 - q$, or $1 - q^2$ might be used as an index of independence, while we might use q, q^2 or $1 - z$ as a measure of relationship between the two sets. The work of Wilks alluded to in Section 1 provides an exact distribution of z on the hypothesis of complete independence, a distribution which may thus be used to test this hypothesis.

If we regard the elements of A, B and C as sample covariances, we have in case $s = t$ a simple interpretation of q. Consider the two matrices of observations on the two sets of variates in N individuals, in which each row corresponds to a variate and each column to an individual observed. From Section 2 it is evident that the square of the sum of the products of corresponding s-rowed determinants in the two matrices is $(-1)^s N^{2s} C$; also that the sums of squares of the s-rowed determinants in the two matrices are $N^s A$ and $N^s B$. Therefore q is simply the product-moment correlation coefficient between corresponding s-rowed determinants.

The generalized variance of a set of variates may be defined as the determinant of their ordered covariances, such as A or B. Let $\xi_1, \xi_2, \ldots, \xi_s$ be estimates respectively of $x_1, x_2, \ldots, x_s$ obtained from $x_{s+1}, \ldots, x_{s+t}$ by least squares, and let the regression equations be

$$\xi_\alpha = \sum_i b_{\alpha i} x_i. \tag{4.7}$$

The appropriateness of Q as a generalization of the correlation coefficient, and of $\sqrt{Z}$ as a generalization of the alienation coefficient, will be apparent from the following theorem:

The ratio of the generalized variance of $\xi_1, \ldots, \xi_s$ *to that of* $x_1, \ldots, x_s$ *is* Q^2. *The ratio of the generalized variance of* $x_1 - \xi_1, x_2 - \xi_2, \ldots, x_s - \xi_s$ *to that of* $x_1, \ldots, x_s$ *is* Z.

This theorem is expressed in terms of the population, but an exactly parallel one holds for a sample.

(*Editors' note*: The proof has been omitted.)

A further property of the vector correlation is obvious from the final paragraph of Section 3:

A necessary and sufficient condition that the number of components in an uncorrelated set of components common to two sets of variates be less than the number of variates in either set is that the vector correlation be zero.

When $s = 2$ the canonical correlations not only determine the vector correlation and alienation coefficients but are determined by them. If as usual we take q positive, (4.6) becomes $q = r_1 r_2$, $z = (1 - r_1^2)(1 - r_2^2)$, whence

$$r_1^2 + r_2^2 = 1 - z + q^2, \qquad r_1^2 r_2^2 = q^2. \tag{4.8}$$

Solving, and denoting the greater canonical correlation by r_1, we have

$$\left.\begin{aligned} r_1 &= \tfrac{1}{2}[\sqrt{(1+q)^2 - z} + \sqrt{(1-q)^2 - z}] \\ r_2 &= \tfrac{1}{2}[\sqrt{(1+q)^2 - z} - \sqrt{(1-q)^2 - z}] \end{aligned}\right\}. \tag{4.9}$$

Since the canonical correlations are real, $(r_1 - r_2)^2$ is positive; therefore

$$z \leqslant (1 - q)^2. \tag{4.10}$$

In like manner, the vector correlation and alienation coefficients in the population are subject not only to the inequalities $0 \leqslant Q^2 \leqslant 1, 0 \leqslant Z \leqslant 1$, but also, when $s = 2$, to

$$Z \leqslant (1 - Q)^2.$$

These inequalities become equalities when the roots are equal.

The fundamental equation (3.6), regarded as an equation in λ^2, has as roots the squares of the canonical correlations. Hence, by (4.8), it reduces it to the form

$$\lambda^4 - (1 - z + q^2)\lambda^2 + q^2 = 0, \tag{4.11}$$

where $s = 2$.

5. Standard Errors

The canonical correlations and the coefficients of the canonical variates are defined in Section 3 in such a way that they are continuous functions of the covariances, with continuous derivatives of all orders, except for certain sets of values corresponding to multiple or zero roots, within the domain of variation for which the covariances are the coefficients of a positive definite quadratic form. This is true for the canonical reduction of a sample as well as for that of a population. The probability that a random sample from a continuous distribution will yield multiple roots is zero; and sample covariances must always be the coefficients of a positive definite form.

We shall in this section derive asymptotic standard errors, variances and covariances for the canonical correlations on the assumption that those in the population are unequal, and that the population has the multiple normal distribution. From these we shall derive standard errors for the vector correlation and alienation coefficients q and z. The deviation of sample from population values in these as in most cases have variances of order n^{-1}, and distributions approaching normality of form as n increases*.

Let $x_1, \ldots, x_p$ be a normally distributed set of variates of zero means and covariances

$$\sigma_{ij} = Ex_i x_j. \tag{5.1}$$

For a sample of N in which x_{if} is the value of x_i observed in the fth individual, the sample covariance of x_i and x_j is

$$s_{ij} = \frac{\sum_f (x_{if} - \bar{x}_i)(x_{jf} - \bar{x}_j)}{N - 1} = \frac{\sum x_{if} x_{jf} - N\bar{x}_i\bar{x}_j}{N - 1}, \tag{5.2}$$

* For a proof of approach to normality for a general class of statistics including those with which we deal, cf. Doob, *op. cit.*

where $\bar{x}_i$ and $\bar{x}_j$ are the sample means. To simplify the later work, we introduce the *pseudo-observations*, x'_{if}, defined in terms of the observations by the equations

$$x'_{if} = \sum_{g=1}^{N} c_{fg} x_{ig}, \tag{5.3}$$

where the quantities c_{fg}, independent of i and therefore the same for all the variates x_i, are the coefficients of an orthogonal transformation, such that

$$c_{N1} = c_{N2} = \cdots = c_{NN} = \frac{1}{\sqrt{N}}. \tag{5.4}$$

Since the transformation is othogonal we must have

$$\sum_h c_{fh} c_{gh} = \delta_{fg}, \tag{5.5}$$

where δ_{fg} is the Kronecker delta, equal to unity if $f = g$, but to zero if $f \neq g$. The coefficients c_{fg} may be chosen in an infinite variety of ways consistently with these requirements, but will be held fixed throughout the discussion. Since linear functions of normally distributed variates are normally distributed, the pseudo-observations are normally distributed. Their population means are, from (5.3),

$$Ex'_{if} = \sum c_{fg} Ex_{ig} = 0,$$

since the original variates were assumed to have zero means. Also, since the expectation of the product of independent variates is zero, and since the different individuals in a sample are assumed independent, so that, by (5.1),

$$Ex_{ih} x_{jk} = \delta_{hk} \sigma_{ij}, \tag{5.6}$$

we have, from (5.3), (5.6) and (5.5),

$$\left.\begin{aligned} Ex'_{if} x'_{jg} &= \sum_{hk}\sum c_{fh} c_{gk} Ex_{ih} x_{jk} \\ &= \sum_{hk}\sum c_{fh} c_{gk} \delta_{hk} \sigma_{ij} \\ &= \sum_h c_{fh} c_{gh} \sigma_{ij} \\ &= \delta_{fg} \sigma_{ij} \end{aligned}\right\}. \tag{5.7}$$

From (5.4) and (5.3) it is clear that

$$x'_{iN} = \frac{\sum_g x_{ig}}{\sqrt{N}} = \sqrt{N} \bar{x}_i. \tag{5.8}$$

The equations (5.3) may, on account of their orthogonality, be solved in the form

$$x_{if} = \pm \sum c_{gf} x'_{ig}.$$

Therefore, by (5.5),

$$\sum_f x_{if}x_{jf} = \sum\sum\sum c_{gf}c_{hf}x'_{ig}x'_{jh} = \sum\sum \delta_{gh}x'_{ig}x'_{jh} = \sum_g x'_{ig}x'_{jg}.$$

Substituting this result and (5.8) in (5.2), we find that the final term of the sum cancels out. Introducing therefore the symbol S for summation from 1 to $N-1$ with respect to the second subscript, and putting also

$$n = N - 1, \tag{5.9}$$

we have the compact result

$$s_{ij} = \frac{Sx'_{ig}x'_{jg}}{n}. \tag{5.10}$$

Since the pseudo-observations are normally distributed with the covariances (5.7) and zero means, they have exactly the same distribution as the observations in a random sample of n from the original population. The equivalence of the mean product (5.10) with the sample covariance (5.2) establishes the important principle that *the distribution of covariances in a sample of $n+1$ is exactly the same as the distribution of mean products in a sample of n*, if the parent population is normally distributed about zero means. Use of this principle will considerably simplify the discussions of sampling.

An important extension of this consideration lies in the use of deviations, not merely from sample means, but from regression equations based on other variates. In such cases the number of degrees of freedom n to be used is the difference between the sample number and the number of constants in each of the regression equations, which number must be the same for all the deviations. The estimate of covariance of deviations in the ith and jth variates to be used is then the sum of the products of corresponding deviations, divided by n. This may also be regarded as the mean product of the values of x_i and x_j in n pseudo-observations, as above, without elimination of the means or of the extraneous variates. The sampling distributions with which we shall be concerned will all be expressed in terms of the number of degrees of freedom n, rather than in terms of the number of observations N. This will permit immediately of the extension, which is equivalent to replacing all the correlations, in terms of which our statistics may be defined, by partial correlations representing the elimination of a particular set of variates, the same in all cases.

A variance is of course the covariance of a variate with itself, so that this whole discussion of covariances is equally applicable to variances.

The characteristic function of a multiple normal distribution with zero means is well known to be

$$M(t_1, t_2, \ldots) = Ee^{\sum t_i x_i} = e^{\sum\sum \sigma_{ij}t_it_j/2}.$$

The moments of the distribution are the derivatives of the characteristic function, evaluated for $t_1 = t_2 = \cdots = 0$. From the fourth derivative with respect to t_i, t_j, t_k and t_m it is easy to show in this way that

$$Ex_ix_jx_kx_m = \sigma_{ij}\sigma_{km} + \sigma_{im}\sigma_{jk} + \sigma_{ik}\sigma_{jm}. \tag{5.11}$$

From (5.10) we have

$$Es_{ij}s_{km} = \frac{1}{n^2}SSEx'_{ig}x'_{jg}x'_{kf}x'_{mf}. \tag{5.12}$$

Now if $f \neq g$,

$$Ex'_{ig}x'_{jg}x'_{kf}x'_{mf} = (Ex'_{ig}x'_{jg})(Ex'_{kf}x'_{mf}) = \sigma_{ij}\sigma_{km}, \tag{5.13}$$

since the expectation of the product of *independent* quantities is the product of their expectations. Of the n^2 terms in the double sum in (5.12), $n^2 - n$ are equal to (5.13). The remaining n terms are those for which $f = g$, and each of them equals (5.11). Hence

$$Es_{ij}s_{km} = \sigma_{ij}\sigma_{km} + \frac{1}{n}(\sigma_{im}\sigma_{jk} + \sigma_{ik}\sigma_{jm}).$$

Inasmuch as

$$Es_{ij} = \sigma_{ij},$$

we have, if we put

$$d\sigma_{ij} = s_{ij} - \sigma_{ij},$$

for the deviation of sample from population value, that the sampling covariances is

$$Ed\sigma_{ij}\,d\sigma_{km} = Es_{ij}s_{km} - \sigma_{ij}\sigma_{km},$$

whence

$$Ed\sigma_{ij}\,d\sigma_{km} = \frac{1}{n}(\sigma_{ik}\sigma_{jm} + \sigma_{im}\sigma_{jk}). \tag{5.14}$$

This is a fundamental formula from which may be derived directly a number of more familiar special cases. For example, to obtain the variance of a variance, merely put $i = j = k = m$, which gives

$$\sigma_{sii}^2 = E(d\sigma_{ii})^2 = \frac{2\sigma_{ii}^2}{n}.$$

Returning from these general considerations to the problem of canonical correlations, we recall from (3.2) and (3.3) that for any particular canonical correlation ρ_1,

$$\sum\sum \sigma_{\alpha\beta}a_\alpha a_\beta = 1, \qquad \sum\sum \sigma_{ij}b_ib_j = 1, \qquad \rho_1 = \sum\sum \sigma_{\alpha i}a_\alpha b_i, \tag{5.15}$$

where α and β are summed from 1 to s, and i and j from $s+1$ to $s+t$. Any particular set of sampling errors $d\sigma_{AB}$ in the covariances determines a corresponding set of sampling errors in the a_α and b_i and in ρ_1, for these quantities are definite analytic functions of the covariances except when ρ_1 is a multiple or zero root of (3.6), cases which we now exclude from consideration. In terms of the derivatives of these functions we define

$$da_\alpha = \sum\sum \frac{\partial a_\alpha}{\partial \sigma_{AB}} d\sigma_{AB}, \qquad db_i = \sum\sum \frac{\partial b_i}{\partial \sigma_{AB}} d\sigma_{AB}, \qquad d\rho_1 = \sum\sum \frac{\partial \rho_1}{\partial \sigma_{AB}} d\sigma_{AB}, \tag{5.16}$$

where $d\sigma_{AB} = s_{AB} - \sigma_{AB}$, and the summations are over all values of A and B from 1 to $s+t$. Then differentiating (5.15) we have

$$\left.\begin{aligned} &\sum\sum(2\sigma_{\alpha\beta}a_\alpha\, da_\beta + a_\alpha a_\beta\, d\sigma_{\alpha\beta}) = 0, \qquad \sum\sum(2\sigma_{ij}b_i\, db_j + b_i b_j\, d\sigma_{ij}) = 0, \\ &d\rho_1 = \sum\sum(\sigma_{\alpha i}a_\alpha\, db_i + \sigma_{\alpha i}b_i\, da_\alpha + a_\alpha b_i\, d\sigma_{\alpha i}) \end{aligned}\right\}. \tag{5.17}$$

Let us now suppose that the variates are in the population canonical. This assumption does not entail any loss of generality as regards ρ_1, since ρ_1 is an invariant under transformations of the variates of either set. Since a_α is the coefficient of x_α in the expression for one of the canonical variates, which we take to be x_1, we have in the population $a_1 = 1, a_2 = a_3 = \cdots = a_s = 0$. In the same way,

$$b_{s+1} = 1, \qquad b_{s+2} = \cdots = b_{s+t} = 0.$$

Also, since the covariances among canonical variates are the elements of the determinant in (3.16), we have

$$\sigma_{\alpha\beta} = \delta_{\alpha\beta}, \qquad \sigma_{ij} = \delta_{ij}, \qquad \sigma_{\alpha i} = \delta_{\alpha+s,i}\rho_\alpha, \tag{5.18}$$

the Kronecker deltas being equal to unity if the two subscripts are equal, and otherwise vanishing. When these special values of the a's, b's and σ's are substituted in (5.17) most of the terms drop out, leaving the simple equations

$$\left.\begin{aligned} &2da_1 + d\sigma_{11} = 0, \qquad 2db_{s+1} + d\sigma_{s+1,s+1} = 0, \\ &d\rho_1 = \rho_1\, db_{s+1} + \rho_1\, da_1 + d\sigma_{1,s+1} \end{aligned}\right\}. \tag{5.19}$$

Substituting from the first two in the third of these equations, we get

$$d\rho_1 = d\sigma_{1,s+1} - \tfrac{1}{2}\rho_1(d\sigma_{11} + d\sigma_{s+1,s+1}). \tag{5.20}$$

For any other simple root ρ_2 we have in the same way

$$d\rho_2 = d\sigma_{2,s+2} - \tfrac{1}{2}\rho_2(d\sigma_{22} + d\sigma_{s+2,s+2}). \tag{5.21}$$

Squaring (5.20), taking the expectation, using the fundamental formula (5.14), and finally substituting the canonical values (5.18), we have

$$\left.\begin{aligned} nE(d\rho_1)^2 &= \sigma_{11}\sigma_{s+1,s+1} + \sigma_{1,s+1}^2 - \rho_1(2\sigma_{11}\sigma_{1,s+1} + 2\sigma_{s+1,s+1}\sigma_{1,s+1}) \\ &\quad + \tfrac{1}{4}\rho_1^2(2\sigma_{11}^2 + 4\sigma_{1,s+1}^2 + 2\sigma_{s+1,s+1}^2) \\ &= 1 + \rho_1^2 - \rho_1(2\rho_1 + 2\rho_1) + \tfrac{1}{4}\rho_1^2(2 + 4\rho_1^2 + 2) \\ &= (1 - \rho_1^2)^2 \end{aligned}\right\}. \tag{5.22}$$

Treating the product of (5.20) and (5.21) in the same way we obtain

$$Ed\rho_1\, d\rho_2 = 0. \tag{5.23}$$

A sample canonical correlation r_1 may be expanded about ρ_1 in a Taylor series of the form

$$r_1 = \rho_1 + \sum\sum \frac{\partial \rho_1}{\partial \sigma_{AB}}\, d\sigma_{AB} + \tfrac{1}{2}\sum\sum\sum\sum \frac{\partial^2 \rho_1}{\partial \sigma_{AB}\partial \sigma_{CD}}\, d\sigma_{AB}\, d\sigma_{CD} + \cdots, \tag{5.24}$$

or, by the last of (5.16),

$$r_1 - \rho_1 = d\rho_1 + \cdots. \tag{5.25}$$

The expectation of the product of any number of the sampling deviations $d\sigma_{AB}$ is a fixed function of the σ's divided by a power of n whose exponent increases with the number of the quantities $d\sigma_{AB}$ in the product. Since $Ed\sigma_{AB} = 0$, we have from (5.24) and (5.14) that $E(r_1 - \rho_1)$ is of order n^{-1}. Hence squaring (5.25) and using (5.22), we find that the sampling variance of r_1 is given by $\dfrac{(1-\rho_1^2)^2}{n}$, apart from terms of higher order in n^{-1}. If by the standard error of r_1 we understand the leading term in the asymptotic expansion of the square root of the variance, we have for this standard error

$$\sigma_{r_1} = \frac{1 - \rho_1^2}{\sqrt{n}}. \tag{5.26}$$

It is remarkable that this standard error of a canonical correlation is of exactly the same form as that of a product-moment correlation coefficient calculated directly from data, at least so far as the leading term is concerned.

The covariance of two statistics or their correlation would ordinarily be of order n^{-1}; but from (5.23) it appears that the covariance of r_1 and r_2 is of order n^{-2} at least. All these results hold as between any pair of simple non-vanishing roots. To summarize:

Let $\rho_1, \rho_2, \ldots, \rho_p$ be any set of simple non-vanishing roots of (3.6). *For sufficiently large samples these will be approximated by certain of the canonical correlations $r_1, r_2, \ldots, r_p$ of the samples in such a way that, when $r_\gamma - \rho_\gamma$ is divided by the standard error*

$$\sigma_{r_\gamma} = \frac{1 - \rho_\gamma^2}{\sqrt{n}} \qquad (\gamma = 1, 2, \ldots, p), \tag{5.27}$$

the resulting variates have a distribution which, as n increases, approaches the normal distribution of p independent variates of zero means and unit standard deviations.

For small samples there will be ambiguities as to which root of the determinantal equation for the sample is to be regarded as approximating a particular canonical correlation of the population. As n increases, the sample roots will separately cluster more and more definitely about individual population roots.

If a canonical correlation ρ_γ is zero, and if $s = t$, the foregoing result is applicable with the qualification that sample values r_γ approximating ρ_γ must not all be taken positive, but must be assigned positive and negative values with equal probabilities. Alternatively, if we insist on taking all the sample canonical correlations as positive, the distribution will be that of absolute values of a normally distributed variate.

To prove this, suppose that the determinantal equation has zero as a double root. For sample covariances sufficiently near those in the population, there will be a root r close to zero, which will be very near the value of λ obtained by dropping from the equation all but the term in λ^2 and that independent of λ. The latter is for $s = t$ a perfect square, and the former does not vanish, since the zero root is only a double one. Hence r is the ratio of a polynomial in the s_{AB}'s to a non-vanishing regular function in the neighbourhood. This means that the differential method applicable to non-vanishing roots is also valid here, and that, since the derivatives are continuous, (5.27) holds even when $\rho_\gamma = 0$.

Since a tetrad difference is proportional to a vector correlation, which is the product of the canonical correlations, the question whether the tetrad differs significantly from zero is equivalent to the question whether a canonical correlation is significantly different from zero. This may be tested by means of the standard error (5.27), which reduces in this case to $\frac{1}{\sqrt{n}}$. Since this is independent of unknown parameters, we have here a method of meeting the third of the difficulties mentioned in Section 1 in connection with testing the significance of the tetrad.

For $s = 2$, a zero root is of multiplicity t at least. From the final result in §9 below it may be deduced that if zero is a root of multiplicity exactly t, if r is the corresponding sample canonical correlation, and if $s = 2$, then nr^2 has the χ^2 distribution with $t - 1$ degrees of freedom. This provides a means of testing the significance of a sample canonical correlation in all cases in which $s = 2$.

We shall conclude this section by deriving standard error formulae for the vector correlation and vector alienation coefficients, assuming the canonical correlations in the population all distinct. Differentiating (4.5) and supposing all canonical correlations positive we have

$$dQ = \sum \frac{d\rho_\gamma}{\rho_\gamma}, \qquad dZ = -2Z \sum \frac{\rho_\gamma\, d\rho_\gamma}{1 - \rho_\gamma^2}.$$

Taking the expectations of the squares and products of these expressions and using (5.22) and (5.23), we obtain for the variances and covariance, apart from terms of higher order in n^{-1},

$$\sigma_q = Q\sqrt{\frac{1}{n}\sum_{\gamma=1}^{s}\frac{(1-\rho_\gamma^2)^2}{\rho_\gamma^2}}, \qquad \sigma_z = 2Z\sqrt{\frac{\rho_1^2 + \cdots + \rho_s^2}{n}},$$

$$E\, dQ\, dZ = -\frac{2}{n} QZ \sum (1 - \rho_\gamma^2). \tag{5.28}$$

For the case $s = 2$ these formulae reduce with the help of (4.5) to

$$\sigma_q = \sqrt{\frac{(1 - Q^2)^2 - Z(1 + Q^2)}{n}}, \qquad \sigma_z = 2Z\sqrt{\frac{1 - Z + Q^2}{n}},$$

$$E\, dQ\, dZ = -\frac{2}{n} QZ(1 + Z - Q^2).$$

6. Examples, and an Iterative Method of Solution

The correlations obtained by Truman L. Kelley* among tests in (1) reading speed, (2) reading power, (3) arithmetic speed, and (4) arithmetic power are given by the elements of the following determinant, in which the rows and columns are arranged in the order given:

$$D = \begin{vmatrix} 1.0000 & .6328 & .2412 & .0586 \\ .6328 & 1.0000 & -.0553 & .0655 \\ .2412 & -.0553 & 1.0000 & .4248 \\ .0586 & .0655 & .4248 & 1.0000 \end{vmatrix} = .4129.$$

These correlations were obtained from a sample of 140 seventh-grade school children. Let us inquire into the relations of arithmetical with reading abilities indicated by these tests.

The two-rowed minors of D in the upper left, lower right, and upper right corners are respectively

$$A = .5996, \quad B = 8195, \quad \sqrt{C} = .01904.$$

Hence, by (4.2)

$$q^2 = .0007377, \quad q = .027161, \quad z = .84036. \tag{6.1}$$

* *Op. cit.*, p. 100. These are the raw correlations, not corrected for attenuation.

By means of (4.9) or (4.11) these values give for the canonical correlations

$$r_1 = .3945, \quad r_2 = .0688. \tag{6.2}$$

In this case $n = N - 1 = 139$, and the standard error (5.27) reduces, for the hypothesis of a zero canonical correlation in the population, to $\frac{1}{\sqrt{139}} =$ .0848. It is plain, therefore, that r_2 is not significant, so that we do not have any evidence here of more than one common component of reading and arithmetical abilities.

Whether we have convincing evidence of *any* common component is another question. It is tempting to compare the value of r_1 also with the standard error .0848 for the purpose of answering this question, which would give a decidedly significant value. This however is not a sensitive procedure for testing the hypothesis that there is no common factor; for this hypothesis of complete independence would mean that both canonical correlations would in the population be zero; they would therefore be a quadruple root of the fundamental equation, to which the standard error is not applicable. We conclude that reading and arithmetic involve one common mental factor but, so far as these data show, only one.

Linear functions $a_1x_1 + a_2x_2$ and $b_3x_3 + b_4x_4$ having maximum correlation with each other may be used either to predict arithmetical from reading ability or vice versa. The coefficients will satisfy (3.4) and (3.5); when in these equations we substitute $r_1 = .3945$ for λ and μ, and the given correlations for the covariances, and divide by $-\lambda = -.3945$, we have

$$\begin{aligned} a_1 + .6328a_2 - .6114b_3 - .1485b_4 &= 0, \\ .6328a_1 + a_2 + .1402b_3 - .1660b_4 &= 0, \\ -.6114a_1 + .1402a_2 + b_3 + .4248b_4 &= 0, \\ -.1485a_1 - .1660a_2 + .4248b_3 + b_4 &= 0. \end{aligned}$$

The fourth equation must be dependent on the preceding three, so we ignore it except for a final checking. Replacing b_4 by unity we may solve the first three equations, which are symmetrical, by the usual least-square method. Thus we write the coefficients, without repetition, in the form

1.0000	.6328	−.6114	−.1485	.8729
	1.0000	.1402	−.1660	1.6070
		1.0000	.4248	.9536

the last column consisting of the sums of the elements written or understood in the respective rows. The various divisions, multiplications and subtractions involved in solving the equations are applied to the elements in the rows, including those in the check column, which at every stage gives the sum of the elements written or understood in a row. In the array above, the coefficients of each equation begin in the first row and proceed downward to the diagonal, then across to the right, and this scheme is followed with the reduced set of equations obtained by eliminating an unknown, which is done in such a way as to preserve symmetry. This process yields finally the ratios

$$a_1 : a_2 : b_3 : b_4 = -2.772 : 2.2655 : -2.4404 : 1.$$

Therefore the linear functions of arithmetical and reading scores that predict each other most accurately are proportional to $-2.7772x_1 + 2.2655x_2$ and $-2.4404x_3 + x_4$, respectively. It is for these weighted sums that the maximum correlation .3945 is attained.

From the same individuals, Kelley obtained the correlations in the following table, in which the first two rows correspond to the arithmetic speed and power tests cited above, while the others are respectively memory for words, memory for meaningful symbols, and memory for meaningless symbols:

$$\left\|\begin{array}{cc|ccc} 1.0000 & .4248 & .0420 & .0215 & .0573 \\ .4248 & 1.0000 & .1487 & .2489 & .2843 \\ \hline .0420 & .1487 & 1.0000 & .6693 & .4662 \\ .0215 & .2489 & .6693 & 1.0000 & .6915 \\ .0573 & .2843 & .4662 & .6915 & 1.0000 \end{array}\right\|.$$

From this we find $q^2 = .0003209$, $q = .01792$, $z = .902466$, whence

$$r_1 = .3073, \qquad r_2 = .0583.$$

Since in this case $s \neq t$, we cannot say as before that the standard error of r_2 when $\rho = 0$ is $n^{-1/2} = .0848$. But, putting $\chi^2 = nr_2^2 = .472$, with two degrees of freedom, we find $P = .79$, so that r_2 is far from significant. However r_1 is decidedly significant.

In view of the tests in Section 11, we conclude in this case also that there is evidence of one common component but not of two.

If each of the two sets contains more than two variates, the two invariants q and z do not suffice to determine the coefficients of the various powers of λ in the determinantal equation, so that its roots can no longer be calculated in the foregoing manner. The coefficients in the equation will involve other rational invariants in addition to q and z, but we shall not be concerned with these, and it is desirable to have a procedure that does not require their calculation, or the explicit determination and solution of the equation. It is also desirable to avoid the explicit solution of the sets of linear equations (3.4) and (3.5) when the variates are numerous, since the labour of the direct procedure then becomes excessive. These computational difficulties are analogous to those in the determination of the principal axes of a quadric in n-space, or of the principal components of a set of statistical variates, problems for which an iterative procedure has been found useful, and has been proved to converge to the correct values in all cases*. We shall now show how

* Harold Hotelling, "Analysis of a Complex of Statistical Variables into Principal Components" in *Journal of Educational Psychology*, Vol. XXIV. pp. 417–441 and 498–520 (September and October, 1933), Section 4.

a process partly iterative in character may be applied to determine canonical variates and canonical correlations between two sets.

If in the s equations (3.4) we regard $\lambda a_1, \lambda a_2, \ldots, \lambda a_s$ as the unknowns, we may solve for them in terms of the b's by the methods appropriate for solving normal equations. Indeed, the matrix of the coefficients of the unknowns is symmetrical; and in the solving process it is only necessary to carry along, instead of a single column of right-hand members, t columns, from which the coefficients of $b_{s+1}, \ldots, b_{s+t}$ in the expressions for $a_1, \ldots, a_s$ are to be determined. The entries initially placed in these columns are of course the covariances between the two sets. Let the solution of these equations consist of the s expressions

$$\lambda a_\alpha = \sum_i g_{\alpha i} b_i \qquad (\alpha = 1, 2, \ldots, s). \tag{6.3}$$

In exactly the same way the t equations (3.5), with μ replaced by λ, may be solved for $\lambda b_{s+1}, \ldots, \lambda b_{s+t}$ in the form

$$\lambda b_i = \sum_\beta h_{i\beta} a_\beta \qquad (i = s + 1, \ldots, s + t). \tag{6.4}$$

If we substitute from (6.4) in (6.3) and set

$$k_{\alpha\beta} = \sum_i g_{\alpha i} h_{i\beta}, \tag{6.5}$$

we have

$$\lambda^2 a_\alpha = \sum_\beta k_{\alpha\beta} a_\beta. \tag{6.6}$$

Now if an arbitrarily chosen set of numbers be substituted for $a_1, \ldots, a_s$ in the right-hand members of (6.6), the sums obtained will be proportional to the numbers substituted only if they are proportional to the true values of $a_1, \ldots,$ a_s. If, as will usually be the case, the proportionality does not hold, the sums obtained, multiplied or divided by any convenient constant, may be used as second approximations to solutions $a_1, \ldots, a_s$ of the equations. Substitution of these second approximations in the right-hand members of (6.6) gives third approximations which may be treated in the same way; and so on. Repetition of this process gives repeated sets of trial values, whose ratios will be seen below to approach as limits those among the true values of $a_1, \ldots, a_s$. The factor of proportionality λ^2 in (6.6) becomes r_1^2, the square of the largest canonical correlation. When the quantities $a_1', \ldots, a_s'$ eventually determined as sufficiently nearly proportional to $a_1, \ldots, a_s$ are substituted in the right-hand members of (6.4), there result quantities $b_{s+1}', \ldots, b_{s+t}'$ proportional to $b_{s+1}, \ldots, b_{s+t}$, apart from errors which may be made arbitrarily small by continuation of the iterative process. The factor of proportionality to be applied in order to obtain linear functions with unit variance is the same for the a's and the b's; from (3.2), (3.4), and (3.5) it may readily be shown that if from the quantities obtained we calculate

$$m = \frac{\sqrt{r_1}}{\sqrt{\sum\sum \sigma_{\alpha i} a'_\alpha b'_i}}, \tag{6.7}$$

then the true coefficients of the first pair of canonical variates are $ma'_1, \ldots, ma'_s, mb'_{s+1}, \ldots, mb'_{s+t}$.

In the iterative process, if $a_1, \ldots, a_s$ represent trial values at any stage, those at the next stage will be proportional to

$$a'_\alpha = \sum k_{\alpha\beta} a_\beta. \tag{6.8}$$

Another application of the process gives

$$a''_\gamma = \sum k_{\gamma\alpha} a'_\alpha,$$

whence, substituting, we have

$$a''_\gamma = \sum k^{(2)}_{\gamma\beta} a_\beta,$$

provided we put

$$k^{(2)}_{\gamma\beta} = \sum_\alpha k_{\gamma\alpha} k_{\alpha\beta}.$$

The last equation is equivalent to the statement that the matrix K^2 of the coefficients $k^{(2)}_{\gamma\beta}$ is the square of the matrix K of the $k_{\alpha\beta}$. It follows therefore that one application of the iterative process by means of the squared matrix is exactly equivalent to two successive applications with the original matrix. This means that if at the beginning we square the matrix only half the number of steps will subsequently be required for a given degree of accuracy.

The number of steps required may again be cut in half if we square K^2, for with the resulting matrix K^4 one iteration is exactly equivalent to four with the original matrix. Squaring again we obtain K^8, with which one iteration is equivalent to eight, and so on. This method of accelerating convergence is also applicable to the calculation of principal components*. It embodies the root-squaring principle of solving algebraic equations in a form specially suited to determinantal equations.

After each iteration it is advisable to divide all the trial values obtained by a particular one of them, say the first, so as to make successive values comparable. The value obtained for a_1, if this is the one used to divide the rest at each step, will approach r_1^2 if the matrix K is used in iteration, but will approach r_1^4 if K^2 is used, r_1^8 if K^4 is used, and so forth. When stationary values are reached, they may well be subjected once to iteration by means of K itself, both in order to determine r^2 without extracting a root of high order, and as a check on the matrix-squaring operations.

If our covariances are derived from a sample from a continuous multivariate distribution, it is infinitely improbable that the equation is ω,

* Another method of accelerated iterative calculation of principal components is given by T.L. Kelley in *Essential Traits of Mental Life*, Cambridge, Mass., 1935. A method similar to that given above is applied to principal components by the author in *Psychometrika*, Vol. I. No. 1 (1936).

$$\begin{vmatrix} k_{11} - \omega & k_{12} & \dots\dots k_{1s} \\ k_{21} & k_{22} - \omega & \dots\dots k_{2s} \\ \dots & \dots & \dots \\ k_{s1} & k_{s2} & \dots\dots k_{ss} - \omega \end{vmatrix} = 0,$$

has multiple roots. If we assume that the roots $\omega_1, \omega_2, \dots, \omega_s$ are all simple, and regard $a_1, \dots, a_s$ as the homogeneous coordinates of a point in $s - 1$ dimensions which is moved by the collineation (6.8) into a point $(a_2', \dots, a_s')$, we know* that there exists in this space a transformed system of coordinates such that the collineation is represented in terms of them by

$$\bar{a}_1' = \omega_1 \bar{a}_1, \quad \bar{a}_2' = \omega_2 \bar{a}_2, \dots, \bar{a}_s' = \omega_s \bar{a}_s.$$

Another iteration yields a point whose transformed homogeneous coordinates are proportional to

$$\omega_1^2 \bar{a}_1, \quad \omega_2^2 \bar{a}_2, \dots, \omega_s^2 \bar{a}_s.$$

Continuation of this process means, if ω_1 is the root of greatest absolute value, that the ratio of the first transformed coordinates to any of the others increases in geometric progression. Consequently the moving point approaches as a limit the invariant point corresponding to this greatest root. Therefore the ratios of the trial values of $a_1, \dots, a_s$ will approach those among the coefficients in the expression for the canonical variate corresponding to the greatest canonical correlation. Thus the iterative process is seen to converge, just as in the determination of principal components.

After the greatest canonical correlation and the corresponding canonical variates are determined, it is possible to construct a new matrix of covariances of deviations from these canonical variates. When the iterative process is applied to this new matrix, the second largest canonical correlation and the corresponding canonical variates are obtained. This procedure may be carried as far as desired to obtain additional canonical correlations and variates, as in the method of principal components; but the later stages of the process will yield results which will usually be of diminishing importance. The modification of the matrix is somewhat more complicated than in the case of principal components, and we shall omit further discussion of this extension.

(*Editors' note*: The remainder of this section, as well as sections 7–16 have been omitted.)

* Bôcher, p. 293.

Introduction to Wilcoxon (1945) Individual Comparisons by Ranking Methods

G.E. Nother

When in 1945, Frank Wilcoxon published this unpretentious little paper, he could hardly have guessed that the two techniques he was proposing would soon occupy a central place in a newly developing branch of statistics that became known as nonparametrics. Wilcoxon (1892–1965), a physical chemist by training who was employed by the American Cyanamid Company in Stamford, Connecticut, came to statistics because of a need for analyzing laboratory data. His main motivation in developing the two new techniques seems to have been a desire to replace the endless t-statistics that he needed for the analysis of his laboratory measurements by something computationally simpler. In a subsequent publication [Wilcoxon (1949)], he explained

> It is not always realized that there are available rapid approximate methods which are quite useful in interpreting the results of experiments, even though these approximate methods do not utilize fully the information contained in the data.

The two procedures, now generally known as the Wilcoxon rank sum test (or two-sample test) and the Wilcoxon signed rank test (or one-sample test), are used for the comparison of two treatments involving unpaired and paired sample observations, respectively. In classical normal theory statistics, two treatments are compared with the help of appropriate t-tests. Wilcoxon proposed replacing the actual data values by their ranks to simplify computational effort.

Some statisticians would date the beginning of nonparametric statistics, in particular ranking methods, with the appearance of Wilcoxon's 1945 paper. As we shall see, the Wilcoxon paper provided a strong impetus to research in nonparametrics. But the historical roots of nonparametric statistics are clearly older and may be traced back all the way to the Arbuthnot paper mentioned in the editors' preface. [An account of the early history of nonparametrics is given in Noether (1984).]

Reflecting Wilcoxon's interest in the practical side of statistics, the two new techniques are introduced with the help of actual data sets, quite likely from Wilcoxon's own laboratory work. A short general discussion follows. The remainder of the paper deals with the computation of relevant probability levels for the two test statistics. For unpaired observations, only the equal sample size case is discussed. Two tables of very limited scope accompany the paper. Table 2 for the signed rank test contains some errors.

In my discussion of subsequent research inspired by the Wilcoxon paper, I shall concentrate on the two-sample case. Developments for the one-sample case are quite analogous and are therefore treated more briefly.

For the two-sample case, we use the following notation. There is an x sample of size m from a population with cumulative distribution function (cdf) $F(z)$ and an independent y sample of size n from a population with cdf $G(z)$. Wilcoxon presumably thought in terms of normal, or near normal, populations. There is no indication that at the time of publication, Wilcoxon was aware of or particularly interested in the much greater generality of the new techniques compared to the standard t-procedures. Wilcoxon two-sample statistic W for testing the hypothesis $G(z) = F(z)$ is the sum of the ranks for the y observations, when all $m + n = N$ observations are ranked from 1 to N.

The Mann–Whitney Test

Two years after the Wilcoxon paper, Mann and Whitney (1947) investigated a test based on the statistic $U = \#(y_j < x_i) = \#$ (negative differences $y_j - x_i$), which is a linear function of W, $W = \frac{1}{2}n(n + 1) + U'$, $U' = mn - U = \#(y_j > x_i)$. In view of this linear relationship between the Wilcoxon and Mann–Whitney statistics, it is possible to reinterpret probability statements involving U as probability statements involving W (and vice versa.)

The Mann–Whitney paper provided a careful study of the null distribution of U, including its asymptotic normality. (For many practical applications, the normal approximation is adequate for sample sizes as small as $m = n = 8$.) Additionally, Mann and Whitney showed that the test is consistent against the monotone alternatives $F(z) < G(z)$ or $F(z) > G(z)$, all z. Indeed, the test is consistent against alternatives for which $p \neq \frac{1}{2}$, where $p = Pr(y < x)$. The statistic U furnishes the unbiased estimate of p, U/mn.

According to Kruskal (1956), the statistic U for the comparison of two treatments was proposed by the German psychologist Gustav Deuchler as early as 1914. But apparently, his suggestion never took hold. Deuchler apparently assumed that the distribution of U was normal. He gave the correct mean, $\frac{1}{2}mn$, but made an error in the computation of its variance.

Estimation

In the statistical literature, the Wilcoxon rank sum test is frequently described as the nonparametric equivalent of the two-sample t-test. It can be shown that two- sample t applied to the ranks of the observations produces a test statistic that is equivalent to W. The condition for consistency of the Wilcoxon test shows, however, that the Wilcoxon test is appropriate for alternatives that are considerably more general than the shift model implicit in the normal theory two-sample t-test. Nevertheless, the shift model $G(z) = F(z - \delta)$ according to which y observations have been shifted by an amount δ relative to x observations is often realistic in practical applications. The model raises the problem of how to estimate the unknown shift parameter δ, when nothing is assumed about the distribution function $F(z)$. Hodges and Lehmann (1963) suggested as an estimate of δ the value $\hat{\delta}$, for which the Wilcoxon test applied to the observations $x_1, \ldots, x_m, y_1 - \hat{\delta}, \ldots, y_n - \hat{\delta}$ is least likely to suggest rejection of the hypothesis that the two samples have come from the same population. This value is the median of the mn sample differences $y_j - x_i$, $i = 1, \ldots, m$; $j = 1, \ldots, n$. Confidence intervals for δ are bounded by symmetrically located differences $y_j - x_i$, when all mn differences are arranged from the smallest to the largest [Moses (1953)].

Relative Efficiency of the Wilcoxon Test

Wilcoxon mentioned that the efficiency of his test relative to the t-test was unknown. He very likely suspected that it was considerably less than 1. Relevant information was supplied by Pitman (1948). Until Pitman, it was rather general to limit efficiency investigations to the case of normal populations. Pitman developed a general theory of the relative efficiency of two tests and gave an expression for the efficiency of the Wilcoxon test relative to the two-sample t-test for the general shift model $g(z) = f(z - \delta)$, where $f(z)$ and $g(z)$ are the densities associated with the x and y populations. For $f(z)$ normal, the efficiency of the Wilcoxon test relative to the t-test turned out to have the amazingly high value $3/\pi = 0.955$. For many distributions, particularly distributions with long tails, the efficiency is actually greater than 1. Even more surprising is a result by Hodges and Lehmann (1956). The efficiency of the Wilcoxon test relative to the t-test is never less than 0.864. Since the Wilcoxon test is much less affected by erratic observations than the t-test, these efficiency results support the conclusion that for many practical purposes, the Wilcoxon test is preferable to the t-test.

According to Hodges and Lehmann (1963) and Lehmann (1963), these efficiency results carry over to the comparison of the point estimate $\hat{\delta}$ with the normal theory estimate $\bar{y} - \bar{x}$ and the "nonparametric" confidence interval mentioned earlier compared to the normal theory t interval.

Linear Rank Statistics

The Wilcoxon rank sum test has been generalized in the following way. Let r_j, $j = 1, \ldots, n$, denote the ranks assigned to observations y_j, when the $m + n = N$ x and y observations are ranked from 1 to N, so that the Wilcoxon rank sum W equals Σr_j. Let $a = \{a_1, \ldots, a_N\}$ be a set of scores with $a_1 \leq \cdots \leq a_N$. Linear rank tests for testing the hypothesis that x and y observations are identically distributed are based on statistics $W_{[a]} = \sum_{j=1}^{n} a(r_j)$, where $a(k) = a_k$. For scores $a_k = k$, $W_{[a]} =$ Wilcoxon W. By choosing appropriate scores a_k, the power of the $W_{[a]}$ test can be maximized for a given distribution $F(z)$. Thus, for normal populations, scores a_k that are equal to the expectations of the kth-order statistics in samples of size N from a standard normal population produce a test known as the normal scores' test that has asymptotic efficiency 1 relative to the t-test and is superior to the t-test for all other distributions. Detailed information about linear rank statistics can be found in Hajek (1961).

The Wilcoxon Signed Rank Test

For the paired observation case with observation pairs (x_k, y_k), $k = 1, \ldots, N$, the Wilcoxon signed rank test ranks the N absolute differences $|y_k - x_k|$ from 1 to N and uses as a test statistic the sum of the ranks that correspond to negative (positive) differences. If we set $z_k = y_k - x_k$, $k = 1, \ldots, N$, the Wilcoxon signed rank test applied to the z_k's becomes a test of the hypothesis that the z's constitute a random sample from a symmetric population with the center of symmetry at the origin. More generally, the hypothesis that the center of symmetry of a z population equals some constant η can be tested by applying the Wilcoxon one-sample test to the differences $z_1 - \eta, \ldots, z_N - \eta$. Tukey (1949) pointed out that the two rank sums for the Wilcoxon signed rank test can also be computed as the number of Walsh averages $\frac{1}{2}(z_h + z_k)$, $1 \leq h \leq k \leq N$, that are respectively smaller than or larger than the hypothetical value η. It follows that the Hodges–Lehmann estimate of η equals the median of all Walsh averages, and confidence intervals for η are bounded by symmetrically located Walsh averages. Efficiency results for the one-sample Wilcoxon procedures remain the same as in the two-sample case, and linear rank statistics for the one-sample problem are defined analogously to the two-sample case.

Summing up, it is probably no exaggeration to suggest that Wilcoxon's 1945 paper accomplished for the development of nonparametrics what Student's 1908 paper accomplished for the development of normal theory methods. Wilcoxon proposed simple solutions for two important practical problems at a time when theoretical statisticians were ready for new challenges. It took many additional years before the great majority of statistical practitioners were ready to abandon their preconceived distrust of things nonparametric, a label that to many was synonymous with rough-and-ready and quick-and-dirty.

Biographical information about Frank Wilcoxon may be found in Bradley and Hollander (1978) and (1988).

References

Bradley, R.A., and Hollander, M. (1978). Wilcoxon, Frank, in *International Encyclopedia of Statistics*, Vol. 2. The Free Press, Glencoe, Ill., pp. 1245–1250.

Bradley, R.A., and Hollander, M. (1988). Wilcoxon, Frank, in *Encyclopedia of Statistical Sciences* (S. Kotz, N.L. Johnson, and C.R. Read, eds.) Vol. 8. Wiley, New York, pp. 609–612.

Hájek, J. (1961). *A Course in Nonparametric Statistics*. Holden-Day, San Francisco, Calif.

Hodges, J.L. Jr., and Lehmann, E.L. (1956). The efficiency of some nonparametric competitors to the *t*-test, *Ann. Math. Statist.*, **27**, 324–335.

Hodges, J.L. Jr., and Lehmann, E.L. (1963). Estimates of locations based on rank tests, *Ann. Math. Statist.*, **34**, 598–611.

Kruskal, W.H. (1956). Historical notes on the Wilcoxon unpaired two-sample test, *J. Amer. Statist. Assoc.*, **52**, 356–360.

Lehmann, E.L. (1963). Nonparametric confidence intervals for a shift parameter, *Ann. Math. Statist.*, **34**, 1507–1512.

Mann, H.B., and Whitney, D.R. (1947). On a test of whether one of two random variables is stochastically larger than the other, *Ann. Math. Statist.*, **18**, 50–60.

Moses, L.E. (1953). Non-parametric methods, in *Statistical Inference* (Walker and Lev, eds.). Henry Holt, New York, Chap. 18.

Noether, G.E. (1984). Nonparametrics: The early years—impressions and recollections, *The Amer. Statistician*, **38**, 173–178.

Pitman, E.J.G. (1948). *Lecture Notes on Non-Parametric Statistics*. Columbia University, New York.

Tukey, J.W. (1949). The simplest signed rank tests. Report No. 17, Statistical Research Group, Princeton University.

Wilcoxon, F. (1949). *Some Rapid Approximate Statistical Procedures*. American Cyanamid Co., Stamford Research Laboratories.

Individual Comparisons by Ranking Methods

Frank Wilcoxon
American Cyanamid Co.

The comparison of two treatments generally falls into one of the following two categories: (a) we may have a number of replications for each of the two treatments, which are unpaired, or (b) we may have a number of paired comparisons leading to a series of differences, some of which may be positive and some negative. The appropriate methods for testing the significance of the differences of the means in these two cases are described in most of the textbooks on statistical methods.

The object of the present paper is to indicate the possibility of using ranking methods, that is, methods in which scores 1, 2, 3, ... n are substituted for the actual numerical data, in order to obtain a rapid approximate idea of the significance of the differences in experiments of this kind.

1. Unpaired Experiments

The following table gives the results of fly spray tests on two preparations in terms of percentage mortality. Eight replications were run on each preparation.

Sample A Percent kill	Rank	Sample B Percent kill	Rank
68	12.5	60	4
68	12.5	67	10
59	3	61	5
72	15	62	6
64	8	67	10
67	10	63	7
70	14	56	1
74	16	58	2
Total 542	91	494	45

Rank numbers have been assigned to the results in order of magnitude. Where the mortality is the same in two or more tests, those tests are assigned the mean rank value. The sum of the ranks for B is 45 while for A the sum is 91. Reference to Table I shows that the probability of a total as low as 45 or lower, lies between 0.0104 and 0.021. The analysis of variance applied to these results gives an F value of 7.72, while 4.60 and 8.86 correspond to probabilities of 0.05 and 0.01 respectively.

Table I. For Determining the Significance of Differences in Unpaired Experiments.

No. of replicates	Smaller rank total	Probability for this total or less
5	16	.016
5	18	.055
6	23	.0087
6	24	.015
6	26	.041
7	33	.0105
7	34	.017
7	36	.038
8	44	.0104
8	46	.021
8	49	.050
9	57	.0104
9	59	.019
9	63	.050
10	72	.0115
10	74	.0185
10	79	.052

Paired Comparisons

An example of this type of experiment is given by Fisher (2, section 17). The experimental figures were the differences in height between cross- and self-fertilized corn plants of the same pair. There were 15 such differences as follows: 6, 8, 14, 16, 23, 24, 28, 29, 41, −48, 49, 56, 60, −67, 75. If we substitute rank numbers for these differences, we arrive at the series 1, 2, 3, 4, 5, 6, 7, 8, 9, −10, 11, 12, 13, −14, 15. The sum of the negative rank numbers is −24. Table II shows that the probability of a sum of 24 or less is between 0.019 and

Table II. For Determining the Significance of Differences in Paired Experiments

Number of Paired Comparisons	Sum of rank numbers, + or −, whichever is less	Probability of this total or less
7	0	0.016
7	2	0.047
8	0	0.0078
8	2	0.024
8	4	0.055
9	2	0.0092
9	3	0.019
9	6	0.054
10	3	0.0098
10	5	0.019
10	8	0.049
11	5	0.0093
11	7	0.018
11	11	0.053
12	7	0.0093
12	10	0.021
12	14	0.054
13	10	0.0105
13	13	0.021
13	17	0.050
14	13	0.0107
14	16	0.021
14	21	0.054
15	16	0.0103
15	19	0.019
15	25	0.054
16	19	0.0094
16	23	0.020
16	29	0.053

0.054 for 15 pairs. Fisher gives 0.0497 for the probability in this experiment by the t test.

The following data were were obtained in a seed treatment experiment on wheat. The data are taken from a randomized block experiment with eight replications of treatments A and B. The figures in columns two and three represent the stand of wheat.

Block	A	B	A-B	Rank
1	209	151	58	8
2	200	168	32	7
3	177	147	30	6
4	169	164	5	1
5	159	166	−7	−3
6	169	163	6	2
7	187	176	11	5
8	198	188	10	4

The fourth column gives the differences and the fifth column the corresponding rank numbers. The sum of the negative rank numbers is −3. Table II shows that the total 3 indicates a probability between 0.024 and 0.055 that these treatments do not differ. Analysis of variance leads to a least significant difference of 14.2 between the means of two treatments for 19:1 odds, while the difference between the means of A and B was 17.9. Thus it appears that with only 8 pairs this method is capable of giving quite accurate information about the significance of differences of the means.

Discussion

The limitations and advantages of ranking methods have been discussed by Friedman (3), who has described a method for testing whether the means of several groups differ significantly by calculating a statistic χ_r^2 from the rank totals. When there are only two groups to be compared, Friedman's method is equivalent to the binomial test of significance based on the number of positive and negative differences in a series of paired comparisons. Such a test has been shown to have an efficiency of 63 percent (1). The present method for comparing the means of two groups utilizes information about the magnitude of the differences as well as the signs, and hence should have higher efficiency, but its value is not known to me.

The method of assigning rank numbers in the unpaired experiments requires little explanation. If there are eight replicates in each group, rank numbers 1 to 16 are assigned to the experimental results in order of magnitude and where tied values exist the mean rank value is used.

In the case of the paired comparisons, rank numbers are assigned to the differences in order of magnitude neglecting signs, and then those rank numbers which correspond to negative differences receive a negative sign. This is necessary in order that negative differences shall be represented by negative rank numbers, and also in order that the magnitude of the rank assigned shall correspond fairly well with the magnitude of the difference. It will be recalled that in working with paired differences, the null hypothesis is that we are dealing with a sample of positive and negative differences normally distributed about zero.

The method of calculating the probability of occurrence of any given rank total requires some explanation. In the case of the unpaired experiments, with rank numbers 1 to $2q$, the possible totals begin with the sum of the series 1 to q, that is, $q(q + 1)/2$; and continue by steps of one up to the highest value possible, $q(3q + 1)/2$. The first two and the last two of these totals can be obtained in only one way, but intermediate totals can be obtained in more than one way, and the number of ways in which each total can arise is given by the number of q-part partitions of T, the total in question, no part being repeated, and no part exceeding $2q$. These partitions are equinumerous with another set of partitions, namely the partitions of r, where r is the serial number of T in the possible series of totals beginning with 0, 1, 2, ... r, and the number of parts of r, as well as the part magnitude, does not exceed q. The latter partitions can easily be enumerated from a table of partitions such as that given by Whitworth (5), and hence serve to enumerate the former. A numerical example may be given by way of illustration. Suppose we have 5 replications of measurements of two quantities, and rank numbers 1 to 10 are to be assigned to the data. The lowest possible rank total is 15. In how many ways can a total of 20 be obtained? In other words, how many unequal 5-part partitions of 20 are there, having no part greater than 10? Here 20 is the sixth in the possible series of totals; therefore $r = 5$ and the number of partitions required is equal to the total number of paritions of 5. The one to one correspondence is shown below:

Unequal 5-part partitions of 20	Partitions of 5
1-2-3-4-10	5
1-2-3-5-9	1-4
1-2-3-6-8	2-3
1-2-4-5-8	1-1-3
1-2-4-6-7	1-2-2
1-3-4-5-7	1-1-1-2
2-3-4-5-6	1-1-1-1-1

By taking advantage of this correspondence, the number of ways in which each total can be obtained may be calculated, and hence the probability of occurrence of any particular total or a lesser one.

The following formula gives the probability of occurence of any total or a lesser total by chance under the assumption that the group means are drawn from the same population:

$$P = 2\left\{1 + \sum_{i=1}^{i=r}\sum_{j=1}^{j=q} \sqcap_j^i - \sum_{n=1}^{n=r-q}\left[(r-q-n+1)\sqcap_{q-1}^{q-2+n}\right]\right\}\Big/\frac{\underline{|2q}}{\underline{|q} \times \underline{|q}}$$

$\sqcap_j^i$ represents the number of j-part partitions of i,
r is the serial number of possible rank totals, 0, 1, 2, ... r.
q is the number of replicates, and
n is an integer representing the serial number of the term in the series.

In the case of the paired experiments, it is necessary to deal with the sum of rank numbers of one sign only, + or −, whichever is less, since with a given number of differences the rank total is determined when the sum of + or − ranks is specified. The lowest possible total for negative ranks is zero, which can happen in only one way, namely, when all the rank numbers are positive. The next possible total is −1, which also can happen in only one way, that is, when rank one receives a negative sign. As the total of negative ranks increases, there are more and more ways in which a given total can be formed. These ways for any totals such as $-r$, are given by the total number of unequal partitions of r. If r is 5, for example, such partitions, are 5, 1–4, 2–3. These partitions may be enumerated, in case they are not immediately apparent, by the aid of another relation among partitions, which may be stated as follows:

The number of unequal j-part partitions of r, with no part greater than i, is equal to the number of j-part partitions of $r - \binom{j}{2}$, parts equal or unequal, and no part greater than $i - j + 1$ (4).

For example, if r equals 10, j equals 3, and i equals 7, we have the correspondence shown below:

Unequal 3-part partitions of 10   1-2-7   1-3-6   1-4-5   2-3-5

3-part partitions of 10-3, or 7, no part greater than 5

1-1-5   1-2-4   1-3-3-   2-2-3

The formula for the probability of any given total r or a lesser total is:

$$P = 2\left[1 + \sum_{n}\left(\sum_{i=n}^{i=r-(n/2)} \sqcap_n^i\right)\right]\Big/ 2^q$$

r is the serial number of the total under consideration in the series of possible totals

$$0, 1, 2, \ldots, r,$$

q is the number of paired differences.

In this way probability tables may be readily prepared for the 1 percent level or 5 percent level of significance or any other level desired.

Literature Cited

1. Cochran, W.G., The efficiencies of the binomial series tests of significance of a mean and of a correlation coefficient. *Jour. Roy. Stat. Soc.*, **100**: 69–73, 1937.
2. Fisher, R.A. *The design of experiments.* Third ed., Oliver & Boyd, Ltd., London, 1942.
3. Friedman, Milton. The use ranks to avoid the assumption of normality. Jour. *Amer. Stat. Assn.* **32**: 675–701. 1937.
4. MacMahon, P.A. *Combinatory analysis*, Vol. II, Cambridge University Press. 1916.
5. Whitworth, W.A. *Choice and chance*, G.F. Stechert & Co., New York, 1942.

Introduction to Mosteller (1946) On Some Useful "Inefficient" Statistics

H.A. David
Iowa State University

1. Introduction

The editors have not done the obvious in selecting this paper. Evidently, they were not put off by its opening words: "Several statistical techniques are proposed for economically analyzing large masses of data by means of punched-card equipment." Moreover, very little of the paper survives unimproved in current statistical practice. Nevertheless, the author made a number of significant advances, pointed the way to many more, and showed great prescience. The paper, as a result, has been extremely influential and has stimulated much research on order statistics.

In the situation considered by Mosteller, "experimenters have little interest in milking the last drop of information out of their [abundant] data" and it is more important to produce timely analyses based, if necessary, on the use of "inefficient" statistics. The basic issue involved here is surely still with us even with high-speed computers. Within this context, the paper is concerned with the estimation of the mean, standard deviation, and correlation coefficient in large samples, especially for normal populations. Some small-sample results are also obtained.

The inefficient statistics of choice are the order statistics. In a random sample of n, drawn from a continuous probability density function $f(x)$, the focus is on a fixed number k of order statistics denoted by $x(n_1)$, $x(n_2)$, $\ldots$, $x(n_k)$, where the ranks n_i satisfy condition 1, namely, $\lim_{n \to \infty} (n_i/n) = \lambda_i$, $i = 1, 2, \ldots, k$, with $0 < \lambda_1 < \lambda_2 < \cdots < \lambda_k < 1$. Thus, in current terminology, $x(n_i)$ is essentially the sample quantile of order λ_i. In particular, the extremes $x(1)$ and $x(n)$, which had already received much sophisticated attention culminating in Gnedenko (1943), play no role.

Mosteller acknowledges valuable help from S.S. Wilks, his major professor, and from J.W. Tukey, both of whom had been working actively on nonparametric tolerance intervals and regions as well as other problems involving order statistics. For references, see the review article by Wilks (1948). However, the present paper is quite unrelated to any earlier Princeton publications.

The title "Order Statistics" of Wilks (1948) refers to the field and includes results on both the theory of order statistics and what would now be called rank-order statistics. The use of "order statistics" to denote the ordered observations seems to have made its first appearance in print in Wilks (1943, p. 89), but general acceptance took some time. Mosteller introduced the term "systematic statistics" at the beginning of his paper as convenient shorthand for "functions of order statistics." Elaborating on this term, Godwin (1949) used "linear systematic statistics," now generally known as L-statistics.

2. Asymptotic Joint Distribution of k Sample Quantiles

Special sample quantiles, such as the median and the quartiles, had long been of interest as measures of location and dispersion. As Harter (1978) points out, Yule (1911) gives the correct formulae for the standard error of a general quantile and for the correlation coefficient of two quantiles. Mosteller claims that the bivariate normal limiting form of the joint distribution of $x(n_i)$ and $x(n_j)$ can be obtained from information supplied by K. Pearson (1920). However, Pearson does not deal with the distribution, only with its parameters. Mosteller goes on to write that the limiting form was derived more rigorously by Smirnov (Smirnoff) (1937) "under rather general conditions" [namely, $f(x)$ assumed everywhere continuous and positive] and is also treated in Kendall (1943). Having earlier (condition B) clarified the meaning of joint asymptotic normality, he restates these findings as Theorem 1.

The first major result in the paper (Theorem 2), although described by the author as an obvious generalization of Theorem 1, gives the explicit form of the limiting multivariate normal distribution of the $x(n_i)$, $i = 1, 2, \ldots, k$ under condition 1. The proof is interesting in that the result is first obtained for a uniform (0, 1) distribution, requiring a nice matrix inversion, and then transformed to the general case. As pointed out in Reiss (1980), the theorem was also proved by Smirnov (1944). Theorem 2 implies that, under mild conditions, a linear function of sample quantiles is asymptotically normally distributed. It is the basis for much of the subsequent large-sample theory developed in the paper and of extensions by other authors. A different application, to goodness-of-fit tests based on all n or a subset of k of the order statistics, is given by Hartley and Pfaffenberger (1972).

True, the result has since been proved under weaker conditions. Mosteller (and Smirnov) require that $f(x)$ be finite and positive at the k population quantiles ξ_{λ_i}, and that $f(x)$ be continuous in the neighborhood of ξ_{λ_i}, $i = 1$,

2, ..., k. The continuity requirement may be dropped in view of J.K. Ghosh's (1971) form of Bahadur's (1966) representation of quantiles in large samples. Another proof under these weakened conditions is given by Walker (1968).

3. Moments of Order Statistics in Small Samples

Following the proof of Theorem 2, there comes a short passage worth citing in full for its prophetic quality.

> It would often be useful to know the small sample distribution of the order statistics, particularly in the case where the sample is drawn from a normal. Fisher and Yates' tables [4] give the expected values of the order statistics up to samples of size 50. However it would be very useful in the development of certain small sample statistics to have further information. It is perhaps too much to expect tabulated distribution functions, but at least the variances and covariances would be useful. A joint effort has resulted in the calculation for samples $n = 2, 3, \ldots, 10$ of the expected values to five decimal places, the variances to four decimal places, and the covariances to nearly two decimal places.

The tables referred to [see Hastings, et al. (1947)] are the first in a long line of tables of means, variances, and covariances of order statistics for many different parent distributions [see, e.g., David (1981, p. 288)]. Most often, sample sizes of $n \leq 20$ are covered. In the normal case, Tietjen et al. (1977) give the variances and covariances to 10 decimal places for $n \leq 50$. Most basically, such tables of the first two moments permit the calculation of the mean and variance of any linear function of order statistics in the cases covered. The tables are crucial in the important method of estimation of location and scale parameters by linear functions of order statistics [Lloyd (1952)] for type-II censored or complete samples [Sarhan and Greenberg (1962)].

4. Optimal Spacing of the Order Statistics

Next comes a very influential pioneering section on the "optimal spacing" of the order statistics. Specifically, Mosteller considers the estimation in large samples of the mean θ of a symmetric distribution by the judicious choice of a small number k of the order statistics. He notes that it is easy to find linear combinations of these order statistics that are unbiased estimators of θ. Moreover, such estimators being asymptotically normally distributed under the conditions of Theorem 2, the efficiency of an estimator in this class is the ratio of asymptotic variances of the best estimator to the proposed estimator.

Mosteller restricts himself to the use of the simple average $\hat{\theta}_k = \Sigma x(n_i)/k$ and gives an asymptotic expression for var $\hat{\theta}_k$ in terms of the λ_i and the densities f_i of the standardized distribution at the quantile of order λ_i. The task then is to find, for each k, the values λ_i minimizing var $\hat{\theta}_k$ for the distribution of interest. For symmetric distributions, he confines consideration to symmetric spacings of the order statistics [i.e., if $x(n_i)$ occurs in $\hat{\theta}_k$, so does $x(n_{k-i+1})$]. In the normal case, on which Mosteller concentrates, he solves the problem numerically for $k \leq 3$ and also shows that the simple choice $\lambda_i = (i - \frac{1}{2})/k$ gives good results, reaching an asymptotic efficiency (compared to the sample mean) of 0.973 for $k = 10$.

The paper then turns to the corresponding problem of estimating the standard deviation σ in the normal case, generalizing Pearson's (1920) result for $k = 2$. The estimators are now of the form $\hat{\sigma} = (\Sigma x(n_{k-i+1}) - \Sigma x(n_i))/C$, where C is an unbiasing constant and the sums run from $i = 1$ to $\frac{1}{2}[k]$. Mosteller expresses amazement that "with punched-card equipment available," Pearson's estimator $\hat{\sigma} = (x(0.93n) - x(0.07n))/C$ is "practically never used" in spite of an efficiency of 0.652 as against 0.37 for the interquartile range estimate. For $k = 8$ he finds that with $\lambda_1 = 0.02$, $\lambda_2 = 0.08$, $\lambda_3 = 0.15$, $\lambda_4 = 0.25$, and $\lambda_i = 1 - \lambda_{9-i}$, $i = 5, \ldots, 8$, the resulting estimator, with $C =$ 10.34, has an efficiency of 0.896, actually slightly higher than the efficiency of 0.88 of the mean deviation from mean or median. Mosteller also makes the interesting point that, with some loss in efficiency, one may want to constrain the λ's from being too close to 0 or 1 because (1) the assumption of normality is particularly suspect in the extreme tails of a distribution, and (2) even for truly normal data, the normality of the more extreme order statistics will hold only in very large samples.

In an important follow-up paper, Ogawa (1951) treated the same problem when general linear functions of k order statistics are used to estimate location or scale parameters or both. Eisenberger and Posner (1965) point out that optimal spacing permits useful data compression since a large sample (e.g., of particle counts on a spacecraft) may be replaced by enough order statistics to allow (on the ground) satisfactory estimation of parameters, as well as a test of the assumed underlying distributional form. For a normal population, the last authors tabulate the optimal estimators when $k = 2(2)20$. In comparison with an efficiency of 0.973 for Mosteller's estimator of the mean for $k = 10$, these tables give efficiencies of 0.981 for $k = 10$ and 0.972 for $k = 8$. The tables give an efficiency of 0.929 for the estimation of σ with $k = 8$ as against 0.896 above. A large number of papers on optimal spacing for various parent distributions continue to be published. For a listing up to 1980, see David (1981, p. 200).

Mosteller's original approach is useful in the following situation [O'Connell and David (1976)]. From a large number n of objects, it is desired to estimate the mean μ_Y of a characteristic Y. However, Y is expensive to measure. Instead, inexpensive related measurements x are first made on the n objects followed by y-measurements on k objects. If the distribution $Y|x$ is normal with constant variance, then a simple estimator of μ_Y is $\sum_{i=1}^k y[n_i]/k$, where $y[n_i]$ denotes the y value for the object whose x rank is n_i, and the n_i are the optimal ranks for estimating μ_X by $\sum_{i=1}^k x(n_i)/k$.

5. Quasi-Ranges

Contrary to the main thrust of the paper, Sec. 4 begins with the estimation of σ in normal samples of *small* size n. Mosteller remarks that "it is now common practice in industry" to estimate σ by $R' = c_n(x_n - x_1)$ for small samples, where the rth order statistic is here simply denoted by x_r, and c_n is an unbiasing constant. Noting that the ratio var S'/var R', where S' is the unbiased root-mean-square estimator of σ, falls off as n increases, he raises the question "whether it might not be worthwhile to change the systematic statistic slightly by using $c_{1|n}\,(x_{n-1} - x_2)$ or more generally $c_{r|n}\,(x_{n-r} - x_{r+1})$," where the c's are again unbiasing constants. Because of the limited tables of Hastings et al. (1947), referred to above, he is able to state only that this is not advantageous for $n \leq 10$. This particular point was settled by Cadwell (1953) who showed that in the class of statistics $w_r = x_{n-r} - x_{r+1}$, w_0 is the best for $n \leq 17$ and w_1 for $18 \leq n \leq 31$. Mosteller's term "quasi range" (now usually hyphenated) for w_r $(r = 1, 2, \ldots)$ is universally adopted in the many investigations of the properties of the w_r and of linear combinations thereof [e.g., Godwin (1949) and Harter (1959)].

6. The Author

Frederick Mosteller (1916–   ) received B.S. and M.S. degrees at the Carnegie Institute of Technology where he began his research in statistics under Edwin G. Olds. He then obtained an A.M. degree from Princeton University and became involved in war work with the small New York branch of the Princeton Statistical Research Group. Other members of the group were John Williams, Cecil Hastings, and Jimmie Savage.

Following return to Princeton and completion of his Ph.D. in 1946, Mosteller went on to a highly distinguished career at Harvard University, where he is still very active. He has served as President of the American Statistical Association, the Institute of Mathematical Statistics, and the American Association for the Advancement of Science. He is an Honorary Fellow of the Royal Statistical Society and a Member of the National Academy of Sciences. Other honors conferred on him include the S.S. Wilks Award of the American Statistical Association and the R.A. Fisher Award of the Committee of Presidents of Statistical Societies. He is the author, coauthor, or editor of over 200 scientific papers and of nearly 50 books of great diversity. For interesting personal glimpses see the article "Frederick Mosteller and John Tukey: A Conversation" in *Statistical Science* (1988). A lively book-length account of Mosteller's life and work is given in Fienberg et al. (1990).

Mosteller's paper is reproduced here, except for a lengthy final section on the estimation of the correlation coefficient of a bivariate normal population, for various degrees of knowledge on the means and variances.

References

Bahadur, R.R. (1966). A note on quantiles in large samples, *Ann. Math. Statist.*, **37**, 577–580.

Cadwell, J.H. (1953). The distribution of quasi-ranges in samples from a normal population, *Ann. Math. Statist.*, **24**, 603–613.

David, H.A. (1981). *Order Statistics*, 2nd ed. Wiley, New York.

Eisenberger, I., and Posner, E.C. (1965). Systematic statistics used for data compression in space telemetry, *J. Amer. Statist. Assoc.*, **60**, 97–133.

Fienberg, S.E., Hoaglin, D.C., Kruskal, W.H., and Tanur, J.M. (eds.) (1990). *A Statistical Model: Frederick Mosteller's Contributions to Statistics, Science, and Public Policy*. Springer-Verlag, New York.

Ghosh, J.K. (1971). A new proof of the Bahadur representation of quantiles and an application, *Ann. Math. Statist.*, **42**, 1957–1961.

Gnedenko, B. (1943). Sur la distribution limite du terme maximum d'une série aléatoire, *Ann. Math.*, **44**, 423–453.

Godwin, H.J. (1949). On the estimation of dispersion by linear systematic statistics, *Biometrika*, **36**, 92–100.

Harter, H.L. (1959). The use of sample quasi-ranges in estimating population standard deviation, *Ann. Math. Statist.*, **30**, 980–999. (correction, **31**, 228).

Harter, H.L. (1978). *A Chronological Annotated Bibliography on Order Statistics*, Vol. 1: Pre-1950. U.S. Government Printing Office, Washington, D.C.

Hartley, H.O., and Pfaffenberger, R.C. (1972). Quadratic forms in order statistics used as goodness-of-fit criteria, *Biometrika*, **59**, 605–612.

Hastings, C., Jr., Mosteller, F., Tukey, J.W., and Winsor, C.P. (1947). Low moments for small samples: A comparative study of order statistics, *Ann. Math. Statist.*, **18**, 413–426.

Kendall, M.G. (1943). *The Advanced Theory of Statistics*, Vol. 1. Griffin, London.

Lloyd, E.H. (1952). Least-squares estimation of location and scale parameters using order statistics, *Biometrika*, **39**, 88–95.

O'Connell, M.J., and David, H.A. (1976). Order statistics and their concomitants in some double sampling situations, in *Essay in Probability and Statistics*, (S. Ikedo, et al., eds.). pp. 451–466. Shinko Tsusho, Tokyo.

Ogawa, J. (1951). Contributions to the theory of systematic statistics, I, *Osaka Math. J.*, **3**, 175–213.

Pearson, K. (1920). On the probable errors of frequency constants, Part III, *Biometrika*, **13**, 113–132.

Reiss, R.-D. (1980). Estimation of quantiles in certain nonparametric models., *Ann. Statist.*, **8**, 87–105.

Sarhan, A.E., and Greenberg, B.G. (eds.) (1962). *Contributions to Order Statistics*. Wiley, New York.

Smirnoff, N. (1937). Sur la dépendance des membres d'une série de variations, *Bull. Univ. État Moscou, Série Int., Sec. A*, Vol. 1, Fasc. 4, 1–12.

Smirnov, N.V. (1944). Approximation of distribution laws of random variables by empirical data, *Uspekhi Mat. Nauk*, **10**, 179–206. (in Russian).

Statistical Science (1988). Frederick Mosteller and John Tukey: A Conversation, **3** (1), 136–144.

Tietjen, G.L., Kahaner, D.K., and Beckman, R.J. (1977). Variances and covariances of the normal order statistics for sample sizes 2 to 50, *Selected Tables in Math. Statist.*, **5**, 1–73.

Walker, A.M. (1968). A note on the asymptotic distribution of sample quantiles, *J. Roy. Statist. Soc.*, **30**, 570–575.

Wilks, S.S. (1943). *Mathematical Statistics*. Princeton Univ. Press, Princeton, N.J.

Wilks, S.S. (1948). Order statistics, *Bull. Amer. Math. Soc.*, **5**, 6–50.

Yule, G.U. (1911). *An Introduction to the Theory of Statistics*. Griffin, London.

On Some Useful "Inefficient" Statistics

Frederick Mosteller
Princeton University

Summary. Several statistical techniques are proposed for economically analyzing large masses of data by means of punched-card equipment; most of these techniques require only a counting sorter. The methods proposed are designed especially for situations where data are inexpensive compared to the cost of analysis by means of statistically "efficient" or "most powerful" procedures. The principal technique is the use of functions of order statistics, which we call *systematic statistics.*

It is demonstrated that certain order statistics are asymptotically jointly distributed according to the normal multivariate law.

For large samples drawn from normally distributed variables we describe and give the efficiencies of rapid methods:

i) for estimating the mean by using 1, 2, ..., 10 suitably chosen order statistics;
ii) for estimating the standard deviation by using 2, 4, or 8 suitably chosen order statistics;
iii) for estimating the correlation coefficient whether other parameters of the normal bivariate distribution are known or not (three sorting and three counting operations are involved).

The efficiencies of procedures ii) and iii) are compared with the efficiencies of other estimates which do not involve sums of squares or products.

1. Introduction

The purpose of this paper is to contribute some results concerning the use of order statistics in the statistical analysis of large masses of data. The present results deal particularly with estimation when normally distributed variables are present. Solutions to all problems considered have been especially designed for use with punched-card equipment although for most of the results a counting sorter is adequate.

Until recently mathematical statisticians have spent a great deal of effort developing "efficient statistics" and "most powerful tests." This concentration of effort has often led to neglect of questions of economy. Indeed some may have confused the meaning of technical statistical terms "efficient" and "efficiency" with the layman's concept of their meaning. No matter how much energetic activity is put into analysis and computation, it seems reasonable to inquire whether the output of information is comparable in value to the input measured in dollars, man-hours, or otherwise. Alternatively we may inquire whether comparable results could have been obtained by smaller expenditures. In some fields where statistics is widely used, the collection of large masses of data is inexpensive compared to the cost of analysis. Often the value of the statistical information gleaned from the sample decreases rapidly as the time between collection of data and action on their interpretation increases. Under these conditions, it is important to have quick, inexpensive methods for analyzing data, because economy demands militate against the use of lengthy, costly (even if more precise) statistical methods. A good example of a practical alternative is given by the control chart method in the field of industrial quality control. The sample range rather than the sample standard deviation is used almost invariably in spite of its larger variance. One reason is that, after brief training, persons with slight arithmetical knowledge can compute the range quickly and accurately, while the more complicated formula for the sample standard deviation would create a permanent stumbling block. Largely as a result of simplifying and routinizing statistical methods, industry now handles large masses of data on production adequately and profitably. Although the sample standard deviation can give a statistically more efficient estimate of the population standard deviation, if collection of data is inexpensive compared to cost of analysis and users can compute a dozen ranges to one standard deviation, it is easy to see that economy lies with the less efficient statistic.

It should not be thought that inefficient statistics are being recommended for all situations. There are many cases where observations are very expensive, and obtaining a few more would entail great delay. Examples of this situation arise in agricultural experiments, where it often takes a season to get a set of observations, and where each observation is very expensive. In such cases the experimenters want to squeeze every drop of information out of their data. In these situations inefficient statistics would be uneconomical, and are not recommended.

A situation that often arises is that data are acquired in the natural course of administration of an organization. These data are filed away until the accumulation becomes mountainous. From time to time questions arise which can be answered by reference to the accumulated information. How much of these data will be used in the construction of say, estimates of parameters, depends on the precision desired for the answer. It will however often be less expensive to get the desired precision by increasing the sample size by dipping deeper into the stock of data in the files, and using crude techniques of analysis, than to attain the required precision by restricting the sample size to the minimum necessary for use with "efficient" statistics.

It will often happen in other fields such as educational testing that it is less expensive to gather enough data to make the analysis by crude methods sufficiently precise, than to use the minimum sample sizes required by more refined methods. In some cases, as a result of the type of operation being carried out sample sizes are more than adequate for the purposes of estimation and testing significance. The experimenters have little interest in milking the last drop of information out of their data. Under these circumstances statistical workers would be glad to forsake the usual methods of analysis for rapid, inexpensive techniques that would offer adequate information, but for many problems such techniques are not available.

In the present paper several such techniques will be developed. For the most part we shall consider statistical methods which are applicable to estimating parameters. In a later paper we intend to consider some useful "inefficient" tests of significance.

2. Order Statistics

If a sample $O_n = x'_1, x'_2, \ldots, x'_n$ of size n is drawn from a continuous probability density function $f(x)$, we may rearrange and renumber the observations within the sample so that

$$x_1 < x_2 < \cdots < x_n \tag{1}$$

(the occurrence of equalities is not considered because continuity implies zero probability for such events). The x_i's are sometimes called *order statistics*. On occasion we write $x(i)$ rather than x_i. Throughout this paper the use of primes on subscripted x's indicates that the observations are taken without regard to order, while unprimed subscripted x's indicate that the observations are order statistics satisfying (1). Similarly $x(n_i)$ will represent the n_ith order statistic, while $x'(n_i)$ would represent the n_ith observation, if the observations were numbered in some random order. The notation here is essentially the *opposite* of usual usage, in which attention is called to the order statistics by the device of primes or the introduction of a new letter. The present reversal of usage seems justified by the viewpoint of the article—that in the problems under consideration the use of order statistics is the natural procedure.

An example of a useful order statistic is the median; when $n = 2m + 1$ ($m = 0, 1, \ldots$), x_{m+1} is called the median and may be used to estimate the population median, i.e. u defined by

$$\int_{-\infty}^{u} f(t)\,dt = \tfrac{1}{2}.$$

In the case of symmetric distributions, the population mean coincides with u and x_{m+1} will be an unbiased estimate of it as well. When $n = 2m$ ($m = 1, 2, \ldots$), the median is often defined as $\frac{1}{2}(x_m + x_{m+1})$. The median so defined is an unbiased estimate of the population median in the case of symmetric distributions; however for most asymmetric distributions $\frac{1}{2}(x_m + x_{m+1})$ will only be unbiased asymptotically, that is in the limit as n increases without bound. For another definition of the sample median see Jackson [8, 1921]. When x is distributed according to the normal distribution

$$N(x, a, \sigma^2) = \frac{1}{\sqrt{2\pi}\,\sigma} e^{-(1/2\sigma^2)(x-a)^2},$$

the variance of the median is well known to tend to $\pi\sigma^2/2n$ as n increases.

It is doubtful whether we can accurately credit anyone with the introduction of the median. However for some of the results in the theory of order statistics it is easier to give credit. In this section we will restrict the discussion to the order statistics themselves, as opposed to the general class of statistics, such as the range $(x_n - x_1)$, which are derived from order statistics. We shall call the general class of statistics which are derived from order statistics, and use the value ordering (1) in their construction, *systematic statistics.*

The large sample distribution of extreme values (examples x_r, x_{n-s+1} for r, s fixed and $n \to \infty$) has been considered by Tippett [17, 1925] in connection with the range of samples drawn from normal populations; by Fisher and Tippett [3, 1928] in an attempt to close the gap between the limiting form of the distribution and results tabled by Tippett [17], by Gumbel [5, 1934] (and in many other papers, a large bibliography is available in [6, Gumbel 1939]), who dealt with the more general case $r \geq 1$, while the others mentioned considered the special case of $r = 1$; and by Smirnoff who considers the general case of x_r, in [15, 1935] and also [16] the limiting form of the joint distribution of x_r, x_s, for r and s fixed as $n \to \infty$.

In the present paper we shall not usually be concerned with the distribution of extreme values, but shall rather be considering the limiting form of the joint distribution of $x(n_1), x(n_2), \ldots, x(n_k)$, satisfying

Condition 1. $\lim_{n\to\infty} \frac{n_i}{n} = \lambda_i$; $i = 1, 2, \ldots, k$;

$$\lambda_1 < \lambda_2 < \cdots < \lambda_k.$$

In other words the proportion of observations less than or equal to $x(n_i)$ tends to a fixed proportion which is bounded away from 0 and 1 as n increases. K. Pearson [13, 1920] supplies the information necessary to obtain the limiting distribution of $x(n_1)$, and limiting joint distribution of $x(n_1)$, $x(n_2)$. Smirnoff gives more rigorous derivations of the limiting form of the marginal distribution of the $x(n_i)$ [15, 1935] and the limiting form of the joint distribution of $x(n_i)$ and $x(n_j)$ [16] under rather general conditions. Kendall [10, 1943, pp. 211–14] gives a demonstration leading to the limiting form of the joint distribution.

Since we will be concerned with statements about the asymptotic properties of the distributions of certain statistics, it may be useful to include a short discussion of their implications both practical and theoretical. If we have a statistic $\hat{\theta}(O_n)$ based on a sample $O_n: x_1', x_2', \ldots, x_n'$ drawn from a population with cumulative distribution function $F(x)$ it often happens that the function $(\hat{\theta} - \theta)/\sigma_n = y_n$, where σ_n is a function of n is such that

$$\lim_{n\to\infty} P(y_n < t) = \frac{1}{\sqrt{2\pi}} \int_{-\infty}^{t} e^{-(1/2)x^2}\, dx. \tag{A}$$

When this condition (A) is satisfied we often say: *$\hat{\theta}$ is asymptotically normally distributed with mean θ and variance σ_n^2*. We will not be in error if we use the statement in italics provided we interpret it as synonymous with (A). However there are some pitfalls which must be avoided. In the first place condition (A) may be true even if the distribution function of y_n, or of $\hat{\theta}$, has no moments even of fractional orders for any n. Consequently we do not imply by the italicized statement that $\lim_{n\to\infty} E[\hat{\theta}(O_n)] = \theta$, nor that $\lim_{n\to\infty} \{[E(\hat{\theta}^2) - [E(\hat{\theta})]^2\} = \sigma_n^2$, for, as mentioned, these expressions need not exist for (A) to be true. Indeed we shall demonstrate that Condition (A) is satisfied for certain statistics even if their distribution functions are as momentless as the startling distributions constructed by Brown and Tukey [1, 1946]. Of course it may be the case that all moments of the distribution of $\hat{\theta}$ exist and converge as $n \to \infty$ to the moments of a normal distribution with mean θ and variance σ_n^2. Since this implies (A), but not conversely, this is a strong convergence condition than (A). (See for example J.H. Curtiss [2, 1942].) However the important implication of (A) is that for sufficiently large n each percentage point of the distribution of $\hat{\theta}$ will be as close as we please to the value which we would compute from a normal distribution with mean θ and variance σ_n^2, independent of whether the distribution of $\hat{\theta}$ has these moments or not.

Similarly if we have several statistics $\hat{\theta}_1, \hat{\theta}_2, \ldots, \hat{\theta}_k$, each depending upon the sample $O_n: x_1', x_2', \ldots, x_n'$, we shall say *that the $\hat{\theta}_i$ are asymptotically jointly normally distributed with means θ_i, variances $\sigma_i^2(n)$, and covariances $\rho_{ij}\sigma_i\sigma_j$*, when

$$\lim_{n\to\infty} P(y_1 < t_1, y_2 < t_2, \ldots, y_k < t_k) = K \int_{-\infty}^{t_1} \int_{-\infty}^{t_2} \cdots \int_{-\infty}^{t_k} e^{-(1/2)Q^2}\, dx_1\, dx_2 \cdots dx_k, \tag{B}$$

where $y_i = (\hat{\theta}_i - \theta_i)/\sigma_i$, and Q^2 is the quadratic form associated with a set of k jointly normally distributed variables with variances unity and covariances ρ_{ij}, and K is a normalizing constant. Once again the statistics $\hat{\theta}_i$ may not have moments or product moments, the point that interests us is that the probability that the point with coordinates $(\hat{\theta}_1, \hat{\theta}_2, \ldots, \hat{\theta}_k)$ falls in a certain region in a k-dimensional space can be given as accurately as we please for sufficiently large samples by the right side of (B).

Since the practicing statistician is very often really interested in the probability that a point will fall in a particular region, rather than in the variance or standard deviation of the distribution itself, the concepts of asymptotic normality given in (A) and (B) will usually not have unfortunate consequences. For example, the practicing statistician will usually be grateful that the sample size can be made sufficiently large that the probability of a statistic falling into a certain small interval can be made as near unity as he pleases, and will not usually be concerned with the fact that, say, the variance of the statistic may be unbounded.

Of course, a very real question may arise: how large must n be so that the probability of a statistic falling within a particular interval can be sufficiently closely approximated by the asymptotic formulas? If in any particular case the sample size must be ridiculously large, asymptotic theory loses much of its practical value. However for statistics of the type we shall usually discuss, computation has indicated that in many cases the asymptotic theory holds very well for quite small samples.

For the demonstration of the joint asymptotic normality of several order statistics we shall use the following two lemmas.

Lemma 1. *If a random variable $\hat{\theta}(O_n)$ is asymptotically normally distributed converging stochastically to θ, and has asymptotic variance $\sigma^2(n) \underset{n\to\infty}{\to} 0$, where n is the size of the sample $O_n : x'_1, x'_2, \ldots, x'_n$, drawn from the probability density function $h(x)$, and $g(\hat{\theta})$ is a single-valued function with a nonvanishing continuous derivative $g'(\hat{\theta})$ in the neighborhood of $\hat{\theta} = \theta$, then $g(\hat{\theta})$ is asymptotically normally distributed converging stochastically to $g(\theta)$ with asymptotic variance $\sigma_n^2[g'(\theta)]^2$.*

PROOF. By the conditions of the lemma

$$\lim_{n\to\infty} P\left[\frac{\hat{\theta} - \theta}{\sigma_n} < t\right] = \frac{1}{\sqrt{2\pi}} \int_{-\infty}^{t} e^{-(1/2)u^2}\, du.$$

Now if $t\sigma_n = \Delta\theta$, $\Delta\theta = \hat{\theta} - \theta$, using the mean value theorem there is a θ_1 in the interval $[\theta, \hat{\theta}]$, such that

$$g(\hat{\theta}) = g(\theta) + (\hat{\theta} - \theta)g'(\theta_1),$$

which implies

$$\lim_{n\to\infty} P\left(\frac{\hat{\theta}-\theta}{\sigma_n} < t\right) = \lim_{n\to\infty} P\left(\frac{g(\hat{\theta})-g(\theta)}{\sigma_n g'(\theta_1)} < t\right), \qquad g'(\theta_1) \neq 0,$$

where θ_1 is a function of n. However $\lim_{\Delta\theta\to 0} g'(\theta_1) = g'(\theta)$ so we may write

$$\lim_{n\to\infty} P\left(\frac{\hat{\theta}-\theta}{\sigma_n} < t\right) = \lim_{n\to\infty} P\left(\frac{g(\hat{\theta})-g(\theta)}{\sigma_n g'(\theta)} < t\right), \qquad g'(\theta) \neq 0,$$

where the form of the expression on the right is the one required to complete the proof of the lemma. □

Of course if we have several random variables $\hat{\theta}_1, \hat{\theta}_2, \ldots, \hat{\theta}_k$, we can prove by an almost identical argument that

Lemma 2. *If the random variables $\hat{\theta}_i(O_n)$ are asymptotically jointly normally distributed converging stochastically to θ_i, and have asymptotic variances $\sigma_i^2(n) \underset{n\to\infty}{\to} 0$, and covariances $\rho_i{}^j\sigma_i\sigma_i$, where n is the size of the sample $O_n: x'_1, x'_2, \ldots, x'_n$ drawn from the probability density function $h(x)$, and $g_i(\hat{\theta}_i)$, $i = 1, 2, \ldots, k$, are single-valued functions with nonvanishing continuous derivatives $g'_i(\hat{\theta}_i)$ in the neighborhood of $\hat{\theta}_i = \theta_i$, then the $g_i(\hat{\theta}_i)$ are jointly asymptotically normally distributed with means $g_i(\theta_i)$, variances $\sigma_i^2[g'_i(\theta_i)]^2$ and covariances $\rho_{ij}\sigma_i\sigma_j g'_i(\theta_i)g'_j(\theta_j)$.*

The following condition represents restrictions on the probability density function $f(x)$ sufficient for the derivation of the limiting form of the joint distribution of the $x(n_i)$ satisfying Condition 1.

Condition 2. The probability density function $f(x)$ is continuous, and does not vanish in the neighborhood of u_i, where

$$\int_{-\infty}^{u_i} f(x)\,dx = \lambda_i, \qquad i = 1, 2, \ldots, k.$$

If we recall the discussion of condition (B) above, the theorem of Pearson and Smirnoff may be stated:

Theorem 1. *If a sample $O_n: x_1, x_2, \ldots, x_n$ is drawn from $f(x)$ satisfying Condition 2, and if $x(n_1)$, $x(n_2)$ satisfy Condition 1 as $n \to \infty$, then $x(n_1)$, $x(n_2)$ are asymptotically distributed according to the normal bivariate distribution with means u_1, u_2,*

$$\int_{-\infty}^{u_i} f(x)\,dx = \lambda_i,$$

and variances

$$\sigma_i^2 = \frac{\lambda_i(1-\lambda_i)}{n[f(u_i)]^2}, \qquad i = 1, 2,$$

and covariance

$$\rho_{12}\sigma_1\sigma_2 = \frac{\lambda_1(1-\lambda_2)}{nf(u_1)f(u_2)}.$$

Theorem 1 has an obvious generalization which seems not to have been carried out in the literature. The generalization may be stated:

Theorem 2. *If a sample $O_n: x_1, x_2, \ldots, x_n$ is drawn from $f(x)$ satisfying Condition 2, and if $x(n_1), x(n_2), \ldots, x(n_k)$ satisfy Condition 1 as $n \to \infty$, then the $x(n_i)$, $i = 1, 2, \ldots, k$, are asymptotically distributed according to the normal multivariate distribution, with means u_i,*

$$\int_{-\infty}^{u_i} f(x)\, dx = \lambda_i,$$

and variances

$$\sigma_i^2 = \frac{\lambda_i(1-\lambda_i)}{nf(u_i)^2}, \qquad i = 1, 2, \ldots, k,$$

and covariances

$$\rho_{ij}\sigma_i\sigma_j = \frac{\lambda_i(1-\lambda_j)}{nf(u_i)f(u_j)}, \qquad 1 \le i < j \le k.$$

Proof. We shall carry out the demonstration for the uniform distribution

$$f(x) = \begin{cases} 1, & 0 \le x \le 1, \\ 0, & \text{elsewhere,} \end{cases}$$

and then utilize the fact that by a suitable transformation of the uniform distribution we may get any $f(x)$ satisfying Condition 2. Of course for the particular case of the uniform distribution all moments of the $x(n_i)$ exist and converge to those of the asymptotic theory.

The joint probability density of the $x(n_i)$, satisfying Condition 1 and drawn from $f(x)$, is given by

$$g[x(n_1), x(n_2), \ldots, x(n_k)] = \frac{n!}{(n_1-1)!(n-n_k)!\prod_{i=2}^{k}(n_i - n_{i-1} - 1)!} \tag{2}$$
$$\left(\int_0^{x(n_1)} dt_1\right)^{n_1-1}\left(\int_{x(n_k)}^{1} dt_{k+1}\right)^{n-n_k}\prod_{i=2}^{k}\left[\int_{x(n_{i-1})}^{x(n_i)} dt_i\right]^{n_i-n_{i-1}-1}.$$

Performing the indicated integrations we get from the right of (2)

$$Cx(n_1)^{n_1-1} \prod_{i=2}^{k} [x(n_i) - x(n_{i-1})]^{n_i-n_{i-1}-1} [1 - x(n_k)]^{n-n_k}, \tag{3}$$

where C is the multinomial coefficient on the right of (2). It is well known that for the uniform distribution $E[x(n_i)] = \frac{n_i}{n+1}$, or asymptotically $\frac{n_i}{n}$, $i = 1, 2, \ldots, k$. We make the transformation $y_i = \left(x(n_i) - \frac{n_i}{n}\right)\sqrt{n}$, leading to

$$C_1\left(\frac{n_1}{n} + \frac{y_1}{\sqrt{n}}\right)^{n_1-1} \prod_{i=2}^{k} \left(\frac{n_i - n_{i-1}}{n} + \frac{[y_i - y_{i-1}]}{\sqrt{n}}\right)^{n_1-n_{i-1}-1} \cdot\left(\frac{n - n_k}{n} - \frac{y_k}{\sqrt{n}}\right)^{n-n_k}. \tag{4}$$

Using the usual technique of factoring our expressions like

$$\left(\frac{n_i - n_{i-1}}{n}\right)^{n_i-n_{i-1}-1},$$

we rewrite (4) with C_2 as a new constant, and setting $\lambda_i = \frac{n_i}{n}$

$$C_2\left(1 + \frac{y_1}{\lambda_1\sqrt{n}}\right)^{n_1-1} \cdot\prod_{i=2}^{k}\left(1 + \frac{(y_1 - y_{i-1})}{(\lambda_i - \lambda_{i-1})\sqrt{n}}\right)^{n_i-n_{i-1}-1}\left(1 - \frac{y_k}{(1-\lambda_k)\sqrt{n}}\right)^{n-n_k}. \tag{5}$$

Now taking the logarithm of (5), expanding, neglecting terms $O\left(\frac{1}{\sqrt{n}}\right)$ and higher, collecting terms and taking the antilogarithm we get the approximate asymptotic distribution of the order statistics

$$g(x(n_1), x(n_2), \ldots, x(n_k)) = C_3 \exp\left[-\frac{1}{2}\left\{\sum_{i=1}^{k} y_i^2 \frac{\lambda_{i+1} - \lambda_{i-1}}{(\lambda_{i+1} - \lambda_i)(\lambda_i - \lambda_{i-1})} - 2\sum_{i=2}^{k} \frac{y_i y_{i-1}}{\lambda_i - \lambda_{i-1}}\right\}\right], \tag{6}$$

where $\lambda_0 = 0$, $\lambda_{k+1} = 1$. Now setting up the matrix of the coefficients of the quadratic expression in the exponent

$$A_{ii} = \frac{\lambda_{i+1} - \lambda_{i-1}}{(\lambda_{i+1} - \lambda_i)(\lambda_i - \lambda_{i-1})}; \qquad A_{i,i-1} = A_{i-1,i} = -\frac{1}{\lambda_i - \lambda_{i-1}},$$

$i = 1, 2, \ldots, k$; $A_{ij} = 0$, $|i - j| > 1$. To obtain the variances and covariances we need

$$A^{ij} = \frac{\text{cofactor of } A_{ij} \text{ in } \|A_{ij}\|}{\text{determinant } A_{ij}}$$

(see for example Wilks [18, p. 63 et seq.]). Now

$$|A| = \text{determinant } A_{ij} = \prod_{1}^{k+1} \frac{1}{\lambda_i - \lambda_{i-1}}; \tag{7}$$

$$\text{cofactor of } A_{ii} = \lambda_i(1 - \lambda_i)|A|, \, i = 1, 2, \ldots, k.$$

$$\text{cofactor of } A_{ij} = \begin{cases} \lambda_i(1 - \lambda_j)|A|, & i < j \\ \lambda_j(1 - \lambda_i)|A|, & j < i. \end{cases}$$

This completes the proof for the uniform distribution.

If the uniform distribution is transformed into a probability density function $f(x)$ satisfying Condition 2, by an order preserving transformation, we appeal to Lemma 2. We notice that the $x(n_i)$ are transformed into $g[x(n_i)]$, and that the probability that $x(n_i)$ falls in the interval $[u_i, u_i + \Delta u_i]$ is transformed into the probability that $g[x(n_i)]$ falls in the interval $[g(u_i), g(u_i + \Delta u_i)]$. Using the mean value theorem we may write

$$g(u_i + \Delta u_i) = g(u_i) + \Delta u_i g'(u_i'),$$

where u_i' lies in the interval $[u_i, u_i + \Delta u_i]$. However

$$\lim_{\Delta u_i \to 0} g'(u_i') = g'(u_i).$$

The density for the uniform distribution in the interval $[u_i, u_i + \Delta u_i]$ is just Δu_i, and this same density will tend to $f(u_i)\Delta u_i g'(u_i)$. Therefore $g'(u_i) = 1/f(u_i)$, which completes the proof of Theorem 2. □

It would often be useful to know the small sample distribution of the order statistics, particularly in the case where the sample is drawn from a normal. Fisher and Yates' tables [4] give the expected values of the order statistics up to samples of size 50. However it would be very useful in the development of certain small sample statistics to have further information. It is perhaps too much to expect tabulated distribution functions, but at least the variances and covariances would be useful. A joint effort has resulted in the calculation for samples $n = 2, 3, \ldots, 10$ of the expected values to five decimal places, the variances to four decimal places, and the covariances to nearly two decimal places. It is expected that these tables will be published shortly.

3. Estimates of the Mean of a Normal Distribution

It will be important in what follows to define efficiency and to indicate its interpretation. Then we shall construct some estimates of the means of certain distributions and compute their efficiencies. Except for the tables given, the discussion is applicable to the estimation of the mean of any symmetric distribution; and, of course, the concept of efficiency is still more general in its application. A statistic $\hat{\theta}(O_n)$, where O_n is the sample, is said to be an *efficient* estimate of θ if

i) $\sqrt{n}(\hat{\theta} - \theta)$ is asymptotically normally distributed with zero mean and finite variance, $\sigma^2(\hat{\theta})$, and
ii) for any other statistic $\hat{\theta}'$ with $\sqrt{n}(\hat{\theta}' - \theta)$ asymptotically normally distributed with zero mean and variance $\sigma^2(\hat{\theta}')$, $\sigma^2(\hat{\theta}) \leq \sigma^2(\hat{\theta}')$.

The ratio $\sigma^2(\hat{\theta})/\sigma^2(\hat{\theta}')$ is termed the efficiency of $\hat{\theta}'$ if $\hat{\theta}$ is an efficient estimate of θ. For discussion see Wilks [18, 1943]. The concepts of efficient statistic or estimate and of efficiency were introduced by R.A. Fisher. They serve as one measure of the amount of information a statistic draws from a sample. It is also common practice to speak of relative efficiencies, for example, of the statistics $\hat{\theta}'$ and $\hat{\theta}''$ described in ii) above, we say if $\sigma^2(\hat{\theta}') < \sigma^2(\hat{\theta}'')$ that the efficiency of $\hat{\theta}''$ relative to $\hat{\theta}'$ is the ratio of the smaller variance to the larger. This concept of efficiency has sometimes been used when the normality assumption has been violated by one or both statistics, when one or both are biased, and when small samples are considered. When used under these conditions the concept of efficiency becomes more difficult to interpret, although a comparison of the variation of two statistics about the value they are commonly estimating is often of value.

In the case of estimates of the mean a of a variable which is normally distributed according to $N(x, a, \sigma^2)$ from a sample of n, we can often express the variance of an asymptotically unbiased estimate as $\sigma^2(\hat{\theta}_i) = k_i\sigma^2/n$. The sample mean $\hat{\theta} = \Sigma x_i'/n$ is an efficient estimate of a with variance σ^2/n. Then in such cases the efficiency of $\hat{\theta}_i$ in estimating a is $1/k_i$. The interpretation is merely that to obtain the same precision using $\hat{\theta}_i$ as is possible with $\hat{\theta}$, one must use a sample k_i times as large.

Bearing in mind that we are at present searching for economical methods for analyzing large samples, it is clear that the concept of efficiency offers us a practical way of comparing cost of information with cost of obtaining it.

In the present section and in sections 4 and 5 we shall develop certain systematic estimates of parameters of normally distributed variables. Our procedure then will be to compare the efficiency of the systematic estimates with the efficient statistic for estimating the parameter in question, and also in sections 4 and 5 we compare our estimates with a statistic not involving squares or products. Of course the efficient statistic for estimating the mean of a normal is the sample mean, therefore in this section we will only compare our estimates with the sample mean.

We can construct unbiased estimates of the mean of a normal distribution from linear combinations of suitably chosen order statistics. These systematic statistics will be asymptotically normally distributed if the order statistics from which they are derived satisfy Condition 1. We will restrict ourselves to a useful practical case where equal weights are used. In other words the estimate discussed is just the average of k order statistics $k^{-1}\Sigma x(n_i)$. Suppose $x(n_i)$, $i = 1, 2, \ldots, k$ satisfy Condition 1, that $E[x(n_i)] = E[x(n_{k-i+1})]$, so that $E[\Sigma x(n_i)] = a$. An important unsolved question is to discover what spacing of the $x(n_i)$ will yield minimum variance, and thereafter at what rate does the efficiency of this optimumly spaced estimate increase with k. Computational methods bog down rapidly after $k = 3$. Because so little is known about this problem it seems worthwhile to offer some results for three arbitrary spacings (these results are of course useful in analyzing data).

If the $x(n_i)$ satisfy Theorem 2 we may approximate the variance of the systematic statistic $\hat{\theta}_k = \Sigma x(n_i)/k$ by the usual formula

$$\sigma^2(\hat{\theta}_k) = E[\Sigma x(n_i)/k]^2 - [E(\Sigma x(n_i)/k)]^2. \tag{8}$$

We lose no generality by assuming the mean and variance of the underlying normal to be 0 and 1 respectively. Then using the fact that $\Sigma u_i = 0$, and the result of Theorem 1 we rewrite (8) as

$$\sigma^2(\hat{\theta}_k) = E[\Sigma(x(n_i) - u_i)/k]^2 = \frac{1}{k^2 n}\left[\sum_{i=1}^{k} \frac{\lambda_i(1-\lambda_i)}{f_i^2} + 2\sum_{i<j} \frac{\lambda_i(1-\lambda_j)}{f_i f_j}\right], \tag{9}$$

where $f_m = f(u_m)$.

Using the symmetry which makes $\lambda_i = 1 - \lambda_{k-i+1}$, $f_i = f_{k-i+1}$, and the fact that for $k = 2r + 1$, $f_{r+1} = 1/\sqrt{2\pi}$, $\lambda_{r+1} = \frac{1}{2}$, we may simplify the right side of equation (9) with the following results for $k = 1, 2, \ldots, 7$. The factor $1/k^2$ has not been disturbed. We also write the general formulas for the simplified form of (9), but we omit a rather lengthy combinatorial argument which establishes the generalization.

$$\begin{aligned}
k = 1&: \frac{\pi}{2n} \\
k = 2&: \frac{2\lambda_1}{4nf_1^2} \\
k = 3&: \frac{2}{9n}\left[\frac{\lambda_1}{f_1^2} + \frac{\lambda_1\sqrt{2\pi}}{f_1} + \frac{\pi}{4}\right] \\
k = 4&: \frac{2}{16n}\left[\frac{\lambda_1}{f_1^2} + \frac{2\lambda_1}{f_1 f_2} + \frac{\lambda_2}{f_2^2}\right] \\
k = 5&: \frac{2}{25n}\left[\frac{\lambda_1}{f_1^2} + \frac{2\lambda_1}{f_1 f_2} + \frac{\lambda_2}{f_2^2} + \sqrt{2\pi}\left(\frac{\lambda_1}{f_1} + \frac{\lambda_2}{f_2}\right) + \frac{\pi}{4}\right] \\
k = 6&: \frac{2}{36n}\left[\frac{\lambda_1}{f_1^2} + \frac{\lambda_2}{f_2^2} + \frac{\lambda_3}{f_3^2} + \frac{2\lambda_1}{f_1 f_2} + \frac{2\lambda_1}{f_1 f_3} + \frac{2\lambda_2}{f_2 f_3}\right] \\
k = 7&: \frac{2}{49n}\left[\frac{\lambda_1}{f_1^2} + \frac{\lambda_2}{f_2^2} + \frac{\lambda_3}{f_3^2} + \frac{2\lambda_1}{f_1 f_2} + \frac{2\lambda_1}{f_1 f_3} + \frac{2\lambda_2}{f_2 f_3}\right. \\
&\qquad \left. + \sqrt{2\pi}\left(\frac{\lambda_1}{f_1} + \frac{\lambda_2}{f_2} + \frac{\lambda_3}{f_3}\right) + \frac{\pi}{4}\right]
\end{aligned} \tag{10}$$

$$k = 2r: \frac{2}{(2r)^2 n}\left[\sum_{i=1}^{r} \frac{\lambda_i}{f_i^2} + 2\sum_{1 \le i < j \le r} \frac{\lambda_i}{f_i f_j}\right], \qquad r \ge 1$$

$$k = 2r + 1: \frac{2}{(2r+1)^2 n}\left[\frac{(2r)^2 n}{2}\sigma^2(\hat{\theta}_{2r}) + \sqrt{2\pi}\sum_{i=1}^{r} \frac{\lambda_i}{f_i} + \frac{\pi}{4}\right], \qquad r \ge 1.$$

In addition to the possibility of minimizing the equations of (10) by numerical methods, three other procedures suggest themselves: i) to space the order statistics uniformly in probability; ii) to choose those k order statistics whose expected values are equal to the expected values of the order statistics in a sample of size k drawn from a unit normal; iii) to choose $\lambda_i = (i - \frac{1}{2})/k$. The following table lists for $k = 1$, 2, and 3 the expected values u_i of the order statistics and the probability to the left of the expected values λ_i for each of the procedures. The chosen order statistics are counted from left to right. It will be noticed that the third method gives very good results, and has the value of simplicity of formula. The following table gives a comparison be-

Table I. Comparison of the Order Statistics Which Would be Chosen According to Each of the Four Procedures for Subsamples of $k = 1, 2, 3$.

		Optimum		Equal probability		Expected values		$\lambda_i = (i - \frac{1}{2})/k$	
k	Order statistic	u_i	λ_i	u_i	λ_i	u_i	λ_i	u_i	λ_i
1	First	.0000	.5000	.0000	.5000	.0000	.5000	.0000	.5000
2	First	−.6121	.2702	−.4307	.3333	−.5642	.2863	−.6745	.2500
	Second	.6121	.7298	.4307	.6667	.5642	.7137	.6745	.7500
3	First	−.9056	.1826	−.6745	.2500	−.8463	.1967	−.9674	.1667
	Second	.0000	.5000	.0000	.5000	.0000	.5000	.0000	.5000
	Third	.9056	.8174	.6745	.7500	.8463	.8033	.9674	.8333

Table II. Comparison of the Efficiencies of Four Methods of Spacing k Order Statistics Used in the Construction of an Estimate of the Mean.

k	$\lambda_i = i/(k + 1)$	Expected values*	$\lambda_i = (i - \frac{1}{2})/k$	Optimum
1	.637	.637	.637	.637
2	.793	.809	.808	.810
3	.860	.878	.878	.879
4	.896	.914	.913	
5	.918	.933	.934	
6	.933	.948	.948	
7	.944	.956	.957	
8	.952	.963	.963	
9	.957	.968	.969	
10	.962	.972	.973	

* The u_i are equal to the expected values of the order statistics of a sample of size k.

Table III*. $P(x < u_{i|k}) \times 10^4$, $u_{i|k} = E(x_{i|k})$, $x_{i|k}$ is the ith Order Statistic in a Sample of Size k Drawn from a Normal Distribution $N(x, 0, 1)$.

k \ i	1	2	3	4	5	6	7	8	9	10
1	5000									
2	2863	7137								
3	1987	5000	8013							
4	1516	3832	6168	8484						
5	1224	3103	5000	6897	8776					
6	1025	2605	4201	5799	7395	8975				
7	0881	2244	3622	5000	6378	7756	9119			
8	0773	1971	3182	4394	5606	6818	8030	9227		
9	0688	1756	2837	3919	5000	6082	7163	8244	9312	
10	0619	1584	2559	3536	4512	5488	6464	7441	7416	9381

* The table is given to more places than necessary for the prupose suggested because it may be of interest in other applications. The $E(x_{i|k})$ from which the table was derived were computed to five decimal places.

tween the efficiencies resulting from spacing by the three methods. The three optimum cases are included for completeness.

Statisticians planning to use the method of expected values suggested above will find Fisher and Yates [4, 1943] table of the expected values of the order statistics in samples of size k drawn from a unit normal helpful for computing the λ_i. Alternatively the following table of λ_i might be used.

As an example of the use of Table III, suppose we are using the expected value method for estimating the mean of a large sample drawn from a normal distribution $N(x, a, \sigma^2)$. If we are willing to use 6 observations out of 1000 for this purpose Table III indicates the selection of $x_{103}, x_{261}, x_{421}, x_{580}, x_{740}, x_{898}$. Furthermore Table II indicates that the variance of the estimate of a based on the average of these six observations will be approximately $\sigma^2/.948n$, $n = 1000$.

4. Estimates of the Standard Deviation

The statistic

$$s^2 = \sum_{i=1}^{n} (x_i' - \bar{x})^2/(n - 1),$$

where $\bar{x} = \sum_{i=1}^{n} x_i'/n$ is well known to be an unbiased estimate of the population variance σ^2, for $n > 1$. However s is not in general an unbiased estimate of σ. We are not interested here in the question of when we should estimate

σ and when it is more advantageous to estimate σ^2. All we want is to have an unbiased estimate of σ, based on some of squares, to compare with another unbiased estimate based on order statistics. In the case of observations drawn from a normal distribution

$$s' = \frac{(\frac{1}{2}n)^{1/2}\Gamma(\frac{1}{2}[n-1])}{\Gamma(\frac{1}{2}n)}\sqrt{\frac{\Sigma(x_i'-\bar{x})^2}{n}}, \tag{11}$$

is an unbiased estimate of σ (see for example Kenney [11], with variance

$$\sigma^2(s') = \left\{\frac{1}{2}\left[\frac{\Gamma(\frac{1}{2}[n-1])}{\Gamma(\frac{1}{2}n)}\right]^2(n-1)-1\right\}\sigma^2. \tag{12}$$

For most practical purposes however, when $n > 10$, the bias in s is negligible. For large samples $\sigma^2(s')$ approaches $\sigma^2/2n$.

4A. The Range as an Estimate of σ.

As mentioned in the Introduction, section 1, it is now common practice in industry to estimate the standard deviation by means of a multiple of the range $R' = c_n(x_n - x_1)$, for small samples, where $c_n = 1/[E(y_n) - E(y_1)]$, y_n and y_1 being the greatest and least observations drawn from a sample of size n from a normal distribution $N(y, a, 1)$. Although we are principally interested in large sample statistics, for the sake of completeness, we shall include a few remarks about the use of the range in small samples.

Now R' is an unbiased estimate of σ, and its variance may be computed for small samples, see for example Hartley [7, 1942]. In the present case, although both R' and s' are unbiased estimates of σ, they are not normally distributed, nor are we considering their asymptotic properties; therefore the previously defined concept of efficiency does not apply. We may however use the ratio of the variances as an arbitrary measure of the relative precision of the two statistics. The following table lists the ratio of the variances of the two statistics, as well as the variances themselves expressed as a multiple of the population variance for samples of size $n = 2, 3, \ldots, 10$.

4B. Quasi Ranges for Estimating σ.

The fact that the ratio $\sigma^2(s')/\sigma^2(R')$ falls off in Table IV as n increases makes it reasonable to inquire whether it might not be worthwhile to change the systematic estimate slightly by using the statistic $c_{1|n}[x_{n-1} - x_2]$, or more generally $c_{r|n}[x_{n-r} - x_{r+1}]$ where $c_{r|n}$ is the multiplicative constant which makes the expression an unbiased estimate of σ (in particular $c_{r|n}$ is the constant to be used when we count in $r + 1$ observations from each end of a sample of size n, thus $c_{r|n} = 1/[E(y_{n-r} - y_{r+1})]$ where the y's are drawn from

Table IV. Relative Precision of s' and R', and Their Variances Expressed as a Multiple of σ^2, the Population Variance.

n	$\sigma^2(s')/\sigma^2(R')$	$\sigma^2(s')/\sigma^2$	$\sigma^2(R')/\sigma^2$
2	1.000	.570	.570
3	.990	.273	.276
4	.977	.178	.182
5	.962	.132	.137
6	.932	.104	.112
7	.910	.0864	.0949
8	.889	.0738	.0830
9	.869	.0643	.0740
10	.851	.0570	.0670

$N(y, a, 1)$). This is certainly the case for large values of n, but with the aid of the unpublished tables mentioned at the close of section 2, we can say that it seems not to be advantageous to use $c_{1|n}[x_{n-1} - x_2]$ for $n \leq 10$. Indeed the variance $c_{1|10}[x_9 - x_2]$, for the unit normal seems to be about .10, as compared with $\sigma^2(R')/\sigma^2 = .067$ as given by Table IV, for $n = 10$. The uncertainty in the above statements is due to a question of significant figures.

Considerations which suggest constructing a statistic based on the difference of two order statistics which are not extreme values in small samples, weigh even more heavily in large samples. A reasonable estimate of σ for normal distributions, which could be calculated rapidly by means of punched-card equipment is

$$\hat{\sigma} = \frac{1}{c}[x(n_2) - x(n_1)], \tag{13}$$

where the $x(n_i)$ satisfy Condition 1, and where $c = u_2 - u_1$, u_2 and u_1 are the expected values of the n_2 and n_1 order statistics of a sample of size n drawn from a unit normal. Without loss of generality we shall assume the x_i are drawn from a unit normal. Furthermore we let $\dfrac{n_2}{n} = \lambda_2 = 1 - \lambda_1 = 1 - \dfrac{n_1}{n}$. Of course σ will be asymptotically normally distributed, with variance

$$\sigma^2(\hat{\sigma}) = \frac{2}{nc^2}\left[\frac{\lambda_1(1-\lambda_1)}{[f(u_1)]^2} + \frac{\lambda_2(1-\lambda^2)}{[f(u_2)]^2} - \frac{2\lambda_1(1-\lambda_2)}{f(u_1)f(u_2)}\right]. \tag{14}$$

Because of symmetry $f(u_1) = f(u_2)$; using this and the fact that $\lambda_1 = 1 - \lambda_2$, we can reduce (14) to

$$\sigma^2(\hat{\sigma}) = \frac{2}{nc^2}\frac{\lambda_1(1-2\lambda_1)}{[f(u_1)]^2}. \tag{15}$$

We are interested in optimum spacing in the minimum variance sense. The minimum for $\sigma^2(\hat{\sigma})$ occurs when $\lambda_1 \doteq .0694$, and for that value of λ_1, $\sigma^2(\hat{\sigma}) \doteq .767\ \sigma^2/n$. Asymptotically s' is also normally distributed, with $\sigma^2(s') = \sigma^2/2n$. Therefore we may speak of the efficiency of $\hat{\sigma}$ as an estimate of σ as .652. It is useful to know that the graph of $\sigma^2(\hat{\sigma})$ is very flat in the neighborhood of the minimum, and therefore varying λ_1 by .01 or .02 will make little difference in the efficiency of the estimate $\hat{\sigma}$ (providing of course that c is appropriately adjusted). K. Pearson [13] suggested this estimate in 1920. It is amazing that with punched-card equipment available it is practically never used when the appropriate conditions described in the Introduction are present.

The occasionally used semi-interquartile range, defined by $\lambda_1 = .25$ has an efficiency of only .37 and an efficiency relative to $\hat{\sigma}$ of only .56.

As in the case of the estimate of the mean by systematic statistics, it is pertinent to inquire what advantage may be gained by using more order statistics in the construction of the estimate σ. If we construct an estimate based on four order statistics, and then minimize the variance, it is clear that the extreme pair of observations will be pushed still further out into the tails of the distribution. This is unsatisfactory from two points of view in practice: i) we will not actually have an infinite number of observations, therefore the approximation concerning the normality of the order statistics may not be adequate if λ_1 is too small, even in the presence of truly normal data; ii) the distribution functions met in practice often do not satisfy the required assumption of normality, although over the central portion of the function containing most of the probability, say except for the 5% in each tail normality may be a good approximation. In view of these two points it seems preferable to change the question slightly and ask what advantage will accrue from holding two observations at the optimum values just discussed (say $\lambda_1 = .07, \lambda_2 = .93$) and introducing two additional observations more centrally located.

We define a new statistic

$$\hat{\sigma}' = \frac{1}{c'}[x(n_4) + x(n_3) - x(n_2) - x(n_1)], \tag{16}$$

$c' = E[x(n_4) + x(n_3) - x(n_2) - x(n_1)]$, where the observations are drawn from a unit normal. We take $\lambda_1 = 1 - \lambda_4, \lambda_2 = 1 - \lambda_3, \lambda_1 = .07$. It turns out that $\sigma^2(\hat{\sigma}')$ is minimized for λ_2 in the neighborhood of .20, and that the efficiency compared with s' is a little more than .75. Thus an increase of two observations in the construction of our estimate of σ increases the efficiency from .65 to .75. We get practically the same result for $.16 \leq \lambda_2 \leq .22$.

Furthermore, it turns out that using $\lambda_1 = .02, \lambda_2 = .08, \lambda_3 = .15, \lambda_4 = .25, \lambda_5 = .75, \lambda_6 = .85, \lambda_7 = .92, \lambda_8 = .98$, one can get an estimate of σ based on eight order statistics which has an efficiency of .896. This estimate is more efficient than either the mean deviation about the mean or median for estimating σ. The estimate is of course

$$\hat{\sigma}'' = [x(n_8) + x(n_7) + x(n_6) + x(n_5) - x(n_4) - x(n_3) - x(n_2) - x(n_1)]/C,$$

where $C = 10.34$.

To summarize: in estimating the standard deviation σ of a normal distribution from a large sample of size n, an unbiased estimate of σ is

$$\hat{\sigma} = \frac{1}{c}(x_{n-r+1} - x_r),$$

where $c = E(y_{n-r+1} - y_r)$ where the y's are drawn from $N(y, a, 1)$. The estimate $\hat{\sigma}$ is asymptotically normally distributed with variance

$$\sigma^2(\hat{\sigma}) = \frac{2}{nc^2}\frac{\lambda_1(1-2\lambda_1)}{[f(u_1)]^2},$$

where $\lambda_1 = r/n$, $f(u_1) = N(E(x_r), 0, \sigma^2)$. We minimize $\sigma^2(\hat{\sigma})$ for large samples when $\lambda_1 \doteq .0694$, and for that value of λ_1,

$$\sigma^2_{\text{opt}}(\hat{\sigma}) \doteq \frac{.767\sigma^2}{n}.$$

The unbiased estimate of σ

$$\hat{\sigma}' = \frac{1}{c'}(x_{n-r+1} + x_{n-s+1} - x_s - x_r)$$

may be used in lieu of $\hat{\sigma}$. If $\lambda_1 = r/n$, $\lambda_2 = s/n$ we find

$$\sigma^2(\hat{\sigma}'|\lambda_1 = .07, \lambda_2 = .20) \doteq \frac{.66\sigma^2}{n}.$$

4C. The Mean Deviations About the Mean and Median

The next level of computational difficulty we might consider for the construction of an estimate of σ is the process of addition. The mean deviation about the mean is a well known, but not often used statistic. It is defined by

$$\text{m.d.} = \sum_{i=1}^{n} |x_i' - \bar{x}|/n. \tag{17}$$

For large samples from a normal distribution the expected value of m.d. is $\sqrt{\frac{2}{\pi}}\sigma$, therefore to obtain an unbiased estimate of σ we define the new statistic $A = \sqrt{\frac{\pi}{2}}\text{m.d.}$ Now for large samples A has variance $\sigma^2[\frac{1}{2}(\pi - 2)]/n$, or an efficiency of .884. However there are slight awkwardnesses in the computation of A which the mean deviation about the median does not have.

It turns out that for samples of size $n = 2m + 1$ drawn from a normal distribution $N(y, a, 0)$ the statistic

$$M' = \sqrt{\frac{\pi}{2}}\frac{\sum|x_i - x_{m+1}|}{2m} \tag{18}$$

asymptotically has mean σ and variance

$$\sigma^2(M') = \frac{1}{2m}\left(\frac{\pi - 2}{2}\right)\sigma^2. \tag{19}$$

Thus in estimating the standard deviation of a normal distribution from large samples we can get an efficiency of .65 by the judicious selection of two observations from the sample, an efficiency of .75 by using four observations, and an efficiency of .88 by using the mean deviation of all the observations from either the mean or the median of the sample, and an efficiency of .90 by using eight order statistics.

(*Editor's note*: Section 5, Estimation of the Correlation Coefficient, has been omitted.)

6. Acknowledgements

The author wishes to acknowledge the valuable help received from S.S. Wilks, under whose direction this work was done, and the many suggestions and constructive criticisms of J.W. Tukey. The author also wishes to acknowledge the debt to his wife, Virginia Mosteller, who prepared the manuscript and assisted in the preparation of the tables and figures.

References

[1] G.W. Brown and J.W. Tukey, "Some distributions of sample means," *Annals of Math. Stat.*, Vol. 17 (1946), p. 1.
[2] J.H. Curtiss, "A note on the theory of moment generating functions," *Annals of Math. Stat.*, Vol. 13 (1942), p. 430.
[3] R.A. Fisher and L.H.C. Tippett, "Limiting forms of the frequency distribution of the largest or smallest member of a sample," *Proc. Camb. Phil. Soc.*, Vol. 24 (1928), p. 180.
[4] R.A. Fisher and F. Yates, *Statistical Tables*, Oliver and Boyd, London, 1943.
[5] E.J. Gumbel, "Les valeurs extremes des distribution statistiques," *Annales de l'Institute Henri Poincare*, Vol. 5 (1934), p. 115.
[6] E.J. Gumbel, "Statische Theorie der grössten Werte," *Zeitschrift für schweizerische Statistik und Vokswirtschaft*," Vol. 75, part 2 (1939), p. 250.
[7] H.O. Hartley, "The probability integral of the range in samples of *n* observations from a normal population," *Biometrika*, Vol. 32 (1942), p. 301.
[8] D. Jackson, "Note on the median of a set of numbers," *Bull. Amer. Math. Soc.*, Vol. 27 (1921), p. 160.
[9] T. Kelley, "The selection of upper and lower groups for the validation of test items," *Jour. Educ. Psych.*, Vol. 30 (1939), p. 17.
[10] M.G. Kendall, *The Advanced Theory of Statistics*, J.B. Lippincott Co., 1943.
[11] J.F. Kenney, *Mathematics of Statistics*, Part II, D. Van Nostrand Co., Inc., 1939, Chap. VII.
[12] K.R. Nair and M.P. Shrivastava, "On a simple method of curve fitting," *Sankhya*, Vol. 6, part 2 (1942), p. 121.

[13] K. Pearson, "On the probable errors of frequency constants, Part III," *Biometrika*, Vol. 13 (1920), p. 113.
[14] K. Pearson (Editor), *Tables for Statisticians and Biometricians*, Part II, 1931, p. 78, Table VIII.
[15] N. Smirnoff, "Über die Verteilung des allgemeinen Gliedes in der Variationsreihe," *Metron*, Vol. 12 (1935), p. 59.
[16] N. Smirnoff, "Sur la dependance des membres d'un series de variations," *Bull. Univ. État Moscou, Series Int., Sect. A., Math. et Mécan.*, Vol. 1, fasc. 4, p. 1.
[17] L.H.C. Tippett, "On the extreme individuals and the range of samples taken from normal population," *Biometrika*, Vol. 17 (1925), p. 364.
[18] S.S. Wilks, *Mathematical Statistics*, Princeton University Press, Princeton, N.J., 1943.

Introduction to Durbin and Watson (1950, 1951) Testing for Serial Correlation in Least Squares Regression. I, II

Maxwell L. King
Monash University

Background

Fitting a linear function of some variables (denoted by $x_1, \ldots, x_k$) to a variable denoted y by least squares is an old statistical technique. In the Fisherian revolution of the 1920s and 1930s, statistical methods were mainly applied in the natural sciences where one could often design the experiments that produced the data. The emphasis was largely on how the experiments should be designed to make the least-squares assumptions valid rather than on checking the correctness of these assumptions.

When economists began to apply least-squares methods to economic data, it slowly became evident that the non experimental nature of the data added formidable complications that required new solutions [see, for example, Frisch (1934) and Koopmans (1937)]. Collectively, these solutions distinguish econometrics from its parent discipline of statistics.

The first major center for econometric research was the Cowles Commission, which was founded in 1932, moved to the University of Chicago in 1939, and then to Yale University in 1955. A second center was set up at Cambridge University after World War II when J.M. Keynes successfully argued for the creation of a department of applied economics. Richard Stone, who had worked with Keynes on creating a system of national accounts for the British economy, was appointed director. Both groups were a mixture of economists and statisticians. Around 1950, the statisticians at Chicago included T.W. Anderson, H. Chernoff, and H. Rubin, with T.C. Koopmans as director, while J. Durbin and G.S. Watson were at Cambridge. Both Koopmans and Stone were subsequently awarded Nobel Prizes in economics.

Three major difficulties with regression analysis of economic time series data are the following: the linear regression being fitted is often one of many simultaneous relationships between the variables; the presence of measurement error; and the presence of nonindependent error terms.

The first two were the main preoccupations of the Chicago group. Two members of the Cambridge group, Cochrane and Orcutt (1949) [also see Orcutt and Cochrane (1949)], attacked the problem of possible dependence among the error terms. They used the Monte Carlo method to show that the von Neumann (1941) ratio test for serial correlation applied to least-squares regression residuals is biased toward accepting independence and that this bias increases with the number of regressors. They argued that it would be adequate to assume that the errors followed a low-order autoregressive process and showed how the parameters could be estimated from least-squares residuals.

The von Neumann tables were designed to test for lag-1 serial correlation in the model $y_i = \mu + \varepsilon_i$. Least-squares regression residuals computed from the model $y_i = \beta_1 x_{i1} + \cdots + \beta_k x_{ik} + \varepsilon_i$ must be orthogonal to the k regression vectors that all probably show serial dependence. Thus, the residuals will tend to have less serial correlation than the errors ε_i. In a fundamental paper, T.W. Anderson (1948) first studied the proper design of significance tests for serial correlation. R.L. Anderson and T.W. Anderson (1950) treated the case when short Fourier series are fitted. Here, using a circular model, they were able to provide simple tables because their regression vectors were eigenvectors of a circulant matrix. This use of circulants was a suggestion of H. Hotelling to R.L. Anderson. However, these tables are not appropriate for practical time series since these data do not behave circularly. That the problem simplifies when the regression vectors are certain eigenvectors was the discovery of T.W. Anderson (1948).

This is the background to the Durbin–Watson papers, which give the first treatment of the serial correlation testing problem for noncircular statistics and arbitrary regressors. Nowadays diagnostic tests for regression are commonplace, although they are usually informal and graphic rather than formal significance tests. With current computing power, one can examine the dependence question in both the time and frequency domains. In 1950, only mechanical machines were available, so that only computationally simple statistical procedures had any hope of being used.

The Authors

James Durbin graduated from Cambridge University with a B.A. during World War II and served the remainder of the war years (1943–45) in the Army Operations Research Group. He returned to Cambridge after the war to complete a one-year postgraduate diploma in mathematical statistics. Richard Stone, his diploma supervisor in economics, offered him a job in the new department of applied economics. There he met and collaborated with

G.S. Watson. Their time at Cambridge overlapped for about six months before Durbin took up a lecturing position at the London School of Economics. Most of the work on the Durbin–Watson papers was done while Durbin was in London and Watson was at Cambridge. Durbin remained at the London School of Economics until his retirement in 1988, having become professor of statistics in 1961. From the interview of Durbin by Phillips (1988), the reader may learn about his subsequent research interests.

Geoffrey S. Watson graduated from the University of Melbourne and, after World War II, went to North Carolina State College as a Ph.D. student. There R.L. Anderson, who had computed the distribution of the circular serial correlation in (1942) and just finished a paper (see above) with T.W. Anderson on this topic, suggested to Watson that he should take as a thesis topic the study of serial correlation testing. John Wishart was then visiting North Carolina and arranged for Watson to go to Cambridge in 1949 to take a two-year appointment at the department of applied economics. There he met and collaborated with Durbin. Currently, he is professor of statistics at Princeton University. In an autobiographical article (1986), Watson describes his career and varied interests.

Paper I

In Sec. 1 and 2, the least-squares fitting of n values of y to a linear function of k regression variables was described in the notation that has now become standard

$$\mathbf{y} = \mathbf{X}\boldsymbol{\beta} + \boldsymbol{\varepsilon},$$

where $\mathbf{y}$ is $n \times 1$, $\mathbf{X}$ $n \times k$, $\boldsymbol{\beta}$ $k \times 1$, and $\boldsymbol{\varepsilon}$ $n \times 1$. Here $\mathbf{X}$ is known and fixed, and for simplicity, of full rank k. Durbin and Watson's use of the now familiar symmetric idempotent $\mathbf{M} = \mathbf{I} - \mathbf{X}(\mathbf{X}'\mathbf{X})^{-1}\,\mathbf{X}'$ and the trace operator came from Aitken (1935). Thus, the least-squares residual vector is $\mathbf{z} = \mathbf{My} = \mathbf{M}\boldsymbol{\varepsilon}$. The problem is to test the null hypothesis that the elements of $\boldsymbol{\varepsilon}$ are i. i. d. $N(0, \sigma^2)$ vs. the alternative that they are serially correlated.

T.W. Anderson (1948) had shown that, if an error vector $\boldsymbol{\varepsilon}$ had a multivariate normal density with exponent

$$-\frac{1}{2\sigma^2}((1 + \rho^2)\boldsymbol{\varepsilon}'\boldsymbol{\varepsilon} - 2\rho\boldsymbol{\varepsilon}'\mathbf{A}\boldsymbol{\varepsilon}), \tag{1}$$

then the statistic to test $\rho = 0$ was $r = \mathbf{z}'\,\mathbf{Az}/\mathbf{z}'\mathbf{z}$. The testing problem becomes one of finding the distribution of

$$r = \frac{\mathbf{z}'\mathbf{Az}}{\mathbf{z}'\mathbf{z}} = \frac{\boldsymbol{\varepsilon}'\mathbf{MAM}\boldsymbol{\varepsilon}}{\boldsymbol{\varepsilon}'\mathbf{M}\boldsymbol{\varepsilon}},$$

where $\rho = 0$ in (1), i.e., the elements of $\boldsymbol{\varepsilon}$ are i. i. d. $N(0, \sigma^2)$. An orthogonal transformation reduces r to its canonical form, with the ζ_i i. i. d. $N(0, 1)$,

$$r = \frac{\sum_1^{n-k} v_i \zeta_i^2}{\sum_1^{n-k} \zeta_i^2}. \tag{2}$$

Since **M** and **MAM** commute, **M** has tr**M** $= n - k$ unit eigenvalues and the eigenvalues v_i (in ascending order) of **MAM** are those of **MA.** Thus, the v_i's depend on **X** and will vary from problem to problem, so tabulation is impossible.

In fact, in 1950 and 1951, the computation of the v_i, and the subsequent step of finding the significance points of (2), were both impossible. It was feasible, as was shown in Sec. 3, to approximate the distribution of r by moments, since these depend upon

$$\sum_{i=1}^{n-k} v_i^q = tr(\mathbf{MA})^q. \tag{3}$$

However, the spectral decomposition of suitable matrices **A** was known; let $\mathbf{A} = \sum_1^n \lambda_i \mathbf{a}_i \mathbf{a}_i'$, $\lambda_1 \leq \cdots \leq \lambda_n$. By the end of Sec. 2, bounds for the v_i in terms of the λ_i's were found by a direct argument. In fact, these inequalities follow from the Courant–Fisher inequalities [Courant and Hilbert (1953, p. 33)]. If none of the columns of **X** are eigenvectors, one has

$$\lambda_i \leq v_i \leq \lambda_{i+k}, \qquad i = 1, \ldots, n - k,$$

implying that

$$r_L = \frac{\sum_1^{n-k} \lambda_i \zeta_i^2}{\sum_1^{n-k} \zeta_i^2} \leq \frac{\sum_1^{n-k} v_i \zeta_i^2}{\sum_1^{n-k} \zeta_i^2} \leq \frac{\sum_1^{n-k} \lambda_{i+k} \zeta_i^2}{\sum_1^{n-k} \zeta_i^2} = r_U. \tag{4}$$

Notice that if the columns of **X** are linear combinations of $\mathbf{a}_{n-k+1}, \ldots, \mathbf{a}_n$, then r becomes r_U. Similarly, a regression on $\mathbf{a}_1, \ldots, \mathbf{a}_k$ leads to r_L. The latter is the situation in which r is a circular correlation coefficient and the regression is on Fourier vectors. If column one of X is $\mathbf{a}_1$, then λ_1 must be deleted and the remaining λ's renumbered $\lambda_1, \ldots, \lambda_{n-1}$.

In Sec. 3 the distributions of r, r_L and r_U are studied. Various results are given, but since only moment approximations were then feasible, most of Sec. 3 concerns computing the moments of r from (2).

Now a practical choice must be made for the statistic. In Sec. 4, reasons were given for choosing effectively the von Neumann ratio rather than the lag-1 serial correlation coefficient when the errors were generated by a stationary first-order autoregressive process. Thus, all subsequent numerical work was done with

$$d = \frac{\mathbf{z}'\mathbf{A}\mathbf{z}}{\mathbf{z}'\mathbf{z}} = \frac{\sum_2^n (z_i - z_{i-1})^2}{\sum_1^n z_i^2}, \tag{5}$$

where

$$\mathbf{A} = \begin{bmatrix} 1 & -1 & 0 & \cdots & & 0 \\ -1 & 2 & -1 & & & 0 \\ 0 & -1 & 2 & & & 0 \\ \vdots & \vdots & & \ddots & & \vdots \\ 0 & 0 & & \cdots & 2 & -1 \\ 0 & 0 & & \cdots & -1 & 1 \end{bmatrix}. \tag{6}$$

The common-sense reason for choosing the numerator of (5) is that it will be small if each z_i is like its predecessor z_{i-1}—and this is what should happen if the residuals are reflecting a strong positive serial correlation in the errors ε_i. Conversely, large d suggests negative correlation. The choice of the denominator of (5) implies also that the regression includes a constant term. This is assumed in the tables of significance points for d_U and d_L, where $k' = k - 1 =$ number of regression variables, excluding the constant term.

The eigenvalues and eigenvectors of **A** are, respectively, for $j = 1, 2, \ldots, n$,

$$\lambda_j = 2\left\{1 - \cos\frac{\pi(j-1)}{n}\right\} = 4\sin^2\frac{\pi(j-1)}{2n} \tag{7}$$

and

$$n^{-1/2}[1, 1, \ldots, 1]$$

$$(2/n)^{1/2}\left[\cos\frac{\pi(j-1)}{2n}, \cos\frac{3\pi(j-1)}{2n}, \ldots, \cos\frac{(2n-1)\pi(j-1)}{2n}\right]. \tag{8}$$

Thus, when $j = 1$, $\lambda_1 = 0$ and the first eigenvector corresponds to fitting a constant term. Hence, from the discussion under (4), $\lambda_1 = 0$ is deleted when making the inequalities. Further notice that the values of the elements of the eigenvectors oscillate about zero faster as j and so λ_j increases. Finally, notice from (7) that the λ_i lie in (0, 4), so the approximate beta distribution of $d/4$ on (0, 1) becomes plausible.

Paper II

The second Durbin–Watson paper gives tables of the significance points of d_L and d_U. From (5), it is obvious that too small a value of d suggests positive serial correlation and too large a value suggests negative correlation. The tables cover $n = 15(1)40(5)100$, $k' =$ number of variables, not counting the mean, $= k - 1 = 1, 2, 3, 4, 5$ and significance levels 5, 2.5, and 1%. The gap between d_L and d_U becomes larger as n decreases and k increases. Suppose the alternative is positive serial correlation. If $d < d_L$, the result is significant. In the worked example of Sec. 2, the case $d < d_L$ is seen. Should the observed value of d lie between d_L and d_U, the bounds test is inconclusive. In this case, since $0 \leq d \leq 4$, it is suggested that the distribution of $d/4$ has approximately the density

$$B(p, q)^{-1}\left(\frac{d}{4}\right)^{p-1}\left(1-\frac{d}{4}\right)^{q-1},$$

where p and q are chosen so that $d/4$ has the right mean and variance. This procedure is illustrated in Sec. 4 on the data used in Sec. 2. If $d > d_U$, then d is not significant.

These significance points and approximations were shown by later authors to be surprizingly accurate. Hannan (1957) pointed out that the exact significance point of d will often be very near d_U. To see this, imagine the common case when all the regression variables are slowly varying. (His proof deals with the case when a polynomial is fitted.) Then these vectors will resemble the eigenvectors for smaller eigenvalues, see (8) and below it. Had the eigenvectors for the k least eigenvalues been fitted, $d \equiv d_U$.

In Sec. 5, formulae for the mean and variance of d are derived for the special cases of one- and two-way classifications and illustrated by data in R.L. Anderson and T.W. Anderson (1950). Section 6 gives $E(d)$ and var (d) for polynomial regression. The paper concludes with a short section which points out, e.g., that if $n = 2m$ and $(z_{m+1} - z_m)^2$ is dropped from the sum in the numerator of d, then the resulting quadratic form has paired roots, so that the method of Anderson (1942) could be used to obtain exact significance points.

Impact

This work had a profound effect on both the methodological and applied aspects of Econometrics. For at least 20 years after Stone (1954, Chap. 19) first drew the Durbin–Watson test to the attention of economists, it almost single-handedly played the role of a misspecification test in the avalanche of regression modeling that followed the introduction and increased availability of computers in econometrics. Regression computer programs were regarded as incomplete if they did not calculate d. To the enlightened econometrician, a significant Durbin-Watson statistic indicated the need to completely rethink the model being estimated because of its ability to detect function misspecification, omitted regressors, and incorrectly specified dynamics among other things. It continues to be seen in every journal on empirical problems.

Largely because of this important role, the Durbin-Watson test has sparked an enormous and varied methodological literature. For example, a recent review by King (1987) lists 245 references. Because the bounds test often produced an inconclusive result and users found it difficult to calculate Durbin and Watson's suggested approximate critical value, researchers explored other test procedures that gave exact results. A common suggestion was to calculate residuals whose distribution is independent of **X.** Others proposed nonparametric alternatives. In both cases, the suggested procedures were found to be less powerful than the Durbin-Watson test. In fact, the problem of testing for autocorrelation in the context of the linear model must surely be the most intensely researched testing problem in econometrics. Durbin (1957) considered the use of d in the simultaneous equation context. The literature is particularly rich in empirical power comparisons under differing circumstances.

Improvements in computer hardware and algorithms now make it possible to calculate the distribution function of d, essentially by doing contour integration numerically. Durbin and Watson (1971) reviewed these developments in a third paper. Some current computer packages, such as SHAZAM [see White (1978)], allow the user to calculate exact p values. Durbin and Watson also showed their test to be (approximately) locally best invariant against first-order autoregressive disturbances. Others have found the test also has this property against a number of other processes that cause first-order autocorrelation. There has also been interest in the behavior of the test when the model contains lagged values of y. Durbin (1970) found the large-sample asymptotic distribution of d, while more recently Inder (1986) and King and Wu (1991), working with the small-disturbance distribution of d, have provided justification for the use of the original tables of bounds.

The work was also influential in that it hastened the use of matrix algebra in the treatment of the linear model. Matrix algebra was being used increasingly [see, for example, Koopmans (1937) and T.W. Anderson (1948)] at that time, but the Durbin–Watson papers established the matrix notation that is now standard. Stone (1954, Chap. 19) was a strong advocate of its use, and when textbooks such as Johnston (1963) and Goldberger (1964) used it, it became standard notation.

References

Aitken, A.C. (1935) On least squares and the linear combination of observations, *Proc. Roy. Soc. Edinburgh* **55**, 42–48.

Anderson, R.L. (1942). Distribution of the serial correlation coefficient, *Ann. Math. Statist* **13**, 1–13.

Anderson, R.L., and Anderson, T.W. (1950). Distribution of the circular serial correlation coefficient for residuals from a fitted Fourier series, *Ann. Math. Statist.* **21**, 59–81.

Anderson, T.W. (1948). On the theory of testing serial correlation, *Skand. Aktuarietidskrift*, **31**, 88–116.

Cochrane, D., and Orcutt, G.H. (1949). Application of least squares regression to relationships containing auto-correlated error terms," *J. Amer. Statist Assoc.* **44**, 32–61.

Courant, R., and Hilbert, D. (1953). *Mathematical Methods of Physics*, Vol. I. Wiley-Interscience, New York.

Durbin, J. (1957). Testing for serial correlation in systems of simultaneous regression equations, *Biometrika*, **44**, 370–377.

Durbin, J. (1970). Testing for serial correlation in least squares regression when some of the regressors are lagged dependent variables, *Econometrica*, **38**, 410–421.

Durbin, J., and Watson, G.S. (1971). Testing for serial correlation in least squares regression, III, *Biometrika*, **58**, 1–19.

Frisch, R. (1934). *Statistical Confluence Analysis by Means of Complete Regression Systems.* University Oekonomiske Institute, Oslo.

Goldberger, A.S. (1964). *Econometric Theory.* Wiley, New York.

Hannan, E.J. (1957). Testing for serial correlation in least squares regression, *Biometrika*, **44**, 57–66.

Inder, B.A. (1986). An approximation to the null distribution of the Durbin–Watson statistic in models containing lagged dependent variables, *Econometric Theory*, **2**, 413–428.

Johnston, J. (1963). *Econometric Methods.* McGraw-Hill, New York.

King, M. L. (1987). Testing for autocorrelation in linear regression models: A survey, in *Specification Analysis in the Linear Model* (M.L. King, and D.E.A. Giles, eds.). Routledge and Kegan Paul, London, pp. 19–73.

King, M.L., and Wu, P.X. (1991). Small-disturbance asymptotics and the Durbin-Watson and related tests in the dynamic regression model *J. Econometrics* **47**, 145–152.

Koopmans, T.C. (1937). *Linear Regression Analysis of Economic Time Series.* Erven F. Bohm, Haarlem.

Phillips, P.C.B. (1988). The ET interview: Professor James Durbin, *Econometric Theory*, **4**, 125–157.

Stone, R. (1954). *The Measurement of Consumers' Expenditure and Behaviour in the United Kingdom*, 1920–1938, Vol. I, Cambridge University Press, Cambridge.

von Neumann, J. (1941). Distribution of the ratio of the mean square successive difference to the variance, *Ann. Math. Statist.* **12**, 367–395.

Watson, G.S. (1986). A boy from the bush, in *The Craft of Probabilistic Modelling* (J. Gani, ed.). Springer-Verlag, New York, pp. 43-60.

White, K.J. (1978). A general computer program for econometric methods-SHAZAM, *Econometrica*, **46**, 239–240.

Testing for Serial Correlation in Least Squares Regression. I

J. Durbin
London School of Economics

G.S. Watson
Department of Applied Economics,
University of Cambridge

A great deal of use has undoubtedly been made of least squares regression methods in circumstances in which they are known to be inapplicable. In particular, they have often been employed for the analysis of time series and similar data in which successive observations are serially correlated. The resulting complications are well known and have recently been studied from the standpoint of the econometrician by Cochrane & Orcutt (1949). A basic assumption underlying the application of the least squares method is that the error terms in the regression model are independent. When this assumption—among others—is satisfied the procedure is valid whether or not the observations themselves are serially correlated. The problem of testing the errors for independence forms the subject of this paper and its successor. The present paper deals mainly with the theory on which the test is based, while the second paper describes the test procedures in detail and gives tables of bounds to the significance points of the test criterion adopted. We shall not be concerned in either paper with the question of what should be done if the test gives an unfavourable result.

Since the errors in any practical case will be unknown the test must be based on the residuals from the calculated regression. Consequently the ordinary tests of independence cannot be used as they stand, since the residuals are necessarily correlated whether the errors are dependent or not. The mean and variance of an appropriate test statistic have been calculated by Moran (1950) for the case of regression on a single independent variable. The problem of constructing an exact test has been completely solved only in one special case. R.L. & T.W. Anderson (1950) have shown that for the case of regression on a short Fourier series the distribution of the circular serial correlation coefficient obtained by R.L. Anderson (1942) can be used to obtain exact significance points for the test criterion concerned. This is due to the coincidence of the regression vectors with the latent vectors of the circular serial covariance matrix. Perversely enough, this is the very case in which the test is least needed, since the least squares regression coefficients are best unbiased estimates even in the non-null case, and in addition estimates of their variance can be obtained which are at least asymptotically unbiased.

The latent vector case is in fact the only one for which an elegant solution can be obtained. It does not seem possible to find exact significance points for any other case. Nevertheless, bounds to the significance points can be obtained, and in the second paper such bounds will be tabulated. The bounds we shall give are 'best' in two senses: first they can be attained (with regression vectors of a type that will be discussed later), and secondly, when they are attained the test criterion adopted is uniformly most powerful against suitable alternative hypotheses. It is hoped that these bounds will settle the question of significance one way or the other in many cases arising in practice. For doubtful cases there does not seem to be any completely satisfactory procedure. We shall, however, indicate some approximate methods which may be useful in certain circumstances.

The bounds are applicable to all cases in which the independent variables in the regression model can be regarded as 'fixed'. They do not therefore apply to autoregressive schemes and similar models in which lagged values of the dependent variable occur as independent variables.

A further slight limitation of the tables in the form in which we shall present them is that they apply directly only to regressions in which a constant term or mean has been fitted. They cannot therefore be used as they stand for testing the residuals from a regression through the origin. In order to carry out the test in such a case it will be necessary to calculate a regression which includes a fitted mean. Once the test has been carried out the mean can be eliminated by the usual methods for eliminating an independent variable from a regression equation (e.g. Fisher, 1946).

Introduction to Theoretical Treatment

Any single-equation regression model can be written in the form

$$y = \beta_1 x_1 + \beta_2 x_2 + \cdots + \beta_k x_k + \varepsilon$$

in which y, the dependent variable, and x, the independent variable, are observed, the errors ε being unobserved. We usually require to estimate β_1, $\beta_2, \ldots, \beta_k$ and to make confidence statements about the estimates given only the sample

$$\begin{array}{ccccc} y_1 & x_{11} & x_{21} & \cdots & x_{k1} \\ y_2 & x_{12} & x_{22} & \cdots & x_{k2} \\ \vdots & \vdots & \vdots & & \vdots \\ y_n & x_{1n} & x_{2n} & \cdots & x_{kn} \end{array}$$

Estimates can be made by assuming the errors $\varepsilon_1, \varepsilon_2, \ldots, \varepsilon_n$ associated with the sample to be random variables distributed with zero expectations independently of the x's. If the estimates we make are maximum likelihood estimates, and if our confidence statements are based on likelihood ratios, we can regard the x's as fixed in repeated sampling, that is, they can be treated as known constants even if they are in fact random variables. If in addition $\varepsilon_1, \varepsilon_2, \ldots, \varepsilon_n$ can be taken to be distributed independently of each other with constant variance, then by Markoff's theorem the least squares estimates of $\beta_1, \beta_2, \ldots, \beta_k$ are best linear unbiased estimates whatever the form of distribution of the ε's. Unbiased estimates of the variances of the estimates can also be obtained without difficulty. These estimates of variance can then be used to make confidence statements by assuming the errors to be normally distributed.

Thus the assumptions on which the validity of the least squares method is based are as follows:

(a) The error is distributed independently of the independent variables with zero mean and constant variance;
(b) Successive errors are distributed independently of one another.

In what follows autoregressive schemes and stochastic difference equations will be excluded from further consideration, since assumption (a) does not hold in such cases. We shall be concerned only with assumption (b), that is, we shall assume that the x's can be regarded as 'fixed variables'. When (b) is violated the least squares procedure breaks down at three points:

(i) The estimates of the regression coefficients, though unbiased, need not have least variance.
(ii) The usual formula for the variance of an estimate is no longer applicable and is liable to give a serious underestimate of the true variance.
(iii) The t and F distributions, used for making confidence statements, lose their validity.

In stating these consequences of the violation of assumption (b) we do not overlook the fact, pointed out by Wold (1949), that the variances of the resulting estimates depend as much on the serial correlations of the independent variables as on the serial correlation of the errors. In fact, as Wold showed, when all the sample serial correlations of the x's are zero the estimates of variance given by the least squares method are strictly unbiased whether the errors are serially correlated or not. It seems to us doubtful, however, whether this result finds much application in practice. It will only rarely be the case that the independent variables are serially uncorrelated while the errors are serially correlated. Consequently, we feel that there can be little doubt of the desirability of testing the errors for independence whenever the least squares method is applied to serially correlated observations.

To find a suitable test criterion we refer to some results obtained by T.W. Anderson (1948). Anderson showed that in certain cases in which the regression vectors are latent vectors of matrices $\mathbf{\Psi}$ and $\mathbf{\Theta}$ occurring in the error distribution, the statistic $\frac{\mathbf{z}'\mathbf{\Theta}\mathbf{z}}{\mathbf{z}'\mathbf{\Psi}\mathbf{z}}$, where $\mathbf{z}$ is the column vector of residuals from regression, provides a test that is uniformly most powerful against certain alternative hypotheses. The error distributions implied by these alternative hypotheses are given by Anderson and are such that in the cases that are likely to be useful in practice $\mathbf{\Psi} = \mathbf{I}$, the unit matrix. These results suggest that we should examine the distribution of the statistic $r = \frac{\mathbf{z}'\mathbf{A}\mathbf{z}}{\mathbf{z}'\mathbf{z}}$ (changing the notation slightly) for regression on any set of fixed variables, $\mathbf{A}$ being any real symmetric matrix.

In the next section we shall consider certain formal properties of r defined in this way, and in §3 its distribution in the null case will be examined. Expressions for its moments will be derived, and it will be shown that its distribution function lies between two distribution functions which could be determined. In §4 we return to discuss the question of the choice of an appropriate test criterion with rather more rigour and a specific choice is made. In the final section certain special properties of this test criterion are given.

2. Transformation of r

We consider the linear regression of y on k independent variables $x_1, x_2, \ldots, x_k$. The model for a sample of n observations is

$$\begin{pmatrix} y_1 \\ y_2 \\ \vdots \\ y_n \end{pmatrix} = \begin{pmatrix} x_{11} & x_{21} & \cdots & x_{k1} \\ x_{12} & x_{22} & \cdots & x_{k2} \\ \vdots & \vdots & & \vdots \\ x_{1n} & x_{2n} & \cdots & x_{kn} \end{pmatrix} \begin{pmatrix} \beta_1 \\ \beta_2 \\ \vdots \\ \beta_k \end{pmatrix} + \begin{pmatrix} \varepsilon_1 \\ \varepsilon_2 \\ \vdots \\ \varepsilon_n \end{pmatrix},$$

or in an evident matrix notation,

$$\mathbf{y} = \mathbf{X}\boldsymbol{\beta} + \boldsymbol{\varepsilon}.$$

The least squares estimate of $\boldsymbol{\beta}$ is $\mathbf{b} = \{b_1, b_2, \ldots, b_k\}$ given by $\mathbf{b} = (\mathbf{X}'\mathbf{X})^{-1}\mathbf{X}'\mathbf{y}$.

The vector $\mathbf{z} = \{z_1, z_2, \ldots, z_n\}$ of residuals from regression is defined by

$$\begin{aligned} \mathbf{z} &= \mathbf{y} - \mathbf{X}\mathbf{b} \\ &= \{\mathbf{I}_n - \mathbf{X}(\mathbf{X}'\mathbf{X})^{-1}\mathbf{X}'\}\mathbf{y}, \end{aligned}$$

where $\mathbf{I}_n$ is the unit matrix of order n.

Thus

$$\begin{aligned} \mathbf{z} &= \{\mathbf{I}_n - \mathbf{X}(\mathbf{X}'\mathbf{X})^{-1}\mathbf{X}'\}(\mathbf{X}\boldsymbol{\beta} + \boldsymbol{\varepsilon}) \\ &= \{\mathbf{I}_n - \mathbf{X}(\mathbf{X}'\mathbf{X})^{-1}\mathbf{X}'\}\boldsymbol{\varepsilon} \\ &= \mathbf{M}\boldsymbol{\varepsilon} \text{ say.} \end{aligned}$$

It may be verified that $\mathbf{M} = \mathbf{M}' = \mathbf{M}^2$; that is, $\mathbf{M}$ is idempotent.*

* This matrix treatment of the residuals is due to Aitken (1935).

We now examine the ratio of quadratic forms $r = \dfrac{\mathbf{z}'\mathbf{A}\mathbf{z}}{\mathbf{z}'\mathbf{z}}$, where $\mathbf{A}$ is a real symmetric matrix. Transforming to the errors we have

$$r = \frac{\boldsymbol{\varepsilon}'\mathbf{M}'\mathbf{A}\mathbf{M}\boldsymbol{\varepsilon}}{\boldsymbol{\varepsilon}'\mathbf{M}'\mathbf{M}\boldsymbol{\varepsilon}} = \frac{\boldsymbol{\varepsilon}'\mathbf{M}\mathbf{A}\mathbf{M}\boldsymbol{\varepsilon}}{\boldsymbol{\varepsilon}'\mathbf{M}\boldsymbol{\varepsilon}}.$$

We shall show that there exists an orthogonal transformation which simultaneously reduces the numerator and denominator of r to their canonical forms; that is, there is an orthogonal transformation $\boldsymbol{\varepsilon} = \mathbf{H}\boldsymbol{\zeta}$ such that

$$r = \frac{\sum_{i=1}^{n-k} \nu_i \zeta_i^2}{\sum_{i=1}^{n-k} \zeta_i^2}.$$

It is well known that there is an orthogonal matrix $\mathbf{L}$ such that

$$\mathbf{L}'\mathbf{M}\mathbf{L} = \begin{pmatrix} \mathbf{I}_{n-k} & \vdots & \mathbf{O} \\ \cdots & & \cdots \\ \mathbf{O} & \vdots & \mathbf{O} \end{pmatrix},$$

where $\mathbf{I}_{n-k}$ is the unit matrix of order $n - k$, and $\mathbf{O}$ stands for a zero matrix with appropriate numbers of rows and columns. This corresponds to the result that $\sum_{i=1}^{n} z_i^2$ is distributed as χ^2 with $n - k$ degrees of freedom. Thus

$$\begin{aligned} \mathbf{L}'\mathbf{M}\mathbf{A}\mathbf{M}\mathbf{L} &= \mathbf{L}'\mathbf{M}\mathbf{L}\,.\,\mathbf{L}'\mathbf{A}\mathbf{L}\,.\,\mathbf{L}'\mathbf{M}\mathbf{L} \\ &= \begin{pmatrix} \mathbf{I}_{n-k} & \vdots & \mathbf{O} \\ \cdots & & \cdots \\ \mathbf{O} & \vdots & \mathbf{O} \end{pmatrix} \begin{pmatrix} \mathbf{B}_1 & \vdots & \mathbf{B}_3 \\ \cdots & & \cdots \\ \mathbf{B}_2 & \vdots & \mathbf{B}_4 \end{pmatrix} \begin{pmatrix} \mathbf{I}_{n-k} & \vdots & \mathbf{O} \\ \cdots & & \cdots \\ \mathbf{O} & \vdots & \mathbf{O} \end{pmatrix} \\ &= \begin{pmatrix} \mathbf{B}_1 & \vdots & \mathbf{O} \\ \cdots & & \cdots \\ \mathbf{O} & \vdots & \mathbf{O} \end{pmatrix}, \end{aligned}$$

where $\begin{pmatrix} \mathbf{B}_1 & \vdots & \mathbf{B}_3 \\ \cdots & & \cdots \\ \mathbf{B}_2 & \vdots & \mathbf{B}_4 \end{pmatrix}$ is the appropriate partition of the real asymmetric matrix $\mathbf{L}'\mathbf{A}\mathbf{L}$.

Let $\mathbf{N}_1$ be the orthogonal matrix diagonalizing $\mathbf{B}_1$, i.e.

$$\mathbf{N}_1'\mathbf{B}_1\mathbf{N}_1 = \begin{pmatrix} \nu_1 & & & \\ & \nu_2 & & \\ & & \ddots & \\ & & & \nu_{n-k} \end{pmatrix}$$

the blank spaces representing zeros. Then $\mathbf{N} = \begin{pmatrix} \mathbf{N}_1 & \vdots & \mathbf{O} \\ \cdots & & \cdots \\ \mathbf{O} & \vdots & \mathbf{I}_k \end{pmatrix}$ is orthogonal, so that $\mathbf{H} = \mathbf{L}\mathbf{N}$ is orthogonal.

Consequently

$$\begin{aligned} \mathbf{H}'\mathbf{M}\mathbf{H} &= \mathbf{N}'\mathbf{L}'\mathbf{M}\mathbf{L}\mathbf{N} \\ &= \mathbf{N}'\begin{pmatrix} \mathbf{I}_{n-k} & \vdots & \mathbf{O} \\ \cdots & & \cdots \\ \mathbf{O} & \vdots & \mathbf{O} \end{pmatrix}\mathbf{N} \\ &= \begin{pmatrix} \mathbf{I}_{n-k} & \vdots & \mathbf{O} \\ \cdots & & \cdots \\ \mathbf{O} & \vdots & \mathbf{O} \end{pmatrix}, \end{aligned}$$

so that

$$\mathbf{H'HAMH} = \mathbf{H'MH}.\mathbf{H'AH}.\mathbf{H'MH}$$

$$= \left(\begin{array}{cccc|c} \nu_1 & & & & \\ & \nu_2 & & & \mathbf{O} \\ & & \ddots & & \\ & & & \nu_{n-k} & \\ \hline & & & \mathbf{O} & \mathbf{O} \end{array}\right).$$

Putting $\boldsymbol{\varepsilon} = \mathbf{H}\boldsymbol{\zeta}$, we have

$$r = \frac{\sum_{i=1}^{n-k} \nu_i \zeta_i^2}{\sum_{i=1}^{n-k} \zeta_i^2}.$$

This result can be seen geometrically by observing that $\boldsymbol{\varepsilon}'\mathbf{MAM}\boldsymbol{\varepsilon}$ = constant and $\boldsymbol{\varepsilon}'\mathbf{M}\boldsymbol{\varepsilon}$ = constant are hypercylinders with parallel generators, the cross-section of $\boldsymbol{\varepsilon}'\mathbf{M}\boldsymbol{\varepsilon}$ constant being an $[n-k]$ hypersphere.

Determination of $\nu_1, \nu_2, \ldots, \nu_{n-k}$

By standard matrix theory $\nu_1, \nu_2, \ldots, \nu_{n-k}$ are the latent roots of $\mathbf{MAM}$ other than k zeros; that is, they are the latent roots of $\mathbf{M}^2\mathbf{A}$, since the roots of the product of two matrices are independent of the order of multiplication.* But $\mathbf{M}^2\mathbf{A} = \mathbf{MA}$ since $\mathbf{M}^2 = \mathbf{M}$. Consequently $\nu_1, \nu_2, \ldots, \nu_{n-k}$ are the latent roots of $\mathbf{MA}$ other than k zeros.

Suppose now that we make the real non-singular transformation of the x's, $\mathbf{P} = \mathbf{XG}$. Then $\mathbf{M} = \mathbf{I}_n - \mathbf{P}(\mathbf{P'P})^{-1}\mathbf{P}'$; that is, $\mathbf{M}$ is invariant under such transformations. We choose $\mathbf{G}$ so that the column vectors $\mathbf{p}_1, \mathbf{p}_2, \ldots, \mathbf{p}_k$ of $\mathbf{P}$ are orthogonal and are each of unit length, i.e.

$$\mathbf{p}_i'\mathbf{p}_j = \begin{cases} 1 & (i = j) \\ 0 & (i \neq j) \end{cases}$$

so that

$$\mathbf{P'P} = \mathbf{I}_n.$$

This amounts to saying that we can replace the original independent variables by a normalized orthogonal set without affecting the residuals.

We have, therefore,

$$\begin{aligned} \mathbf{M} &= \mathbf{I}_n - (\mathbf{p}_1\mathbf{p}_1' + \mathbf{p}_2\mathbf{p}_2' + \cdots + \mathbf{p}_k\mathbf{p}_k') \\ &= (\mathbf{I}_n - \mathbf{p}_1\mathbf{p}_1')(\mathbf{I}_n - \mathbf{p}_2\mathbf{p}_2')\ldots(\mathbf{I}_n - \mathbf{p}_k\mathbf{p}_k') \\ &= \mathbf{M}_1\mathbf{M}_2\ldots\mathbf{M}_k, \text{ say.} \end{aligned}$$

* See, for instance, C.C. Macduffee, *The Theory of Matrices* (Chelsea Publishing Company, 1946), Theorem 16·2.

Each factor $\mathbf{M}_i$ has the same form as $\mathbf{M}$, the matrix $\mathbf{P}$ being replaced by the vector $\mathbf{p}_i$; it is idempotent of rank $n-1$ as can be easily verified. From the derivation it is evident that the $\mathbf{M}_i$'s commute. This is an expression in algebraic terms of the fact that we can fit regressions on orthogonal variables separately and in any order without affecting the final result.

Returning to the main argument we have the result that $v_1, v_2, \ldots, v_{n-k}$ are the roots of $\mathbf{M}_k \ldots \mathbf{M}_2\mathbf{M}_1\mathbf{A}$ other than k zeros. From the form of the products we see that any result we establish about the roots of $\mathbf{M}_1\mathbf{A}$ in terms of those of $\mathbf{A}$ will be true of the roots of $\mathbf{M}_2\mathbf{M}_1\mathbf{A}$ in terms of those of $\mathbf{M}_1\mathbf{A}$. This observation suggests a method of building up a knowledge of the roots of $\mathbf{M}_k \ldots \mathbf{M}_2\mathbf{M}_1\mathbf{A}$ in stages starting from the roots of $\mathbf{A}$ which we assume known.

We therefore investigate the latent roots of $\mathbf{M}_1\mathbf{A}$, say $\theta_1, \theta_2, \ldots, \theta_{n-1}, 0$. These are the roots of the determinantal equation

$$|\mathbf{I}_n\theta - \mathbf{M}_1\mathbf{A}| = 0,$$

i.e.

$$|\mathbf{I}_n\theta - (\mathbf{I}_n - \mathbf{p}_1\mathbf{p}_1')\mathbf{A}| = 0. \qquad (1)$$

Let $\mathbf{T}$ be the orthogonal matrix diagonalizing $\mathbf{A}$, i.e.

$$\mathbf{T}'\mathbf{A}\mathbf{T} = \mathbf{\Lambda} = \begin{pmatrix} \lambda_1 & & & \\ & \lambda_2 & & \\ & & \ddots & \\ & & & \lambda_n \end{pmatrix},$$

where $\lambda_1, \lambda_2, \ldots, \lambda_n$ are the latent roots of $\mathbf{A}$. Pre- and post-multiplying (1) by $\mathbf{T}'$ and $\mathbf{T}$, we have

$$|\mathbf{I}_n\theta - (\mathbf{I}_n - \mathbf{1}_1\mathbf{1}_1')\mathbf{\Lambda}| = 0,$$

where $\mathbf{1}_1 = \{l_{11}, l_{12}, \ldots, l_{1n}\}$ is the vector of direction cosines of $\mathbf{p}_1$ referred to the latent vectors of $\mathbf{A}$ as axes. (Complications arising from multiplicities in the roots of $\mathbf{A}$ are easily overcome in the present context.) Dropping the suffix from $\mathbf{1}_1$ for the moment, we have

$$|\mathbf{I}_n\theta - (\mathbf{I}_n - \mathbf{1}\mathbf{1}')\mathbf{\Lambda}| = 0.$$

Writing out the determinant in full,

$$\begin{vmatrix} \theta - \lambda_1 + l_1^2\lambda_1, & l_1l_2\lambda_2 & \cdots & l_1l_n\lambda_n \\ l_2l_1\lambda_1 & \theta - \lambda_2 + l_2^2\lambda_2 & \cdots & \vdots \\ \vdots & \vdots & \cdots & \vdots \\ \vdots & \vdots & \cdots & \vdots \\ l_nl_1\lambda_1 & \cdots & \cdots & \theta - \lambda_n + l_n^2\lambda_n \end{vmatrix} = 0.$$

Subtracting l_2/l_1 times the first row from the second row, l_3/l_1 times the first from the third, and so on, we can expand the determinant to give the equation

$$\prod_{j=1}^{n}(\theta-\lambda_j)+\sum_{i=1}^{n} l_i^2\lambda_i\prod_{j\neq i}^{n}(\theta-\lambda_j)=0.$$

Reducing and taking out a factor θ corresponding to the known zero root of $\mathbf{M}_1\mathbf{A}$ gives

$$\sum_{i=1}^{n} l_i^2\prod_{j\neq i}^{n}(\theta-\lambda_j)=0. \tag{2}$$

$\theta_1, \theta_2, \ldots, \theta_{n-1}$ are the roots of this equation.

We notice that when $l_r = 0$, $\theta - \lambda_r$ is a factor of (2) so that $\theta = \lambda_r$ is a solution. Thus when $\mathbf{p}_1$ coincides with a latent vector of $\mathbf{A}$, $\theta_1, \theta_2, \ldots, \theta_{n-1}$ are equal to the latent roots associated with the remaining $n-1$ latent vectors of $\mathbf{A}$. In the same way if $\mathbf{p}_2$ also coincides with a latent vector of $\mathbf{A}$, the roots of $\mathbf{M}_2\mathbf{M}_1\mathbf{A}$ other than two zeros are equal to the latent roots associated with the remaining $n-2$ latent vectors of $\mathbf{A}$. Thus, in general, if the k regression vectors coincide with k of the latent vectors of $\mathbf{A}$, $v_1, v_2, \ldots, v_{n-k}$ are equal to the roots associated with the remaining $n-k$ latent vectors of $\mathbf{A}$. This result remains true if the regression vectors are (linearly independent) linear combinations of k of the latent vectors of $\mathbf{A}$.

For other cases it would be possible to write down an equation similar to (2) giving the roots of $\mathbf{M}_2\mathbf{M}_1\mathbf{A}$ in terms of $\theta_1, \theta_2, \ldots, \theta_{n-1}$, and so on. In this way it would be theoretically possible to determine $v_1, v_2, \ldots, v_{n-k}$. The resulting equations would, however, be quite unmanageable except in the latent vector case just mentioned.

Inequalities on $v_1, v_2, \ldots, v_{n-k}$*

We therefore seek inequalities on $v_1, v_2, \ldots, v_{n-k}$. For the sake of generality, we suppose that certain of the regression vectors, say s of them, coincide with latent vectors of $\mathbf{A}$ (or linear combinations of them). From the results of the previous section, it follows immediately that the problem is reduced to the consideration of $k-s$ arbitrary regression vectors, while $\mathbf{A}$ may be supposed to have s zero roots, together with the roots of $\mathbf{A}$ not associated with the s latent vectors mentioned above. These roots may be renumbered so that

$$\lambda_1 \leq \lambda_2 \leq \cdots \leq \lambda_{n-s}.$$

We proceed to show that

$$\lambda_i \leq v_i \leq \lambda_{i+k-s} \quad (i = 1, 2, \ldots, n-k).$$

* We are grateful to Professor T.W. Anderson for pointing out an error in the section entitled "Inequalities on v_1 $v_2, \ldots, v_{n-k}$." While part (b) of the lemma and its corollary were incorrectly stated in the original paper, they have been correctly stated here.

It is convenient to establish first an analogous result for the full sets of ν's and λ's. We therefore arrange the suffixes so that

$$\nu_1 \leqslant \nu_2 \leqslant \cdots \leqslant \nu_{n-k},$$
$$\lambda_1 \leqslant \lambda_2 \leqslant \cdots \leqslant \lambda_n.$$

We also arrange the θ's so that

$$\theta_1 \leqslant \theta_2 \leqslant \cdots \leqslant \theta_{n-1}.$$

It was noted above that if $l_r = 0$, λ_r is a root of (2). Also if any two of the λ's, say λ_r and λ_{r+1}, are equal, then $\lambda_r = \lambda_{r+1}$ is a root of (2). These are the only two cases in which any of the λ's is a root of (2).

For the remaining roots let

$$f(\theta) = \sum_{i=1}^{n} l_i^2 \prod_{j \neq i}^{n} (\theta - \lambda_i).$$

Then

$$f(\lambda_r) = l_r^2 \prod_{j \neq r}^{n} (\lambda_r - \lambda_j),$$

so that if $f(\lambda_r) > 0$ then $f(\lambda_{r+1}) \leqslant 0$, and if $f(\lambda_r) < 0$ then $f(\lambda_{r+1}) \geqslant 0$. Since $f(\theta)$ is continuous there must therefore be a root in every interval $\lambda_r \leqslant \theta \leqslant \lambda_{r+1}$. Thus

$$\lambda_i \leqslant \theta_i \leqslant \lambda_{i+1} \qquad (i = 1, 2, \ldots, n-1). \tag{4}$$

To extend this result we recall that $\mathbf{M}_1\mathbf{A}$ has one zero root in addition to $\theta_1, \theta_2, \ldots, \theta_{n-1}$. Suppose $\theta_l \leqslant 0 \leqslant \theta_{l+1}$; then the roots of $\mathbf{M}_1\mathbf{A}$ can be arranged in the order

$$\theta_1 \leqslant \theta_2 \leqslant \cdots \leqslant \theta_l \leqslant 0 \leqslant \theta_{l+1} \leqslant \cdots \leqslant \theta_{n-1}.$$

Let the roots of $\mathbf{M}_2(\mathbf{M}_1\mathbf{A})$ be $\phi_1, \phi_2, \ldots, \phi_{n-1}$ together with one zero root. Then by (4)

$$\theta_1 \leqslant \phi_1 \leqslant \theta_2 \leqslant \phi_2 \leqslant \cdots \leqslant \theta_l \leqslant \phi_l \leqslant 0 \leqslant \phi_{l+1} \leqslant \cdots.$$

But $\mathbf{M}_2\mathbf{M}_1\mathbf{A}$ certainly has two zero roots, since $\mathbf{M}_2\mathbf{M}_1\mathbf{A}$ has rank at most $n-2$. Thus either ϕ_l or ϕ_{l+1} must be zero. Rejecting one of them and renumbering we have

$$\lambda_i \leqslant \phi_i \leqslant \lambda_{i+2} \qquad (i = 1, 2, \ldots, n-2).$$

Applying the same argument successively we have

$$\lambda_i \leqslant \nu_i \leqslant \lambda_{i+k} \qquad (i = 1, 2, \ldots, n-k).$$

Deleting cases due to regression vectors coinciding with latent vectors of $\mathbf{A}$ we have (3).

The results of this section will be gathered into a lemma.

Lemma. *If* $\mathbf{z}$ *and* $\boldsymbol{\varepsilon}$ *are* $n \times 1$ *vectors such that* $\mathbf{z} = \mathbf{M}\boldsymbol{\varepsilon}$, *where* $\mathbf{M} = \mathbf{I}_n - \mathbf{X}(\mathbf{X}'\mathbf{X})^{-1}\mathbf{X}'$, *and if* $r = \dfrac{\mathbf{z}'\mathbf{A}\mathbf{z}}{\mathbf{z}'\mathbf{z}}$, *where* $\mathbf{A}$ *is a real symmetric matrix, then*

(a) *There is an orthogonal transformation* $\boldsymbol{\varepsilon} = \mathbf{H}\boldsymbol{\zeta}$, *such that*

$$r = \frac{\sum_{i=1}^{n-k} \nu_i \zeta_i^2}{\sum_{i=1}^{n-k} \zeta_i^2},$$

where $\nu_1, \nu_2, \ldots, \nu_{n-k}$ *are the latent roots of* $\mathbf{MA}$ *other than k zeros;*

(b) *Suppose s of the columns of* $\mathbf{X}$ *are linear combinations of s of the latent vectors of* $\mathbf{A}$. *If the roots of* $\mathbf{A}$ *associated with the remaining n − s latent vectors of* $\mathbf{A}$ *are renumbered so that*

$$\lambda_1 \leqslant \lambda_2 \leqslant \cdots \leqslant \lambda_{n-s}$$

then

$$\lambda_i \leqslant \nu_i \leqslant \lambda_{i+k-s} \qquad (i = 1, 2, \ldots, n-k).$$

We deduce the following corollary:

Corollary.

$$r_L \leq r \leq r_U$$

where

$$r_L \sim \frac{\sum_{i=1}^{n-k} \lambda_i \zeta_i^2}{\sum_{i=1}^{n-k} \zeta_i^2}$$

and

$$r_U = \frac{\sum_{i=1}^{n-k} \lambda_{i+k-s} \zeta_i^2}{\sum_{i=1}^{n-k} \zeta_i^2}.$$

This follows immediately by appropriate numbering of suffixes, taking $\lambda_{n-s+1} \ldots \lambda_n$ as the latent roots corresponding to the latent regression vectors and arranging the remainder so that $\lambda_i \leqslant \lambda_{i+1}$. The importance of this result is that it sets bounds upon r which do not depend upon the particular set of regression vectors. r_L and r_U are the best such bounds in that they can be attained, this being the case when the regression vectors coincide with certain of the latent vectors of $\mathbf{A}$.

3. Distribution of r

It has been pointed out that when the errors are distributed independently of the independent variables the latter can be regarded as fixed. There is one special case, however, in which it is more convenient to regard the x's as varying. We shall discuss this first before going on to consider the more general problem of regression on 'fixed variables'.

The case we shall consider is that of a multivariate normal system. In such a system the regressions are linear and the errors are distributed independently of the independent variables. It will be shown that if $y, x_1, \ldots, x_k$ are distributed jointly normally such that the regression of y on the x's passes through the origin, and if successive observations are independent, then r is distributed as if the residuals $z_1, \ldots, z_n$ were independent normal variables. That is, the regression effect disappears from the problem. Similarly, when the regression does not pass through the origin r is distributed as if the z's were residuals from the sample mean of n normal independent observations.

This is perhaps not a very important case in practice, since it will rarely happen that we shall wish to test the hypothesis of serial correlation in the errors when it is known that successive observations of the x's are independent. Nevertheless, it is convenient to deal with it first before going on to discuss the more important case of regression on 'fixed variables'.

To establish the result we consider the geometrical representation of the sample and observe that the sample value of r depends only on the direction in space of the residual vector $\mathbf{z}$. If the x's are kept fixed and the errors are normal and independent, $\mathbf{z}$ is randomly directed in the $[n-k]$ space orthogonal to the space spanned by the x vectors. If the x's are allowed to vary $\mathbf{z}$ will be randomly directed in the $[n]$ space if and only if the x's are jointly normal and successive observations are independent (Bartlett, 1934). In this case the direction of $\mathbf{z}$ is distributed as if $z_1, z_2, \ldots, z_n$ were normal and independent with the same variance. Thus when $y, x_1, \ldots, x_k$ are multivariate normal such that the regression of y on x passes through the origin, r is distributed as if the residuals from the fitted regression through the origin were normal and independent variables.

In the same way it can be shown that if we fit a regression including a constant term, r is distributed as if the z's were residuals from a sample mean of normal independent variables whether the population regression passes through, the origin or not.

Regression on 'Fixed Variables'

To examine the distribution of r on the null hypothesis in the 'fixed variable' case we assume that the errors $\varepsilon_1, \varepsilon_2, \ldots, \varepsilon_n$ are independent normal variables with constant variance, i.e. they are independent $N(0, \sigma^2)$ variables. Transforming as in §2 we have

$$r = \frac{\sum_{i=1}^{n-k} \nu_i \zeta_i^2}{\sum_{i=1}^{n-k} \zeta_i^2}.$$

Since the transformation is orthogonal, $\zeta_1, \zeta_2, \ldots, \zeta_{n-k}$ are independent $N(0, \sigma^2)$ variables. It is evident that the variation of r is limited to the range (ν_1, ν_{n-k}).

Assuming the v's known, the exact distribution of r has been given by R.L. Anderson (1942) for two special cases: first for $n-k$ even, the v's being equal in pairs, and second for $n-k$ odd, the v's being equal in pairs with one value greater or less than all the others. Anderson's expressions for the distribution function are as follows:

$$P(r > r') = \sum_{i=1}^{m} \frac{(\tau_i - r')^{(1/2)(n-k)-1}}{\alpha_i} \qquad (\tau_{m+1} \leqslant r' \leqslant \tau_m),$$

where

$n-k$ *even*: the v_i's form $\frac{1}{2}(n-k)$ distinct pairs denoted by

$$\tau_1 > \tau_2 > \cdots > \tau_{(n-k)/2}$$

$$\alpha_i = \prod_{j \neq i}^{(n-k)/2} (\tau_i - \tau_j),$$

$n-k$ *odd*: the v_i's form $\frac{1}{2}(n-k-1)-1$ distinct pairs as above together with one isolated root τ less than all the others and

$$\alpha_i = \prod_{j \neq i}^{(n-k-1)/2} (\tau_i - \tau_j)\sqrt{(\tau_i - \tau)}.$$

The expression for $n-k$ odd, $\tau > \tau_1$ is obtained by writing $-r$ for r. Formulae for the density function are also given by Anderson.

For the case in which the v's are all different and $n-k$ is even the $[\frac{1}{2}(n-k)-1]$th derivative of the density function has been given by von Neumann (1941), but up to the present no elementary expression for the density function itself has been put forward. Von Neumann's expression for the derivative is as follows:

$$\frac{d^{(1/2)(n-k)-1}}{dr^{(1/2)(n-k)-1}} f(r) = 0 \quad m \text{ even}$$

$$= \frac{(-1)^{(n-k-m-1)/2}\left(\dfrac{n-k}{2}-1\right)!}{\pi\sqrt{(-\prod_{j=1}^{n-k}(r-v_i))}}, \qquad m \text{ odd}$$

for $v_m < r < v_{m+1}$, $m = 1, 2, \ldots, n-k-1$.

To use these results in any particular case the v's would need to be known quantities, which means in practice that the regression vectors must be latent vectors of **A**. In addition, the roots associated with remaining $n-k$ latent vectors of **A** must satisfy R.L. Anderson's or von Neumann's conditions.

The results can also be applied to the distributions of r_L and r_U, the lower and upper bounds of r, provided the appropriate λ's satisfy the conditions. Using the relations

$$F_L(r) \geqslant F(r) \geqslant F_U(r), \tag{5}$$

where F_L and F_U are the distribution functions of r_L and r_U we would then have limits to the distribution function of r. The truth of the relations (5) can be seen by noting that r_L and r are in (1, 1) correspondence and $r_L \leqslant r$ always.

Approximations

R.L. Anderson's distribution becomes unwieldy to work with when $n - k$ is moderately large, and von Neumann's results can only be used to give an exact distribution when $n - k$ is very small. For practical applications, therefore, approximate methods are required.

We first mention the result, pointed out by T.W. Anderson (1948), that as $n - k$ becomes large r is asymptotically normally distributed with the mean and variance given later in this paper. For moderate values of $n - k$, however, it appears that the distributions of certain statistics of the type r are better approximated by a β-distribution, even when symmetric.* One would expect the advantage of the β over the normal approximation to be even greater when the v's are such that the distribution of r is skew. For better approximations various expansions in terms of β-functions can be used. One such expansion was used for most of the tabulation of the distribution of von Neumann's statistic (Hart 1942). Another method is to use a series expansion in terms of Jacobi polynomials using a β-distribution expression as weight function. (See, for instance, Courant & Hilbert,† 1931, p. 76.) The first four terms of such a series will be used for calculating some of the bounds to the significance points of r tabulated in our second paper.

Moments of r

To use the above approximations we require the moments of r. First we note that since r is independent of the scale of the ζ's we can take σ^2 equal to unity. We therefore require the moments of $r = u/v$, where $u = \sum_{i=1}^{n-k} v_i \zeta_i^2$ and $v = \sum_{i=1}^{n-k} \zeta_i^2$, $\zeta_1, \zeta_2, \ldots, \zeta_{n-k}$ being independent $N(0, 1)$ variables.

It is well known (Pitman, 1937; von Neumann, 1941) that r and v are distributed independently. Consequently

$$E(u^s) = E(r^s v^s) = E(r^s)E(v^s),$$

so that

$$E(r^s) = \frac{E(u^s)}{E(v^s)},$$

that is, the moments of the ratio are the ratios of the moments.

* See, for instance, Rubin (1945), Dixon (1944), R.L. Anderson and T.W. Anderson (1950).

† Note, however, the misprint: $x^q(1 - x)^{p-q}$ should read $x^{q-1}(1 - x)^{p-q}$.

The moments of u are most simply obtained by noting that u is the sum of independent variables $v_i\zeta_i^2$, where ζ_i^2 is a χ^2 variable with one degree of freedom. Hence the sth cumulant of u is the sum of sth cumulants, that is

$$\kappa_s(u) = 2^{s-1}(s-1)! \sum_{i=1}^{n-k} v_i^s,$$

since

$$\kappa_s(v_i\zeta_i^2) = 2^{s-1}(s-1)!\, v_i^s.$$

In particular

$$\kappa_1(u) = \sum v_i, \qquad \kappa_2(u) = 2\sum v_i^2.$$

The moments of u can then be obtained from the cumulants.

The moments of v are simply those of χ^2 with $n-k$ degrees of freedom, i.e.

$$E(v) = n - k,$$
$$E(v^2) = (n-k)(n-k+2), \text{ etc.}$$

Hence

$$E(r) = \mu_1' = \frac{1}{n-k}\sum_{i=1}^{n-k} v_i = \bar{v} \text{ say.} \tag{6}$$

To obtain the moments of r about the mean we have

$$r - \mu_1' = \frac{\sum (v_i - \bar{v})\zeta_i^2}{\sum \zeta_i^2} = \frac{u'}{v} \quad \text{say.}$$

As before the moments of $r - \mu_1'$ are the moments of u' divided by the moments of v. The moments of u' are obtained from the cumulants

$$\kappa_s(u') = 2^{s-1}(s-1)! \sum_{i=1}^{n-k} (v_i - \bar{v})^s.$$

In this way we find

$$\left.\begin{aligned} \operatorname{var} r = \mu_2 &= \frac{2\sum (v_i - \bar{v})^2}{(n-k)(n-k+2)}, \\ \mu_3 &= \frac{8\sum (v_i - \bar{v})^3,}{(n-k)(n-k+2)(n-k+4)}, \\ \mu_4 &= \frac{48\sum (v_i - \bar{v})^4 + 12\{\sum (v_i - \bar{v})^2\}^2}{(n-k)(n-k+2)(n-k+4)(n-k+6)}. \end{aligned}\right\} \tag{7}$$

It must be emphasized at this point that the moments just given refer to regression through the origin on k independent variables. If the regression model includes a constant term, that is, if the calculated regression includes a fitted mean, and if, as is usual, we wish to distinguish the remaining independent variables from the constant term, then k must be taken equal to $k' + 1$ in the above expressions, k' being the number of independent variables in addition to the constant. We emphasize this point, since it is k' that is usually referred to as the number of independent variables in such a model.

The expressions given will enable the moments of r to be calculated when the v's are known. In most cases that will arise in practice, however, the v's will be unknown and it will be impracticable to calculate them. We therefore require means of expressing the power sums $\sum v_1^2$ in terms of known quantities, namely the matrix $\mathbf{A}$ and the independent variables.

To do this we make use of the concept of the trace of a matrix, that is, the sum of its leading diagonal elements. This is denoted for a matrix $\mathbf{S}$ by tr $\mathbf{S}$, $\mathbf{S}$ being of course square. It is easy to show that the operation of taking a trace satisfies the following simple rules:

(a) $\operatorname{tr}(\mathbf{S} + \mathbf{T}) = \operatorname{tr} \mathbf{S} + \operatorname{tr} \mathbf{T}$,
(b) $\operatorname{tr} \mathbf{ST} = \operatorname{tr} \mathbf{TS}$ whether $\mathbf{S}$ and $\mathbf{T}$ are square or rectangular.

From these rules we deduce a third:

(c) $\operatorname{tr}(\mathbf{S} + \mathbf{T})^q = \operatorname{tr} \mathbf{S}^q + \binom{q}{1} \operatorname{tr} \mathbf{S}^{q-1}\mathbf{T} + \binom{q}{2} \operatorname{tr} \mathbf{S}^{q-2}\mathbf{T}^2 + \cdots + \operatorname{tr} \mathbf{T}^q$,

when $\mathbf{S}$ and $\mathbf{T}$ are square. In addition, we note that $\operatorname{tr} \mathbf{S} = \sum_{i=1}^{m} \sigma_i$, where $\sigma_1, \sigma_2, \ldots, \sigma_m$ are the latent roots of $\mathbf{S}$, and in general that $\operatorname{tr} \mathbf{S}^q = \sum_{i=1}^{m} \sigma_i^q$.

Thus we have immediately

$$\sum_{i=1}^{n-k} v_i^q = \operatorname{tr}(\mathbf{MA})^q,$$

since $v_1, v_2, \ldots, v_{n-k}$ together with k zeros are the latent roots of $\mathbf{MA}$.

In cases in which the independent variables are known constants it is sometimes possible to construct the matrix $\mathbf{MA}$ directly and hence to obtain the mean and variance of r in a fairly straightforward way.

For models of other types in which the independent variables can take arbitrary values further reduction is needed. For the mean we require

$$\begin{aligned} \sum v_i &= \operatorname{tr} \mathbf{MA} = \operatorname{tr}\{\mathbf{I}_n - \mathbf{X}(\mathbf{X}'\mathbf{X})^{-1}\mathbf{X}\}\mathbf{A} \\ &= \operatorname{tr} \mathbf{A} - \operatorname{tr} \mathbf{X}(\mathbf{X}'\mathbf{X})^{-1}\mathbf{X}'\mathbf{A} \quad \text{by rule (a)} \\ &= \operatorname{tr} \mathbf{A} - \operatorname{tr} \mathbf{X}'\mathbf{A}\mathbf{X}(\mathbf{X}'\mathbf{X})^{-1} \quad \text{by rule (b).} \end{aligned} \tag{8}$$

The calculation of this expression is not as formidable an undertaking as might at first sight appear, since $(\mathbf{X}'\mathbf{X})^{-1}$ will effectively have to be calculated in any case for the estimation of the regression coefficients. It is interesting to note incidentally that the matrix $\mathbf{X}'\mathbf{A}\mathbf{X}(\mathbf{X}'\mathbf{X})^{-1}$ in the expression is a direct multivariate generalization of the statistic r.

For the variance we require

$$\begin{aligned}\sum v_i^2 &= \operatorname{tr}(\mathbf{MA})^2 = \operatorname{tr}\{\mathbf{A} - \mathbf{X}(\mathbf{X}'\mathbf{X})^{-1}\mathbf{X}'\mathbf{A}\}^2 \\ &= \operatorname{tr}\mathbf{A}^2 - 2\operatorname{tr}\mathbf{X}'\mathbf{A}^2\mathbf{X}(\mathbf{X}'\mathbf{X})^{-1} + \operatorname{tr}\{\mathbf{X}'\mathbf{A}\mathbf{X}(\mathbf{X}'\mathbf{X})^{-1}\}^2,\end{aligned} \tag{9}$$

by rules (b) and (c).

Similarly

$$\begin{aligned}\sum v_i^3 = \operatorname{tr}\mathbf{A}^3 &- 3\operatorname{tr}\mathbf{X}'\mathbf{A}^3\mathbf{X}(\mathbf{X}'\mathbf{X})^{-1} \\ &+ 3\operatorname{tr}\{\mathbf{X}'\mathbf{A}^2\mathbf{X}(\mathbf{X}'\mathbf{X})^{-1}\mathbf{X}'\mathbf{A}\mathbf{X}(\mathbf{X}'\mathbf{X})^{-1}\} - \operatorname{tr}\{\mathbf{X}'\mathbf{A}\mathbf{X}(\mathbf{X}'\mathbf{X})^{-1}\}^3,\end{aligned} \tag{10}$$

$$\begin{aligned}\sum v_i^4 = \operatorname{tr}\mathbf{A}^4 &- 4\operatorname{tr}\mathbf{X}'\mathbf{A}^4\mathbf{X}(\mathbf{X}'\mathbf{X})^{-1} + 6\operatorname{tr}\{\mathbf{X}'\mathbf{A}^3\mathbf{X}(\mathbf{X}'\mathbf{X})^{-1}\mathbf{X}'\mathbf{A}\mathbf{X}(\mathbf{X}'\mathbf{X})^{-1}\} \\ &- 4\operatorname{tr}[\mathbf{X}'\mathbf{A}^2\mathbf{X}(\mathbf{X}'\mathbf{X})^{-1}\{\mathbf{X}'\mathbf{A}\mathbf{X}(\mathbf{X}'\mathbf{X})^{-1}\}^2] + \operatorname{tr}\{\mathbf{X}'\mathbf{A}\mathbf{X}(\mathbf{X}'\mathbf{X})^{-1}\}^4,\end{aligned} \tag{11}$$

and so on.

When the independent variables are orthogonal these expressions can be simplified somewhat since $\mathbf{X}'\mathbf{X}$ is then a diagonal matrix. Thus

$$\operatorname{tr}\mathbf{X}'\mathbf{A}\mathbf{X}(\mathbf{X}'\mathbf{X})^{-1} = \sum_{i=1}^{k}\frac{\mathbf{x}_i'\mathbf{A}\mathbf{x}_i}{\mathbf{x}_i'\mathbf{x}_i},$$

$\mathbf{x}_i$ standing for the vector of sample values of the ith independent variable. Each term in the summation has the form r in terms of one of the independent variables. Similarly for $\operatorname{tr}\mathbf{X}'\mathbf{A}^2\mathbf{X}(\mathbf{X}'\mathbf{X})^{-1}$, $\operatorname{tr}\mathbf{X}'\mathbf{A}^3\mathbf{X}(\mathbf{X}'\mathbf{X})^{-1}$, etc. We have also

$$\operatorname{tr}\{\mathbf{X}'\mathbf{A}\mathbf{X}(\mathbf{X}'\mathbf{X})^{-1}\}^2 = \sum_{i=1}^{k}\left(\frac{\mathbf{x}_i'\mathbf{A}\mathbf{x}_i}{\mathbf{x}_i'\mathbf{x}_i}\right)^2 + 2\sum_{i\neq j}^{k}\frac{(\mathbf{x}_i'\mathbf{A}\mathbf{x}_j)^2}{\mathbf{x}_i'\mathbf{x}_i\mathbf{x}_j'\mathbf{x}_j}.$$

Thus when the regression vectors are orthogonal the following formulae enable us to calculate the mean and variance of r:

$$\left.\begin{aligned}\sum v_i &= \operatorname{tr}\mathbf{A} - \sum_{i=1}^{k}\frac{\mathbf{x}_i'\mathbf{A}\mathbf{x}_i}{\mathbf{x}_i'\mathbf{x}_i}, \\ \sum v_i^2 &= \operatorname{tr}\mathbf{A}^2 - 2\sum_{i=1}^{k}\frac{\mathbf{x}_i'\mathbf{A}^2\mathbf{x}_i}{\mathbf{x}_i'\mathbf{x}_i} + \sum_{i=1}^{k}\left(\frac{\mathbf{x}_i'\mathbf{A}\mathbf{x}_i}{\mathbf{x}_i'\mathbf{x}_i}\right)^2 + 2\sum_{i<j}^{k}\frac{(\mathbf{x}_i'\mathbf{A}\mathbf{x}_j)^2}{\mathbf{x}_i'\mathbf{x}_i\mathbf{x}_j'\mathbf{x}_j}.\end{aligned}\right\} \tag{12}$$

The mean and variance are obtained by substituting these values in (6) and (7).

Similar results apply when $\mathbf{X}$ is partitioned into two or more orthogonal sets of variables. For instance, when $\mathbf{X}$ consists of the constant vector $\{c, c, \ldots, c\}$ together with the matrix $\dot{\mathbf{X}}$ of deviations from the means of the remaining $k-1$ variables, i.e.

$$\dot{\mathbf{x}}_{ij} = \mathbf{x}_{ij} - \bar{\mathbf{x}}_i \qquad (i = 2, 3, \ldots, k; j = 1, 2, \ldots, n),$$

then

$$\operatorname{tr}\mathbf{X}'\mathbf{A}\mathbf{X}(\mathbf{X}'\mathbf{X})^{-1} = \frac{\mathbf{i}'\mathbf{A}\mathbf{i}}{n} + \operatorname{tr}\dot{\mathbf{X}}'\mathbf{A}\dot{\mathbf{X}}(\dot{\mathbf{X}}'\dot{\mathbf{X}})^{-1},$$

$$\operatorname{tr}\{\mathbf{X}'\mathbf{A}\mathbf{X}(\mathbf{X}'\mathbf{X})^{-1}\}^2 = \left(\frac{\mathbf{i}'\mathbf{A}\mathbf{i}}{n}\right)^2 + \frac{2\mathbf{i}'\mathbf{A}\dot{\mathbf{X}}(\dot{\mathbf{X}}'\dot{\mathbf{X}})^{-1}\dot{\mathbf{X}}'\mathbf{A}\mathbf{i}}{n} + \operatorname{tr}\{\dot{\mathbf{X}}'\mathbf{A}\dot{\mathbf{X}}(\dot{\mathbf{X}}'\dot{\mathbf{X}})^{-1}\}^2,$$

where $\mathbf{i}$ is the equiangular vector $\{1, 1, \ldots, 1\}$. When this is a latent vector of $\mathbf{A}$ corresponding to a latent root of zero, $\mathbf{i}'\mathbf{A} = \mathbf{O}$. We then have the important result that (8)–(11) apply without change except that the original variables $\mathbf{X}$ are replaced by the deviations from their means $\dot{\mathbf{X}}$. This result holds whenever $\mathbf{x}'\mathbf{A}\mathbf{x}$ is invariant under a change of origin of $\mathbf{x}$.

Before closing this treatment of moments we should mention one difficulty in using them for obtaining approximations in terms of β-distributions and associated expansions. In constructing such approximations one usually knows the range within which the variable is distributed. In the present problem, however, the range is (v_1, v_{n-k}), which will often be unknown and impracticable to determine. In such cases it will accordingly be necessary to use approximations to v_1 and v_{n-k} before the distributions can be fitted.

(*Editors' note*: The section "Characteristic Function of u and v" has been omitted.)

4. Choice of Test Criterion*

To decide upon a suitable test criterion an important consideration is the set of alternative hypotheses against which it is desired to discriminate. The kind of alternative we have in mind in this paper is such that the correlogram of the errors diminishes approximately exponentially with increasing separation of the observations. A convenient model for such hypotheses is the stationary Markoff process

$$\varepsilon_i = \rho\varepsilon_{i-1} + u_i \quad (i = \ldots -1, 0, 1, \ldots), \tag{16}$$

where $|\rho| < 1$ and u_i is normal with mean zero and variance σ^2 and is independent of $\varepsilon_{i-1}, \varepsilon_{i-2}, \ldots$ and $u_{i-1}, u_{i-2}, \ldots$. The null hypothesis is then the hypothesis that $\rho = 0$ in (16).

It has been shown by T.W. Anderson (1948) that no test of this hypothesis exists which is uniformly most powerful against alternatives (16). Anderson also showed, however, that for certain regression systems with error distributions close to that given by (16) tests can be obtained which are uniformly most powerful against one-sided alternatives (16) and which give type B_1 regions for two-sided alternatives (16).

These regression systems include cases in which the regression vectors are constant vectors coinciding with latent vectors of a matrix $\mathbf{\Theta}$ (or with linear combinations of k of them) and in which the error distributions have density functions of the form

$$K \exp\left[-\frac{1}{2\sigma^2}\{(1+\rho^2)\boldsymbol{\varepsilon}'\boldsymbol{\varepsilon} - 2\rho\boldsymbol{\varepsilon}'\mathbf{\Theta}\boldsymbol{\varepsilon}\}\right]. \tag{17}$$

For such cases the uniformly most powerful test of the hypothesis $\rho = 0$ against alternatives $\rho > 0$ is given by $r > r_0$, where $r = \dfrac{\mathbf{z}'\mathbf{\Theta}\mathbf{z}}{\mathbf{z}'\mathbf{z}}$, $\mathbf{z}$ being the

* This section is based on the treatment given by T.W. Anderson (1948).

vector of residuals from least squares regression, and r_0 being determined to give a critical region of appropriate size. For two-sided alternatives to $\rho = 0$ the type B_1 test is given by $r < r_2$, $r > r_3$, where r_2 and r_3 are determined so as to give a critical region of appropriate size and to satisfy the relation

$$\int_{r_2}^{r_3} rp(r)\,dr = E(r)\int_{r_2}^{r_3} p(r)\,dr,$$

$p(r)$ being the density function of r in the null case.

We recall that whatever the regression vectors,

$$r_L \leqslant r \leqslant r_U, \tag{18}$$

where r_L and r_U are defined in the Corollary, §2. Now r_L and r_U have distributions in the null case identical with distributions of r obtained from residuals from regressions on certain latent vectors of the matrix $\mathbf{A}$. Thus if we put $\mathbf{\Theta} = \mathbf{A}$ in (17) we can say that when the lower bound r_L in (18) is attained (or the upper bound), the statistic $r = \dfrac{\mathbf{z}'\mathbf{A}\mathbf{z}}{\mathbf{z}'\mathbf{z}}$ gives a test which is uniformly most powerful against one-sided alternatives (16) and which is of type B_1 against two-sided alternatives.

The error distribution for the stationary Markoff process (16) has the density function

$$K \exp\left[-\frac{1}{2\sigma^2}\left\{(1+\rho^2)\sum_{i=1}^{n} \varepsilon_i^2 - \rho^2(\varepsilon_1^2 + \varepsilon_n^2) - 2\rho\sum_{i=2}^{n} \varepsilon_i\varepsilon_{i-1}\right\}\right]. \tag{19}$$

Taking $\boldsymbol{\varepsilon}'\mathbf{\Theta}\boldsymbol{\varepsilon} = \sum_{i=2}^{n} \varepsilon_i\varepsilon_{i-1}$ in (17) gives a density function

$$K \exp\left[-\frac{1}{2\sigma^2}\left\{(1+\rho^2)\sum_{i=1}^{n} \varepsilon_i^2 - 2\rho\sum_{i=2}^{n} \varepsilon_i\varepsilon_{i-1}\right\}\right], \tag{20}$$

while taking $\boldsymbol{\varepsilon}'\mathbf{\Theta}\boldsymbol{\varepsilon} = \sum_{i=1}^{n} \varepsilon_i^2 - \frac{1}{2}\sum_{i=2}^{n} (\varepsilon_i - \varepsilon_{i-1})^2$ in (17) gives a density function

$$K \exp\left[-\frac{1}{2\sigma^2}\left\{(1+\rho^2)\sum_{i=1}^{n} \varepsilon_i^2 - \rho(\varepsilon_1^2 + \varepsilon_n^2) - 2\rho\sum_{i=2}^{n} \varepsilon_i\varepsilon_{i-1}\right\}\right]. \tag{21}$$

These are both close to (19). Thus following Anderson we conjecture that either value of $\mathbf{\Theta}$ would give a good statistic r for testing against alternatives (16). Between the two statistics there is not much to choose. We ourselves have adopted a slight modification of the second, partly for reasons of computational convenience and partly because of similarity to von Neumann's statistic δ^2/s^2 (1941) already well known to research workers.

The statistic we have adopted is defined by

$$d = \frac{\sum_{i=2}^{n} (z_i - z_{i-1})^2}{\sum_{i=1}^{n} z_i^2},$$

which is related to $\frac{\delta^2}{s^2}$ by $\frac{\delta^2}{s^2} = \frac{nd}{n-1}$. This is a special case of the general statistic $r = \frac{\mathbf{z}'\mathbf{A}\mathbf{z}}{\mathbf{z}'\mathbf{z}}$ discussed in §2 and §3, in which

$$\mathbf{A} = \mathbf{A}_d = \tfrac{1}{2}\begin{pmatrix} 1 & -1 & 0 & 0 & \cdots & \cdots & 0 \\ -1 & 2 & -1 & 0 & \cdots & \cdots & \cdots \\ 0 & -1 & 2 & -1 & \cdots & \cdots & \cdots \\ 0 & 0 & -1 & 2 & \cdots & \cdots & \cdots \\ \cdots & \cdots & \cdots & \cdots & \cdots & \cdots & \cdots \\ \cdots & \cdots & \cdots & \cdots & \cdots & 2 & -1 \\ 0 & \cdots & \cdots & \cdots & 0 & -1 & 1 \end{pmatrix}.$$

In the notation of the previous paragraph we would take $\mathbf{\Theta} = \mathbf{I} - \frac{1}{2}\mathbf{A}_d$ to give the density (21). Now the latent vectors of the matrices $\mathbf{A}_d$ and $\mathbf{\Theta}$ in this equation are the same. Thus when the regression vectors are latent vectors of $\mathbf{A}_d$ the statistic d provides a uniformly most powerful test against one-sided alternatives (21). In particular the test given by d when the bounds r_L and r_U are attained is uniformly most powerful.

The main alternative to using d or a related statistic as a test criterion would be to use one of the circular statistics such as

$$r_c = \frac{\sum_{i=1}^{n} z_i z_{i-1}}{\sum_{i=1}^{n} z_i^2}$$

or

$$d_c = \frac{\sum_{i=1}^{n} (z_i - z_{i-1})^2}{\sum_{i=1}^{n} z_i^2},$$

where we define $z_0 \equiv z_n$ in each case. T.W. Anderson (1948) has shown that r_c and d_c give uniformly most powerful tests against one-sided alternatives in the circular population having a density function

$$K \exp\left[-\frac{1}{2\sigma^2}\left\{(1+\rho^2)\sum_{i=1}^{n} \varepsilon_i^2 - 2\rho \sum_{i=1}^{n} \varepsilon_i \varepsilon_{i-1}\right\}\right], \tag{22}$$

where $\varepsilon_0 \equiv \varepsilon_n$. r_c was the statistic adopted by R.L. Anderson & T.W. Anderson (1950) for testing the residuals from regression on a Fourier series.

The disadvantage of r_c and d_c is that (22) is not so close to (19) as (20) or (21). The advantage is that since the latent roots of the associated values of $\mathbf{A}$ are equal in pairs, the results of R.L. Anderson (1942) can sometimes be used to obtain exact distributions in the null case. The roots of $\mathbf{A}_d$, on the other hand, are all distinct. We conclude that d or a related non-circular statistic would seem to be preferable whenever an approximation to the distribution is sufficient, but that a circular statistic would seem to be preferable if exact results are required at the expense of some loss of power. We mention that the computations involved in using Anderson's exact distribution become very tedious as the number of degrees of freedom increases.

The next question that arises is how good these statistics are as test criteria in cases in which the regression vectors are not latent vectors. Such cases are of course by far the more frequent in practice. It is evident that we can expect the power of the test to diminish as the regression vectors depart from the latent vectors, since the least squares regression coefficients are not then maximum likelihood estimates in the non-null case. Thus any test based on least squares residuals cannot even be a likelihood ratio test. Against this three points can be made. The first is that we still have a valid test, though possibly of reduced power. Secondly, it is desirable on grounds of convenience to have a test based on least squares residuals even though it is not an optimal test. Thirdly, the statistic r necessarily lies between the bounds r_L and r_U and when these bounds are attained the test is optimal. We note also that it is only for the latent vector case that the distribution problems have been approached with any success.

5. Some Special Results

To obtain the moments of d we need the powers of $\mathbf{A}_d$. Because of the symmetry of these matrices they are completely specified by the top left-hand triangle. Thus we can write

$$\mathbf{A}_d \doteq \begin{array}{rrr} 1 & -1 & 0 \\ & 2 & -1 \\ & & 2 \end{array}$$

We find

$$\mathbf{A}_d^2 \doteq \begin{array}{rrrr} 2 & -3 & 1 & 0 \\ & 6 & -4 & 1 \\ & & 6 & -4 \\ & & & 6 \end{array}$$

$$\mathbf{A}_d^3 \doteq \begin{array}{rrrrrr} 5 & -9 & 5 & -1 & 0 & 0 \\ & 19 & -15 & 6 & -1 & 0 \\ & & 20 & -15 & 6 & -1 \\ & & & 20 & -15 & 6 \end{array}$$

$$\mathbf{A}_d^4 \doteq \begin{array}{rrrrrrr} 14 & -28 & 20 & -7 & 1 & 0 & 0 \\ & 62 & -55 & 28 & -8 & 1 & 0 \\ & & 70 & -56 & 28 & -8 & 1 \\ & & & 70 & -56 & 28 & 1 \end{array}$$

Rather than use these matrices as they stand, however, it will probably be more convenient to proceed by finding the sums of squares of the successive differences of the z's. Denoting the sth differences by $\Delta^s z$ we have

$$\left.\begin{aligned} z'\mathbf{A}_d z &= \sum_{i=1}^{n-1} (\Delta z_i)^2, \\ z'\mathbf{A}_d^2 z &= \sum (\Delta^2 z_i)^2 + (z_1 - z_2)^2 + (z_{n-1} - z_n)^2, \\ z'\mathbf{A}_d^3 z &= \sum (\Delta^3 z_i)^2 + 4z_1^2 + 9z_2^2 + z_3^2 - 12z_1z_2 - 6z_2z_3 + 4z_1z_3 \\ &\quad + \text{a similar expression in } z_n, z_{n-1}, z_{n-2}, \\ z'\mathbf{A}_d^4 z &= \sum (\Delta^4 z_i)^2 + 13z_1^2 + 45z_2^2 + 17z_3^2 + z_4^2 - 48z_1z_2 - 54z_2z_3 - 8z_3z_4 \\ &\quad + 28z_1z_3 + 12z_2z_4 - 6z_1z_4 \\ &\quad + \text{a similar expression in } z_n, z_{n-1}, z_{n-2}, z_{n-3}. \end{aligned}\right\} \tag{23}$$

For the circular definition of d, i.e.

$$d_c = \frac{\sum_{i=1}^n (z_i - z_{i-1})^2}{\sum_{i=1}^n z_i^2} = \frac{\mathbf{z}'\mathbf{A}_{dc}\mathbf{z}}{\mathbf{z}'\mathbf{z}} \quad \text{with } z_0 \equiv z_n,$$

the correction terms disappear, giving

$$\mathbf{z}'\mathbf{A}_{dc}^s\mathbf{z} = \sum_{i=1}^n (\Delta^s z_i)^2, \quad \text{where} \quad z_{-i} \equiv z_{n-i}.$$

The latent roots of $\mathbf{A}_d$ are given by

$$\lambda_j = 2\left\{1 - \cos\frac{\pi(j-1)}{n}\right\} \quad (j = 1, 2, \ldots, n) \tag{24}$$

(von Neumann 1941). The first four power sums are:

$$\left.\begin{aligned} \sum_{j=1}^n \lambda_j &= 2(n-1), \\ \sum \lambda_j^2 &= 2(3n-4), \\ \sum \lambda_j^3 &= 4(5n-8), \\ \sum \lambda_j^4 &= 2(35n-64). \end{aligned}\right\} \tag{25}$$

The latent vector corresponding to the zero root λ_1 is $\{1, 1, \ldots, 1\}$, which is the regression vector corresponding to a constant term in the regression model. For regressions with a fitted mean, therefore, we need only consider the remaining $n-1$ λ's which we renumber accordingly so that

$$\lambda_j = 2\left(1 - \cos\frac{\pi j}{n}\right) \quad (j = 1, 2, \ldots, n-1).$$

With these λ's we have from the Corollary, §2,

$$d_L \leqslant d \leqslant d_U, \tag{26}$$

where

$$d_L = \frac{\sum_{i=1}^{n-k'-1} \lambda_i \zeta_i^2}{\sum_{i=1}^{n-k'-1} \zeta_i^2}, \tag{27}$$

$$d_U = \frac{\sum_{i=1}^{n-k'-1} \lambda_{i+k'} \zeta_i^2}{\sum_{i=1}^{n-k'-1} \zeta_i^2}, \tag{28}$$

k' being the number of independent variables in the model in addition to the constant term.

With the error distribution assumed in §3 the limits of the mean of d are given by

$$E(d) \leqslant E(d_U) = 2 - \frac{2}{n-k'-1} \sum_{j=k'+1}^{n-1} \cos \frac{\pi j}{n}$$

$$\geqslant E(d_L) = 2 - \frac{2}{n-k'-1} \sum_{j=1}^{n-k'-1} \cos \frac{\pi j}{n}.$$

We state without proof the limits of the variance of d:

$$\operatorname{var}(d) \leqslant \frac{16}{(n-k'-1)(n-k'+1)} \sum_{j=1}^{(n-k'-1)/2} \cos^2 \frac{\pi j}{n} \quad (n-k' \text{ odd}),$$

$$\leqslant \frac{16}{(n-k'-1)(n-k'+1)} \sum_{j=1}^{(n-k')/2-1} \cos^2 \frac{\pi j}{n}$$

$$+ \frac{8(n-k'-2)}{(n-k'-1)^2(n-k'+1)} \cos^2 \frac{(n-k')\pi}{2n} \quad (n-k' \text{ even}),$$

$$\geqslant \frac{16}{(n-k'-1)(n-k'+1)} \sum_{i=k'+1}^{(n-1)/2} \cos^2 \frac{\pi j}{n} \quad (n \text{ odd}),$$

$$\geqslant \frac{16}{(n-k'-1)(n-k'+1)} \sum_{i=k'+1}^{(n-2)/2} \cos^2 \frac{\pi j}{n} \quad (n \text{ even}).$$

To give some idea of how the distribution of d can vary for different regression vectors we give a short table of the limiting means and variances.

		$k'=1$		$k'=3$		$k'=5$	
		Mean	Variance	Mean	Variance	Mean	Variance
$n=20$	Lower	1.89	0.157	1.65	0.101	1.38	0.048
	Upper	2.11	0.200	2.35	0.249	2.62	0.313
$n=40$	Lower	1.95	0.090	1.84	0.077	1.72	0.063
	Upper	2.05	0.100	2.16	0.111	2.28	0.124
$n=60$	Lower	1.97	0.062	1.89	0.057	1.82	0.051
	Upper	2.03	0.067	2.11	0.071	2.18	0.077

We wish to record our indebtedness to Prof. R.L. Anderson for suggesting this problem to one of us.

References

Aitken, A.C. (1935). *Proc. Roy. Soc. Edinb.* **55**, 42.

Anderson, R.L. (1942). *Ann. Math. Statist.* **13**, 1.

Anderson, R.L. & Anderson, T.W. (1950). *Ann. Math. Statist.* **21**, 59.

Anderson, T.W. (1948). *Skand. Aktuar Tidetr.* **31**, 88.

Bartlett, M.S. (1934). *Proc. Camb. Phil. Soc.* **30**, 327.

Cochrane, D. & Orcutt, G.H. (1949). *J. Amer. Statist. Soc.* **44**, 32.

Courant, R. & Hilbert, D. (1931). *Methoden der Mathematischen Physik.* Julius Springer.

Dixon, W.J. (1944). *Ann. Math. Statist.* **15**, 119.

Fisher, R.A. (1946). *Statistical Methods for Research Workers*, 10th ed. Oliver and Boyd.

Hart, B.I. (1942). *Ann. Math. Statist.* **13**, 207.

Moran, P.A.P. (1950). *Biometrika* **37**, 178.

von Neumann, J. (1941). *Ann. Math. Statist.* **12**, 367.

Pitman, E.J.G. (1937). *Proc. Camb. Phil. Soc.* **33**, 212.

Rubin, H. (1945). *Ann. Math. Statist.* **16**, 211.

Wold, H. (1949). 'On least squares regression with auto-correlated variables and residuals.' (Paper read at the 1949 Conference of the International Statistical Institute.)

Testing for Serial Correlation in Least Squares Regression. II

J. Durbin
London School of Economics

G.S. Watson
Department of Applied Economics,
University of Cambridge

1. Introduction

In an earlier paper (Durbin & Watson, 1950) the authors investigated the problem of testing the error terms of a regression model for serial correlation. Test criteria were put forward, their moments calculated, and bounds to their distribution functions were obtained. In the present paper these bounds are tabulated and their use in practice is described. For cases in which the bounds do not settle the question of significance an approximate method is suggested. Expressions are given for the mean and variance of a test statistic for one- and two-way classifications and polynomial trends, leading to approximate tests for these cases. The procedures described should be capable of application by the practical worker without reference to the earlier paper (hereinafter referred to as Part I).

It should be emphasized that the tests described in this paper apply only to regression models in which the independent variables can be regarded as 'fixed variables'. They do not, therefore, apply to autoregressive schemes and similar models in which lagged values of the dependent variable occur as independent variables.

2. The Bounds Test

Throughout the paper the procedures suggested will be illustrated by numerical examples. We begin by considering some data from a demand analysis study.

Editor's note: Selected sections of this paper are reproduced.

EXAMPLE 1. (Annual consumption of spirits from 1870 to 1938.) The data were compiled by A.R. Prest, to whose paper (1949) reference should be made for details of the source material. As is common in econometric work the original observations were transformed by taking logarithms:

y = log consumption of spirits per head;
x_1 = log real income per head;
x_2 = log relative price of spirits (i.e. price of spirits deflated by a cost-of-living index).

We suppose that the observations satisfy the regression model

$$y = \beta_0 + \beta_1 x_1 + \beta_2 x_2 + \varepsilon, \tag{1}$$

where β_0 is a constant, β_1 is the income elasticity, β_2 is the price elasticity, and ε is a random error with zero mean and constant variance.

It was shown in Part I that exact critical values of this kind cannot be obtained. However, it is possible to calculate upper and lower bounds to the critical values. These are denoted by d_U and d_L. If the observed d is less than d_L we conclude that the value is significant, while if the observed d is greater than d_U we conclude that the value is not significant at the significance level concerned. If d lies between d_L and d_U the test is inconclusive.

Significance points of d_L and d_U are tabulated for various levels in Tables 4, 5 and 6. In addition, a diagram is given to facilitate the test procedure in the most usual case of a test against positive serial correlation at the 5% level (Fig. 1). k' is the number of independent variables.

In the present example $n = 69$ and $k' = 2$, so that at the 5% level $d_L = 1 \cdot 54$ approximately. The observed value $0 \cdot 25$ is less than this and therefore indicates significant positive serial correlation at the 5% level. In fact, the observed value is also significant at the 1% level.

The procedure for other values of k' is exactly similar.

Tests Against Negative Serial Correlation and Two-Sided Tests

Tests against negative serial correlation may sometimes be required. For instance, it is a common practice in econometric work to analyse the first differences of the observations rather than the observations themselves, on the ground that the serial correlation of the transformed errors is likely to be less than that of the original errors. We may wish to ensure that the transformation has not overcorrected, thus introducing negative serial correlation into the transformed errors. To make a test against negative serial correlation, d is calculated as above and subtracted from 4. The quantity $4 - d$ may now be treated as though it were the value of a d-statistic to be tested for positive serial correlation. Thus if $4 - d$ is less than d_L, there is significant evidence of negative serial correlation, and if $4 - d$ is greater than d_U, there is not significant evidence; otherwise the test is inconclusive.

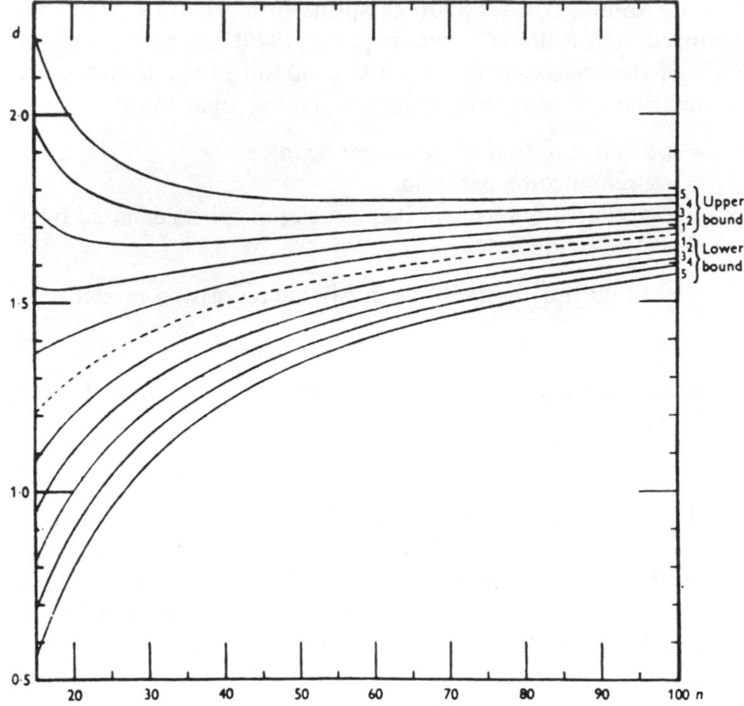

Figure 1. Graphs of 5% values of d_L and d_U against n for $k' = 1, 2, 3, 4, 5$.

When there is no prior knowledge of the sign of the serial correlation, two-sided tests may be made by combining single-tail tests. Using only equal tails, d will be significant at level α if either d is less than d_L or $4 - d$ is less than d_L, non-significant if d lies between d_U and $4 - d_U$ and inconclusive otherwise; the level α may be 2, 5 and 10%. Thus, by using the 5% values of d_L and d_U from Table 4, a two-sided test at the 10% level is obtained.

4. Approximate Procedure when the Bounds Test is Inconclusive

No satisfactory procedure of general application seems to be available for cases in which the bounds test is inconclusive. However, an approximate test can be made, and this should be sufficiently accurate if the number of degrees of freedom is large enough, say greater than 40. For smaller numbers this test can only be regarded as giving a rough indication.

The method used is to transform d so that its range of variation is approxi-

Table 4. Significance Points of d_L and d_U: 5%.

	$k' = 1$		$k' = 2$		$k' = 3$		$k' = 4$		$k' = 5$	
n	d_L	d_U	d_L	d_U	d_L	d_U	d_L	d_U	d_L	d_U
15	1.08	1.36	0.95	1.54	0.82	1.75	0.69	1.97	0.56	2.21
16	1.10	1.37	0.98	1.54	0.86	1.73	0.74	1.93	0.62	2.15
17	1.13	1.38	1.02	1.54	0.90	1.71	0.78	1.90	0.67	2.10
18	1.16	1.39	1.05	1.53	0.93	1.69	0.82	1.87	0.71	2.06
19	1.18	1.40	1.08	1.53	0.97	1.68	0.86	1.85	0.75	2.02
20	1.20	1.41	1.10	1.54	1.00	1.68	0.90	1.83	0.79	1.99
21	1.22	1.42	1.13	1.54	1.03	1.67	0.93	1.81	0.83	1.96
22	1.24	1.43	1.15	1.54	1.05	1.66	0.96	1.80	0.86	1.94
23	1.26	1.44	1.17	1.54	1.08	1.66	0.99	1.79	0.90	1.92
24	1.27	1.45	1.19	1.55	1.10	1.66	1.01	1.78	0.93	1.90
25	1.29	1.45	1.21	1.55	1.12	1.66	1.04	1.77	0.95	1.89
26	1.30	1.46	1.22	1.55	1.14	1.65	1.06	1.76	0.98	1.88
27	1.32	1.47	1.24	1.56	1.16	1.65	1.08	1.76	1.01	1.86
28	1.33	1.48	1.26	1.56	1.18	1.65	1.10	1.75	1.03	1.85
29	1.34	1.48	1.27	1.56	1.20	1.65	1.12	1.74	1.05	1.84
30	1.35	1.49	1.28	1.57	1.21	1.65	1.14	1.74	1.07	1.83
31	1.36	1.50	1.30	1.57	1.23	1.65	1.16	1.74	1.09	1.83
32	1.37	1.50	1.31	1.57	1.24	1.65	1.18	1.73	1.11	1.82
33	1.38	1.51	1.32	1.58	1.26	1.65	1.19	1.73	1.13	1.81
34	1.39	1.51	1.33	1.58	1.27	1.65	1.21	1.73	1.15	1.81
35	1.40	1.52	1.34	1.58	1.28	1.65	1.22	1.73	1.16	1.80
36	1.41	1.52	1.35	1.59	1.29	1.65	1.24	1.73	1.18	1.80
37	1.42	1.53	1.36	1.59	1.31	1.66	1.25	1.72	1.19	1.80
38	1.43	1.54	1.37	1.59	1.32	1.66	1.26	1.72	1.21	1.79
39	1.43	1.54	1.38	1.60	1.33	1.66	1.27	1.72	1.22	1.79
40	1.44	1.54	1.39	1.60	1.34	1.66	1.29	1.72	1.23	1.79
45	1.48	1.57	1.43	1.62	1.38	1.67	1.34	1.72	1.29	1.78
50	1.50	1.59	1.46	1.63	1.42	1.67	1.38	1.72	1.34	1.77
55	1.53	1.60	1.49	1.64	1.45	1.68	1.41	1.72	1.38	1.77
60	1.55	1.62	1.51	1.65	1.48	1.69	1.44	1.73	1.41	1.77
65	1.57	1.63	1.54	1.66	1.50	1.70	1.47	1.73	1.44	1.77
70	1.58	1.64	1.55	1.67	1.52	1.70	1.49	1.74	1.46	1.77
75	1.60	1.65	1.57	1.68	1.54	1.71	1.51	1.74	1.49	1.77
80	1.61	1.66	1.59	1.69	1.56	1.72	1.53	1.74	1.51	1.77
85	1.62	1.67	1.60	1.70	1.57	1.72	1.55	1.75	1.52	1.77
90	1.63	1.68	1.61	1.70	1.59	1.73	1.57	1.75	1.54	1.78
95	1.64	1.69	1.62	1.71	1.60	1.73	1.58	1.75	1.56	1.78
100	1.65	1.69	1.63	1.72	1.61	1.74	1.59	1.76	1.57	1.78

mately from 0 to 1 and to fit a Beta distribution with the same mean and variance. The mean and variance of d vary according to the values of the independent variables, so the first step is to calculate them for the particular case concerned.

The description of the computing procedure is greatly facilitated by the introduction of matrix notation. Thus the set $\{y_1, y_2, \ldots, y_n\}$ of observations of the independent variable is denoted by the column vector $\mathbf{y}$. In the same way the set

$$\begin{bmatrix} x_{11} & x_{21} & \cdots & x_{k'1} \\ \vdots & \vdots & & \vdots \\ x_{1n} & x_{2n} & \cdots & x_{k'n} \end{bmatrix}$$

of observations of the independent variables is denoted by the matrix $\mathbf{X}$. We suppose that all these observations are measured from the sample means. The corresponding sets of first differences of the observations are denoted by $\Delta\mathbf{y}$ and $\Delta\mathbf{X}$.

$\mathbf{A}$ is the real symmetric matrix

$$\begin{bmatrix} 1 & -1 & 0 & \cdots & \cdots & \cdots & 0 \\ -1 & 2 & -1 & \cdots & \cdots & \cdots & \cdots \\ 0 & -1 & 2 & \cdots & \cdots & \cdots & \cdots \\ \cdots & \cdots & \cdots & \cdots & \cdots & \cdots & \cdots \\ \cdots & \cdots & \cdots & \cdots & \cdots & \cdots & 0 \\ \cdots & \cdots & \cdots & \cdots & \cdots & 2 & -1 \\ 0 & \cdots & \cdots & \cdots & 0 & -1 & 1 \end{bmatrix}$$

The moments of d are obtained by calculating the traces of certain matrices. The trace of a square matrix is simply the sum of the elements in the leading diagonal. For example, the trace of $\mathbf{A}$, denoted by tr $\mathbf{A}$, is $2(n-1)$, where n is the number of rows or columns in $\mathbf{A}$.

It was shown in Part I that the mean and variance of d are given by

$$E(d) = \frac{P}{n - k' - 1}, \tag{5}$$

$$\operatorname{var}(d) = \frac{2}{(n - k' - 1)(n - k' + 1)}\{Q - PE(d)\}, \tag{6}$$

where

$$P = \operatorname{tr} \mathbf{A} - \operatorname{tr}\{\mathbf{X}'\mathbf{A}\mathbf{X}(\mathbf{X}'\mathbf{X})^{-1}\}, \tag{7}$$

and

$$Q = \operatorname{tr} \mathbf{A}^2 - 2\operatorname{tr}\{\mathbf{X}'\mathbf{A}^2\mathbf{X}(\mathbf{X}'\mathbf{X})^{-1}\} + \operatorname{tr}[\{\mathbf{X}'\mathbf{A}\mathbf{X}(\mathbf{X}'\mathbf{X})^{-1}\}^2]. \tag{8}$$

The elements of $(\mathbf{X}'\mathbf{X})^{-1}$ will have been obtained for the calculation of the regression coefficients, and the elements of $\mathbf{X'AX}$ for the calulation of d; for $\mathbf{X'AX} = (\mathbf{\Delta X})'(\mathbf{\Delta X})$, so that the (i, j)th element of $\mathbf{X'AX}$ is simply the sum of products $\Sigma \Delta x_i \Delta x_j$. Thus the only new matrix requiring calculation is $\mathbf{X'A^2X}$. Now $\mathbf{X'A^2X}$ is very nearly equal to $(\mathbf{\Delta^2 X})'(\mathbf{\Delta^2 X})$, where $\mathbf{\Delta^2 X}$ represents the matrix of second differences of the independent variables. Thus the (i, j)th element of $\mathbf{X'A^2X}$ will usually be given sufficiently closely by $\Sigma(\Delta^2 x_i)\,(\Delta^2 x_j)$, where $\Delta^2 x_i$ stands for the second difference of the ith independent variable. (More exactly

$$(\mathbf{X'A^2X})_{ij} = \Sigma(\Delta^2 x_i)(\Delta^2 x_j) + (x_{i1} - x_{i2})(x_{j1} - x_{j2})$$
$$+ (x_{in-1} - x_{in})(x_{jn-1} - x_{jn}).) \ldots$$

We now assume that $\frac{1}{4}d$ is distributed in the Beta distribution with density

$$\frac{1}{B(p, q)}\left(\frac{d}{4}\right)^{p-1}\left(1 - \frac{d}{4}\right)^{q-1}.$$

This distribution gives

$$\mathbf{E}(d) = \frac{4p}{p + q},$$

$$\mathbf{var}\ d = \frac{16pq}{(p + q)^2(p + q + 1)},$$

from which we find p and q by the equations

$$\left.\begin{aligned} p + q &= \frac{\mathbf{E}(d)\{4 - \mathbf{E}(d)\}}{\mathbf{var}\ d} - 1, \\ p &= \tfrac{1}{4}(p + q)\mathbf{E}(d). \end{aligned}\right\} \qquad (9)$$

To test against positive serial correlation we require the critical value of $\frac{1}{4}d$ at the lower tail of the distribution. If $2p$ and $2q$ are integers, this can be obtained from Catherine Thompson's tables (1941), or indirectly from tables of the variance ratio or Fisher's z, such as those in the Fisher-Yates tables (1948); if $2p$ and $2q$ are not both integers, a first approximation may be found using the nearest integral values. Thus $F = \dfrac{p(4 - d)}{qd}$ is distributed as the variance ratio and $z = \frac{1}{2}\log_e F$ is Fisher's z, both with $n_1 = 2q$, $n_2 = 2p$ degrees of freedom.

References

Anderson, R.L. (1942). *Ann. Math. Statist.* **13**, 1.

Anderson, R.L. & Anderson, T.W. (1950). *Ann. Math. Statist.* **21**, 59.

Anderson, R.L. & Houseman, E.E. (1942). Tables of orthogonal polynomial values extended to $N = 104$. *Res. Bull. Iowa St. Coll.* no. 297.

Carter, A.H. (1947). *Biometrika*, **34**, 352.
Cochran, W.G. (1938). *J. R. Statist. Soc., Suppl.*, **5**, 171.
Courant, R. & Hilbert, D. (1931). *Methoden der Mathematischen Physik*. Berlin: Julius Springer.
Durbin, J. & Watson, G.S. (1950). *Biometrika*, **37**, 409.
Fisher, R.A. & Yates, F. (1948). *Statistical Tables*. Edinburgh: Oliver and Boyd.
Pearson, K. (1948). *Tables of the Incomplete Beta Function*. Cambridge University Press.
Prest, A.R. (1949). *Rev. Econ. Statist.* **31**, 33.
Scmultz, Henry (1938). *Theory and Measurement of Demand*, pp. 674–7. University of Chicago Press.
Snedecor, G.W. (1937). *Statistical Methods*. Collegiate Press.
Thompson, Catherine (1941). *Biometrika*, **32**, 151.
Watson, G.S. & Durbin, J. (1951). Exact tests of serial correlation using non-circular statistics.
Wise, M.E. (1950). *Biometrika*, **37**, 208.

Introduction to Box and Wilson (1951) On the Experimental Attainment of Optimum Conditions

Norman R. Draper
University of Wisconsin-Madison

Imperial Chemical Industries (ICI) was one of the first companies to recognize the value of statisticians and they had several, spread over the various divisions. ICI Dyestuffs Division, in particular, had the good fortune and foresight to have George E.P. Box as its statistical head and K.B. Wilson as one of its chemists. Their cooperation in this particular study set the foundation for the entire present field of response surface methodology.

A problem of great importance to a chemical company is that of finding optimal operating conditions for its processes. For example, what combination of input levels of temperature, pressure, reaction concentration, and so on delivers a process's maximum yield? Box and Wilson began their study by estimating first derivatives of a response surface, and moving up the unknown surface based on what their estimates showed. After early successes with laboratory-scale experiments on previously unexplored systems, they found, on the plant processes, that steepest ascent methods alone were unlikely to give much more improvement after a while. The presence of obvious interactions between inputs soon led to a study of curved surfaces and the realization that many of the surfaces that actually occured in practice were ridgelike ones, with multiple optima, affording choices in the experimental conditions that could be used.

Several of the ideas and techniques now in common use originated in this paper.

1. Instead of attention being concentrated only on effects estimated by a factorial design, which permitted only some of the higher-order derivatives to be estimated, thought was given to the *order* of the Taylor's series approximation and *all* derivatives of a specified order were estimated instead.

2. The alias (or bias) matrix was used to study the way estimates of lower-order derivatives were biased by higher-order derivatives that had been omitted from the fitted model.
3. Sequential experimentation was used and emphasized as a feasible technique for complicated studies of this type.
4. Canonical analysis (reduction) of response surfaces not only enabled one to grasp the type of response surface that existed, but also made it clear that, in many practical cases, redundancy existed and the surfaces were often not "onions" but multidimensional ridges. Such ridges typically provide multiple optimal points, giving a way to find satisfactory conditions for more than one response variable simultaneously.
5. The central composite design was introduced for the first time. Many years of investigation have since established this design as an economical and robust way to conduct sequential experimentation.

The paper itself follows a now classic pattern. Initially, fit a first-order model and make a steepest ascent or descent. Then fit a higher-order surface. The example in Sect. 3.1 shows the fitting of a second-order surface. Section 3.2 illustrates how the third order biases in the model may be associated with the second-order model coefficients.

In subsequent sections, the authors discuss designs suitable for first-order fitting and second-order fitting. We see the three-factor composite design in Fig. 3. Then follows an explanation of canonical reduction in order to appreciate what the model represents. The comprehensive example of Sec. 5.3 lays out the whole process for us. A Royal Statistical Society discussion follows the paper itself.

One of the remarkable qualities of this paper is the freshness of it after 40 years! It would still be a good place to send a beginner in response surface methodology. In fact, it might be the best place, because the calculations (which were then so difficult to carry out) are laid out fully for the student to verify.

The impact of this paper has been long-lasting. It influenced the way in which industrial and university experiments were planned and performed, it led to many practical design innovations, and it spawned a massive amount of research on the theoretical aspects of response surface methodology. The three books by Box et al. (1978), Box and Draper (1987), and Khuri and Cornell (1987) provide a comprehensive view of these developments.

In 1986, George Box was interviewed by Maurice DeGroot. The subsequent article, DeGroot (1987), contains many personal insights and is recommended for the light it casts on the development of response surface methods and of other topics. (see also the introduction to Box and Jenkins (1962) in this volume.)

I am grateful to George Box for some clarifying conversations.

References

Box, G.E.P., and Draper, N.R. (1987). *Empirical Model-Building and Response Surfaces.* Wiley, New York.

Box, G.E.P., Hunter, W.G., and Hunter, J.S. (1978). *Statistics for Experimenters.* Wiley, New York.

DeGroot, M.H. (1987). An interview with George Box, *Statist. Sci.*, **2**, 239–258.

Khuri, A.J., and Cornell, J.A. (1987). *Response Surfaces.* Marcel Dekker, New York and Quality Press, Milwaukee.

On the Experimental Attainment of Optimum Conditions

G.E.P. Box and K.B. Wilson
Imperial Chemical Industries
Blackley, Manchester, UK

1. Introduction

The work described is the result of a study extending over the past few years by a chemist and a statistician. Development has come about mainly in answer to problems of determining optimum conditions in chemical investigations, but we believe that the methods will be of value in other fields where experimentation is sequential and the error fairly small.

1.1. Statement of the Problem

The problem of experimental attainment of optimum conditions has been stated and discussed by Hotelling (1941) and by Friedman and Savage (1947). We define the problem as follows:

A *response* η is supposed dependent on the levels of k quantitative *factors* or *variables* $x_1, \ldots, x_t, \ldots, x_k$ capable of exact measurement and control. Thus for the uth combination of factor levels ($u = 1, \ldots N$)

$$\eta_u = \varphi(x_{1u}, \ldots x_{ku})$$

Owing to unavoidable uncontrolled factors, the observed response y_u varies in repeated observations, having mean η_u and variance σ^2.

In the whole k dimensional factor space, there is a region R, bounded by practical limitation to change in the factors, which we call the *experimental region.* The problem is to find, in the smallest number of experiments, the point $(x_1^0, \ldots x_t^0, \ldots x_k^0)$ within R at which η is a maximum or a minimum. In our field, yield, purity or cost of product are the responses which have to be

Editor's note: This paper could be read in conjunction with J.C. Kiefer's (1959) paper which is reproduced in part in Volume I of this series.

maximized (or minimized in the case of cost). The factors affecting these responses are variables such as temperature, pressure, time of reaction, proportions of the reactants.

2. Outline of the Methods Adopted

2.1. Experimental Strategy

The following circumstances influence the strategy of the experimenter:

(i) The magnitude of experimental error,
(ii) The complexity of the response surface,
(iii) Whether or not experiments may be conducted sequentially so that each set may be designed using the knowledge gained from the previous sets.

A sure way of finding optimum conditions would be to explore the whole experimental region. In practice this would have to be done by carrying out experiments on a grid of points extending through R, and would in principle be possible whatever were the circumstances i, ii, iii, above. Now for a response surface of given complexity there must exist a grid of minimum density to allow adequate approximation, but it is easy to show that the number of points on such a grid would usually be far too large, in the investigations with which we have been concerned. On the other hand, in these investigations the experimental error is usually rather small and the experiments are conducted sequentially. A shorter method is therefore required which will exploit these advantages.

Since the experimental error is small, small changes can be determined accurately, and the experimenter may explore adequately a small *sub-region* of the whole region R with only a few experiments. Also since the experiments are sequential the possibility arises of using the results obtained in one sub-region to move to a second in which the response is higher. By successive application of such a procedure, a maximum or at least a near-stationary point of high response should be reached. By restricting the region in which experiments are conducted, we may find only a local maximum and miss a higher ultimate maximum; however, where fuller exploration is impracticable this possibility must be accepted, and we have not found the implied risk troublesome.

2.2. Sequential Procedures for Attaining Higher Responses

A rule is required, therefore, which will lead from one sub-region to another where the response is improved. The most obvious procedure of this sort is the "one factor at a time" method. It has in one form or another been used for a long time, and is formalized by Friedman and Savage. A more efficient procedure is developed below.

2.2.1. *Maximum Gain Methods and Steepest Ascent*

We shall assume that within the region considered the derivatives of the response function are continuous. It is desired to proceed from a point O in the k-dimensional space to a point P distant r from O, at which the gain in response is a maximum. For convenience make O the origin so that the response there is $\varphi(O)$ and at P is $\varphi(P) = \varphi(x_1 \dots x_k)$. Since OP is to be equal to r,

$$r^2 = \sum x_t^2 \tag{1}$$

We require $\varphi(P) - \varphi(O)$ to be a maximum subject to condition (1). Using Lagrange's method of undetermined multipliers we construct the function:

$$\psi = \varphi(P) - \varphi(O) - \tfrac{1}{2}\mu \sum x_t^2 \tag{2}$$

The required maximum is where $\partial\psi/\partial x_t$ are all zero, that is at the point where the k equations

$$\mu x_t = \varphi_t(P) \qquad t = 1, 2, \dots, k \tag{3}$$

are satisfied. (The notation $\varphi_t(P)$ denotes that the function is differentiated with respect to x_t and the value at P inserted.) From (1) and (3)

$$\mu = \pm\{\sum[\varphi_t(P)]^2\}^{1/2}/r \tag{4}$$

The equations show that for P to be the point distance r from O at which the gain is a maximum, the co-ordinates at P must be proportional to the first order derivatives at P (assumed not all zero). This is equivalent to saying that the point of maximum gain will be one of the points at which a hypersphere radius r and centre O touches a contour surface. Now derivatives at P are in general unknown. Assuming however that φ can be represented about the origin by its Taylor's series in which terms of degree greater than d are ignored, the derivatives at P may be expressed in terms of those at O by the equation

$$\varphi_t(P) = \left[D_t \sum_{s=0}^{d-1} \{(D_1 x_1 + D_2 x_2 + \cdots + D_k x_k)^s/s!\} \right] \varphi(O), \tag{5}$$

where D_t denotes differentiation with respect to x_t and the expression within the square brackets is expanded and allowed to operate on φ, the values of the derivatives at O being inserted in the equation. Formulae which specify the co-ordinates of the point P of maximum gain distance r from O in terms of the derivatives at O may now be obtained by substituting (5) in (3) and choosing μ so that the solutions of (3) give a maximum. In particular if in the region considered second and higher degree terms may be ignored we have k equations,

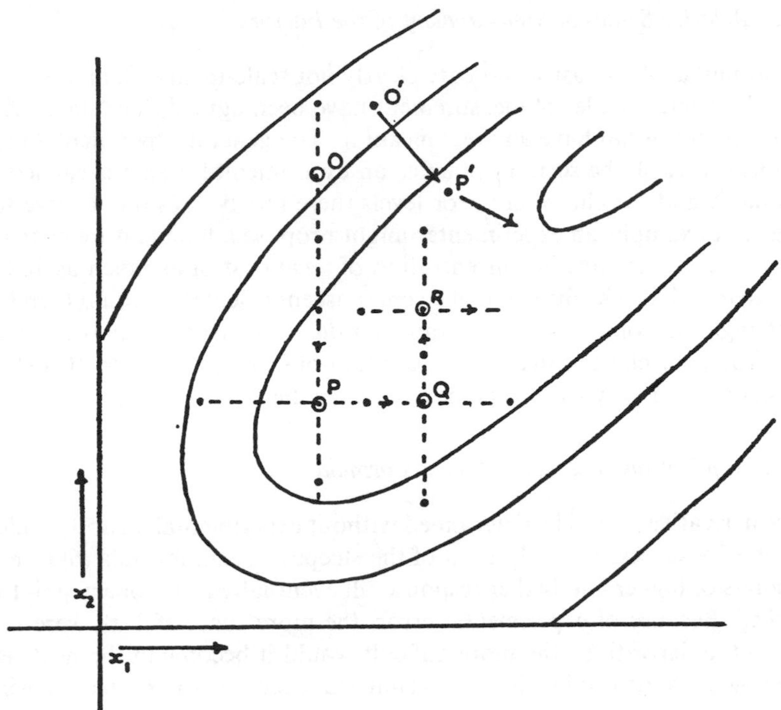

Figure 1. The "one factor at a time" path - - - and the steepest ascent path ——.

$$\mu x_t = \varphi_t(O) \qquad t = 1, 2, \ldots k. \tag{6}$$

These are the equations of "steepest ascent" or greatest slope from O. By varying the factors in proportion to their first order partial derivatives at O, we move in a direction at right angles to contour planes assumed to be locally parallel and equidistant. If, therefore, the first order derivatives at O can be determined, we may use them to move to a better response at P, provided that the distance r is not taken so large as to seriously strain the linear approximation. Fig. 1 illustrates for two factors x_1 and x_2 the "one factor at a time" approach via the path $OPQR$ and the steepest ascent approach along the path $O'P'$. The curved lines are response contours.

If the experimental error were larger and experimentation could therefore be conducted more economically by covering a wider subregion with rather more experimental points, the second degree approximation could be used. The locus of P would then be obtained by solving the k linear equations,

$$\mu x_t = \varphi_t + x_1 \varphi_{1t} + x_2 \varphi_{2t} + \cdots + x_k \varphi_{kt} \tag{7}$$

for a suitable set of values of μ, the values of the derivatives being taken at the point O.

2.2.2. *Relative Scales of Measurement of the Factors*

The formulae of the last section are clearly not scale-invariant, and it is only after the relative scales of measurement have been agreed that "distance" in the factor space can have any real meaning. The general experimental implications can easily be seen. In practice an experimental design is carried out around O, and the choice of factor levels there tacitly fixes the relative scale units. For example, an experimenter might propose a temperature change of say 5°C., but a change in concentration of a catalyst of as much as 100 per cent. to match it. The direction of steepest ascent calculated would then be at right angles to contour surfaces drawn on *this scale*. Experience is necessary, and clearly the choice can affect the amount of subsequent work. It is shown in §(4) how a grossly wrong choice can be rectified.

2.2.3. *Application of Steepest Ascent Formula*

If the derivatives could be determined without experimental error it would be possible by successive application of the steepest ascent formula (6) to move to points of higher and higher response till eventually a stationary point was reached. Because of experimental error, the more successful we were in reducing the derivatives, the more difficult would it become in the next set of experiments to determine them with sufficient accuracy to proceed further. In practice, therefore, the steepest ascent technique is employed to move from a point remote from a stationary value at which the surface slopes are large, to a point closer to it at which they are small compared with the experimental error.

2.3. Exploration of the Surface in a Near-Stationary Region

When the experimenter is in a near-stationary region (either as the result of successive application of the steepest ascent procedure, or because he is working to improve a process which has already received much previous attention), he will wish to conduct more detailed experiments in the limited region in order to determine the local nature of the surface.

In particular for example he will desire to know whether it probably contains a true maximum (in which case he will wish to estimate its position), a minimax or col (in which case he will wish to know how to "climb out of it"), or a ridge (when he will wish to know its direction and slope). By determining effects of second and possibly higher order in the region by suitably designed experiments it will be possible to answer these questions.

2.3.1. *Arrangement of Sections 3, 4 and 5*

In what follows, we shall be concerned with: (i) the application of the steepest ascent technique to move in the space to a near-stationary point; (ii) the examination of a near-stationary region.

For (i) we shall mainly require designs for the determination of first-order effects, and for (ii) designs to determine effects of higher order. Section (3), which follows, is concerned with the derivation of these designs, (4)* with the application of the first order steepest ascent process and (5) with the examination of a near-stationary region.

3. Experimental Designs to Determine Differential Coefficients

We assume that the response η at any point $(x_1, \ldots x_t, \ldots x_k)$ in the factor space can be represented by a regression equation of the form

$$\eta = \beta_0 + \beta_1 x_1 + \beta_2 x_2 + \cdots + \beta_{11} x_1^2 + \cdots + \beta_{12} x_1 x_2 + \cdots + \beta_{111} x_1^3 + \cdots \tag{8}$$

In general the coefficient of $x_1^\alpha x_2^\gamma \ldots x_k^\pi$ is denoted by $\beta_{1^\alpha 2^\alpha \ldots k^\pi}$, the notation implying that the subscript 1 is to be repeated α times; the subscript 2, γ times, etc.

Now suppose that *all* terms of degree d and less are included and within a given region a perfect fit is thus obtained. Then equation (8) will correspond with the Taylor expansion of the response function $\varphi(x_1, x_2, \ldots x_k)$ about the origin O of the variables and the derivatives at O will be simple multiples of the β's. For example, $\varphi_0 = \beta_0$, $\varphi_1 = \beta_1$, $\varphi_{11} = 2\beta_{11}$, $\varphi_{12} = \beta_{12}$, $\varphi_{111} = 6\beta_{111}$ and in general

$$\varphi_{1^\alpha 2^\gamma \ldots k^\pi} = \alpha! \gamma! \ldots \pi! \beta_{1^\alpha 2^\gamma \ldots k^\pi}, \tag{9}$$

where the quantity on the left denotes the value of the derivative obtained by differentiating the function α times with respect to x_1, γ times with respect to x_2, etc., at the origin O.

If responses are observed at a set of points suitably numerous and suitably placed within the region concerned, estimates b_0, b_1, b_2, etc., of β_0, β_1, β_2, etc., can be obtained by fitting the regression equation. Using (9) to obtain the appropriate multiplier, the b's will provide estimates f_0, f_1, f_2, etc., of the derivatives at O; φ_0, φ_1, φ_2, etc. We define an $N \times k$ matrix $\mathbf{D}$ called the *design matrix* whose elements are the levels of the k variables employed in each of the N trials. The elements $x_{1u}, x_{2u}, \ldots, x_{ku}$ of the uth row of $\mathbf{D}$ are the co-ordinates of a point in the k dimensional factor space. The N points thus defined constitute the *experimental design.* The N elements of a vector $\mathbf{Y}$ are observations of the responses obtained at these points and $\boldsymbol{\eta} = E(\mathbf{Y})$ is the corresponding vector of expected values. Suppose equation (8) contains L terms then it may be written

$$\eta = \sum_{i=1}^{L} \beta_i X_i, \tag{10}$$

* *Editor's note*: Section 4 has not been reproduced in this volume.

where the X_i are called the independent variables. In the case we are considering they are products and powers of the co-ordinates of the experimental points. Denoting by $\mathbf{X}$ the $N \times L$ *matrix of independent variables*, showing the N values of the L functions of the co-ordinates, (10) becomes in matrix notation

$$\boldsymbol{\eta} = \mathbf{X}\boldsymbol{\beta}, \tag{11}$$

where $\boldsymbol{\beta}$ is the $L \times 1$ vector of unknown constants. To fit the surface we employ the method of least squares. Assuming that the observational errors have constant variance σ^2 and are uncorrelated and also that $\mathbf{X}$ has rank L it is well known (Gauss, 1821; Markoff, 1912; Aitken, 1935; David and Neyman, 1938; Plackett, 1949) that

(i) Unbiased estimates of the elements of the vector $\boldsymbol{\beta}$, linear in the observations, with smallest variance are provided by the L x 1 vector of estimates $\mathbf{B} = \mathbf{TY}$ where the *transforming matrix* $\mathbf{T}$ is $(\mathbf{X}'\mathbf{X})^{-1}\,\mathbf{X}'$.

(ii) The $L \times L$ matrix of variances and covariances of these estimates is $\mathbf{C}^{-1}\sigma^2$ where $\mathbf{C} = \mathbf{X}'\mathbf{X}$ is the matrix of sums of squares and products of the independent variables. (We call $\mathbf{C}^{-1}$ the *precision matrix*.)

(iii) An unbiased estimate of $(N - L)\sigma^2$ is provided by the residual sum of squares

$$(N - L)s^2 = (\mathbf{Y} - \mathbf{XB})'(\mathbf{Y} - \mathbf{XB}) = \mathbf{Y}'\mathbf{Y} - \mathbf{B}'\mathbf{CB} = \mathbf{Y}'\mathbf{Y} - \mathbf{Y}'\mathbf{XB}.$$

We note that once a design for determining derivatives up to a given order has been decided on, the matrices $\mathbf{T}$ and $\mathbf{C}^{-1}$ can be calculated and tabled once for all. The computations necessary to calculate the estimates and their standard errors from any given set of observations are then extremely simple.

3.1. An Example

Suppose there were two variables only and it was desired to determine the derivatives up to the second order in a neighbourhood where terms of higher order could be ignored. We should fit the equation

$$\eta = \beta_0 + \beta_1 x_1 + \beta_2 x_2 + \beta_{11} x_1^2 + \beta_{22} x_2^2 + \beta_{12} x_1 x_2 \tag{12}$$

to an arrangement of six or more points in the two-dimensional space. One suitable arrangement would be a factorial design in which each of the two factors was varied at three levels. This design is rather specialized, however, and as a more general illustration we consider a design not of the factorial type.

One such arrangement consists of five points in the shape of a regular pentagon with one point in the centre. If the figure chosen had its uppermost side horizontal and the distance of any point from the centre was one unit, the co-ordinates of the point would be given by the design matrix

$$\mathbf{D} = \begin{array}{c} \\ 1 \\ 2 \\ 3 \\ 4 \\ 5 \\ 6 \end{array} \begin{array}{c} \begin{array}{cc} \quad x_1 \quad & \quad x_2 \quad \end{array} \\ \left[\begin{array}{rr} 0.5878 & 0.8090 \\ 0.9511 & -0.3090 \\ 0.0000 & -1.0000 \\ -0.9511 & -0.3090 \\ -0.5878 & 0.8090 \\ 0.0000 & 0.0000 \end{array}\right] \end{array}. \tag{13}$$

The matrix **X** of independent variables has six columns giving the corresponding values of $x_0, x_1, x_2, x_1^2, x_2^2$ and $x_1 x_2$ at the experimental points (x_0 is the variable corresponding with β_0 and is always unity). From **X** we calculate in turn* $\mathbf{X'X} = \mathbf{C}$, $\mathbf{C}^{-1}$ and $\mathbf{T}'$; the last two matrices are given below:

$$\mathbf{C}^{-1} = (\mathbf{X'X})^{-1} = \begin{array}{c} \\ b_0 \\ b_1 \\ b_2 \\ b_{11} \\ b_{22} \\ b_{12} \end{array} \begin{array}{c} \begin{array}{cccccc} b_0 & b_1 & b_2 & b_{11} & b_{22} & b_{12} \end{array} \\ \left[\begin{array}{cccccc} 1 & . & . & -1 & -1 & . \\ . & 0.4 & . & . & . & . \\ . & . & 0.4 & . & . & . \\ -1 & . & . & 1.6 & 0.8 & . \\ -1 & . & . & 0.8 & 1.6 & . \\ . & . & . & . & . & 1.6 \end{array}\right] \end{array}. \tag{14}$$

$\mathbf{T}' = \mathbf{X}(\mathbf{X'X})^{-1}$

$$= \begin{array}{c} \\ 1 \\ 2 \\ 3 \\ 4 \\ 5 \\ 6 \end{array} \begin{array}{c} \begin{array}{cccccc} b_0 & b_1 & b_2 & b_{11} & b_{22} & b_{12} \end{array} \\ \left[\begin{array}{rrrrrr} 0 & 0.2351 & 0.3236 & 0.0764 & 0.3236 & 0.7608 \\ 0 & 0.3804 & -0.1236 & 0.5236 & -0.1236 & -0.4702 \\ 0 & 0.0000 & -0.4000 & -0.2000 & 0.6000 & 0.0000 \\ 0 & -0.3804 & -0.1236 & 0.5256 & -0.1236 & 0.4702 \\ 0 & -0.2351 & 0.3236 & 0.0764 & 0.3226 & -0.7608 \\ 1 & 0.0000 & 0.0000 & -1.0000 & -1.0000 & 0.0000 \end{array}\right] \end{array}. \tag{15}$$

To use the design, observations would be made at levels of the variables proportional to the elements in **D**. The sums of products of the observations with the columns of the matrix $\mathbf{T}'$ supply the estimates b_0, b_1, etc., of the β's. The variances and covariances of the estimates are given by the elements of $\mathbf{C}^{-1}$ multiplied by σ^2 (assumed known).

The relations $f_0 = b_0$, $f_1 = b_1$, $f_2 = b_2$, $f_{11} = 2b_{11}$, $f_{22} = 2b_{22}$, $f_{12} = b_{12}$, would supply unbiased estimates f_0, f_1, etc., of the derivatives φ_0, φ_1, etc., at the origin of the variables, providing it could be assumed that over the region considered the second degree approximation was adequate.

* In this particular example there are as many constants to estimate as there are observations, consequently **X** is a square matrix and $\mathbf{T} = \mathbf{X}^{-1}$ can be obtained directly.

3.2. Effect of Lack of Fit of Assumed Equation. General Theory of Aliases

In fitting an equation like (12) to a set of experimental points the estimates of the β's differ from the true values on account of: (i) experimental error in determining the responses; (ii) biases arising when it is impossible to represent the function by an equation of the type fitted.

An experimental design should therefore be judged, partly by the apparent precision with which the desired constants are estimated (as shown by the matrix $\mathbf{C}^{-1}$), and partly by the magnitude of the possible biases in the estimates.

The nature of the biases for any given design may be determined as follows. Suppose that, within a given region of the factor space, the response function may be represented exactly by an equation such as (8) involving L constants; but that the experimenter assumes an adequate fit to be possible using an equation involving only $M < L$ of these constants, and performs $N \geq M$ experiments to estimate them. Then, it is wrongly assumed that—

$$\boldsymbol{\eta} = \mathbf{X}_1 \boldsymbol{\beta}_1, \tag{16}$$

when in fact

$$\boldsymbol{\eta} = \mathbf{X}_1 \boldsymbol{\beta}_1 + \mathbf{X}_2 \boldsymbol{\beta}_2, \tag{17}$$

and $\mathbf{X}_1$ is $(N \times M)$; $\mathbf{X}_2$, $(N \times S)$; $\boldsymbol{\beta}_1$, $(M \times 1)$; $\boldsymbol{\beta}_2$, $(S \times 1)$ and $M + S = L$. The least squares estimates obtained assuming (16) is true are

$$\mathbf{B}_1 = (\mathbf{X}_1'\mathbf{X}_1)^{-1}\mathbf{X}_1'\mathbf{Y} \tag{18}$$

and will in general be biased, since

$$\begin{aligned} E(\mathbf{B}_1) &= (\mathbf{X}_1'\mathbf{X}_1)^{-1}\mathbf{X}_1'\boldsymbol{\eta} && (19)\\ &= (\mathbf{X}_1'\mathbf{X}_1)^{-1}\mathbf{X}_1'\mathbf{X}_1\boldsymbol{\beta}_1 + (\mathbf{X}_1'\mathbf{X}_1)^{-1}\mathbf{X}_1'\mathbf{X}_2\boldsymbol{\beta}_2 && \text{(using (17))} \end{aligned}$$

and consequently

$$\mathbf{B}_1 \to \boldsymbol{\beta}_1 + \mathbf{T}_1\mathbf{X}_2\boldsymbol{\beta}_2 \tag{20}$$

The arrow notation is used to indicate that the quantity on the left is an unbiased estimate of the quantity on the right. The $M \times S$ matrix $\mathbf{T}_1\mathbf{X}_2$ will be called the *alias matrix* and be denoted by $\mathbf{A}$. The expression (20) defines M relations of the type

$$b_i \to \beta_i + \sum_{j=1}^{S} a_{ij}\beta_j, \tag{21}$$

where a_{ij} is the element of the ith row and jth column of $\mathbf{A}$. Thus if S extra constants β_j are needed accurately to describe the function, these may bias the estimates of the M constants which the experimenter is attempting to estimate. The extent of the biases will depend on the magnitude of the elements

a_{ij} of the matrix **A** which themselves depend upon the arrangement of points chosen. In particular, if a certain a_{ij} is zero then the estimate b_i will not be biased by β_j.

As an illustration the nature of the biases arising from third-order effects is found for the pentagonal design. The four third-order derivatives are proportional to $\beta_{111}, \beta_{222}, \beta_{112}, \beta_{122}$, which are the coefficients of x_1^3, x_2^3, $x_1^2x_2$ and $x_1x_2^2$ respectively. Therefore

$$\mathbf{X}_2 = \begin{array}{c} \\ 1 \\ 2 \\ 3 \\ 4 \\ 5 \\ 6 \end{array} \begin{array}{c} \begin{array}{cccc} x_1^3 & x_2^3 & x_1^2x_2 & x_1x_2^2 \end{array} \\ \left[\begin{array}{rrrr} 0.2031 & 0.5295 & 0.2795 & 0.3847 \\ 0.8602 & -0.0295 & -0.2795 & 0.0908 \\ 0.0000 & -1.0000 & 0.0000 & 0.0000 \\ -0.8602 & -0.0295 & -0.2795 & -0.0908 \\ -0.2031 & 0.5295 & 0.2795 & -0.3847 \\ 0.0000 & 0.0000 & 0.0000 & 0.0000 \end{array}\right] \end{array}. \tag{22}$$

Multiplying this matrix by **T** in (15) we obtain the alias matrix

$$\mathbf{A} = \mathbf{TX}_2 = \begin{array}{c} \\ 0 \\ 1 \\ 2 \\ 11 \\ 22 \\ 12 \end{array} \begin{array}{c} \begin{array}{cccc} 111 & 222 & 112 & 122 \end{array} \\ \left[\begin{array}{cccc} \cdot & \cdot & \cdot & \cdot \\ 0.75 & \cdot & \cdot & 0.25 \\ \cdot & 0.75 & 0.25 & \cdot \\ \cdot & 0.25 & -0.25 & \cdot \\ \cdot & -0.25 & 0.25 & \cdot \\ -0.50 & \cdot & \cdot & 0.50 \end{array}\right] \end{array}. \tag{23}$$

whence it follows that

$$\left.\begin{aligned} &b_0 \to \beta_0 \\ &b_1 \to \beta_1 + 0.75\beta_{111} + 0.25\beta_{122} \\ &b_2 \to \beta_2 + 0.75\beta_{222} + 0.25\beta_{112} \\ &b_{11} \to \beta_{11} + 0.25\beta_{222} - 0.25\beta_{112} \\ &b_{22} \to \beta_{22} - 0.25\beta_{222} + 0.25\beta_{112} \\ &b_{12} \to \beta_{12} - 0.50\beta_{111} + 0.50\beta_{122} \end{aligned}\right\}. \tag{24}$$

In general the coefficients in A and consequently the magnitude of the possible biases will depend upon the choice of experimental design. It can be shown, however, that if more constants are needed to represent the function than there are experiments, the extra constants will either bias the estimates or else appear in the residual degrees of freedom and bias the error estimate. The only exception to this rule occurs when a design is such that one or more columns of $\mathbf{X}_2$ consists entirely of zeros. In this case the corresponding "extra" constants will not appear.

3.3. Comparison of Designs

The matrices $\mathbf{C}^{-1}$ and $\mathbf{A}$ depend only on the arrangement of the experimental points, and not on any particular set of observations $\mathbf{Y}$ which might be made at these points. They supply therefore an objective basis for comparing designs for precision and bias. If the number of experiments in two designs which it is desired to compare are not the same, the precision of the estimates may be brought to a common basis by considering the relative variance of the estimates *per observation*. Thus we compare $N_1\mathbf{C}_1^{-1}$ with $N_2\mathbf{C}_2^{-1}$.

A more difficult problem arises because both $\mathbf{C}^{-1}$ and $\mathbf{A}$ are dependent on scale factors. If therefore the comparison between designs is to have any meaning they must first be brought to the same "size". The conclusions we draw will depend to some extent on how we define "size". A design is of course a k dimensional distribution of points and the measures for size adopted in this paper are the marginal second moments of these points about their means. Thus if there are k factors and N observations s_t is called the *spread* for the tth variable where

$$s_t^2 = \left\{\sum_{u=1}^{N} (x_{tu} - \bar{x}_t)^2\right\}/N,$$

and two designs are regarded as of comparable size when they are measured so that the spread for each of the factors is the same in the two designs. For designs in which each linear effect is uncorrelated with every other effect the convention amounts to choosing the units so that the linear effects are determined with equal accuracy in the designs to be compared. Other conventions could of course be used, and would possibly be more appropriate for certain types of arrangement. The equating of second moments does in the cases so far studied seem to provide reasonable results. In any case changes of scale have opposite effects on $\mathbf{C}^{-1}$ and $\mathbf{A}$, for example choosing wider ranges for the factors in a design will result in the reduction of the variances of the estimates but an increase in the coefficients of the aliases. It is unlikely therefore that we should be seriously misled by this convention.

For example, suppose the merits of the 3^2 factorial and pentagonal design in determining the derivatives up to second order are to be compared and the levels of the factorial design are $-1, 0$, and 1 and the levels for the pentagonal design are those shown in (13). Then for the factorial design, $s_1^2 = s_2^2 = 2/3$ and for the pentagonal, $s_1^2 = s_2^2 = 5/12$, consequently the dimensions of the pentagonal design must be increased by the factor

$$g = s(F)/s(P) = (2/3 \times 12/5)^{1/2} = 1.265 \tag{25}$$

The appearances of the two designs scaled on this basis are shown in Fig. 2.

The relative precision of the estimates when the designs are scaled on this basis may be judged from the matrices $N\mathbf{C}^{-1}$, the elements of which are proportional to the variances and covariances per observation.

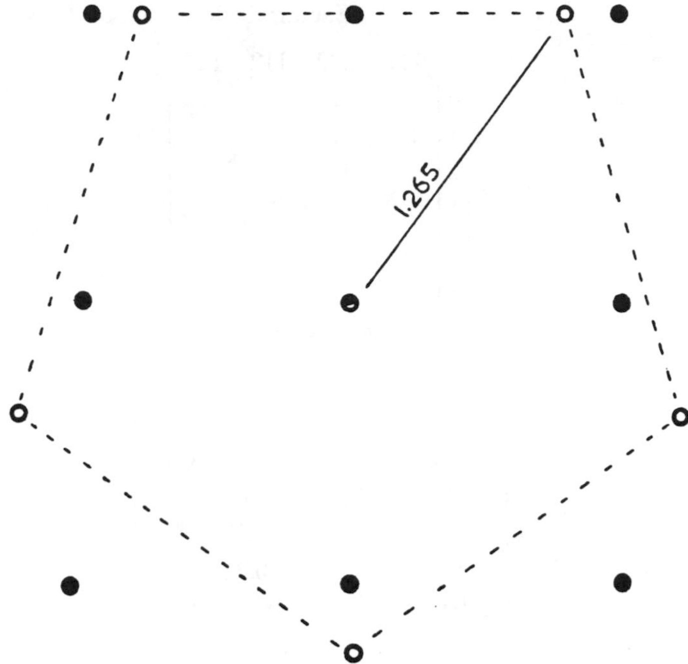

Figure 2. The pentagonal design and the 3^2 factorial with marginal second moments equated.

Factorial

$$
9\mathbf{C}_F^{-1} = \begin{array}{c} \\ b_0 \\ b_1 \\ b_2 \\ b_{11} \\ b_{22} \\ b_{12} \end{array}
\begin{array}{c} \begin{array}{cccccc} b_0 & b_1 & b_2 & b_{11} & b_{22} & b_{12} \end{array} \\
\left[\begin{array}{cccccc}
5.0 & . & . & -3.0 & -3.0 & . \\
. & 1.5 & . & . & . & . \\
. & . & . & . & . & . \\
-3.0 & . & . & 4.5 & . & . \\
-3.0 & . & . & . & 4.5 & . \\
. & . & . & . & . & 2.25
\end{array}\right] \end{array}
$$

Pentagonal

$$
6\mathbf{C}_P^{-1} = \begin{array}{c} \begin{array}{cccccc} b_0 & b_1 & b_2 & b_{11} & b_{22} & b_{12} \end{array} \\
\left[\begin{array}{cccccc}
6.0 & . & . & -3.75 & -3.75 & . \\
. & 1.5 & . & . & . & . \\
. & . & 1.5 & . & . & . \\
-3.75 & . & . & 3.75 & 1.875 & . \\
-3.75 & . & . & 1.875 & 3.75 & \\
. & . & . & . & . & 3.75
\end{array}\right] \end{array}. \tag{26}
$$

The biases due to possible third order effects are found from the alias matrices:

$$
\text{Factorial}\qquad
\mathbf{A}_F = \begin{array}{c} \\ 0 \\ 1 \\ 2 \\ 11 \\ 22 \\ 12 \end{array}
\begin{array}{c}
\begin{array}{cccc} 111 & 222 & 112 & 122 \end{array} \\
\left[\begin{array}{cccc}
\cdot & \cdot & \cdot & \cdot \\
1.0 & \cdot & \cdot & \cdot \\
\cdot & 1.0 & \cdot & \cdot \\
\cdot & \cdot & \cdot & \cdot \\
\cdot & \cdot & \cdot & \cdot \\
\cdot & \cdot & \cdot & \cdot
\end{array}\right]
\end{array}.
$$

$$
\text{Pentagonal}\qquad
\mathbf{A}_P = \begin{array}{c}
\begin{array}{cccc} 111 & 222 & 112 & 122 \end{array} \\
\left[\begin{array}{cccc}
\cdot & \cdot & \cdot & \cdot \\
1.2 & \cdot & \cdot & 0.4 \\
\cdot & 1.2 & 0.4 & \cdot \\
\cdot & 0.32 & -0.32 & \cdot \\
\cdot & -0.32 & 0.32 & \cdot \\
-0.63 & \cdot & \cdot & 0.63
\end{array}\right]
\end{array}. \tag{27}
$$

There is little to choose in the relative precision of the designs. The pentagonal design estimates the quadratic effects rather more accurately and interaction effects rather less accurately. The quadratic estimates are, however, correlated in the pentagonal design. The alias matrices show that if third order terms were not negligible the estimates would be more heavily biased in the case of the pentagonal design. This is to be expected in view of the fewer experiments used.

Two types of design are of particular importance, and will be referred to as designs of types A and B. Designs of type A of order d will be defined as those which give unbiased estimates of all derivatives of order 1 to d, providing the assumption is strictly true that all terms of higher degree may be ignored. In such designs N can be as small as M, the number of constants to be determined. Designs of type B of order d will be defined as those which provide unbiased estimates of all derivatives of order 1 to d, even though terms of order $d + 1$ exist. In such designs the number of experiments N must be larger than the number of constants M to be estimated, so that $N - M$ residual degrees of freedom exist in which to accommodate the aliases of order $d + 1$. Such designs need not, however, allow for separate estimation of these higher order effects, and S the number of extra constants accommodated may be considerably greater than $N - M$, the number of residual degrees of freedom.

Although φ_0, the value of the function at the point $(0, 0, \ldots 0)$ which can be regarded as the derivative of order zero, appears in the set-up, we do not need to know its value in order to apply the formula for steepest ascent or to study the nature of the response surface. From this point of view bias in its estimate will be immaterial.

3.4. Designs for Determining First Order Differential Coefficients

If higher order differential coefficients can be ignored, the first order differential coefficients at a point O, the origin for the variables, can be obtained by fitting the plane

$$\eta = \beta_0 + \beta_1 x_1 + \beta_2 x_2 + \cdots + \beta_k x_k \tag{28}$$

to a set of points arranged in a suitable manner about O. The estimates b_0, $b_1, \ldots b_k$ of the β's will also be estimates $f_0, f_1, \ldots, f_k$ of $\varphi_0, \varphi_1, \ldots, \varphi_k$, the required derivatives at the point O. Designs suitable for this purpose can be obtained by varying each of the factors $x_1, x_2, \ldots, x_k$ at only two levels. For convenience these levels are taken to be -1 and 1.

If all possible combinations of levels are included, the design will be a complete factorial design, and the 2^k points in space will be at the vertices of a k dimensional hypercube of side 2 units with its centre at the point $(0, 0, \ldots 0)$. If not all the 2^k experiments are performed, the design will be called an *incomplete factorial.* Particular classes of incomplete factorial designs are the fractional factorial designs introduced by Finney (1945) and the multifactorial designs of Plackett and Burman (1946).

First order designs of type A are provided by the multifactorial designs of Plackett and Burman. These authors supply designs for $k = 3, 7, 11, 15, \ldots, 4m - 1, \ldots 99$ factors in $N = 4, 8, 12, 16, \ldots 4m \ldots, 100$ experiments. For intermediate values of k the next higher design may be used and the appropriate number of columns omitted from the design matrix. When N is a power of two these arrangements are identical with the fractional factorial designs.

First order designs of type B can always be obtained by duplicating the appropriate Plackett and Burman design of type A and order 1, with reversed signs. This means that if a design of type A had been completed it would always be possible to make a partial check on the assumptions by modifying it in this simple way to a design of type B. Any variable held constant in the first set of experiments could be held at a different level* on duplicating the design. Thus corresponding to a type A design for $N - 1$ factors in N experiments we would obtain a type B design for N factors in $2N$ experiments. The above assertion is a special case of a more general theorem proved in Appendix (2). All these designs are orthogonal (in the sense that $\mathbf{X}'\mathbf{X} = N\mathbf{I}$). The transforming matrix $\mathbf{T}$ is therefore $N^{-1}\mathbf{X}'$, and the estimates f_0, f_1, f_2, etc., can be obtained very simply and are uncorrelated. The alias matrix is of the

* The intention here is to indicate the method for constructing the design rather than the practical advisability of introducing an extra factor in this way.

form $N^{-1}\mathbf{X}_1'\mathbf{X}_2$, and since for each factor the levels $+1$ and -1 occur equal numbers of times none of the first-order estimates are biased by quadratic derivatives φ_{11}, φ_{22}, etc. Thus, if the curvature of a surface, however great, could be expressed in terms of quadratic effects alone, the estimates of the first order effects would remain completely unaffected. On the other hand, the existence of mixed second order derivatives φ_{12}, etc., corresponding to the two-factor interactions, may bias these estimates.

3.4.1. *Two Level Factorial and Fractional Factorial Designs*

Most of the designs of order 1 used in this work have been two level factorial and fractional factorial arrangements. It is not our intention to describe their construction and properties in detail. They have been the subject of study by Fisher (1935), Yates (1935 and 1937), Hotelling (1944), Plackett and Burman (1946), Rao (1947), Kempthorne (1947), Davies and Hay (1950), and others. We illustrate here certain features which are of particular interest in the study of the present problem. Consider for simplicity, the two-level factorial design for the variables x_1, x_2, x_3.

The design matrix **D** consists of the vertices of a 3-dimensional cube. Suppose the function

$$\eta = \beta_0 x_0 + \beta_1 x_1 + \beta_2 x_2 + \beta_3 x_3 + \beta_{12} x_1 x_2 + \beta_{13} x_1 x_3 + \beta_{23} x_2 x_3 + \beta_{123} x_1 x_2 x_3$$

is fitted to the responses found at these eight points.

The matrix of independent variables is

$$\begin{array}{cccccccc} & (1) & (2) & (3) & (5) & (6) & (7) & (4) \\ x_0 & x_1 & x_2 & x_3 & x_1x_2 & x_1x_3 & x_2x_3 & x_1x_2x_3 \end{array}$$

$$\mathbf{X}_1 = \begin{bmatrix} 1 & -1 & -1 & -1 & 1 & 1 & 1 & -1 \\ 1 & -1 & -1 & 1 & 1 & -1 & -1 & 1 \\ 1 & -1 & 1 & -1 & -1 & 1 & -1 & 1 \\ 1 & -1 & 1 & 1 & -1 & -1 & 1 & -1 \\ 1 & 1 & -1 & -1 & -1 & -1 & 1 & 1 \\ 1 & 1 & -1 & 1 & -1 & 1 & -1 & -1 \\ 1 & 1 & 1 & -1 & 1 & -1 & -1 & -1 \\ 1 & 1 & 1 & 1 & 1 & 1 & 1 & 1 \end{bmatrix}. \qquad (29)$$

The matrix **T** is $\frac{1}{8}\mathbf{X}_1'$. The estumate of each β is found, therefore, by forming the sum of products of the observed responses with the appropriate column of matrix $\mathbf{X}_1$ and dividing the result by 8. The estimates $b_0, b_1, b_2, b_3, b_{12}, b_{13}, b_{23}, b_{123}$ thus obtained are the mean $\bar{y}$, and one half of the quantities usually referred to as main effects and interactions (see for example Yates, 1937).

From the approach of the previous section, the nature of possible biases in the estimates for this design may now be determined. We assume that the response function can be represented exactly by a Taylor series, including all terms* up to degree d, involving the fitting of L constants. The matrix $\mathbf{X}_1$ of independent variables then corresponds to $\mathbf{X}_1$ of equation (16), whilst $\mathbf{X}_2$ is the matrix of powers and products of degree d and less not included in $\mathbf{X}_1$. Now (Fisher, 1942; Finney, 1945) the columns of the matrix $\mathbf{X}_1$ may be regarded as the elements of a finite Abelian group. In particular they form a set closed with respect to multiplication (in the sense that if the elements of the columns are multiplied by themselves or by the elements of any other column, the new column resulting will always be identical with one or other of those already contained in $\mathbf{X}_1$). It follows that every column of the matrix $\mathbf{X}_2$ must coincide with one or other of the columns in $\mathbf{X}_1$. The alias matrix, $\mathbf{A} = \frac{1}{8}\mathbf{X}_1'\mathbf{X}_2$, is particularly simple therefore. Every column in $\mathbf{X}_2$ will have a non-zero inner product with one and only one column in $\mathbf{X}_1$, and consequently each of the extra constants will be associated with one and only one of the estimates b_0, b_1, etc. The group property arises because for any row of the matrix (29) $x_0 = 1 = x_1^2 = x_2^2 = x_3^2$; consequently, for example, $x_1^2 x_2 = x_2$, $x_1^4 x_2^3 x_3 = x_2 x_3$, etc. It follows from the nature of $\mathbf{A}$ that $\beta_{11}, \beta_{22}, \beta_{1111}, \beta_{1122}$, and in fact all the constants containing an even number of subscripts will appear in the expected value of b_0 and nowhere else. Similarly β_{112} will be associated with b_2, $\beta_{11112223}$ with b_{23} and so on. The defining relation for the group may be written

$$I = 11 = 22 = 33. \tag{30}$$

The identity I corresponds with the column x_0 and consequently with b_0. Since this also implies that $I = 1111 = 1122$, etc., the nature of the biases associated with a particular estimate may be obtained by multiplying the defining relation by the subscripts of the estimate. For example, $1 = 111 = 122 = 133 = 11111 = 11122$, etc., and consequently

$$b_1 \to \beta_1 + \beta_{111} + \beta_{122} + \beta_{133} + \beta_{11111} + \text{etc.} \tag{31}$$

or

$$b_1 \to \varphi_1 + \tfrac{1}{6}\varphi_{111} + \tfrac{1}{2}\varphi_{122} + \tfrac{1}{2}\varphi_{133} + \tfrac{1}{120}\varphi_{11111} + \text{etc.} \tag{32}$$

where the multipliers of the derivatives are given by (9). The expected values,

* The logic of including all derivatives up to a given degree d, rather than only those which correspond to interaction terms, is seen if it is supposed that experiments are being performed in the neighbourhood of a maximum. Since, for a maximum the matrix of second order derivatives φ_{st}, must be negative definite—and this in turn implies that $\varphi_{ss}\varphi_{tt} > \varphi_{st^2}$—the geometric mean of any pair of quadratic terms would be greater than the corresponding interaction term. Thus if the interaction terms could not be ignored then certainly the corresponding quadratic terms could not be negligible.

including terms up to order 3, for the eight estimates from this complete factorial design are thus

$$\left.\begin{aligned} &b_0 \to \varphi_0 + \tfrac{1}{2}\varphi_{11} + \tfrac{1}{2}\varphi_{22} + \tfrac{1}{2}\varphi_{33} \\ &b_1 \to \varphi_1 + \tfrac{1}{6}\varphi_{111} + \tfrac{1}{2}\varphi_{122} + \tfrac{1}{2}\varphi_{133} \\ &b_2 \to \varphi_2 + \tfrac{1}{6}\varphi_{222} + \tfrac{1}{2}\varphi_{112} + \tfrac{1}{2}\varphi_{233} \\ &b_3 \to \varphi_3 + \tfrac{1}{6}\varphi_{333} + \tfrac{1}{2}\varphi_{113} + \tfrac{1}{2}\varphi_{223} \\ &b_{12} \to \varphi_{12} \\ &b_{13} \to \varphi_{13} \\ &b_{23} \to \varphi_{23} \\ &b_{123} \to \varphi_{123} \end{aligned}\right\}. \tag{33}$$

Rao (1947) obtained fractional factorial designs with the properties he desired by associating further factors with the independent variables corresponding with the interactions in the full factorial design, a method used previously by Yates (1935). Thus to obtain the design of type B and order 1 from which unbiased estimates of the "main effects" could be derived even though two-factor interactions existed, Rao associated the factor x_4 with the elements in the column corresponding to $x_1x_2x_3$. This device is also used in the derivation of fractional factorial designs by Davies and Hay (1950), who show in addition how the association will determine the alias relationships. Thus x_4 may be put equal to $x_1x_2x_3$ or to $-x_1x_2x_3$. If the former relationship is used we have in shortened notation

$$4 = 123,$$

whence since $44 = I$ we can multiply both sides by 4 and we have

$$I = 1234.$$

This relation indicates that if the elements of the columns corresponding to the variables x_1, x_2, x_3 and x_4 are multiplied together the result will be a column of $+1$'s. Thus the particular factor combinations used correspond to that half of a full 2^4 factorial design for which the elements of the independent variable $x_1x_2x_3x_4$ are all equal to $+1$. The design under investigation is thus a one-half replicate of the full 2^4 factorial "split" along the interaction having subscript 1234. (The relationship $-4 = 123$ defines the other half.) Therefore the defining relation is

$$I = 11 = 22 = 33 = 44 = 1234\ (= 1111 = 1122 = 111234, \text{etc.}) \tag{34}$$

and consequently to terms of third order

$$\left.\begin{aligned}
b_0 &\to \varphi_0 + \tfrac{1}{2}\varphi_{11} + \tfrac{1}{2}\varphi_{22} + \tfrac{1}{2}\varphi_{33} + \tfrac{1}{2}\varphi_{44} \\
b_1 &\to \varphi_1 + \tfrac{1}{6}\varphi_{111} + \tfrac{1}{2}\varphi_{122} + \tfrac{1}{2}\varphi_{133} + \tfrac{1}{2}\varphi_{144} + \varphi_{234} \\
b_2 &\to \varphi_2 + \tfrac{1}{6}\varphi_{222} + \tfrac{1}{2}\varphi_{112} + \tfrac{1}{2}\varphi_{233} + \tfrac{1}{2}\varphi_{244} + \varphi_{134} \\
b_3 &\to \varphi_3 + \tfrac{1}{6}\varphi_{333} + \tfrac{1}{2}\varphi_{113} + \tfrac{1}{2}\varphi_{223} + \tfrac{1}{2}\varphi_{344} + \varphi_{124} \\
b_{12} &\to \varphi_{12} + \varphi_{34} \\
b_{13} &\to \varphi_{13} + \varphi_{24} \\
b_{23} &\to \varphi_{23} + \varphi_{14} \\
b_{123} &\to \varphi_4 + \tfrac{1}{6}\varphi_{444} + \tfrac{1}{2}\varphi_{114} + \tfrac{1}{2}\varphi_{224} + \tfrac{1}{2}\varphi_{334} + \varphi_{123}
\end{aligned}\right\}. \tag{35}$$

This is a design of type B for, if third and higher order effects were absent but second order effects were not, b_1, b_2, b_3 and b_4 would supply estimates of $\varphi_1, \varphi_2, \varphi_3, \varphi_4$ unbiased by the effects of second order.

3.4.2. *Fractional Factorial Designs of Type A*

For 3 factors in 4 experiments a third factor may be associated with the independent variable $x_1 x_2$ in the 2^2 factorial arrangement. The matrix of independent variables is thus

$$\mathbf{X} = \begin{array}{c} \begin{matrix} x_0 & x_1 & x_2 & x_3 = x_1 x_2 \end{matrix} \\ \begin{bmatrix} 1 & 1 & 1 & 1 \\ 1 & 1 & -1 & -1 \\ 1 & -1 & 1 & -1 \\ 1 & -1 & -1 & 1 \end{bmatrix} \end{array} \tag{36}$$

and since $I = 11 = 22 = 33 = 123$

$$\left.\begin{aligned}
b_0 &\to \varphi_0 + \tfrac{1}{2}\varphi_{11} + \tfrac{1}{2}\varphi_{22} + \tfrac{1}{2}\varphi_{33} \\
b_1 &\to \varphi_1 + \varphi_{23} \\
b_2 &\to \varphi_2 + \varphi_{13} \\
b_3 &\to \varphi_3 + \varphi_{12}
\end{aligned}\right\}. \tag{37}$$

For 7 factors in 8 experiments to order 2, the design is formed by associating each of the interaction columns in (29) with new factors. If we put

$$x_4 = x_1 x_2 x_3, \quad x_5 = x_1 x_2, \quad x_6 = x_1 x_3, \quad x_7 = x_2 x_3$$

$$\left.\begin{aligned}
b_0 &\to \varphi_0 + \tfrac{1}{2}\varphi_{11} + \tfrac{1}{2}\varphi_{22} + \cdots + \tfrac{1}{2}\varphi_{77} \\
b_1 &\to \varphi_1 + \varphi_{25} + \varphi_{36} + \varphi_{47}
\end{aligned}\right\} \tag{38}$$

$$\left.\begin{aligned} b_2 &\to \varphi_2 + \varphi_{15} + \varphi_{37} + \varphi_{46} \\ b_3 &\to \varphi_3 + \varphi_{16} + \varphi_{27} + \varphi_{45} \\ b_{12} &\to \varphi_5 + \varphi_{12} + \varphi_{34} + \varphi_{67} \\ b_{13} &\to \varphi_6 + \varphi_{13} + \varphi_{24} + \varphi_{57} \\ b_{23} &\to \varphi_7 + \varphi_{14} + \varphi_{23} + \varphi_{56} \\ b_{123} &\to \varphi_4 + \varphi_{35} + \varphi_{26} + \varphi_{17} \end{aligned}\right\} \tag{38}$$

When the factors are intermediate in number between 3, 7, etc., Rao, and Plackett and Burman, suggest that the next higher design be used, treating the remaining factors as dummies. The aliases may then be obtained from the table of aliases for the full design by omitting all terms containing dummy subscripts. Some caution is required here, since in some cases not all such designs are equally satisfactory. If four factors are to be tested in eight experiments it would usually be most desirable to omit a set of columns such as (5), (6), (7) in the design for seven variables. If other sets of 3 subscripts were omitted the designs would of course be of type A, but not necessarily of type B. In general the design would only be of type B if the 4 columns selected in (29) were such that, for every point on the design, the product of the co-ordinates was equal to unity, thus ensuring a sub-group relation of type $I = 1234$.

With 5 or 6 factors the arrangements are of the same type whichever subscripts are omitted. For 5 factors one of the first order effects has two second order aliases and the remaining four have only one, while the two dummy comparisons each have two second order aliases. For 6 factors each of the first order effects have two second-order aliases and the dummy comparison has three second-order aliases.

The location of the aliases is often of considerable importance both in planning the sets of experiments and in their subsequent interpretation. Thus the experimenter might feel that if appreciable interactions did occur they would be more likely, between certain pairs of factors, than between other pairs. In such instances this information could be used in arranging the experiment so that, as far as possible, these more suspect second-order effects were not associated with first order effects, but were isolated in comparisons corresponding to "dummy" factors. The possibilities can be examined by altering the identity of the dummy factors and rearranging the subscripts among themselves.

Certain limitations will be found to exist. For example in the design for determining first order effects for 5 factors in 8 experiments, it is not possible to arrange that two interactions which do not have one subscript in common should appear one in each of the two residual comparisons. From the nature of the alias relations it is clear that type A designs must always be used with extreme caution.

3.4.3. *Fractional Factorial Designs of Type B*

The design of type B for 4 factors in 8 experiments has already been derived (35). It could also have been obtained by duplicating with reversed signs the matrix of independent variables (36) for the design of type A, associating a fourth factor x_4 with x_0. It is easy to see that the new defining relation is obtained by reproducing terms containing an even number of elements as they stand, and multiplying the terms containing an odd number of elements by the subscripts of the newly introduced variable. In this case $I = 1234$.

If any one of the subscripts is omitted the type B design for 3 factors is obtained, which is of course the complete factorial. This design is one of a class which (on the assumption that effects of higher order than the second may be ignored) allows unbiased estimates to be obtained not only of effects of first order but also of mixed derivatives φ_{12}, φ_{13}, etc. To distinguish these designs they will be said to be of type B' and order 1.

The design for 8 factors in 16 experiments is obtained by duplicating with reversal of signs the matrix of independent variables (29) for the type A design for 7 factors in 8 experiments. The relations below, (39), may be compared with those obtained for the type A design. The groups of second order effects which were associated with first order effects in the original design, keep their identity, but are associated with the extra degrees of freedom arising from the duplication.

$$\left.\begin{aligned}
b_0 &\to \varphi_0 + \tfrac{1}{2}\varphi_{11} + \tfrac{1}{2}\varphi_{22} + \cdots + \tfrac{1}{2}\varphi_{88}\\
b_1 &\to \varphi_1\\
b_2 &\to \varphi_2\\
b_3 &\to \varphi_3\\
b_4 &\to \varphi_4\\
b_5 &\to \varphi_5\\
b_6 &\to \varphi_6\\
b_7 &\to \varphi_7\\
b_8 &\to \varphi_8\\
b_{12} &\to \varphi_{12} + \varphi_{34} + \varphi_{58} + \varphi_{67}\\
b_{13} &\to \varphi_{13} + \varphi_{24} + \varphi_{57} + \varphi_{68}\\
b_{14} &\to \varphi_{14} + \varphi_{23} + \varphi_{56} + \varphi_{78}\\
b_{15} &\to \varphi_{15} + \varphi_{28} + \varphi_{37} + \varphi_{46}\\
b_{16} &\to \varphi_{16} + \varphi_{27} + \varphi_{38} + \varphi_{45}\\
b_{17} &\to \varphi_{17} + \varphi_{26} + \varphi_{35} + \varphi_{48}\\
b_{18} &\to \varphi_{18} + \varphi_{25} + \varphi_{36} + \varphi_{47}
\end{aligned}\right\} . \qquad (39)$$

The designs for 7 and for 6 factors may be obtained as before by omitting subscripts in the design for 8 factors or by duplicating the corresponding type A design with reversed signs. A type B design for 5 factors could be obtained in a similar way, the design being a half replicate of the full 2^5 factorial typified by the sub-group relation $I = 1234$. In this case, however, a design of type B' exists, a half replicate with defining relation $I = 12345$, which would normally be used in practice. It is easily obtained by writing down the full 2^4 factorial design for four factors x_1, x_2, x_3, x_4, and adding the column $x_5 = x_1x_2x_3x_4$ obtained by multiplying these elements together. The sub-group relation $I = 12345$ together with the group relations $I = 11 = 22 = 33 = 44 = 55$ allow the aliases to be easily written down. If third and higher order effects are ignored, the 15 estimates corresponding with the 5 main effects and 10 two-factor interactions are all unbiased.

3.5. Designs to Estimate Effects of First and Second Order

First order designs of both types A and B were supplied by two level factorial and incomplete factorial arrangements. When it is desired to estimate effects of higher order, factorial and fractional factorial designs are less satisfactory, (1) because the number of experiments necessary is usually very much larger than the number of effects it is desired to estimate, and (2) because the relative precision with which the effects are estimated is often not that desired.

3.5.1. *Effects Estimated Using Factorial Designs*

As an example, suppose that $k = 3$ variables are being considered. A table is shown below setting out the subscripts of effects up to order 6 (to save space the subscripts are not fully enumerated for $d = 4$, $d = 5$ and $d = 6$).

Subscripts for Terms of Order d when $k = 3$

$d = 0$	$d = 1$	$d = 2$	$d = 3$	$d = 4$	$d = 5$	$d = 6$
0	1	12	123	1123	11223	112233
	2	13	122	1223	11233	111111
	3	23	112	1233	12233	222222
		11	133	1122	11111	333333
		22	113	1133	22222	122222
		33	223	2233	33333	111112
			233	1111	12222	133333
			111	2222	11112	111113

222	3333	13333	222223
332	1222	11113	233333
	1112	23333	112222
	to 15 terms	to 21 terms	to 28 terms

Consider a factorial experiment with each factor varied at three levels. With a basic equation in which the independent variables consisted of all combinations of powers up to the second of x_1, x_2 and x_3, 27 coefficients b_0, b_1, b_2, $\ldots b_{112233}$ could be estimated from the 27 observations. Their subscripts would be those enclosed by the dotted lines above. b_1, b_2, b_3 would be proportional to what are usually called the linear effects, b_{11}, b_{22}, b_{33} to the quadratic effects, and the remaining coefficients to the interactions between these effects.

If, however, it was reasonable to approximate to the response function by a series containing *all* terms up to some specified degree d, then with $d = 2$, the ten coefficients of order 2 and less would provide unbiased estimates of the derivatives of order 2 and less, the 17 remaining coefficients being estimates of experimental error only. On the other hand, if d were higher than 2, these remaining coefficients would provide estimates of some but not all of the extra constants, and the constants not included would inevitably bias some at least of the estimates.

In general, from a complete factorial design in which each of the k factors is varied at p levels, it is possible to form estimates from the k^p observations of k^p quantities*. The quantities may be chosen to be the coefficients of the k^p independent variables formed by expanding the expression

$$\pi_{t=1}^{k} (x_t^0 + x_t^1 + \cdots + x_t^{p-1}). \tag{40}$$

The estimates will include all the b's of order $p - 1$ and less together with certain of those of order $p, p + 1, \ldots, k(p - 1)$; the latter will have subscripts repeated not more than $p - 1$ times for any single variable. On the assumption that only the $(k + p - 1)!/k!(p - 1)!$ effects of order $p - 1$ and less exist, the design supplies unbiased estimates of these effects and an estimate of experimental error based on $p^k - (k + p - 1)!/k!(p - 1)!$ degrees of freedom. If it is necessary to use a series with all terms up to order d greater than $p - 1$ it will be possible to estimate those of the extra constants in which no subscript is repeated more than $p - 1$ times; but the remaining "extra constants" will bias some of the estimates.

In practice it would often be useful to employ designs which, while supplying estimates of all effects up to order d, also estimated "sample members" of the hierarchy of higher order effects. For if such effects were not small then

* *Editor's note*: The values k^p appearing in the original text have been changed to p^k.

the experimenter would be warned of the danger of proceeding on that assumption. However, unless p and k were small the number of experiments required by the full factorial design would be unreasonably large compared with the number of constants to be estimated and the number of "extra" effects determined in the complete factorial design would often be far larger than was necessary to check the fit of the equation. These designs could not be used efficiently therefore when experimental error was small. Fractional replication, in which the comparisons measuring effects of high order are used to estimate the lower order effects of additional factors, can help to remedy this defect. Unfortunately, except when the number of levels is 2, the comparisons in the fractional factorials do not directly correspond with derivatives which it is desired to estimate and, although some saving is possible, even these designs would often require an excessive number of experiments. For example, for the 3^k designs, no useful fractional factorial exists when $k = 3$, so that it would be necessary to carry out all the 27 experiments of the complete factorial design in order to determine the 10 effects of second order or less. Similarly when $k = 4$, 81 experiments are needed to determine 15 effects. When $k = 5$ a one-third replicate may be used to determine the 21 effects of order 2 and less; but even this involves the carrying out of 81 experiments, a number which would be excessive if the experimental error were small.

3.5.2. *Relative Precision of Effects in Factorial Designs*

Consider the designs to supply estimates of effects up to the second order. These estimates may be used to indicate a path along which maximum increase might be expected; more usually, when the design has been performed in the neighbourhood of a stationary point, they may be used to estimate the position of that point by solving equations (7) with $\mu = 0$. Now the coefficients of the derivatives φ_{st} in (7) are of the same degree, irrespective of whether $t = s$ or $t \neq s$ it would seem reasonable therefore to require that all second order effects (i.e., both the quadratic effects and interaction effects) should be determined with about the same precision. Since f_{st}, $s \neq t$, is estimated by b_{st} but f_{ss} is estimated by $2b_{ss}$, it is easily shown that 3^k factorial designs provide estimates of the quadratic derivatives having variances eight times as great as those for the interaction derivatives.

3.5.3. *Composite Designs*

The problem of finding "best" designs of types A and B when d is greater then unity is being investigated. In the meantime, designs, which did not require an excessively large number of experiments and which would estimate effects with reasonably high precision and low bias, were needed for use in current investigations. In particular, designs of type A and order 2 were required.

Factorial or fractional designs at two levels are easily found which allow

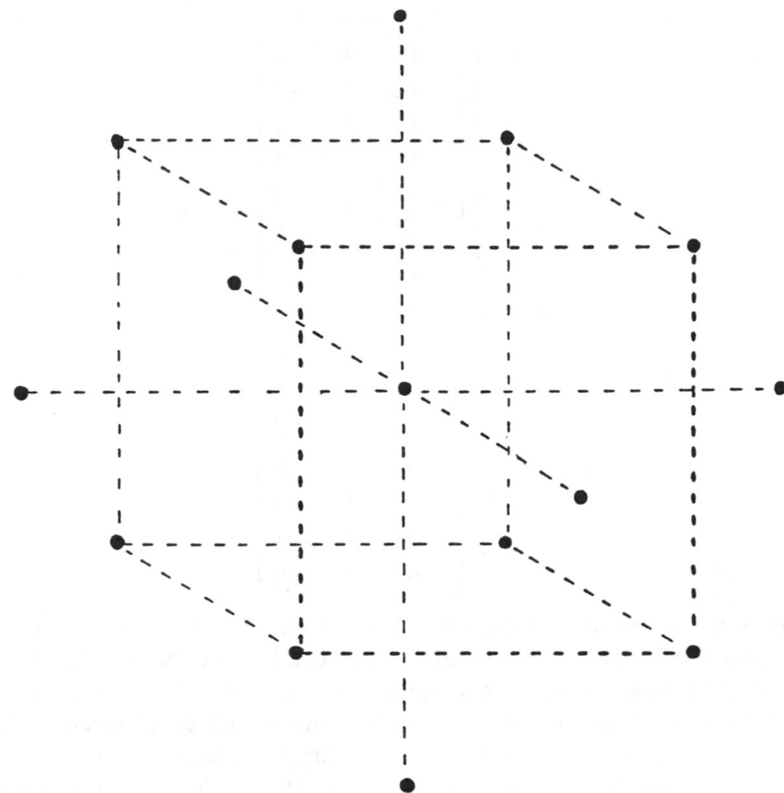

Figure 3. A three factor composite design.

the efficient estimation, without confusion, of first order derivatives and mixed derivatives not higher than the second order. These are what we have called type B' designs, and a list of such designs is given by Rao (1947). In constructing an arrangement which will estimate *all* derivatives up to the second order, we can start with a design of this sort, as a basis, and add further points which in conjunction with the type B' design make possible the estimation of the quadratic effects also. These designs have the advantage that they can be performed in stages. The first order design can first be completed. Then, if fairly large first order effects are found, while the effects of the two factor interactions, which can be estimated, are small, the experimenter may proceed to a new base by means of the first order steepest ascent formula. But if the relative magnitude of effects of first order and the two factor interactions show that it will be necessary to determine all second order derivatives, the extra points may be added to form the composite design. A design of this sort is illustrated for three variables in Fig. 3 below. The type B' arrangement forming the nucleus is in this case the two level factorial design. The design matrix $\mathbf{D}$ for such an arrangement can be written

$$\mathbf{D} = \begin{array}{c} \\ 1 \\ 2 \\ 3 \\ 4 \\ 5 \\ 6 \\ 7 \\ 8 \\ 9 \\ 10 \\ 11 \\ 12 \\ 13 \\ 14 \\ 15 \end{array} \begin{array}{c} \begin{array}{ccc} x_1 & x_2 & x_3 \end{array} \\ \begin{bmatrix} 1 & 1 & 1 \\ 1 & 1 & -1 \\ 1 & -1 & 1 \\ 1 & -1 & -1 \\ -1 & 1 & 1 \\ -1 & 1 & -1 \\ -1 & -1 & 1 \\ -1 & -1 & -1 \\ \alpha & 0 & 0 \\ -\alpha & 0 & 0 \\ 0 & \alpha & 0 \\ 0 & -\alpha & 0 \\ 0 & 0 & \alpha \\ 0 & 0 & -\alpha \\ 0 & 0 & 0 \end{bmatrix} \end{array}. \tag{41}$$

where α is the distance of the axial points from the centre. It is shown in Appendix 4 that designs built up in this way can be very effective. In particular α may be chosen so that the design is orthogonal or, alternatively, so that the derivatives of second order are all determined with equal precision. The precision of the estimates and the extent of the possible biases show that the designs compare favourably with 3 level factorials, when the fewer experiments required is taken into account.

In examples where the first order design has indicated the probability of an imminent stationary point in a particular direction, the extra points may be added in that direction. In favour of this is the fact that the distance between the centroid of the design and the stationary point will probably be reduced and the fit of the second degree approximating equation improved. A number of interesting possibilities arise, one of which is demonstrated in the example in section 5.

(*Editor's note*: Section 4 has been omitted.)

5. Exploration of the Surface in a Near-Stationary Region

In this section it is assumed that the experimenter is using a sub-region of the factor space, within which first order derivatives are small. He may have arrived at this region by successive application of the steepest ascent technique, or already have found it at the commencement of his investigation. Examples of the latter kind commonly occur when the problem is further to improve some well established process; for often, large first order effects will have been eliminated in previous work. In either case, initially at any rate, only the immediate neighbourhood need be explored and this may be done without an excessively large number of experiments.

Here we shall be concerned only with approximating equations of the second degree. We assume that a suitable design has been performed, and that estimates $f_0, f_1, f_2, \ldots f_{11}, f_{22}, \ldots f_{12}$, etc., of reasonable precision are available for second and lower orders.

5.1 Analysis of the Fitted Equation

In the neighbourhood of the design the response surface is given approximately by a $k+1$ dimensional paraboloid in the variables $y, x_1, x_2 \ldots x_k$.

$$y = f_0 + f_1 x_1 + f_2 x_2 + \cdots + \tfrac{1}{2} f_{11} x_1^2 + \tfrac{1}{2} f_{22} x_2^2 + \cdots f_{12} x_1 x_2 + f_{13} x_1 x_3 + \cdots. \tag{44}$$

Associated with the surface are k dimensional contour surfaces such as

$$y_c = f_0 + f_1 x_1 + f_2 x_2 + \cdots + \tfrac{1}{2} f_{11} x_1^2 + \tfrac{1}{2} f_{22} x_2^2 + \cdots f_{12} x_1 x_2 + f_{13} x_1 x_3 + \text{etc.} \ldots \tag{45}$$

on which the response is equal to y_c.

Differentiating (44) with respect to $x_1, x_2 \ldots x_k$ in turn, we obtain k linear simultaneous equations, such as (7) but with $\mu = 0$. On solving these we obtain $x_1^0, x_2^0, \ldots x_k^0$, the co-ordinates of S the stationary point on the fitted surface. On substituting these values in (44) we obtain the predicted response at S,

$$y^0 = f_0 + \tfrac{1}{2}(f_1 x_1^0 + f_2 x_2^0 + \cdots + f_k x_k^0). \tag{46}$$

The system of conics corresponding to the contour surfaces may take a large variety of forms; it would usually be quite impossible to appreciate the nature of the fitted surfaces by inspection of the values of the coefficients in (45). The nature of the system is, however, readily made apparent by reduction of the conic to canonical form.

This consists essentially of shifting the origin to the stationary point S, the centre of the system of curves representing the contour surfaces, and rotating the co-ordinate axes so that they correspond to the axes of these conics. Equation (45) then reduces to

$$y_c - y^0 = \lambda_1 X_1^2 + \lambda_2 X_2^2 + \cdots + \lambda_k X_k^2 *. \tag{47}$$

* Theoretically some of the λ's may be exactly zero, in which case no unique stationary point S will exist. With experimental data the λ's may approach zero; the behaviour of the contour surfaces, in these important limiting cases, are mentioned in the text. For a full account of the reduction of a conic to canonical form, the reader is referred to text-books such as Turnbull and Aitken (1932). An example of the reduction is given in the next section.

The expression shows the loss of response on moving from S, so that for example, if λ_1 is negative, S is a maximum for the response curve drawn along X_1. Large values of λ correspond with rapid changes in the response whilst small values indicate slow changes.

(i) If $\lambda_1, \lambda_2, \ldots \lambda_k$ were all negative, the fitted surface would have a true maximum at the stationary point, and the fitted contour surfaces would be ellipsoids.
(ii) If one or more of the λ's were positive, there would be a col or minimax with the contour surfaces elliptic hyperboloids.
(iii) If one or more of the λ's approached zero, the curves would be attenuated along their corresponding axes and the resulting surfaces would approach elliptic or hyperbolic cylinders; or the fitted response surface would possess a ridge.

Suppose that in a particular example the λ's and the directions of $X_1, X_2 \ldots X_k$ had been determined. Suppose also that on differentiating (44) a stationary point had been found in the immediate neighbourhood of the design. In case (i), above, the co-ordinates of the stationary point would provide an estimate of the position of the maximum. In case (ii) the directions corresponding to the positive λ's would indicate how the experimenter should move from the col to points of higher yield; further experiments would then be made to explore these possibilities. In case (iii) the directions of the line, plane, or space, in which the response was nearly constant could be determined. This is very important in practice; for it is thus possible to indicate alternative sets of conditions at which almost equal responses would be expected and so to find the most satisfactory compromise for auxiliary responses in the chosen process. The analysis of the surface in this way will also serve to indicate where further experiments are needed. Thus where some of the λ's are near to zero, it may be important to discover whether the true values are positive or negative. In such cases extra experiments may best be performed along the corresponding estimated axes; hence re-estimates of the constants and λ's can be made. In some cases strong divergencies between the values obtained at these extra points and the values already found may indicate the necessity for fitting an equation containing higher order terms.

Sometimes the stationary point of the fitted function will be remote from the neighbourhood of the design near which it was expected. In this remote region the fitted surface could not provide any accurate information about the true surface. The fitted contour surface in the immediate neighbourhood of the design will, however, supply valuable information. In a commonly occurring case of this kind, the near-stationary region to which the experimenter has been led is not in fact close to a stationary point, but is in the neighbourhood of a slowly rising ridge which leads to higher responses. One axis (say X_1) of the fitted conic will lie in the direction of the ridge and the corresponding λ_1 will be small. Referred to a local origin on X_1, the equations of the fitted contour surfaces are

$$y_c - y^0 = B_1 X_1 + \lambda_1 X_1^2 + \lambda_2 X_2^2 + \cdots \lambda_k X_k^2, \tag{48}$$

where y^0 is the response at the origin. The limiting case is when $\lambda_1 = 0$. The contour surfaces are then paraboloids with their axes along X_1 and B_1 measures the slope of the ridge. In practice the experimenter would explore the axis X_1 with further experiments; if the slope was confirmed he would then follow this direction.

5.2 The Inclusion of Additional Observations

Thus it is frequently necessary by the use of additional points to explore further the possibilities suggested by the analysis of the second degree equation. The matrix $\mathbf{C}^{-1}$ would be known for the basic design but not for the arrangement of points including the additional observations; the calculation of a new set of constants *ab initio* would be laborious. Fortunately, however, a complete recalculation is unnecessary. Plackett (1950) shows how "corrections" can be obtained, which allow new estimates to be formed from the old with a minimum of recalculation.

In our notation Plackett's results may be stated as follows: In an experiment with a design $\mathbf{D}_1$ let the N_1 resulting observations be given by the vector $\mathbf{Y}$; let $\mathbf{X}$ be the $N_1 \times L$ matrix of rank L for the L independent variables (functions of the co-ordinates in $\mathbf{D}$). Suppose that the matrix $(\mathbf{X}'\mathbf{X})^{-1} = \mathbf{C}^{-1}$ is known; and that the vector $\mathbf{B}$ of the estimates of the elements of $\boldsymbol{\beta}$, as well as the residual sums of squares $\mathbf{S}$ (based on $N - L$ degrees of freedom) have been calculated. Let a further N_2 observations $\mathbf{Z}$ become available as a result of experiments performed at a set of points whose co-ordinates are given by $\mathbf{D}_2$ and for which the matrix of independent variables is $\mathbf{W}$. The new values of $\mathbf{C}_0^{-1}$, $\mathbf{B}_0'$ and S_0 for the whole of the $N_1 + N_2$ points are given by

$$\mathbf{C}_0^{-1} = \mathbf{C}^{-1} - \mathbf{J}'\mathbf{G}\mathbf{J} \tag{49}$$

$$\mathbf{B}_0' = \mathbf{B}' + \boldsymbol{\Delta}'\mathbf{G}\mathbf{J} \tag{50}$$

$$S_0 = S + \boldsymbol{\Delta}'\mathbf{G}\boldsymbol{\Delta} \tag{51}$$

where

$$\mathbf{J} = \mathbf{W}\mathbf{C}^{-1}; \quad \mathbf{G} = (\mathbf{I} + \mathbf{R})^{-1}; \quad \mathbf{R} = \mathbf{W}\mathbf{C}^{-1}\mathbf{W}'; \quad \boldsymbol{\Delta} = \mathbf{Z} - \mathbf{W}\mathbf{B}. \tag{52}$$

In each case the new values are obtained by adding appropriate correction terms to the old. For formulae (50) and (51) these correction terms contain the vector $\mathbf{Z} - \mathbf{WB}$. The elements of this vector are the discrepancies between the values which the equation first fitted predicts for the responses at the additional points and the responses actually found. Each of the correction terms contains the matrix $(\mathbf{I} + \mathbf{R})^{-1}$. It is thus necessary to invert a $N_2 \times N_2$ matrix only; providing N_2, the number of additional observations, is not too great, the inverse can be found fairly readily. For systematic methods for inverting matrices the reader is referred to papers by Dwyer (1942), Hotelling (1943), and Fox *et al.* (1948).

5.3 An Example

This example is chosen not because the results obtained were at all spectacular; but because it provides a good illustration of the devices discussed in previous sections and, since only three variables are involved, it is possible to demonstrate the conclusions geometrically.

The investigation concerned one stage of a particular chemical process in which a reaction of type,

$$A + B \rightarrow C + \text{other products},$$

was carried out. This process was already well established on the manufacturing scale, but it was thought that slightly better conditions might possibly be found. (The experiments were carried out in the laboratory. The reaction, however, was of the type in which plant conditions were almost exactly reproducible, and it was fairly certain that any changes in the laboratory scale would be applicable in the larger scale operation.) The yield of the product C could be accurately determined and the experiments were very reproducible. They were not easy to perform, however, and were time-absorbing; each one engaged a chemist and his assistants for a number of days. It was essential therefore to reduce the number of experiments to a minimum.

The factors varied were temperature (x_1), concentration (x_2), and the molar ratio of B to A (x_3). The most economic process was not necessarily that corresponding with the highest yield; the response considered was the calculated cost per pound of final product. This was a function not only of the yield obtained but also of the experimental conditions employed. The object of the experiments was to estimate the point at which the cost would be a *minimum.*

For convenience in calculation an arbitrary quantity has been subtracted from the cost per pound of product assessed in tenths of a penny; in what follows the resulting amount y is termed "cost". There was some evidence that the standard deviation of individual observations would be about 1 unit.

The normal conditions were:

Temperature °C.	*Concentration %*	*Molar Ratio B/A*
145	20	6

The region of practical variation for the factors was fairly extensive. However, from the plant-handling point of view at the next stage of the process, it was essential that the concentration should not fall below 18 per cent.

When the first set of experiments was begun, it was uncertain whether the known conditions would be sufficiently far from the optimum to make necessary the preliminary application of the technique of §4 to move to the region of a stationary point. Therefore a complete 2^3 factorial was performed. This is a design of type B' which allows all the first order derivatives and mixed derivatives of second order to be estimated. The levels used were:

Temperature °C.		*Concentration* %		*Molar Ratio A/B*	
140	145	20	24	6	7
	x_1		x_2		x_3
-1	1	-1	1	-1	1

The requirement that the concentration should not be less than 18 per cent. corresponds with the condition that x_2 should not be less than -2. The co-ordinates of the experimental points and corresponding costs were (1, 1, 1) 17; (1, 1, -1) 9; (1, -1, 1) 12; (1, -1, -1) 15; (-1, 1, 1) 24; (-1, 1, -1)11; (-1, -1, 1) 7; (-1, -1, -1); 5. The effects and their expected values, including terms up to third order, are shown below:

$$\left.\begin{aligned}
b_0 &\rightarrow \varphi_0 + \tfrac{1}{2}\varphi_{11} + \tfrac{1}{2}\varphi_{22} + \tfrac{1}{2}\varphi_{33} &&= 12.50 \pm 0.4\\
b_1 &\rightarrow \varphi_1(+\tfrac{1}{6}\varphi_{111} + \tfrac{1}{2}\varphi_{122} + \tfrac{1}{2}\varphi_{133}) &&= 0.75 \pm 0.4\\
b_2 &\rightarrow \varphi_2(+\tfrac{1}{6}\varphi_{222} + \tfrac{1}{2}\varphi_{112} + \tfrac{1}{2}\varphi_{233}) &&= 2.75 \pm 0.4\\
b_3 &\rightarrow \varphi_3(+\tfrac{1}{6}\varphi_{333} + \tfrac{1}{2}\varphi_{113} + \tfrac{1}{2}\varphi_{223}) &&= 2.50 \pm 0.4\\
b_{12} &\rightarrow \varphi_{12} &&= -3.00 \pm 0.4\\
b_{13} &\rightarrow \varphi_{13} &&= -1.25 \pm 0.4\\
b_{23} &\rightarrow \varphi_{23} &&= 2.75 \pm 0.4\\
b_{123} &\rightarrow \varphi_{123} &&= 0.00 \pm 0.4
\end{aligned}\right\}. \tag{53}$$

The estimates are followed by their standard errors based on the assumption that σ the experimental error is about 1 unit. From the relative size of the interaction terms, it is clear that the steepest descent* relaxation process is unlikely to be very effective here; the design must be augmented to determine all effects of second order. This example is interesting, because owing to a misunderstanding the steepest descent path was in fact calculated and followed.

The steepest descent path will be followed when the factors are varied in proportion to the first order effects with reversed signs, i.e., -0.75, -2.75, -2.50.

Two experiments were performed on this path. The levels for the factors are given below, in units of the design, together with the estimated costs:

$$\begin{array}{cccc}
\textit{Experiment} & x_1 \quad\; x_2 \quad\; x_3 & & y\\
(9) & \left[\begin{array}{ccc} -0.54 & -2.00 & -1.82\\ -0.81 & -3.00 & -2.73\end{array}\right] & & \left[\begin{array}{c}13\\27\end{array}\right].\\
(10) & & &
\end{array} \tag{54}$$

* "Descent", since we are seeking a minimum.

As might have been expected, no reduction in cost was in fact found in these experiments. However, since the first order effects for each of the factors is positive, it seemed likely that if a minimum existed it would be towards the vertex $(-1\ -1\ -1)$. Therefore the factorial design was augmented with three further points in the manner discussed at the end of §3.53. The levels used in these experiments, together with the costs found, are shown below:

$$\begin{array}{c} \textit{Experiment} \\ (11) \\ (12) \\ (13) \end{array} \quad \begin{array}{c} \begin{array}{ccc} x_1 & x_2 & x_3 \end{array} \\ \begin{bmatrix} -3 & -1 & -1 \\ -1 & -3 & -1 \\ -1 & -1 & -3 \end{bmatrix} \end{array} \quad \begin{array}{c} y \\ \begin{bmatrix} 19 \\ 12 \\ 12 \end{bmatrix} \end{array}. \tag{55}$$

The matrix T' for the composite design is:

	b_0	b_1	b_2	b_3	b_{11}
1	0.2656	0.1250	0.1250	0.1250	−0.0468
2	−0.0156	0.1250	0.1250	−0.1250	0.0468
3	−0.0156	0.1250	−0.1250	0.1250	0.0468
4	0.1406	0.1250	−0.1250	−0.1250	0.0781
5	−0.0156	−0.1250	0.1250	0.1250	0.0468
6	0.1406	−0.1250	0.1250	−0.1250	−0.0468
7	0.1406	−0.1250	−0.1250	0.1250	−0.0468
8	0.7344	−0.1250	−0.1250	−0.1250	−0.2031
11	−0.1250	.	.	.	0.1250
12	−0.1250	.	.	.	.
13	−0.1250	.	.	.	.

	b_{22}	b_{23}	b_{12}	b_{13}	b_{23}
1	−0.0468	−0.0468	0.1250	0.1250	0.1250
2	0.0468	0.0468	0.1250	−0.1250	−0.1250
3	0.0468	0.0468	−0.1250	0.1250	−0.1250
4	−0.0468	−0.0468	−0.1250	−0.1250	0.1250
5	0.0468	0.0468	−0.1250	−0.1250	0.1250
6	0.0781	−0.0468	−0.1250	0.1250	−0.1250
7	−0.0468	0.0781	0.1250	−0.1250	−0.1250
8	−0.2031	−0.2031	0.1250	0.1250	0.1250
11	.	.	.	.	.
12	0.1250	.	.	.	.
13	.	0.1250	.	.	.

(56)

It will be noted from the above matrix that the estimates of the linear effects and first order interactions remain the same as before. Their standard errors and aliases are also undisturbed. For b_0 and the quadratic effects the estimates obtained were:

$$\left.\begin{aligned} b_0 &\to \varphi_0 + \tfrac{1}{2}(\varphi_{111} + \varphi_{222} + \varphi_{333}) + \tfrac{1}{2}(\varphi_{122} + \varphi_{133} + \varphi_{112} \\ &\qquad + \varphi_{233} + \varphi_{113} + \varphi_{223}) + \tfrac{9}{8}\varphi_{123} = 6.750 \pm 0.8 \\ b_{11} &\to \tfrac{1}{2}\varphi_{11} - \tfrac{1}{2}(\varphi_{111} + \varphi_{112} + \varphi_{113}) - \tfrac{3}{8}\varphi_{123} = 3.000 \pm 0.3 \\ b_{22} &\to \tfrac{1}{2}\varphi_{22} - \tfrac{1}{2}(\varphi_{222} + \varphi_{122} + \varphi_{223}) - \tfrac{3}{8}\varphi_{123} = 1.625 \pm 0.3 \\ b_{33} &\to \tfrac{1}{2}\varphi_{33} - \tfrac{1}{2}(\varphi_{333} + \varphi_{133} + \varphi_{233}) - \tfrac{3}{8}\varphi_{123} = 1.125 \pm 0.3 \end{aligned}\right\}. \tag{57}$$

On the assumption that third and higher order effects might be ignored, the estimated equation of best fit, in the neighbourhood of the design, was therefore:

$$\begin{aligned} y = 6.750 &+ 0.750x_1 + 2.750x_2 + 2.500x_3 + 3.000x_1^2 + 1.625x_2^2 + 1.125x_3^2 \\ &- 3.000x_1x_2 - 1.250x_1x_3 + 2.750x_2x_3. \end{aligned} \tag{58}$$

The residual sum of squares $\mathbf{S}$ based on one degree of freedom was zero. For the composite design the matrix $\mathbf{C}^{-1}$ is:

	0	1	2	3	11
0	0.7168	−0.0781	−0.0781	−0.0781	−0.1816
1	−0.0781	0.1250	.	.	0.0468
2	−0.0781	.	0.1250	.	0.0156
3	−0.0781	.	.	0.1250	0.0156
11	−0.1816	0.0468	0.0156	0.0156	0.0762
22	−0.1816	0.0156	0.0468	0.0156	0.0449
33	−0.1816	0.0156	0.0156	0.0468	0.0449
12	0.1094	.	.	.	−0.0468
13	0.1094	.	.	.	−0.0468
23	0.1094	.	.	.	−0.0156

	22	33	12	13	23
0	−0.1816	−0.1816	0.1094	0.1094	0.1094
1	0.0156	0.0156	.	.	.
2	0.0468	0.0156	.	.	.
3	0.0156	0.0468	.	.	.
11	0.0449	0.0449	−0.0468	−0.0468	−0.0156
22	0.0762	0.0449	−0.0468	−0.0156	−0.0468
33	0.0449	0.0762	−0.0156	−0.0468	−0.0468
12	−0.0468	−0.0156	0.1250	.	.
13	−0.0156	−0.0468	.	0.1250	.
23	−0.0468	−0.0468	.	.	0.1250

(59)

Thus the estimates of the quadratic effects were rather strongly correlated. The point at which the two steepest ascent experiments were performed was reasonably close to the design and the information from these was now included. By substituting the independent variables corresponding to the experiments (9) and (10) in the fitted equation (58), the predicted values 12.94 and 28.51 were obtained; these may be compared with the values found, 13 and 27. The agreement is remarkably good, and we are encouraged to hope that the assumptions made are satisfactory within the region considered. With the aid of Plackett's equations, discussed in §5.2, these discrepancies, between the values predicted and those actually found, were now used to modify the estimates of the b's. A new matrix $\mathbf{C}_0^{-1}$ appropriate for the whole set of 13 points was also obtained. The matrix $\mathbf{W}$, for the two new sets of independent variables, is

$$\mathbf{W} = \begin{array}{c} \\ (9) \\ (10) \end{array} \begin{array}{c} \begin{array}{ccccc} 0 & 1 & 2 & 3 & 11 \end{array} \\ \left[\begin{array}{ccccc} 1 & -0.5400 & -2.0000 & -1.8200 & 0.2916 \\ 1 & -0.8100 & -3.0000 & -2.7300 & 0.6561 \end{array}\right. \end{array}$$

$$\begin{array}{c} \begin{array}{ccccc} 22 & 33 & 12 & 13 & 23 \end{array} \\ \left.\begin{array}{ccccc} 4.0000 & 3.3124 & 1.0800 & 0.9828 & 3.6400 \\ 9.0000 & 7.4529 & 2.4300 & 2.2113 & 8.1900 \end{array}\right]. \end{array} \tag{60}$$

The vector $\mathbf{Z}$ which gives the costs found at these points, is $\mathbf{Z} = \begin{bmatrix} 13 \\ 27 \end{bmatrix}$. Whence $\mathbf{J} = \mathbf{W}\mathbf{C}^{-1}$ is given by

$$\mathbf{J} = \begin{array}{c} (9) \\ (10) \end{array} \left[\begin{array}{ccccc} 0.3000 & -0.0177 & -0.0843 & -0.0833 & -0.0695 \\ -0.4765 & 0.1084 & 0.0054 & 0.0808 & 0.1344 \end{array}\right.$$

$$\left.\begin{array}{ccccc} -0.0823 & -0.0951 & -0.0085 & 0.0008 & 0.2170 \\ 0.1399 & 0.1068 & -0.1560 & -0.1349 & 0.3516 \end{array}\right] \tag{61}$$

and

$$\mathbf{R} = \mathbf{W}\mathbf{C}^{-1}\mathbf{W}' = \mathbf{J}\mathbf{W}' = \begin{array}{c} \\ (9) \\ (10) \end{array} \begin{array}{c} \begin{array}{cc} (9) & (10) \end{array} \\ \begin{bmatrix} 0.7469 & 1.0582 \\ 1.0582 & 3.2742 \end{bmatrix}. \end{array} \tag{62}$$

$\mathbf{I} + \mathbf{R}$ is obtained by adding unity to each of the diagonal elements in $\mathbf{R}$; and $(\mathbf{I} + \mathbf{R})^{-1} = \mathbf{G}$, by solving the equations $(\mathbf{I} + \mathbf{R})\mathbf{G} = \mathbf{I}$. Thus,

$$\left\{\begin{array}{l} 1.7469g_{11} + 1.0582g_{21} = 1 \\ 1.0582g_{11} + 4.2742g_{21} = 0 \end{array}\right\} \quad \left\{\begin{array}{l} 1.7469g_{12} + 1.0582g_{22} = 0 \\ 1.0582g_{12} + 4.2742g_{22} = 1 \end{array}\right\}. \tag{63}$$

Solving for the g's we have

$$\mathbf{G} = \begin{array}{c} \\ (9) \\ (10) \end{array} \begin{array}{c} \begin{array}{cc} (9) & (10) \end{array} \\ \begin{bmatrix} 0.6734 & -0.1667 \\ -0.1667 & 0.2752 \end{bmatrix}. \end{array} \tag{64}$$

We now pre-multiply $\mathbf{J}$ by $\mathbf{G}$ and obtain

$$\mathbf{GJ} = \begin{array}{c} \\ (9) \\ (10) \end{array} \begin{array}{cccccccccc} 0 & 1 & 2 & 3 & 11 & 22 & 33 & 12 & 13 & 23 \\ 0.2815 & -0.0300 & -0.0727 & -0.0696 & -0.0692 & -0.0787 & -0.0819 & 0.0202 & 0.0230 & 0.0875 \\ -0.1812 & 0.0328 & 0.0403 & 0.0361 & 0.0486 & 0.0522 & 0.0453 & -0.0415 & -0.0373 & 0.0606 \end{array}. \tag{65}$$

The new values for the b's can now be calculated simply from (50). The vector of the discrepancies is

$$\boldsymbol{\Delta} = \mathbf{Z} - \mathbf{WB} = \begin{array}{c} (9) \\ (10) \end{array} \begin{bmatrix} 0.0623 \\ -1.5137 \end{bmatrix}. \tag{66}$$

Thus, for example, the value for b_0 is

$$6.7500 + \{0.0623 \times 0.2815\} + \{(-1.5137) \times (-0.1812)\} = 7.0418$$

The calculations are carried out as a single operation on the machine. In this way, we find

$$b_0 = 7.0418$$

$$\begin{array}{lll} b_1 = 0.6985 & b_{11} = 2.9221 & b_{12} = -2.9359 \\ b_2 = 2.6844 & b_{22} = 1.5410 & b_{13} = -1.1921 \\ b_3 = 2.4410 & b_{33} = 1.0510 & b_{23} = 2.6637 \end{array}$$

The correction for $\mathbf{C}^{-1}$ may now be obtained by pre-multiplying $\mathbf{GJ}$ by $\mathbf{J}'$. Again the new matrix can be obtained from the matrices $\mathbf{C}^{-1}$, $\mathbf{GJ}$, and $\mathbf{J}$, by a series of operations on the machine. For example, the new first element in the matrix is

$$0.7168 - \{0.3000 \times 0.2815\} - \{(-0.4765) \times (-0.1812)\} = 0.5460.$$

The new matrix $\mathbf{C}_0^{-1}$ obtained in this way is:

	0	1	2	3	11
0	0.5460	−0.0535	−0.0371	−0.0400	−0.1377
1	−0.0535	0.1209	−0.0057	−0.0051	0.0404
2	−0.0371	−0.0057	0.1150	−0.0093	0.0051
3	−0.0400	−0.0051	−0.0093	0.1163	0.0059
11	−0.1377	0.0404	0.0051	0.0059	0.0648
22	−0.1331	0.0086	0.0352	0.0048	0.0324
33	−0.1355	0.0093	0.0044	0.0364	0.0331
12	0.0835	0.0048	0.0057	0.0050	−0.0399
13	0.0847	0.0044	0.0055	0.0049	−0.0403
23	0.1120	−0.0050	0.0016	0.0024	−0.0177

$$\begin{array}{rrrrr} 22 & 33 & 12 & 13 & 23 \\ -0.1331 & -0.1355 & 0.0835 & 0.0847 & 0.1120 \\ 0.0086 & 0.0093 & 0.0048 & 0.0044 & -0.0050 \\ 0.0352 & 0.0044 & 0.0057 & 0.0055 & 0.0016 \\ 0.0048 & 0.0064 & 0.0050 & 0.0049 & 0.0024 \\ 0.0324 & 0.0331 & -0.0399 & -0.0403 & -0.0177 \\ 0.0624 & 0.0319 & -0.0394 & -0.0085 & -0.0481 \\ 0.0318 & 0.0636 & -0.0093 & -0.0407 & -0.0450 \\ -0.0394 & -0.0093 & 0.1187 & -0.0056 & 0.0102 \\ -0.0085 & -0.0407 & -0.0056 & 0.1199 & 0.0081 \\ -0.0481 & -0.0450 & 0.0102 & 0.0081 & 0.0847 \end{array} \Bigg] . \qquad (67)$$

The additional points have slightly reduced the variance for the estimates (cf. (67) and (59)). (The matrix (67) is required later, otherwise it might not have been worthwhile to calculate it.)

Using (51), the new value for the residual sum of squares, now based on three degrees of freedom, is

$$S_0 = 0.00 + 0.66 = 0.66 \qquad (68)$$

Whence, if the fit were perfect, the estimated standard deviation would be

$$s = (0.66/3)^{1/2} = 0.47.$$

In order to study the nature of the fitted surface we must reduce it to standard form, in the manner discussed in §5.1.

The discriminating cubic is

$$\begin{vmatrix} 2.9221 - \lambda & -1.4679 & -0.5960 \\ -1.4679 & 1.5410 - \lambda & 1.3318 \\ 0.5960 & 1.3318 & 1.0510 - \lambda \end{vmatrix} = 0 \qquad (69)$$

i.e.,

$$\lambda^3 - 5.5141\lambda^2 + 4.9114\lambda + 0.9290 = 0$$

which has the roots

$$\lambda_1 = -0.1597 \quad \lambda_2 = 1.3434 \quad \lambda_3 = 4.3304. \qquad (70)$$

The equations for the co-ordinates of the centre are:

$$\begin{aligned} 5.8842x_1 - 2.9359x_2 - \ \ 1.921x_3 &= -0.6985 \\ -2.9359x_1 + 3.0820x_2 + 2.6637x_3 &= -2.6844 \\ -1.1921x_1 + 2.6637x_2 + 2.1020x_3 &= -2.4410 \end{aligned}$$

the solutions of which give the co-ordinates of the stationary point S.

$$x_1^0 = -0.3365 \quad x_2^0 = 0.2411 \quad x_3^0 = -1.6576. \tag{71}$$

By substituating these values in the fitted equation, or using (46), the predicted value for the cost at this point is found to be $y^0 = 5.2247$. The equation of the contour surfaces may therefore be written in some set of orthogonal co-ordinates as

$$y_c - 5.2247 = -0.1597X_1^2 + 1.3434X_2^2 + 4.3304X_3^2 \tag{72}$$

The fitted contour surfaces are hyperboloids of one sheet; thus the sections by the planes $X_2 = 0$ and $X_3 = 0$ are hyperbolas, and that by X_1 is an ellipse. From the smallness of λ_1 compared with the other coefficients, it is clear that the surface is attenuated along the X_1 axis, i.e., that there is a ridge running in this direction.

To find the direction of the new axes we require the orthogonal matrix which transforms the old variables to the new, that is the matrix **M** for which

$$\mathbf{X} = \mathbf{M}(\mathbf{x} - \mathbf{x}^0). \tag{73}$$

If $\mathbf{M}_t$ is the vector corresponding to the tth row of **M**, it is proved in text books that

$$\mathbf{M}_t(\tfrac{1}{2}\mathbf{F} - \mathbf{I}\lambda_t) = \mathbf{O}, \tag{74}$$

where **F** is the matrix $\{f_{st}\}$ of second order derivatives. Substituting λ_1 in this equation we obtain the set of homogeneous equations:

$$3.0818M_{11} - 1.4679M_{12} - 0.5960M_{13} = 0$$

$$-1.4679M_{11} + 1.7007M_{12} + 1.3318M_{13} = 0$$

$$-0.5960M_{11} + 1.3318M_{12} + 1.2107M_{13} = 0$$

the solutions of which are proportional to the elements of the first row of **M**. Putting $M_{11} = 1$ we have a consistent set of three equations in two unknowns. Solving any two of these we obtain the values $M_{11} = 1$, $M_{12} = 3.4318$, $M_{13} = -3.2813$. Since **M** is orthogonal, the sum of squares of the elements in any row or column is unity. Thus division of the elements M_{11}, etc., by the square root of their sum of squares gives the first row of **M**.

$$m_{11} = 0.2061 \quad m_{12} = 0.7073 \quad m_{13} = -0.6763.$$

Substituting λ_2 in (74) we find in a similar way the elements

$$m_{21} = 0.6385 \quad m_{22} = 0.4265 \quad m_{23} = 0.6406.$$

Finally substituting λ_3 we obtain

$$m_{31} = 0.7416 \quad m_{32} = -0.5638 \quad m_{33} = -0.3636.$$

The co-ordinates of the centre and any point along the three principle axes are thus given by

New Variables		
0	0	0
X_1	0	0
0	X_2	0
0	0	X_3

Old Variables		
-0.3365	0.2411	-1.6576
$-0.3365 + 0.2061X_1$	$0.2411 + 0.7073X_1$	$-1.6576 - 0.6763X_1$
$-0.3365 + 0.6385X_2$	$0.2411 + 0.4265X_2$	$-1.6576 + 0.6406X_2$
$-0.3365 + 0.7416X_3$	$0.2411 - 0.5638X_3$	$-1.6576 - 0.3636X_3$

The appearance of the fitted surface can be appreciated from Fig. 4, which shows the arrangement of the points in space, looking down the axis X_1. The contour lines for costs of 10 and 20 are drawn as they would be seen on the plane $X_1 = 0$. Owing to the small value of λ_1, the estimated contour surfaces in the neighbourhood of the design are approximately cylinders having the elliptic cross section shown.

From the figure it will be seen that the agreement between the contours of the fined surface and the actual experimental points is good. In the circumstances considered this arrangement of points would be expected to supply fairly accurate estimates of λ_2 and λ_3. However, a range of hyperboloids of one sheet and ellipsoids attenuated along the X_1 axis having this elliptic section at $X_1 = 0$ would give almost as satisfactory a fit. In order to determine λ_1 more accurately, it was decided to carry out two further experiments at points where the X_1 axis intersected the two planes $x_3 = 1$, $x_3 = -3$.

The co-ordinates for these two points may be obtained from (75). In the new system they are $(-3.9296, 0, 0)$ and $(1.9849, 0, 0)$. The co-ordinates in the old system, and the costs found when experiments were performed at these points, are given below:

$$\begin{array}{c} \\ (14) \\ (15) \end{array} \begin{array}{c} \begin{array}{ccc} x_1 & x_2 & x_3 \end{array} \\ \begin{bmatrix} -1.1465 & -2.5383 & 1 \\ 0.0727 & 1.6450 & -3 \end{bmatrix} \end{array} \begin{array}{c} y \\ \begin{bmatrix} 6 \\ 19 \end{bmatrix} \end{array}$$

The first point confirms the fairly stationary nature of the cost moving up the axis X_1, but the second point suggests that the contours do in fact close in the opposite direction.

The information supplied by these two additional points was now included, as described before. The adjusted values for the constants were:

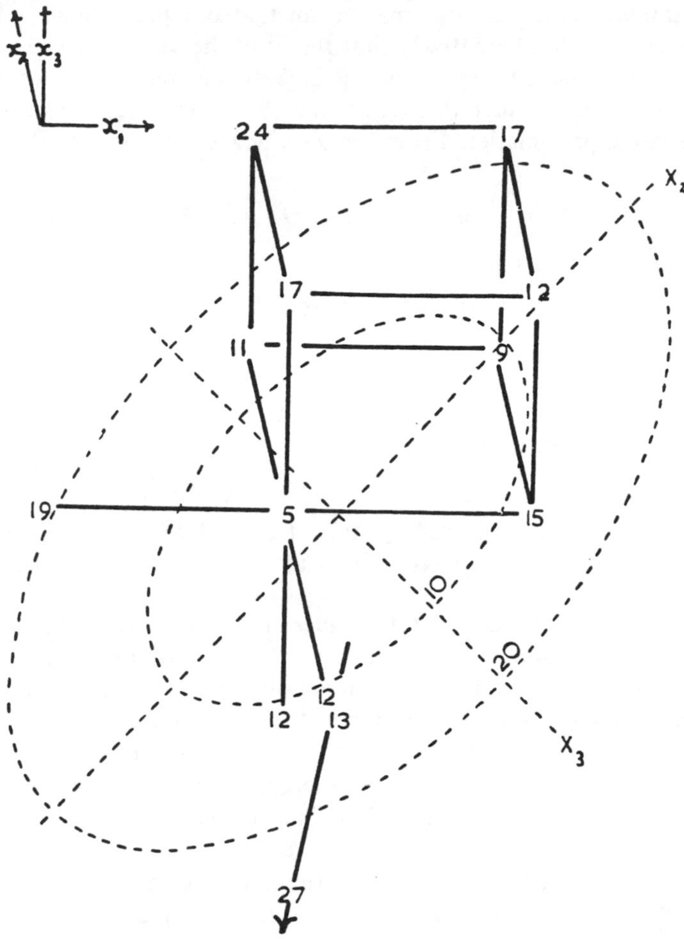

Figure 4. The arrangement of experiments and fitted contours "looking down".

$$b_0 = 5.277$$

$$\begin{array}{lll} b_1 = 0.874 & b_{11} = 3.339 & b_{12} = -3.233 \\ b_2 = 3.220 & b_{22} = 2.239 & b_{13} = -1.561 \\ b_3 = 2.177 & b_{33} = 1.778 & b_{23} = \quad 1.7451 \end{array}$$

and the new residual sum of squares, based now on 5 degrees of freedom, was 19.93. Whence if the fit were perfect, the estimated standard deviation would be

$$s = (19.93/5)^{1/2} = 2.00.$$

This is significantly greater than the value of one unit which we assumed for the standard deviation; this may mean that our preliminary value was an underestimate, or alternatively that the fit of the second degree equation was imperfect. The latter explanation is a likely one, but the departure from expectation is small enough to suggest that the second degree equation is at least a useful approximation. From the b's a new set of λ's was calculated as follows:

$$\lambda_1 = 0.9461 \quad \lambda_2 = 1.4999 \quad \lambda_3 = 4.9113$$

and the new centre was

$$x_1^0 = -0.7601 \quad x_2^0 = -1.1093 \quad x_3^0 = -0.4014$$

at which the predicted cost was

$$y^0 = 2.6719.$$

The new transformation was

$$\mathbf{M} = \begin{bmatrix} 0.3786 & 0.7933 & -0.4768 \\ 0.5424 & 0.2271 & 0.8088 \\ 0.7500 & -0.5649 & -0.3442 \end{bmatrix}. \tag{76}$$

The directions of the axes and the values of λ_2 are not greatly changed but the value for λ_1 is now positive, indicating ellipsoidal contours with the centre a true maximum. Two further experiments were performed at this centre, the costs found being 4 and 3 respectively. Four additional experiments, two performed on the newly calculated axis of X_1 and two elsewhere, confirmed the general nature of the surface; by including these the constants of the surface were not greatly changed. The final conclusion drawn was that a minimum for cost occurred at about the levels 141°C., 20 per cent. concentration and molar ratio (A/B) 6.2; here the estimated cost was about 3.2. The final estimate of the standard deviation based on the 11 residual degrees of freedom was 1.7; thus the fitted equation was probably a satisfactory approximation.

6. Summary and Discussion

The general problem is discussed of finding experimentally the levels of a number of quantitative variables at which some dependent response has a maximum value. The problem can be solved by exploring the whole experimental region; but the number of experiments necessary to do this would usually be prohibitively large. When the experimental error is small and experiments are conducted sequentially, the derivatives determined in a given sub-region may be used to locate another in which the response is greater. By repetition of this process a near-stationary region is found. This region may then be explored by determining derivatives of higher order. The derivatives are deduced by performing experiments in the sub-region and fitting an equation of suitable degree to the experimental points. Two possible sources of error arise, that due to errors of observation, and that due to bias which might occur if the response function were not capable of representation by an equation of the type assumed. The extent to which the estimates will be affected by these errors is completely determined by two matrices which depend only on the arrangement of the experimental points. Using these matrices it is possible to judge the suitability of possible designs and to compare different types of design.

When it is desired to estimate derivatives of first order only, the two-level factorial and incomplete factorial designs will provide efficient estimates. When effects of higher order are needed, multi-level factorial and incomplete factorial designs are less satisfactory and new types of design are proposed. Particular examples of these are "composite" designs which are used to determine the effects up to order 2 and which are formed by adding further points to two-level designs.

When an equation of second degree has been fitted in a particular region the nature of the local surface can be deduced approximately, by reducing the conic, of which the fitted surface is part, to canonical form. This will usually suggest points at which confirmatory experiments should be performed; the information from these may be incorporated without undue recalculation. The methods discussed below have been used in a number of investigations with considerable success; parts of two of these investigations are used as examples.

The procedure described can be regarded as an extension and rationalization of the methods of the experienced experimenter in situations where the experimental error is fairly small and sequential experimentation is possible. When confronted with problems of the sort discussed he will, as a rule, consider carefully the result of one set of experiments to plan the next, move the levels of factors in the direction in which increases are indicated, and carry out extra experiments at points where the information is inadequate. Any rigid systemization of his procedure may lose more by hindering power of manoeuvre than it gains by increased precision, and will (quite rightly in our opinion) not appeal to him. Much work remains to be done, in particular in the study of designs of second and higher orders, and in the determining of confidence regions for a predicted stationary point. Work is proceeding on both these problems, and it is hoped to discuss them in later communications. In conclusion we wish to acknowledge our indebtedness to many colleagues whose helpful criticisms and suggestions have been of the greatest value, in particular to Dr. O.L. Davies, Dr. T.S. Kenney, Dr. P.V. Youle and Dr. R.J. Benzie.

(*Editor's note*: Appendices 1–4 have been omitted.)

References

Aitken, A.C. (1935), *Proc. Roy. Soc. Edin.*, **55**, 42.
——— (1937), *ibid.*, **57**, 269.
Banerjee, K.S. (1949), *Ann. Math. Statist.*, **20**, 300.
Barnard, G.A. (1946), *J.R. Statist. Soc. Suppl.*, **8**, 1.
Booth, A.D. (1949), *Quart. J. Appl. Math.*, **1**, 237.
Brownlee, K.A., Kelly, B.K., and Loraine, P.K. (1948), *Biometrika*, **35**, 268.
David, F.N., and Neyman, J. (1938), *Statist. Res. Mem.*, **2**, 105.
Davies O.L., and Hay, W. (1950), *Biometrics*, **6**, 233.
Dwyer, P.S. (1942), *J. Amer. Statist. Assn.*, **37**, 441.
Finney, D.J. (1945), *Ann. Eugen. Lond.*, **12**, 291.
Fisher, R.A. (1935), *The Design of Experiments*. Edinburgh: Oliver and Boyd.
——— (1942), *Ann. Eugen. Lond.*, **11**, 341.
Fox. L., Huskey, H.O. and Wilkinson, J.H. (1948), *Quart. J. Mech. Appl. Math.*, **1**, 149.
Friedman, M., and Savage, L.J. (1947), *Selected Techniques of Statistical Analysis*. New York: McGraw-Hill.
Gauss, C.F. (1821), *Werke*, 4 Göttingen.
Hotelling, H. (1941), *Ann. Math. Statist.*, **12**, 20.
——— (1944), *ibid.*, **15**, 297.
Kempthorne, O. (1947), *Biometrika*, **34**, 255.
——— (1948), *Ann. Math. Statist.*, **19**, 238.
Kendall, M.G. (1946), *The Advanced Theory of Statistics*, Vol. II. London: Griffin.
Markoff, A.A. (1912), *Wahrscheinlichkeitsrechnung*. Leipzig: Tembner.
Plackett, R.L., and Burman, J.P. (1946), *Biometrika*, **33**, 305.
Plackett, R.L. (1949), *Biometrika*, **36**, 458.
——— (1950), *ibid.*, **37**, 149.
Rao, C.R. (1947), *J.R. Statist. Soc., Suppl.*, **9**, 128.
Southwell, R.V. (1946), *Relaxation Methods in Theoretical Physics*. Oxford.
——— (1940), *Relaxation Methods in Engineering Science*. Oxford.
Synge, J.L. (1944), *Quart. J. Appl. Math.*, **1**, 237.
Temple, G. (1939), *Proc. Roy. Soc.*, A, **169**, 476.
Yates, F. (1935), *J.R. Statist. Soc., Suppl.*, **2**, 181.
——— (1937), *Imp. Bur. Soil Sci. Tech. Comm.*, No. 35.
Wald, A. (1945), *Ann. Math. Statist.*, **16**, 117.

Introduction to Kaplan and Meier (1958) Nonparametric Estimation from Incomplete Observations

N.E. Breslow
University of Washington

During the late 1940s and early 1950s, stimulated in part by Sir Bradford Hill's (1951) advocacy of the randomized clinical trial as a cornerstone of modern scientific medicine, medical research workers began to undertake numerous follow-up studies of patient populations in order to assess the effectiveness of medical treatment. A challenging problem in the analysis of data from such investigations, whose typical endpoint was the death of a patient, stemmed from the fortunate fact that not all the patients had died at the end of the study. Thus, some of the observed lifetimes were *censored* on the right, meaning simply that they were known only to exceed an observation limit equal to the time interval from the patient's entry on study to its close or, if follow-up was incomplete, to the time of loss from further observation. This precluded the use of classical statistical procedures based on averages.

The standard approach to the presentation and analysis of such censored survival data at the time was the construction of life tables such as had been used by actuaries and demographers for centuries [Glass (1950)]. The state of the art was well represented in the work of Berkson and Gage (1950) on the survival of cancer patients at the Mayo Clinic. Table 1, adapted from their paper, shows the grouping of patient lifetimes into yearly intervals. If n_j denotes the number known to be alive at the start of the *jth* interval, d_j the number who died, and w_j the number withdrawn alive during the interval, they estimated the conditional probability of death during the interval as $\hat{q}_j = d_j/n'_j$. The utilization of $n'_j = n_j - w_j/2$ as the effective number at risk was based on the somewhat vague notion that withdrawals occurred, on average, at the interval midpoint. The probability of survival at the end of the *kth* interval was estimated as the product $\hat{P}_k = \prod_{j=1}^{k} \hat{p}_j$ of the conditional probabilities $\hat{p}_j = 1 - \hat{q}_j$ of survival over each of the preceding intervals. A formula

Table 1. Calculation of Survival Rates by the Actuarial Method[a]

Interval (years) $j-1$	Last report: Dead d_j	Last report: Living w_j	No. alive at start n_j	Effective no. at risk n'_j	Probability of death q_j	Cumulative survival (%) P_j
0–	90	—	374	374.0	0.2406	100.0
1–	76	—	284	284.0	0.2676	75.9
2–	51	—	208	208.0	0.2452	55.6
3–	25	12	157	151.0	0.1656	42.0
4–	20	5	120	117.5	0.1702	35.0
5–	7	9	95	90.5	0.0773	29.1
6–	4	9	79	74.5	0.0537	26.8
7–	1	3	66	64.5	0.0155	25.4
8–	3	5	62	59.5	0.0504	25.0
9–	2	5	54	51.5	0.0388	23.7
10–	21	26	47	—	—	22.8

[a] Adapted from Berkson and Gage (1950).

for the standard error of $\hat{P}_k$, namely,

$$\mathrm{SE}(\hat{P}_k) = \hat{P}_k \sqrt{\sum_{j=1}^{k} \frac{\hat{q}_j}{n'_j \hat{p}_j}}, \tag{1}$$

was available based on approximations derived by Greenwood (1926) and Irwin (1949) under the assumption that the $\hat{q}_j$ were independent binomial proportions. However, the accuracy of the various ad hoc adjustments and approximations that entered into this standard methodology never had been investigated fully.

Kaplan and Meier (1958) proposed a refined estimate of the survival curve based on the notion that one was free, when carrying out the standard life table operations, "to take the intervals as short and as numerous as one pleases, and to regard each death as occupying an interval by itself." They termed the result the product-limit (PL) estimate. Kaplan and Meier (hereafter KM) traced the origin of this idea back to Böhmer (1912), and others [Andersen and Borgan 1985)] have suggested that it may even go back as far as Karup (1893). As pointed out by Kaplan (personal communication), there is some slight ambiguity in the formal definition of the PL estimate given on pp. 463–465 of their paper. This may be resolved quite simply, however, by defining the division points u_j between the runs of deaths and losses to occur at the *latest* loss rather than "anywhere among" them so that n_{j+1}, the number still on study at the start of the next interval, excludes the losses that occur at the end of the preceding one.

Nowadays a formal definition of the PL estimate would likely be given as follows. Denote by $t_1 < t_2 < \cdots < t_k$ the distinct times at which deaths occur.

Redefine d_j and n_j to denote, respectively, the number of deaths that occur at t_j and the number at risk, i.e., the number alive and under observation, just prior to that time. Then the probability of survival beyond any time t, namely, $1 - F(t)$ where F is the cumulative distribution of (true) lifetimes, is estimated by

$$1 - \hat{F}(t) = \prod_{t_j \le t} \left(1 - \frac{d_j}{n_j}\right).$$

KM called attention to an important feature of this definition, namely, its accommodation of left truncation as well as right censorship. In other words, the numbers n_j at risk may be augmented by the entry of new patients on study during the course of follow-up, as well as reduced through deaths and losses. If everyone at risk at the maximum observed true lifetime t_k succumbs, then $1 - \hat{F}(t) = 0$ for $t \ge t_k$. Otherwise, KM regarded the estimate as undefined beyond t_k.

KM demonstrated that the PL estimate is (1) nearly unbiased; (2) consistent; (3) the nonparametric maximum likelihood estimate of $1 - F(t)$; and (4) has approximate standard error given by eq. (1) with the summation taken over $t_j < t$ and with n_j' replaced by n_j as redefined above. Thus, the ingenious device of allowing each distinct lifetime to occupy its own time interval resolved the major concerns about the bias and standard error of the grouped life table estimate. The reader may find the KM derivations of these key properties to be somewhat incomplete in places. This is not surprising since a completely rigorous mathematical treatment of censorship raises fundamental issues in probability theory that have been fully resolved only recently. Indeed, the problem of the distributional properties of the PL estimate and related quantities has challenged a new generation of statisticians to develop probabilistic techniques specifically to handle it. KM's approach *via* successive conditional expectations, conditioning on the past history of deaths and losses, anticipated the modern approach based on martingale theory.

Efron (1967) stimulated much further work by his utilization of the PL estimate as a stepping stone to a generalized Wilcoxon test for comparing two samples of censored data. Suppose that $\{x_i^0 : i = 1, \cdots, n\}$ is a random sample from F and that u_i are the corresponding observation limits, which are best regarded as fixed numbers. One observes $x_i = \min(x_i^0, u_i)$ together with censoring indicators δ_i that take the value 1 if $x_i = x_i^0$ (no censorship) and 0 otherwise. Let $Y(t) = \#\{i : x_i \ge t\}$ denote the process that keeps track of the number of persons at risk at time t and, anticipating later developments, let $N(t) = \#\{i : x_i \le t, \delta_i = 1\}$ denote the counting process of true observed lifetimes. Assuming that an initial estimate of the lifetime distribution was available, Efron suggested that one estimate the number of survivors to time t as the number known to survive that long plus the sum of the estimated conditional probabilities that those censored earlier also survived to t. He thus defined a *self-consistent* estimate to be one that satisfied

$$n[1 - \hat{F}(t-)] = Y(t) + \sum_{x_i<t,\delta_i=0} \frac{1 - \hat{F}(t-)}{1 - \hat{F}(x_i)}$$

and proved that the PL estimate uniquely had this property.

Supposing F to be absolutely continuous and $\{u_i : i = 1, \ldots, n\}$ to be an independent random sample from a cdf G, Efron also stated without proof an important result regarding the asymptotic normality of the PL estimate regarded as a stochastic process, namely, that the "KM process" $Z = \sqrt{n}(\hat{F} - F)/(1 - F)$ converges in distribution to a Gaussian process Z^∞ having covariance function

$$\operatorname{cov}[Z^\infty(s), Z^\infty(t)] = \int_0^{s\wedge t} \frac{dF(u)}{[1 - F(u-)]^2[1 - G(u-)]}, \tag{2}$$

which he recognized to be that of a Brownian bridge on a transformed time scale. [Following Gill (1983), this expression has been written to show the proper extension for noncontinuous F.] Since (2) becomes infinite if G approaches 1 too quickly in relation to F, some regularity conditions on the distribution of the censoring limits are needed.

Continuing the work with Efron's random censorship model, Breslow and Crowley (1974) established the weak convergence of Z to Z^∞ on any finite interval $[0, T]$ such that $[1 - H(T)] \overset{\text{def}}{=} [1 - F(T)][1 - G(T)] > 0$. Their proof involved the approximation of Z by an expression that was linear in N and Y and, in the words of Gill (1983) who corrected an error in the evaluation of one of the remainder terms, was "unavoidably complex" in regard to the derivation of the covariance function (2). Nonetheless, it provided the machinery needed for the construction of simultaneous asymptotic confidence bands for the survival curve $1 - F(t)$, considered a function of t over the interval $[0, T]$, work that was undertaken by Gillespie and Fisher (1979) and by Hall and Wellner (1980). Meier (1975), in a paper that elaborated on the original KM work, outlined how the weak convergence result applied also to the case of fixed censoring limits. He pointed out that the limiting distribution Z^∞ could contain a Poisson component, as well as or instead of a Gaussian component, if censorship was severe in a region where F changed little. Wellner (1985) subsequently investigated precise conditions leading to a fully Poisson limit.

At the same time that Breslow and Crowley were working on the random censorship model using standard tools of weak convergence [Billingsley (1968), Pyke and Shorack (1968)], O. Aalen (1975, 1978) was developing an approach based on martingale theory as part of his Ph.D. dissertation at Berkeley under Lucien LeCam. This work provided a more natural framework for the study of survival processes under quite general censorship models by regarding them as counting processes whose conditional distribution at any point in time was determined by the entire past history of events (deaths) and censorship. It was a "decisive breakthrough" that forms the cornerstone of the modern theory of life history analysis [Andersen and Borgan (1985)], and it greatly simplified the rigorous evaluation of the asymptotic properties of the KM estimate and of the Cox (1972) proportional hazards model that is considered elsewhere in this volume.

Aalen's point of departure was estimation of the cumulative hazard function

$$\Lambda(t) = \int_0^t \lambda(s)\, ds = \int_0^t \frac{dF(s)}{[1 - F(s-)]},$$

where $\lambda(t)$ denotes the instantaneous hazard or mortality function, i.e., $\lambda(t)\, dt = \Pr\{x_i \varepsilon [t, t + dt) | x_i \geq t\}$. The analogue of the KM estimate for the cumulative hazard,

$$\hat{\Lambda}(t) = \sum_{t_j \leq t} \frac{d_j}{n_j} = \int_0^t \frac{dN(s)}{Y(s)},$$

where $1/Y(s)$ is defined to be 0 whenever $Y(s)$ is 0, is known as the Nelson–Aalen estimate. Nelson (1969) had earlier proposed its use for the analysis of incomplete failure data in an industrial setting. Aalen observed that $M(t) = N(t) - \int_0^t Y(s)\, d\Lambda(s)$ was a martingale with respect to a nested family $\{\mathscr{F}_{t-}\}$ of sigma algebras that specify the history of events and losses prior to time t, and furthermore, that $\hat{\Lambda} - \Lambda$ could be expressed as the stochastic integral $\int YdM$ of the *predictable* process Y with respect to M. Using work by Rebolledo (1980) on martingale central limit theorems, he concluded that $\sqrt{n}(\hat{\Lambda} - \Lambda)$ converged in distribution to a Gaussian process with covariance function $\int_0^{s \wedge t} g^2(u)\, du$, provided only that a nonnegative, square integrable function g existed such that

$$\int_0^t g^2(s)\, ds = \lim_{n \uparrow \infty} n^{-1} \int_0^t \frac{d\Lambda(s)}{Y(s)}.$$

This limit equals the covariance function (2) of the limiting KM process Z^∞ in the case of random censorship. In view of the approximate relation $\hat{\Lambda} \approx -\ln(1 - \hat{F})$ and the Taylor expansion $-\ln(1 - \hat{F}) - \Lambda \approx (1 - F)^{-1}(\hat{F} - F)$, it is clear that the limiting processes must indeed be the same. Gill (1983) used this approach to investigate the asymptotic behavior of the PL estimate on the entire real line.

Other investigators have considered the small sample performance of the PL estimate and the adequacy of the Greenwood–Irwin standard error (1). For example, using a random censorship model in which the censoring and survival hazards were proportional $[(1 - G) = (1 - F)^\alpha]$, Chen, et al. (1982) demonstrated numerically the utility of the KM bound

$$0 \leq E\hat{F}(t) - F(t) \leq \exp[-EY(t + 0)]$$

for establishing the exponentially small bias of the PL estimate. They found that the Greenwood-Irwin formula often slighty underestimated the true standard error, a result that also followed from Wellner's (1985) investigations. For practical purposes, however, the standard formula is generally adequate.

Although a simple plot of the survival curve $1 - \hat{F}(t)$ against t is quite adequate for estimating the proportion of survivors at various points following treatment, it often is not very informative in regard to the shape of the instantaneous hazard function. Smoothed estimates of λ based on integrating an appropriate kernel with respect to the Nelson–Aalen cumulative hazard estimate have been proposed by Ramlau–Hansen (1983), Yandell (1983), and Tanner and Wong (1954). Greater use of smoothed hazard plots as an adjunct to the simple PL estimate would encourage more detailed study of the shape of survival distributions.

The impact of KM's work on statistical practice is perhaps even greater than its role in stimulating basic research. In the early 1970s the PL estimate was adopted by statisticians, epidemiologists, and clinicians as the method of choice for reporting results of clinical trials of chronic disease and other follow-up studies in the medical literature. According to data compiled by the Institute for Scientific Information in Philadelphia, the citation rate rose dramatically throughout the 1970s and 1980s. By 1988, KM's paper was receiving nearly 800 citations annually in the scientific literature, with a cumulative total approaching 4000.

Edward Kaplan and Paul Meier were graduate students of John Tukey at Princeton University, where they completed their doctoral dissertations in 1951. Although neither of their theses involved the analysis of survival data, both students were led to consider the problem shortly after they left Princeton: Meier at the department of biostatistics at The Johns Hopkins University and Kaplan in the mathematics research department at the Bell Telephone Laboratories. Initially, they submitted separate manuscripts to the *Journal of the American Statistical Association*, Kaplan's oriented toward industrial life-testing applications and Meier's drawing from the classical biostatistical literature. Both were encouraged by Tukey, to whom they had circulated their manuscripts, and *JASA* editor Allan Wallis to collaborate on a joint version. This occupied the next three or four years as each reworked the other's material. When the joint paper was finally submitted, the referees were divided on whether it should be published. Fortunately for the sake of posterity, the journal's editor decided to go with the positive review.

Following the publication of their paper in 1958, Kaplan went on to pursue interests in probability theory and mathematical programming. He is the author of a 1982 text, *Mathematical Programming and Games*, as well as unpublished works on poetry and music theory. He retired in 1981 from his position as professor of mathematics at Oregon State University, Corvallis. Meier has pursued a varied and influential career as a biostatistician, at Hopkins until 1957 and since then at the University of Chicago. He is well known for his early work with the Salk polio vaccine trial and as a strong advocate for the role of statistics in medicine. He is currently Ralph and Mary Otil Isham Professor of Statistics and Medicine at the University of Chicago.

References

Aalen, O.O. (1975). Statistical inference for a family of counting processes. Ph.D. dissertation, Department of Statistics, University of California, Berkeley.

Aalen, O.O. (1978). Non-parametric inference for a family of counting processes, *Ann. Statist.*, **6**, 701–726

Andersen, P.K., and Borgan, O. (1985). Counting process models for life history data: A review, *Scand. J. Statist.*, **12**: 97–158.

Berkson, J., and Gage, R.P. (1950). Calculation of survival rates for cancer, *Proc. Staff Meetings Mayo Clinic*, **25**, 270–286.

Billingsley, P. (1968). *Weak Convergence of Probability Measures*. Wiley, New York.

Böhmer, P.E. (1912). Theorie der unabhängigen Wahrscheinlichkeiten, *Rapp., Mém. et Procés-verbaux 7e Congr. Internat. Act.*, Amsterdam, **2**, 327–343.

Breslow, N., and Crowley, J. (1974). A large sample study of the life table and product limit estimates under random censorship, *Ann. Statist.*, **2**, 437–453.

Chen, Y.Y., Hollander, M., and Langberg, N.A. (1982). Small sample results for the Kaplan–Meier estimator, *J. Amer. Statist. Assoc.*, **77**, 141–144.

Cox, D.R. (1972). Regression models and life tables, *J. Roy. Statist. Soc., Ser. B*, **34**, 187–220 (with discussion).

Efron, B. (1967). The two sample problem with censored data, in *Proceedings of 5th Berkeley Symposium on Mathematical Statistics and Probability*, Vol. 4 University of California Press. pp. 831–853.

Gill, R (1983). Large sample behaviour of the product-limit extimator on the whole line, *Ann. Statist.*, **11**, 49–58.

Gillespie, M.J., and Fisher, L. (1979). Confidence bands for the Kaplan–Meier survival curve estimate, *Ann. Statist.*, **7**, 920–924.

Glass, D.V. (1950). Graunt's life table, *J. Inst. Actuaries*, **76**: 60–64.

Greenwood, M. (1926). The natural duration of cancer, in *Reports on Public Health and Medical Subjects*, vol. 33. H. M. Stationery Office, London.

Hall, W.J., and Wellner, J.A. (1980). Confidence bands for a survival curve from censored data, *Biometrika*, **67**, 133–143.

Hill, A.B. (1951). The clinical trial, *Brit. Med. Bull.*, **7**, 278–282.

Irwin, J.O. (1949). The standard error of an estimate of expectational life, *J. Hygiene*, **47**, 188–189.

Karup, I. (1893). *Die Finanzlage der Gothaischen Staatsdiener- Wittwen-Societät*. Dresden.

Meier, P. (1975). Estimation of a distribution function from incomplete observations, in *Perspectives in Probability and Statistics* (J. Gani, ed.). Applied Probability Trust, Sheffield, pp. 67–87.

Nelson, W. (1969). Hazard plotting for incomplete failure data, *J. Qual. Technol.*, **1**, 27–52.

Pyke, R., and Shorack, G. (1968). Weak convergence of a two-sample empirical process and a new approach to the Chernoff–Savage theorems, *Ann. Math. Statist.*, **39**, 755–771.

Ramlau-Hansen, H. (1983). Smoothing counting process intensities by means of kernel functions, *Ann. Statist.*, **11**, 453–466.

Rebolledo, R. (1980). Central limit theorems for local martingales, *Z. Wahr. verw. Geb.*, **51**, 269–286.
Tanner, M.A., and Wong, W.H. (1984). Data-based nonparametric estimation of the hazard function with applications to model diagnostics and exploratory analysis, *J. Amer. Statist. Assoc.*, **79**: 174–182.
Wellner, J.A. (1985). A heavy censoring limit theorem for the product limit estimator. *Ann. of Statist.* **13**: 150–162.
Yandell, B.S. (1983). Nonparametric inference for rates and densities with censored survival data, *Ann. Statist.*, **11**, 1119–1135.

Nonparametric Estimation from Incomplete Observations*

E. L. Kaplan
University of California Radiation Laboratory

Paul Meier
University of Chicago

Abstract

In lifetesting, medical follow-up, and other fields the observation of the time of occurrence of the event of interest (called a *death*) may be prevented for some of the items of the sample by the previous occurrence of some other event (called a *loss*). Losses may be either accidental or controlled, the latter resulting from a decision to terminate certain observations. In either case it is usually assumed in this paper that the lifetime (age at death) is independent of the potential loss time; in practice this assumption deserves careful scrutiny. Despite the resulting incompleteness of the data, it is desired to estimate the proportion $P(t)$ of items in the population whose lifetimes would exceed t (in the absence of such losses), without making any assumption about the form of the function $P(t)$. The observation for each item of a suitable initial event, marking the beginning of its lifetime, is presupposed.

For random samples of size N the product-limit (PL) estimate can be defined as follows: List and label the N observed lifetimes (whether to death or loss) in order of increasing magnitude, so that one has $0 \leq t_1' \leq t_2' \leq \cdots \leq t_N'$. Then $\hat{P}(t) = \Pi_r\, [(N - r)/(N - r + 1)]$, where r assumes those values for which $t_r' \leq t$ and for which t_r' measures the time to death. This estimate is the distribution, unrestricted as to form, which maximizes the likelihood of the observations.

Other estimates that are discussed are the actuarial estimates (which are also products, but with the number of factors usually reduced by grouping); and reduced-sample (RS) estimates, which require that losses not be accidental, so that the limits of observation (potential loss times)

* Prepared while the authors were at Bell Telephone Laboratories and Johns Hopkins University respectively. The work was aided by a grant from the Office of Naval Research.

are known even for those items whose deaths are observed. When no losses occur at ages less than t, the estimate of $P(t)$ in all cases reduces to the usual binomial estimate, namely, the observed proportion of survivors.

1. Introduction

1.1 Formulation

In many estimation problems it is inconvenient or impossible to make complete measurements on all members of a random sample. For example, in medical follow-up studies to determine the distribution of survival times after an operation, contact with some individuals will be lost before their death, and others will die from causes it is desired to exclude from consideration. Similarly, observation of the life of a vacuum tube may be ended by breakage of the tube, or a need to use the test facilities for other purposes. In both examples, incomplete observations may also result from a need to get out a report within a reasonable time.

The type of estimate studied here can be briefly indicated as follows. When a random sample of N values, $T_1, T_2, \ldots, T_N$ of a random variable is given, the sample distribution function $\hat{F}(t)$ is naturally defined as that which assigns a probability of $1/N$ to each of the given values, so that $\hat{F}(t)$ equals $1/N$ times the number of sample values less than the argument t. Besides describing the sample, this $\hat{F}(t)$ is also a nonparametric estimate of the population distribution, in the sense indicated in 1.2 below. When the observations are incomplete, the corresponding estimate is still a step-function with discontinuities at the ages of observed deaths, but it can no longer be obtained as a mere description of the sample.

The samples considered in this paper are incomplete in the sense that one has given, not a random sample $T_1, \ldots, T_N$ of values of the random variable T itself (called the lifetime), but the *observed lifetimes*

$$t_i = \min(T_i, L_i), \qquad i = 1, 2, \ldots, N. \tag{1a}$$

Here the L_i, called *limits of observation*, are constants or values of other random variables, which are assumed to be independent of the T_i unless otherwise stated (in Sections 3.2 and 7). For each item it is known whether one has

$$T_i \leqq L_i, \qquad t_i = T_i \text{ (a } \textit{death}\text{)} \tag{1b}$$

or

$$L_i < T_i, \qquad t_i = L_i \text{ (a } \textit{loss}\text{)}.$$

Ordinarily the T_i and L_i are so defined as to be necessarily nonnegative.

The items in the sample are thus divided into two mutually exclusive classes, namely deaths and losses. A loss by definition always precludes the desired knowledge of T_i. On the other hand, a death does not always preclude the knowledge of the corresponding L_i, in case the limits of observation are nonrandom and foreseeable. Such knowledge of the L_i may have value; for example, it makes available the reduced-sample estimate (Section 3), if one chooses to use it.

The type of sample described is a generalization of the censored sample defined by Hald [17], and a specialization of the situation considered by Harris, Meier, and Tukey [18].

The term death has been adopted as being at least metaphorically appropriate in many applications, but it can represent any event susceptible of random sampling. In particular, the roles of death and (random) loss may be interchangeable. By redefining the classification of events into deaths and losses, it may be possible to approach the same data from various points of view and thus to estimate the survivorship functions $P(t)$ that would be appropriate to various categories of events in the absence of the others. This is familiar enough; see for example [6], [13], [16].

1.2 Nonparametric Estimation

Most general methods of estimation, such as maximum likelihood or minimum chi-square, may be interpreted as procedures for selecting from an admissible class of distributions one which, in a specified sense, best fits the observations. To estimate a characteristic (or parameter) of the true distribution one uses the value that the characteristic has for this best fitting distribution function. It seems reasonable to call an estimation procedure *nonparametric* when the class of admissible distributions from which the best-fitting one is to be chosen is the class of all distributions. (Wolfowitz [28] has used the term similarly in connection with the likelihood ratio in hypothesis testing). With a complete sample, it is easy to see that the sample distribution referred to in 1.1 is the nonparametric estimate on the maximum likelihood criterion. The same result is true of the product-limit estimate for incomplete samples, as will be demonstrated in Section 5.

The most frequently used methods of parametric estimation for distributions of lifetimes are perhaps the fitting of a normal distribution to the observations or their logarithms by calculating the mean and variance, and fitting an exponential distribution $e^{-t/\mu}dt/\mu$ by estimating the mean life μ alone. Such assumptions about the form of the distribution are naturally advantageous insofar as they are correct; the estimates are simple and relatively efficient, and a complete distribution is obtained even though the observations may be restricted in range. However, nonparametric estimates have the important functions of suggesting or confirming such assumptions, and of supplying the estimate itself in case suitable parametric assumptions are not known. An important property of these nonparametric estimates is that if the age scale is transformed from t to $t^* = f(t)$, where f is a strictly increasing function, then the corresponding estimated distribution functions are simply related by $\hat{F}^*(f(t)) = \hat{F}(t)$.

1.3 Examples of the RS and PL Estimates

We will consider the following situation. A random sample of 100 items is put on test at the beginning of 1955; during the year 70 items die and 30 survive. At the end of the year, a larger sample is available and 1000 additional items are put on test. During 1956, 15 items from the first sample and 750 from the second die, leaving 15 and 250 survivors respectively. As of the end of 1956, it is desired to estimate the proportion $P(2)$ of items in the population surviving for two years or more.

The survival probabilities are supposed to depend on the age (the duration of the test) rather than on the calendar year, and hence the data are arranged as in Table 460.*

Table 460.

Samples	I	II
Initial numbers	100	1000
Deaths in first year of age	70	750
One-year survivors	30	250
Deaths in second year of age	15	
Two-year survivors	15	

This particular example is such that it is easy to form an estimate $P^*(2) = 15/100 = 0.15$ from the first sample alone. This is called the *reduced-sample* (RS) estimate because it ignores the 1000 items tested only during 1956. It is a legitimate estimate only when the reduced sample is itself a random sample; this will be the case only when (as assumed here) the observation limits (two years for the first sample, and one year for the second) are known for all items, deaths as well as losses. In the absence of this information, one would have no basis for discriminating among the 835 deaths observed before the age of two years. One cannot simply ignore the 250 losses at age one year; since only 15 items have survived for two years, $P(2)$ would then be estimated as $15/850 = .018$, an absurd result. The point is discussed further in 3.1 below.

We now inquire whether the second sample, under test for only one year, can throw any light on the estimate of $P(2)$. Clearly it will be necessary to assume that both samples have been drawn from the same population, an

* *Editor's note*: The number 460 and the numbers of the subsequent tables correspond to the pages of the original articles on which these tables appeared.

assumption that the RS estimate $P^*(2)$ avoided. At any rate, the estimates of $P(1)$ from the two samples, namely 0.30 and 0.25, are not sufficiently different to contradict the assumption. By combining the two samples, the estimate

$$\hat{P}(1) = P^*(1) = (30 + 250)/(100 + 1000) = .255$$

is obtained for $P(1)$. (In this case the RS has the same value as the other estimate to be discussed, the product-limit or PL.) This result exhausts the usefulness of the second sample for the present purposes; how does it help to estimate $P(2)$?

The answer is that there are advantages to using the first (the smaller) sample for estimating $P(2)/P(1)$, the conditional probability of survival for two years given survival for one year, rather than $P(2)$ itself. This estimate is

$$\hat{P}(2)/\hat{P}(1) = 15/30 = 0.50, \text{ whence}$$

$$\hat{P}(2) = 0.255 \times 0.50 = 0.127,$$

a very simple example of the *product-limit* (PL) estimate. The outstanding advantage of this strategy is that it works just as well if we are not privileged to know that the 750 deaths in the second sample had observation limits of one year, because these items are irrelevant to the estimation of $P(2)/P(1)$ in any case. Other considerations for deciding between the two estimates will be set forth in Section 3.

The discussion of the PL estimate will be continued shortly, in Sections 2 and 3. Section 3 is equally concerned with the RS estimate, while Section 4 is devoted to the actuarial estimates.* The remaining three sections consist principally of mathematical derivations. Though much older, the actuarial estimates are essentially approximations to the PL; they are products also, but typically they aim to reduce the number of factors by grouping. (Grouping may or may not be possible for the PL itself.) The distinguishing designation product-*limit* was adopted because this estimate is a limiting case of the actuarial estimates. It was proposed as early as 1912 by Böhmer [6] (referred to by Seal [26]), but seems to have been lost sight of by later writers and not further investigated.

1.4 Notation

The survival function

$$P(t) = \Pr(T > t), \tag{1c}$$

giving the population probability of surviving beyond t, will be used in place of the distribution function $F(t) = 1 - P(t)$ because of its convenience where the product-limit estimate and its actuarial approximations are concerned. In addition the following functions are defined:

$\hat{P}(t)$ = product-limit (PL) estimate of $P(t)$.

* *Editor's note*: Section 4 is not reproduced in this volume.

$P^*(t)$ = reduced-sample (RS) estimate of $P(t)$.

$n(t)$ = the number of items observed and surviving at age t, when deaths (but not losses) at t itself are subtracted off.

$N(t)$ = the expectation of $n(t)$, for fixed observation limits.

$N^0(t)$ = the number of items having observation limits L such that $L \geq t$. In practice this function is not necessarily known.

For the first reading of the paper it may be desirable to suppose that the death of one item and the loss of the same or any other item never occur at the same age, and never coincide with an age t at which any of the above functions are to be evaluated. This condition can always be met by fudging the ages a little when necessary. On the other hand, a regular user of the techniques will probably come to regard overt fudging as naive; he will prefer to formalize his notation and record-keeping by adopting the conventions already insinuated into the definitions of death and loss, $P(t)$, and $n(t)$ above. These conventions may be paraphrased by saying that deaths recorded as of an age t are treated as if they occurred slightly before t, and losses recorded as of an age t are treated as occurring slightly after t. In this way the fudging is kept conceptual, systematic, and automatic.

The convention that deaths precede losses in case of ambiguity is based on the following sequence of operations, which is clearly more efficient than the reverse sequence: Examine a group of items of age t_0, observe the number δ of deaths since the last examination, and then remove (or lose contact with) a number λ of the survivors. It may then be convenient simply to record t_0 as the age of death or loss of the $\delta + \lambda$ items, especially if t_0 is always an integral multiple of a fundamental time interval; in fact, however, the deaths will have preceded the losses.

The chief exception to the immediate applicability of this convention occurs when the losses are random but cannot affect items that have already died. Then the possible sequences of occurrence of deaths and losses between examination times (assumed to be close together) are approximately equally likely, and a reasonable compromise is to assume that half the losses in the interval precede and half follow the deaths, as in (4b) below. If the loss of an item is compatible with the possibility of its having died (unknown to the experimenter) between the time when it was last examined and the time of loss, an item lost in this way is effectively lost just after it was last examined, and the convention is entirely appropriate. The disappearance of individuals subject to medical follow-up is a case in point.

The remainder of the conventions concern the treatment of discontinuities in the functions listed above. Should the value assigned to $n(t)$, say, for an argument t for which it is discontinuous be the right-hand limit $n(t + 0) = \lim n(t + h)$ as $h \to 0$ with $h > 0$, the left-hand limit $n(t - 0)$ (the same but with $h < 0$), or something else? The superior expressiveness of the notation adopted is illustrated by the relations

$$n(t-0) - n(t) = \text{the number of deaths at } t,$$
$$n(t) - n(t+0) = \text{the number of losses at } t, \tag{1d}$$

or equivalently by the formula (3a): $P^*(t) = n(t)/N^0(t)$, which otherwise would not be valid at discontinuities. Analogous conventions are adopted for all the above-mentioned functions of t, so that $P(t)$ and $\hat{P}(t)$ are right-continuous, $N^0(t)$ is left-continuous, and $n(t)$ and $P^*(t)$ are neither. Other advantages of the convention for $P(t)$ and its estimates are the following: (a) According to the sequence of operations assumed above, one may record deaths as of age t although they actually occurred slightly earlier. (b) It makes $P(0) = 1$ if and only if no item dies at birth (age zero). This is convenient and natural.

One other possibility may be mentioned briefly. The assumption that $P(\infty) = 0$, so that the lifetimes are finite with probability one, is necessary only for parts of Section 7 and for the calculation of a finite mean lifetime. However, in practice there is no apparent need to contradict the assumption either. If half of a sample dies in one day and the other half is still alive after 1000 days, one should still report $\hat{P}(1000)$ (not $\hat{P}(\infty)$) $= 0.50$, since the argument 1000 is not an arbitrary large number, but the actual duration of the test.

2. The Product-Limit Estimate

2.1 Definition and Calculation

Both the PL and the actuarial estimates of Section 4 are based on the following general procedure:

(a) The age scale is divided into suitably chosen intervals, $(0, u_1)$, (u_1, u_2), ..., as described below. (In the example of 1.3, there were only two such intervals, namely (0, 1) and (1, 2).)
(b) For each interval (u_{j-1}, u_j), one estimates $p_j = P_j/P_{j-1}$, the proportion of items alive just after u_{j-1} that survive beyond u_j.
(c) If t is a division point (it may be introduced specially if necessary), the proportion $P(t)$ in the population surviving beyond t is estimated by the product of the estimated p_j for all intervals prior to t.

If step (a) is left relatively arbitrary and approximations or parametric assumptions are accepted in step (b), one arrives at the actuarial estimates. The PL is obtained by selecting the intervals in (a) so that the estimation in (b) is a simple binomial, without any recourse to assumptions of functional form. The condition for this is that within each interval, deaths and losses be segregated in a known fashion. As a beginning, it may be assumed that no interval contains both deaths and losses. Then if the number under observation just after u_{j-1} is denoted by n_j, and δ_j deaths are observed in the interval (u_{j-1}, u_j), the estimate is clearly

$$\hat{p}_j = (n_j - \delta_j)/n_j = n_j'/n_j, \tag{2a}$$

where n_j' is the number under observation just after the δ_j deaths. However, if the interval contains only losses (but at least one item survives throughout the interval), the estimate is $\hat{p}_j = 1$.

In the product of conditional probabilities formed in step (c), unit factors may as well be suppressed; and we need not be concerned with the manner in which the losses are distributed among the intervals, so long as n_j and n_j' are correctly evaluated in (2a), and no losses occur at ages intermediate to the δ_j deaths. The situation is illustrated by the following scheme:

Table 463.

No. of items	N		n_i		n_1'		n_2		n_2'		....
No. of deaths or losses		λ_0		δ_1		λ_1		δ_2		λ_2	
Division points	$u_0 = 0$					u_1				u_2	

Here N is the initial number of items, and the braces join the numbers whose ratios are the conditional probabilities (2a). The numbers in the second line are the differences of those in the first, the λ's counting losses and the δ's deaths; some of these could be zero. The division points u_j are placed in the third line to show that the λ_j deaths occur between u_{j-1} and u_j, while u_j is located anywhere among the λ_j losses. The relation $n_j' \leq n(u_j) \leq n_{j+1}$ holds.

The PL estimate is now given by

$$\hat{p}(t) = \prod_{j=1}^{k} (n_j'/n_j), \quad \text{with} \quad u_k = t, n_j' = n_j - \delta_j. \tag{2b}$$

If the greatest observed lifetime t^* corresponds to a loss, (2b) should not be used with $t > t^*$; in this case $\hat{P}(t)$ can be regarded as lying between 0 and $\hat{P}(t^*)$, but is not more closely defined.

If it is desired to permit the entrance of items into the sample after the commencement of their lifetimes, this can be done by treating such entrances as "losses" that are counted negatively in λ_j. The same items can of course disappear again at a later age and so yield ordinary losses as well. It is assumed that nothing is known of the existence of any such item that dies before it becomes available for observation; that is, the observation is censored on the right but truncated on the left, in the terminology of Hald [17].

The form (2b) was selected for the PL estimate to give the minimum number of elementary factors and the maximum grouping of the observations. Nevertheless, the number of deaths δ_j in an interval can easily be as small as unity. The resulting estimate $\hat{p}_j$, though of limited value by itself, is none the less acceptable as a component of $\hat{P}(t)$. In fact, one is at liberty to take the intervals as short and as numerous as one pleases, and to regard each death as occupying an interval by itself. To specify the resulting expression, one relabels the N ages t_i of death or loss in order of increasing magnitude, and denotes them by $t'_1 \le t'_2 \le \cdots \le t'_N$. Then

$$\hat{P}(t) = \prod_r [(N - r)/(N - r + 1)], \tag{2c}$$

where r runs through those positive integers for which $t'_r \le t$ and t'_r is the age of death (not loss). The cancellation of like integers in numerator and denominator where they occur reduces (2c) to (2b). If there are no losses, everything cancels except the first denominator N and the last numerator $n(t)$, say, and the PL reduces to the usual binomial estimate $n(t)/N$. (2c) shows that $\hat{P}(t)$ is a step-function which changes its value only at the observed ages of death, where it is discontinuous.

In analyzing data on lifetimes by the multiplication of conditional probabilities, one of the following three procedures will usually suffice:

(1) If the number of deaths is relatively small, these deaths may be arranged in order of age without grouping, and the numbers of losses in the intervening age intervals counted. The PL estimate is calculated by (2c).
(2) If (1) is too time-consuming but the number of distinct ages of loss is relatively small, these ages may be arranged in order, additional division points inserted as desired, and the numbers of deaths in the resulting age intervals counted. If some of these intervals are shorter than necessary and are found to contain no deaths, they can be combined with adjacent intervals. The PL estimate is calculated by (2b).
(3) If neither (1) nor (2) is compact enough, then division points are chosen without close consideration of the sample, deaths and losses are counted in each interval, and an actuarial approximation to the PL, such as (4b), is used.*

As a miniature example of case (1), suppose that out of a sample of 8 items the following are observed:

Deaths at 0.8, 3.1, 5.4, 9.2 months.

Losses at 1.0, 2.7, 7.0, 12.1 months.

The construction of the function $\hat{P}(t)$ then proceeds as follows:

Table 464.

u_j	n_j	n'_j	λ_j	$\hat{P}(u_j)$
0.8	8	7	2	7/8
3.1	5	4	0	7/10
5.4	4	3	1	21/40
9.2	2	1	0	21/80
(12.1)	1	1	1	21/80

* *Editor's note*: (46) is given by $P^{(1)} = \dfrac{n - \lambda/2 - \delta}{n - \lambda/2}$. Here n is the number of items at the beginning of the interval known to be depleted by δ deaths and λ losses within the interval.

Each value of $\hat{P}(u_j)$ is obtained by multiplying n_j'/n_j by the preceding value $\hat{P}(u_{j-1})$. The age 12.1 is recorded in the last line to show the point at which $\hat{P}(t)$ becomes undefined; since it is a loss time, the 12.1 is enclosed in parentheses. It is to be inferred from the table that $\hat{P}(5.3) = 7/10$, for example. The third and fourth columns could be omitted since $n_j' = n_j - 1$ (except in the last line, which corresponds to a loss) and $\lambda_j = n_j' - n_{j+1}$.

A rudimentary illustration of case (2) has already been given in 1.3. A little more elaborate example of a similar sort with $N = 100$ is given in Table 465. Here 1.7, 3.6, and 5.0 are assumed to be the only ages at which losses occur; they are prescribed as division points. The other division points (1, 2, 3, 4) are selected at pleasure, with the object of interpolating additional points on the curve of $\hat{P}(t)$ vs. t. $\hat{N}_{Ej}$ is the effective sample size defined in (2j) below. In practice four columns headed u_j, n_j, n_j', $\hat{P}_j$ will suffice. From the table one infers that $.74 < \hat{P}(2.5) < .87$, for example.

Table 465.

Interval	Factor						
u_{j-1}, u_j	j	n_j	δ_j	λ_j	$\hat{p}_j$	$\hat{P}(u_j)$	$\hat{N}_{Ej}$
0–1	1	100	3	0	97/100	.97	100
1–1.7	2	97	5	20	92/97	.92	100
1.7–2	3	72	4	0	68/72	.87	88
2–3	4	68	10	0	58/68	.74	83
3–3.6	5	58	9	12	49/58	.63	80
3.6–4	6	37	6	0	31/37	.52	73
4–5.0	7	31	15	16	16/31	.27	51

2.2 Mean and Variance of $\hat{P}(t)$

The important facts here, derived in Section 6.1, are that $\hat{P}(t)$ is consistent and of negligible bias (unless excessive averaging is done) and that an asymptotic expression for its variance can be obtained. Like the estimate itself, the sample approximation to its variance proves to be independent of the limits of observation of items not actually lost. However, the variance derived from population values does depend on all the limits of observation, which are assumed to be fixed during the sampling.

It has been noted that if the greatest observed lifetime t^* corresponds to a loss, then for $t > t^*$, $\hat{P}(t)$ is undefined though bounded by 0 and $\hat{P}(t^*)$. Unless the probability of this ambiguous situation is quite small, however, a non-parametric estimate of $P(t)$ will not be very informative in any case. The ambiguity cannot occur unless the $N^0(t)$ items observable to t all die at ages less than t. The probability of this event is

$$[1 - P(t)]^{N^0(t)} \leq e^{-N^0(t)P(t)} = e^{-N(t)} \tag{2d}$$

This is already less than 0.01 when $N(t)$ is only five.

It is shown in Section 6.1 that if one can supplement the ambiguous case by ascertaining the age of death of the item lost at t^*, or of one or more other randomly selected items alive at t^*, and defines $\hat{P}(t)$ for $t > t^*$ as $\hat{P}(t^*)$ times the survival function for the supplementary sample, then the expected value of $\hat{P}(t)$ is precisely the population value $P(t)$. In practice this supplementation would often be neither feasible nor worthwhile, but with otherwise adequate data the resulting bias in one or a few samples will be too small to have any practical importance.

It will be shown in Section 6.1 that the variance of $P(t)$ is given approximately by

$$V[\hat{P}(t)] \doteq P^2(t) \sum_1^k (q_j/N_j p_j), \tag{2e}$$

where the distinct limits of observation L_j' are now used as the division points; L_{k-1}' is the greatest preceding t; and $p_j = 1 - q_j = P(L_j')/P(L_{j-1}')$, with $L_0' = 0$, $L_k' = t$. After dividing (2e) by $P^2(t)$, one sees that the square of the coefficient of variation (CV) of $\hat{P}(t)$ is set equal to the sum of the squares of the CV's of the estimates of the p_j, the usual approximation for the variance of a product.

If the sample estimates are inserted in (2e) one obtains

$$\hat{\mathrm{V}}\,[\hat{P}(t)] \doteq \hat{P}^2(t) \sum_1^k [\delta_j/n_j(n_j - \delta_j)] = \hat{P}^2(t) \sum_1^k \left(\frac{1}{n_j'} - \frac{1}{n_j}\right). \tag{2f}$$

A very similar formula was derived by Greenwood [15] and later by Irwin [19] in connection with actuarial estimates. It is easily verified that (2f) remains valid when the number of intervals is reduced to those used in (2b), or expanded to one interval for each death as in (2c). In the latter case it may be written

$$\hat{V}[\hat{P}(t)] \doteq \hat{P}^2(t) \sum_r [(N - r)(N - r + 1)]^{-1}, \tag{2g}$$

where r runs through the positive integers for which $t_r' \leq t$ and t_r' corresponds to a death.

In terms of integrals (2e) can be written

$$\begin{aligned} V[\hat{P}(t)] &\doteq P^2(t) \int_0^t \frac{|dP(u)|}{N^0(u)P^2(u)} \\ &\doteq P^2(t) \int_0^t \frac{|dP(u)|}{N(u)P(u)} \end{aligned} \tag{2h}$$

In case of ambiguity the first form should be referred to and interpreted in accordance with 1.4. If $P(u)$ alone is discontinuous within the range of integration, it should be regarded as a (continuous) independent variable, so that $\int P(u)^{-2}|dP(u)| = 1/P(u)$.

Since $N^0(t) \le N^0(u) \le N$, it is clear that (2h) is greater than the complete sample variance $P(t)[1 - P(t)]/N$ but less than the reduced-sample variance $P(t)[1 - P(t)]/N^0(t)$ of (3b). If losses are random and the instantaneous rates of death and loss among survivors are in the ratio 1 to ρ at all ages, then one has $E[N^0(u)^{-1}] \doteq [NP^{\rho}(u)]^{-1}$ for large samples, and (2h) reduces to

$$V[\hat{P}(t)] \doteq [P^{1-\rho}(t) - P^2(t)]/(1 + \rho)N \tag{2i}$$

It is often instructive to estimate an effective sample size $\hat{N}_E(t)$, which in the absence of losses would give the same variance $\hat{V}[\hat{P}(t)]$. Evidently

$$\hat{N}_E(t) = \hat{P}(t)[1 - \hat{P}(t)]/\hat{V}[\hat{P}(t)]. \tag{2j}$$

It can be shown that $\hat{N}_E(t)$ is nonincreasing as t increases, and that

$$n(t)/\hat{P}(t) \le \hat{N}_E(t) \le D(t)/[1 - \hat{P}(t)], \tag{2k}$$

where $D(t)$ is the number of deaths observed at ages not exceeding t. The upper bound was proposed as an approximation to $\hat{N}_E(t)$ by Cornfield [8]. The lower bound corresponds to the reduced-sample variance.

In Table 464 of 2.1 we had $\hat{P}(6) = 21/40 = 0.525$. By (2f) the variance of $P(6)$ is estimated as

$$\hat{V}[\hat{P}(6)] = (0.525)^2\left(\frac{1}{8 \times 7} + \frac{2}{5 \times 3}\right) = 0.042.$$

By (2g) the same result appears in the form

$$(0.525)^2\left(\frac{1}{7 \times 8} + \frac{1}{4 \times 5} + \frac{1}{3 \times 4}\right).$$

The effective sample size is estimated as

$$\hat{N}_E(6) = (0.525)(0.475)/0.042 = 6.0.$$

The bounds for $\hat{N}_E(6)$ in (2k) are 5.7 and 6.3. Values of $\hat{N}_E$ are also indicated in Table 465.

2.3 Mean Lifetime

The PL estimate $\hat{\mu}$ of the mean lifetime μ is defined as the mean of the PL estimate of the distribution. It is well-known (and easily proved by integrating by parts) that the mean of a nonnegative random variable is equal to the area under the corresponding survivorship function. Hence

$$\hat{\mu} = \int_0^\infty \hat{P}(t)\, dt.$$

Of course, if $\hat{P}(t)$ is not everywhere determined, $\hat{\mu}$ is undefined. In cases where the probability of an indeterminate result is small, $\hat{P}(t)$ is practically unbiased and the same is true of $\hat{\mu}$.

If we "complete" $\hat{P}(t)$ in the example of Table 464 by following the longest observed individual (with $t_r' = 12.1$) to death at $t = 14.3$ we have

$$\hat{\mu} = (1.000)(0.8) + (0.875)(3.1 - 0.8) + (0.700)(5.4 - 3.1)$$
$$+ (0.525)(9.2 - 5.4) + (0.2625)(14.3 - 9.2)$$
$$= 0.800 + 2.012 + 1.610 + 1.995 + 1.339 = 7.76.$$

If $\hat{P}(t)$ were "completed" by setting $\hat{P}(t) = 0$ for the indeterminate range, the last term in the above sum would be replaced by (0.2625) (12.1 − 9.2) = 0.761, and $\hat{\mu}$ would be estimated as 7.18.

Of course, if the probability of an indeterminate result is high, there is no satisfactory way to estimate μ. In such cases Irwin [19] has suggested that in place of estimating the mean itself, one should estimate the "mean life limited to a time L," say $\mu_{[L]}$. This is the mean of $\min(T_i, L)$, with L chosen at the investigator's convenience. Naturally, one would choose L to make the probability of an indeterminate result quite small. If one chooses to use this procedure he should give an estimate of $P(L)$ along with $\hat{\mu}_{[L]}$. If we take $L = 10$ in our example,

$$\hat{\mu}_{[10]} = 0.800 + 2.012 + 1.610 + 1.995 + (0.2625)(10 - 9.2) = 6.63,$$

and $\hat{P}(10) = 0.2625$.

In Section 6.2 an approximate formula is given for the variance of $\hat{\mu}$:

$$V(\hat{\mu}) \doteq \int_0^\infty \frac{A^2(t)|dP(t)|}{N(t)P(t)} = \int_0^\infty \frac{A^2(t)|dP(t)|}{N^0(t)P^2(t)}, \tag{2l}$$

where $A(t) = \int_t^\infty P(u)du$. Upon making the obvious substitutions we find, after some reduction, the following estimate of $V(\hat{\mu})$:

$$\hat{V}(\hat{\mu}) = \sum_r \frac{A_r^2}{(N-r)(N-r+1)} \tag{2m}$$

where r runs over those integers for which t_r corresponds to a death, and $A_r = \int_{t_a}^\infty \hat{P}(u)du$. If there are no losses,

$$A_r = \sum_{i=r+1}^{N} (t_i - t_r)/N,$$

and it can be shown that $\hat{V}(\hat{\mu})$ reduces to $\sum(t_i - \bar{t})^2/N^2$. This fact, plus the impossibility of estimating the variance on the basis of only one observed death, suggests that (2*m*) might be improved by multiplication by $D/(D-1)$, where D is the number of deaths observed.

For our first estimate of $\hat{\mu}$ above we have

$$A_1 = 2.012 + 1.610 + 1.995 + 1.339 = 6.956,$$
$$A_4 = 1.610 + 1.995 + 1.339 = 4.994,$$
$$A_5 = 1.995 + 1.339 = 3.334,$$
$$A_7 = 1.339 = 1.339.$$

(Obviously the A_r and $\hat{\mu}$ are best calculated in reverse order.) We then have for the estimated variance,

$$\hat{V}(\hat{\mu}) = \frac{(6.956)^2}{7 \times 8} + \frac{(4.944)^2}{4 \times 5} + \frac{(3.334)^2}{3 \times 4} + \frac{(1.339)^2}{1 \times 2} = 3.91$$

The estimated standard deviation of $\hat{\mu}$ is $\sqrt{3.91} = 1.98$. If the factor $D/(D-1) = 4/3$ is included, these results become 5.21 and 2.28 respectively.

If one is limited to grouped data as in Table 465 it is necessary to use actuarial-type assumptions (e.g., the trapezoidal rule) to estimate $\hat{\mu}$ and its variance. Thus, we may estimate the mean life limited to 5 time units as follows:

$$\hat{\mu}_{[5]} = (1/2)[(1.00 + 0.97)(1.0 - 0.0) + 0.97 + 0.92)(1.7 - 1.0)$$
$$+ \cdots + (0.52 + 0.27)(5{,}0 - 4.0)] = 3.76.$$

We observe in conjunction with this estimate that the estimated proportion surviving the limit is $\hat{P}(5) = 0.27$.

(*Editor's note*: Sections 3–5 (pp. 469–476) have been omitted.)

6. Means and Variances

In this section the mean of $\hat{P}(t)$ and the variances of $\hat{P}(t)$ and of the meanlife estimates $\hat{\mu}$ and μ^* are derived. The observation limits are regarded as fixed from sample to sample. In principle, unconditional results for random loss times could be obtained by integration.

6.1 The PL Estimate $\hat{P}(t)$

Let $L_1 < L_2 < \ldots < L_{k-1}$ be the distinct observation limits that are less than the age $t = L_k$ at which $P(t)$ is being estimated. Let $n_{j+1} = n(L_j + 0)$ be the number of survivors[1] observed beyond L_j; $N_{j+1} = En_{j+1} = N^0(L_j + 0)P(L_j)$; $p_j = 1 - q_j = P(L_j)/P(L_{j-1})$; and δ_j the number of deaths observed in the interval $(L_{j-1}, L_j]$, excluding nonzero values of the left endpoint, with $L_0 = 0$. Then

$$\hat{P}(t) = \prod_{j=1}^{k} \hat{p}_j \quad \text{with} \quad \hat{p}_j = (n_j - \delta_j)/n_j. \tag{6a}$$

Let E_j denote a conditional expectation for $n_1, \ldots, n_j$ (and hence also $\delta_1, \ldots, \delta_{j-1}$) fixed. Then $E_j\hat{p}_j = p_j$, provided that $n_j > 0$. If the condition $n_j > 0$ could be ignored, one could write

[1] The expression $L_j + 0$ means that losses at L_j itself have been subtracted off.

$$
\begin{aligned}
E\hat{P}(t) &= E[\hat{p}_1 \dots \hat{p}_{k-1} E_k \hat{p}_k] = E[\hat{p}_1 \dots \hat{p}_{k-1} p_k] \\
&= p_k E[\hat{p}_1 \dots \hat{p}_{k-2} E_{k-1} \hat{p}_{k-1}] \\
&= \dots = p_k p_{k-1} \dots p_1 = P(t),
\end{aligned} \tag{6b}
$$

so that $\hat{P}(t)$ would be an unbiased estimate. The flaw in this demonstration is the indeterminacy resulting when (for some j) $n_{j+1} = 0$ but $\hat{P}(L_j) > 0$, an event whose probability is bounded by (2d), and which could be obviated in principle by obtaining one or more supplementary observations. In the derivations of approximate formulas that follow, any bias that $\hat{P}(t)$ may have is neglected.

Both Greenwood [15] and Irwin [19] when giving formulas for the variance of an actuarial estimate of $P(t)$ actually treat a special case in which the actuarial estimator and $\hat{P}(t)$ coincide. Translated into our notation their common argument is the following.

Suppose we consider conditional variances for n_j held fixed at the values N_j. Then the $\hat{p}_j$ may be treated as independent quantities and

$$
\begin{aligned}
E[\hat{P}^2(t) &= \prod_{j=1}^{k} \left(p_j^2 + \frac{p_j q_j}{N_j} \right) = p_1^2 \dots p_k^2 \prod_{j=1}^{k} \left(1 + \frac{q_j}{N_j p_j} \right) \\
&= p^2(t) \prod_{j=1}^{k} \left(1 + \frac{q_j}{N_j p_j} \right).
\end{aligned} \tag{6c}
$$

If we ignore terms of order N_j^{-2} we have as an approximation for the variance

$$
V[\hat{P}(t)] \doteq P^2(t) \sum_{j=1}^{k} \frac{q_j}{N_j p_j}. \tag{6d}
$$

This last expression we will refer to as "Greenwood's formula." The calculation of the variance with $n_1, \dots, n_k$ fixed can apparently be justified only on the ground that it doesn't matter a great deal; the fixed n_j are unrealistic, and inconsistent with the previous assumption of a sample of fixed size and fixed limits of observation.

The authors have therefore rederived (6d) by means of successive conditional expectations. The procedure is to verify by induction the approximate relation

$$
E_{j+1}[\hat{P}^2(t)] \doteq \hat{p}_1^2 \dots \hat{p}_j^2 \left(1 + \sum_{i=j+1}^{i=k} \frac{q_i}{p_i E_{j+1} n_i} \right) p_{j+1}^2 \dots p_k^2 \tag{6e}
$$

which reduces to (6d) when $j = 0$. Applying E_j to (6e) has the effect of replacing j by $j - 1$ provided that one sets

$$
\begin{aligned}
E_j \frac{\hat{p}_j^2}{E_{j+1} n_i} &\doteq E_j \left[\hat{p}_j \cdot \frac{E_j \hat{p}_j}{E_j E_{j+1} n_i} \right] \\
&\doteq p_j^2 / E_j n_i.
\end{aligned} \tag{6f}
$$

wherein $\hat{p}_j / E_{j+1} n_i$ has been replaced by the ratio of the expectations of numerator and denominator. The fact that $\hat{p}_j$ and $E_{j+1}n_i$ are positively correlated improves the approximation.

In the special case in which $k = 2$ (all observation limits less than t are equal to L_1) and any indeterminacy is resolved by one supplementary observation, the exact variance of $\hat{P}(t)$ can be calculated to be

$$V[\hat{P}(t)] = P^2(t)\left(\frac{q_1}{Np_1} + \frac{q_2}{N'p_1p_2}\left[1 + \left(1 - \frac{N'}{N}\right)^2 \frac{N'p_1 - [N', p_1]}{[N', p_1]} + \frac{N'}{N}\left(1 - \frac{N'}{N}\right)\left(\frac{q_1}{[N', p_1]} - 2q_1^{N'}\right)\right]\right), \tag{6g}$$

where N is the total sample size; $N' = N^0(t)$, the number of items observable to t or beyond; and

$$[N', p_1] = 1/E(X^{-1}) \tag{6h}$$

with $X = \max(Y, 1)$ and Y distributed binomially (N', p_1). The methods used by Stephan [27] show that $[N', p_1] = N'p_1 + 0(1)$ for N' large. Thus the term in square brackets is of order $1 + 0(1/N')$, and (6g) approaches Greenwood's formula (6d) for large N'.

6.2 Covariances and Mean Lifetimes

The method of Irwin, [19] can be used to derive the variance of $\hat{\mu} = \int_0^\infty \hat{P}(t)dt$ by writing

$$\begin{aligned} E(\hat{\mu}^2) &= E\int_0^\infty\int_0^\infty \hat{P}(u)\hat{P}(v)\,dv\,du \\ &= 2\int_0^\infty\int_u^\infty E[\hat{P}(u)\hat{P}(v)]\,dv\,du. \end{aligned} \tag{6i}$$

If $u < v$ and L_h and L_k denote u and v respectively, then apart from bias due to indeterminate cases one has

$$\begin{aligned} E[\hat{P}(u)\hat{P}(v)] &= E[\hat{P}^2(u)E_{h+1}(\hat{p}_{h+1}\hat{p}_{h+2}\cdots\hat{p}_k)] \\ &\doteq p_{h+1}p_{h+2}\cdots p_k E[\hat{P}^2(u)] \\ &= [P(v)/P(u)]E[\hat{P}^2(u)] \doteq P(u)P(v)[1 + U(u)], \end{aligned} \tag{6j}$$

where by (2h) and (2g)

$$U(u) = \int_0^u \frac{|dP(t)|}{N(t)P(t)} \doteq \sum[(N - r)(N - r + 1)]^{-1};$$

the summation is over deaths at ages not exceeding u, and gives the sample estimate. Thus one has approximately

$$
\begin{aligned}
V(\hat{\mu}) &\doteq 2\int_0^\infty \int_u^\infty P(u)P(v)U(u)\,dv\,du \\
&= 2\int_0^\infty A(u)P(u)U(u)\,du = \int_0^\infty A^2(u)\,dU(u),
\end{aligned}
\tag{6k}
$$

where $A(u) = \int_u^\infty P(v)dv$. The result is discussed in 2.3.

To obtain analogous (but in this case exact) results for the RS estimates $P^*(t)$ and μ^*, let $u < v$ as before, and split off two independent subsamples consisting of the $N' = N^0(v)$ items observable to v or beyond, and the $N'' = N^0(u) - N^0(v)$ additional items observable to u or beyond, but not to v. Let the first subsample yield δ' deaths in $(0, u)$ and ε' in (u, v), while the second yields δ'' deaths in $(0, u)$. Also let $P_1 = P(u)$ and $P_2 = P(v)$. Then

$$
\begin{aligned}
&E[P^*(u)P^*(v)] \qquad (\text{with } u \leq |) \\
&= E[N' + N'' - \delta' - \delta'')(N' - \delta' - \varepsilon')]/N'(N' + N'') \\
&= E[(N' + N'' - \delta' - \delta'')(N' - \delta')]P_2/P_1N'(N' + N'') \\
&= [E(N' - \delta')^2 + E(N'' - \delta'')\cdot E(N' - \delta')]P_2/P_1N'(N' + N'') \\
&= [N'^2P_1^2 + N'P_1(1 - P_1) + N'N''P_1^2]P_2/P_1N'(N' + N'') \\
&= P(u)P(v) + P(v)[1 - P(u)]/N^0(u).
\end{aligned}
\tag{6l}
$$

Substituting this result in the analogue of (6i) and subtracting μ^2 gives $V(\mu^*)$ as in (3d).

7. Consistency; Testing with Replacement

The RS estimate $P^*(t)$, being binomial, is of course consistent. In view of the approximate nature of the formula for the variance of $\hat{P}(t)$, however, a proof of its consistency is in order. In fact, by giving closer attention to the errors of the approximations in the variance derivation, it can be shown that $\hat{P}(t)$ does have the limit $P(t)$ in probability provided only that $N^0(t) \to \infty$. However, the proof is too lengthy to be given here. Of course, the consistency is obvious enough in the special case in which the number k of conditional probabilities to be estimated remains bounded as $N^0(t) \to \infty$.

(*Editor's note*: The discussion of Consistency in Testing with Replacement (p. 479) has been omitted.)

Acknowledgment

The viewpoint adopted in this paper owes much to discussions with John W. Tukey. It also incorporates a number of suggestions kindly forwarded by readers of a preliminary draft.

References

[1] Bailey, W.G., and Haycocks, H.W., "A synthesis of methods of deriving measures of decrement from observed data," *Journal of the Institute of Actuaries*, 73 (1947), 179–212.

[2] Bartlett, M.S., "On the statistical estimation of mean lifetime," *Philosophical Magazine* (Series 7), 44 (1953), 249–62.

[3] Berkson, J., and Gage, R.P., "Calculation of survival rates for cancer," *Proceedings of the Staff Meetings of the Mayo Clinic*, 25 (1950), 270–86.

[4] Berkson, J., and Gage, R.P., "Survival curve for cancer patients following treatment," *Journal of the American Statistical Association*, 47 (1952), 501–15.

[5] Berkson, J., "Estimation of the interval rate in actuarial calculations: a critique of the person-years concept," (Summary) *Journal of the American Statistical Association* 49 (1954), 363.

[6] Böhmer, P.E., "Theorie der unabhängigen Wahrscheinlichkeiten," Rapports, Mémoires et Procès-verbaux de Septième Congrès International d'Actuaires, Amsterdam, 2 (1912), 327–43.

[7] Brown, G.W. and Flood, M.M., "Tumbler mortality," *Journal of the American Statistical Association*, 42 (1947), 562–74.

[8] Cornfield, J., "Cancer illness among residents in Atlanta, Georgia," Public Health Service Publication No. 13, Cancer Morbidity Series No. 1, 1950.

[9] Davis, D.J., "An analysis of some failure data," *Journal of the American Statistical Association* 47 (1952), 113–150.

[10] Epstein, Benjamin, and Sobel, Milton, "Lifetesting," *Journal of the American Statistical Association*, 48 (1953), 486–502.

[11] Epstein, Benjamin, and Sobel, Milton, "Some theorems relevant to lifetesting from an exponential distribution," *Annals of Mathematical Statistics*, 25 (1954), 373–81.

[12] Fix, Evelyn, "Practical implications of certain stochastic models on different methods of follow-up studies," paper presented at 1951 annual meeting of the Western Branch, American Public Health Association, November 1, 1951.

[13] Fix, Evelyn, and Neyman, J., "A simple stochastic model of recovery, relapse, death and loss of patients," *Human Biology*, 23 (1951), 205–41.

[14] Goodman, L.A., "Methods of measuring useful life of equipment under operational conditions," *Journal of the American Statistical Association*, 48 (1953) 503–30.

[15] Greenwood, Major, "The natural duration of cancer," Reports on Public Health and Medical Subjects, No. 33 (1926), His Majesty's Stationery Office.

[16] Greville, T.N.E., "Mortality tables analyzed by cause of death," *The Record*, American Institute of Actuaries, 37 (1948), 283–94.

[17] Hald, A., "Maximum likelihood estimation of the parameters of a normal distribution which is truncated at a known point," *Skandinavisk Aktuarietidskrift*, 32 (1949) 119–34.

[18] Harris, T.E., Meier, P., and Tukey, J.W., "Timing of the distribution of events between observations," *Human Biology*, 22 (1950), 249–70.

[19] Irwin, J.O., "The standard error of an estimate of expectational life," *Journal of Hygiene*, 47 (1949), 188–9.

[20] Jablon, Seymour, "Testing of survival rates as computed from life tables," unpublished memorandum, June 14, 1951.

[21] Kahn, H.A., and Mantel, Nathan, "Variance of estimated proportions withdrawing in single decrement follow-up studies when cases are lost from observation," unpublished manuscript.

[22] Littell, A.S., "Estimation of the T-year survival rate from follow-up studies over a limited period of time," *Human Biology*, 24 (1952), 87–116.
[23] Merrell, Margaret, "Time-specific life tables contrasted with observed survivorship," *Biometrics*, 3 (1947), 129–36.
[24] Sartwell, P.E., and Merrell, M., "Influence of the dynamic character of chronic disease on the interpretation of morbidity rates," *American Journal of Public Health*, 42 (1952), 579–84.
[25] Savage, I.R., "Bibliography of nonparametric statistics and related topics," *Journal of the American Statistical Association*, 48 (1953), 844–906.
[26] Seal, H.L., "The estimation of mortality and other decremental probabilities," *Skandinavisk Aktuarietidskrift*, 37 (1954), 137–62.
[27] Stephan, F.F., "The expected value and variance of the reciprocal and other negative powers of a positive Bernoullian variate," *Annals of Mathematical Statistics*, 16 (1945), 50–61.
[28] Wolfowitz, J., "Additive partition functions and a class of statistical hypotheses," *Annals of Mathematical Statistics*, 13 (1942), 247–79.

Introduction to Chernoff (1959) Sequential Design of Experiments

P.K. Sen
University of North Carolina at Chapel Hill

Herman Chernoff was born in New York City on July 1, 1923. His parents immigrated from Russia. Herman graduated from Townsend Harris High School in New York City in 1939 and entered the New York University to study mathematics (with minor in physics). He received a B.S. in 1943 and went to Brown University for graduate studies in applied mathematics. With a brief interruption (as a result of the draft in 1945), he completed his non-thesis requirement at Brown and came to Columbia University, in January 1947, to work on his dissertation under the supervision of Abraham Wald. He was awarded a Ph.D. from Brown University in 1948. Herman served on the faculty at the University of Illinois, Stanford University, and MIT; currently, he is at Harvard University. He is a member of the National Academy of Sciences, Fellow (and past president) of the Institute of Mathematical Statistics, Fellow of the American Statistical Association, elected member of the International Statistical Institute, and a member of a number of other professional socities.

While in the undergraduate program, Herman's interest in statistics was aroused by the fundamental work of J. Neyman and E.S. Pearson (on hypothesis testing), and in his doctoral dissertation, he worked on asymptotic studentization in the testing of (statistical) hypothesis. During his stay at Columbia University and Chicago (at the Cowles Commission for Research in Economics), Herman met with a number of mathematicians, statisticians, and economists whose influence on his later professional work is quite noticeable. In particular, his acquaintance with Henry Scheffé in 1950 influenced him in pursuing research work on the duality of some optimization problems and some generalizations of the work of Abraham Wald and Stanley Isaacson on type-A and type-D critical (similar) regions [Chernoff and Scheffé (1952)]. For quite some time, Herman was deeply routed in these optimization problems, and these optimization findings have great impact in several other areas of research where he has made outstanding contributions. Some of these works, completed before 1959, have great bearing on the paper under commentary, and will be mentioned briefly. Chernoff (1952) introduced a measure of the asymptotic efficiency of statistical tests incorporating the fruitful use of probabilities of large deviations; both these aspects have contributed a lot to the future research in this area. In collaboration with I.R. Savage [Chernoff and Savage (1958)], he formulated a novel approach to the study of the asymptotic normality (and efficiency) of a class of linear rank statistics when the underlying distributions are not necessarily the same, and the past 30 years have witnessed the phenomenal growth of literature on research work in this area. His optimization background haunted him, resulting in a novel formulation of locally optimal designs for estimating parameters [Chernoff (1953)]. All these, in turn, initiated his prolonged interest in optimal designs, both in the traditional fixed-sample size case and sequential setups, for which he can be undoubtedly regarded as the founder. This quest culminated in the paper under commentary: For the sequential design of experiments, he clearly motivated the problems, posed some novel and workable solutions, and demonstrated their close relations with his earlier works. Before we proceed to discuss the significance of this work, perhaps it would be more appropriate to discuss the basic motivation in a nontechnical way. Not surprisingly, I located a description of this motivation in even more clarity in a follow-up publication [Chernoff (1960)], and hence, I could not resist the temptation of quoting a few lines from this supplement:

> Considerable scientific research may be characterized in the following way. A scientist is interested in a problem. He performs an experiment to obtain information. This information not only serves to illuminate the problem but is used to design a more informative experiment. As information is accumulated, he continues designing more and more effective experiments until he reaches the point where he feels that further experimentation seems no longer necessary. Then he announces his result. The procedure outlined may reasonably be entitled *sequential design of experiments.* In spite of its apparent usefulness there seems to be no formal theory of sequential design of experiments in the literature of statistics....

It is indeed Chernoff's own creative thoughtfulness that establishes the foundation for the sequential design of experiments. The basic feature of optimization (in terms of extractable information) through sequential designing of an experiment (incorporating appropriate stopping rules) constitutes the most salient aspect of this paper, and based on this novel feature, it well qualifies as a breakthrough in statistics. Prior to this work, a variation of this problem, termed the *two-armed bandit problem* [Robbins (1952)] was considered in the literature, although with a limited amount of success. A bandit problem (in statistical decision theory) involves (sequential) selections from several stochastic processes. In contrast, in this paper, Chernoff presented a procedure for the sequential design of experiments in which the problem is one of testing a hypothesis. It is assumed that there are only two possible actions (terminal decisions) and a class of available experiments. After each observation, the statistician has to decide on whether to continue experimentation or not. In the event of continuation, he has to select one of the available experiments, whereas in case of termination, he must select one of the two terminal actions. A general feature of the two-arm bandit problem (as well as other related problems) is that while the goal is to prescribe an optimal solution, such a prescription may not work out neatly in an exact sense. Chernoff rightfully characterized the role of asymptotic methods in these problems, and with his expertise in this general domain, he came up with simple solutions. For the special case in which there are only a finite number of states of nature and a finite number of available experiments, Chernoff's procedure is "asymptotically optimal" as the cost of sampling approaches zero. Although a similar asymptotic consideration underlies Wald's (1947) sequential probability ratio test (with respect to the excess over the boundaries), Chernoff managed to incorporate the cost function neatly along with the Kullback–Leibler information numbers to characterize the asymptotic risks corresponding to a sequential strategy and therby to formulate an asymptotically optimal strategy.

These details are provided in Sec. 2 for the case of two simple hypotheses, and a special problem of special practical interest involving composite hypotheses is studied in detail in Sec. 3. Suppose that one wants to compare the efficacy of two drugs. The experiment E_j consists of using the jth drug, $j = 1, 2$. The response is quantal with the probability p_j for the jth drug, $j = 1, 2$. It is intended to test the hypothesis $H_1: p_1 > p_2$ against $H_2: p_1 \leq p_2$. Then the unknown state of nature is represented by $\theta = (p_1, p_2)$. It is desired to adopt the first drug if H_1 is true and the second one if H_2 is true. The procedure can be outlined as follows. After n observations compute the maximum likelihood estimate (MLE) $\hat{\theta}_n$ of θ. Act as though $\hat{\theta}_n$ were the true value of θ by selecting the next experiment $E^{(n+1)}$ to maximize

$$\min_{\phi \in a(\hat{\theta}_n)} I(\hat{\theta}_n, \phi, E), \tag{1}$$

where $I(\theta, \Phi, E)$ is the Kullback–Leibler information number of θ with respect to ϕ for the experiment E and $a(\theta)$ is the set alternative to θ. This simple motivation as formalized in Sec. 4 for a description of the general procedures when one wants to test for $H_1: \theta \in \omega_1$ vs. $H_2: \theta \in \omega_2$ and there is an available set of experiments $\{E\}$, each of which may be replicated. Let $f(x, \theta, E)$ be the density of the outcome x of experiment E with respect to a measure μ_E, let $E^{(n)}$ represent the experiment used for the nth trial, x_n being the corresponding outcome, and let

$$z_n(\theta, \phi) = \log \frac{[f(x_n, \theta, E^{(n)})}{f(x_n, \phi, E^{(n)})]}, \qquad n \geq 1, \tag{2}$$

$$S_n(\theta, \phi) = \textstyle\sum_{i \leq n} z_i(\theta, \phi) \qquad \text{and} \qquad S_n = \textstyle\sum_{i \leq n} z_i(\hat{\theta}_n, \tilde{\theta}_n), \tag{3}$$

where $\hat{\theta}_n$ is the MLE of θ based on the first observations and $\tilde{\theta}_n$ the MLE under the hypothesis alternative to $\hat{\theta}_n$. Then the Chernoff procedure can be described as follows. Stop sampling at the nth observation and select the hypothesis of $\hat{\theta}_n$ if $S_n > -\log c$, where c is the cost per unit sampling. If sampling is to be continued, let $E^{(n+1)} = E(\hat{\theta}_n)$ be selected so that (1) is maximized. In Sec. 5 and 6, the asymptotic characteristics of this procedure are studied. In this context, it is assumed that there are only a finite number of states of nature and a finite number of available (pure) experiments. It is shown that as $c \to 0$, for this procedure the probability of making the wrong decision is $O(c)$ and the expected sample size is (asymptotically) not larger than $(-\log c)/I(\theta)$, where

$$I(\theta) = \inf_{\phi \in a(\theta)} I(\theta, \phi, E(\theta)) = \sup_E \inf_{\phi \in a(\theta)} I(\theta, \phi, E). \tag{4}$$

The risk or expected cost $R(\theta)$ for using this procedure has the asymptotic form

$$R(\theta) \approx -c \log c/I(\theta), \tag{5}$$

and further, the procedure is asymptotically optimal in the following way: For any other procedure to do better for some θ, i.e., to decrease $R(\theta)$ by some factor for some θ, it must do worse by a greater order of magnitude for some other θ. The concluding section deals with a number of pertinent remarks that clarify the basic appeal of this procedure and explain its limitations too.

The designing aspect (in a sequential setup) of this approach is reflected in the choice of the experiments $E^{(n)}$ at the successive stages, while the stopping rule is framed in a manner analogous to Wald's (1947) sequential probability ratio test. To make this point clear, let us look back at the problem treated in Sec. 3. Wald (1947) considered a sequential probability ratio test in which at each stage, the same experiment is conducted (which consists of taking a pair of observations, one from each of the two populations). In this Chernoff procedure, at the successive stages, the choice of the experiments $E^{(n)}$ depends on the outcome of the preceding ones, and this adaptive procedure results in a substantial savings of the expected sample size. This convincing argument pertains to the general scheme outlined in Sec. 4 through 6 of Chernoff's paper and explains the basic importance of the designing aspect in the context of the sequential testing of statistical hypothesis. In depicting such a picture, Chernoff displays true and novel, mathematical ingenuity and statistical foresight. The technicalities, limited to Sec. 5 and 6, do not obscure the main flow of the ideas, and in this respect, his earlier work on the asymptotic efficiency and probability of large deviations has provided him with the necessary ingredients for a simple enough presentation. This designing aspect pertains also to the case that involves selecting one of k (≥ 2) mutually exclusive hypotheses. Moreover, the cost per unit sampling may also be made to depend on the experiments, and in (3) in the definition of S_n, the $Z_i(\hat{\theta}_n, \tilde{\theta}_n)$ may also be replaced by $Z_i(\hat{\theta}_i, \tilde{\theta}_i)$. In this sense, Chernoff's sequential design of experiments pertains to a general class of problems arising in statistical inference (although he mainly emphasized the hypothesis testing problem). One particular aspect deserves some detailed discussion. The optimality picture is depicted here in an asymptotic shade where c, the cost per unit sampling, is made to converge to zero. The main advantage of this asymptotic solution is its simplicity. However, in actual practice, an important question may arise: How small should c actually be in order that such an asymptotic solution provides a reasonably good approximation to the actual one? Being well aware of this query, Chernoff has not hesitated to offer some remarks concerning the adequacy of his proposed asymptotic solution. In passing, it may be remarked that as $c \to 0$, the asymptotic optimality may not be especially relevant for the initial stage of experimentation. It is desirable to first apply experiments that are informative for a broad range of parameter values and then gradually restrict the course leading to the proposed asymptotically optimal one. The probability of reaching a terminal decision at a very early stage becomes very small as $c \to 0$, and hence, this modification causes little damage to the proposed asymptotic procedure. In this context, it may be noted that maximizing the Kullback–Leibler information number (as is needed in the Chernoff method) may yield experiments that are efficient only for θ close to the estimated value, and hence, the limitation of the Chernoff procedure for small samples (i.e., c is not so small) should not be overlooked. This point was raised by Chernoff in his later works too [Chernoff (1972)]. Also, numerical studies made later on indicate that some alternative procedures that appear to be less theoretically appealing may perform better than the Chernoff asymptotically optimal solution when c is not so small. In this respect, the asymptotic results (mostly, of the first order) presented in a simple fashion may not be good enough to provide adequate approximations for moderate values of c, and a "second-order" asymptotic approximation may be needed to encompass such situations. Theoretically, this is a harder problem, and only recently, Robert Keener, one of Chernoff's own advisees, has considered an elegant approximation that works out well for moderate values of c as well Keener (1984). In Keener's treatment too, it is assumed that there are only finitely many states of nature and a finite number of available experiments. Nevertheless, Chernoff's original solution and its extensions would continue to provide additional challenges in the future. Elimination of the finiteness of the nature space and the space of experiments may be the first step in this direction.

In conclusion, I would like to commend Herman Chernoff for a job well done, for opening up a new avenue of research in a very important area of statistics, and for his vast insights. I am honored to write this commentary on this historic paper.

References

Chernoff, H. (1952). A measure of asymptotic efficiency for tests of a hypothesis based on sums of observations, *Ann. Math. Statist.*, **23**, 493–507.

Chernoff, H. C1953). Locally optimal designs for estimating parameters, *Ann. Math. Statist.*, **24**, 586–602.

Chernoff, H. (1954). On the distribution of the likelihood ratio, *Ann. Math. Statist.*, **25**, 573–578.

Chernoff, H. (1956). Large sample theory: Parametric case, *Ann. Math. Statist.*, **27**, 1–22.

Chernoff, H. (1960). Motivation for an approach to the sequential design of experiments, in *Information and Decision Processes* (R.E. Machol, ed.), McGraw-Hill, New York.

Chernoff, H. (1972). *Sequential Analysis and Optimal Design*. SIAM, Philadelphia, Pa.

Chernoff, H., and Savage, I.R. (1958). Asymptotic normality and efficiency of certain nonparametric test statistics, *Ann. Math. Statist.*, **29**, 972–994.

Chernoff, H., and Scheffé, H. (1952). A generalization of the Neyman–Pearson fundamental lemma, *Ann. Math. Statist.*, **23**, 213–225.

Keener, R.W. (1984). Second order efficiency in the sequential design of experiments, *Ann. Statist.*, **12**, 510–532.

Rizvi, M.H., Rustagi, J.S., and Siegmund, D. (eds.) (1983). *Recent Advances in Statistics: Papers in Honor of Herman Chernoff on His Sixtieth Birthday*. Academic Press, New York.

Robbins, H.E. (1952). Some aspects of the sequential design of experiments, *Bull. Amer. Math. Soc.*, **58**, 527–535.

Wald, A. (1947). *Sequential Analysis*. Wiley, New York.

Sequential Design of Experiments

Herman Chernoff[1]
Stanford University

1. Introduction

Considerable scientific research is characterized as follows. The scientist is interested in studying a phenomenon. At first he is quite ignorant and his initial experiments are preliminary and tentative. As he gathers relevant data, he becomes more definite in his impression of the underlying theory. This more definite impression is used to construct more informative experiments. Finally after a certain point he is satisfied that his evidence is sufficient to allow him to announce certain conclusions and he does so.

While this sequential searching for relevant and informative experiments is common, very little statistical theory has been directed in this direction. The general problem may reasonably be called that of sequential design of experiments. A truncated variation of this problem called the two-armed bandit problem has attracted some attention (see [1] and [5]). Up to now an optimal solution for the two-armed bandit problem has not been attained. The failure to solve the two-armed bandit problem and certain obvious associated results indicate strongly that while optimal strategies are difficult to characterize, *asymptotically* optimal results should be easily available. Here the term asymptotic refers to large samples. For the sequential design problems, large samples and small cost of experimentation are roughly equivalent.

In this paper we present a procedure for the sequential design of experiments where the problem is one of testing a hypothesis. Formally, we assume that there are two possible actions (terminal decisions) and a class of available experiments. After each observation, the statistician decides on whether to continue experimentation or not. If he decides to continue, he must select one

Received July 5, 1958.

[1] This work was sponsored by the Office of Naval Research under Contract N6 onr-25140 (NR-342-022).

of the available experiments. If he decides to stop he must select one of the two terminal actions.

For the special case where there are only a finite numer of states of nature and a finite number of available experiments this procedure will be shown to be "asymptotically optimal" as the cost of sampling approaches zero. The procedure can be partially described by saying that at each stage the experimenter acts as though he is almost convined that $\hat{\theta}$, the current maximum likelihood estimate of the state of nature, is actually equal to or very close to the true state of nature.

In problems were the cost of sampling is not small, this procedure may leave something to be desired. More specifically, until enough data are accumulated, the procedure may suggest very poor experiments because it does not sufficiently distinguish between the cases where $\hat{\theta}$ is a poor estimate and where $\hat{\theta}$ is a good estimate. For small cost of experimentation initial bungling is relatively unimportant. It is hoped and expected that with minor modifications the asymptotically optimal procedure studied here can be adapted to problems with relatively large cost of experimentation.

The procedures studied make extensive use of the Kullback-Leibler information numbers (see [2] and [4]).

2. Preliminaries Involving Two Simple Hypotheses

The procedure presented in this paper may be motivated by an asymptotic study of the classical problem of sequentially testing a simple hypothesis vs. a simple alternative with only one available experiment. Suppose $H_0 : \theta = \theta_0$ and $H_1 : \theta = \theta_1$ are two simple hypotheses, and the experiment yields a random variable x whose density is $f_i(x)$ under H_i, $i = 0, 1$. The Bayes strategies are the Wald sequential likelihood-ratio tests. These are characterized by two numbers A and B and consist of reacting to the first n observations $x_1, x_2, \ldots,$ x_n

by rejecting H_0 if $\quad S_n \geqq A,$
accepting H_0 if $\quad S_n \leqq B,$

and continuing sampling as long as $B < S_n < A$ where

$$S_n = \sum_{i=1}^{n} \log[f_1(x_i)/f_0(x_i)]. \tag{2.1}$$

The appropriate numbers A and B are determined by the *a priori* probability w of H_1, and the costs. These are the cost c per observation (which is assumed fixed), the loss r_0 due to rejecting H_0 when it is true and the loss r_1 due to accepting H_0 when it is false. The risks corresponding to a sequential strategy are given by

$$\begin{aligned} R_0 &= r_0\alpha + c\varepsilon(N|H_0) \\ R_1 &= r_1\beta + c\varepsilon(N|H_1) \end{aligned} \tag{2.2}$$

where α and β are the two probabilities of error, and N is the possibly random sample size. Of course A and B are determined so as to minimize

$$(1 - w)R_0 + wR_1.$$

Suppose that c approaches zero. Then A and $-B$ are large and Wald's approximations [6] give

$$\begin{gathered} \alpha \approx e^{-A}, \qquad\qquad \beta \approx e^{B}, \\ \mathrm{E}(N|H_0) \approx -B/I_0, \quad \text{and} \quad \mathrm{E}(N|H_1) \approx A/I_1 \end{gathered} \tag{2.3}$$

where I_0 and I_1 are the Kullback-Leibler information numbers given by

$$I_0 = \int \log[f_0(x)/f_1(x)]f_0(x)\,dx \tag{2.4}$$

and

$$I_1 = \int \log[f_1(x)/f_0(x)]f_1(x)\,dx \tag{2.5}$$

and are assumed to exist finite and positive. Minimizing the approximation to $(1 - w)R_0 + wR_1$ we find that

$$A \approx -\log c + \log[I_1 r_0(1 - w)/w] \approx -\log c, \tag{2.6}$$

$$B \approx \log c + \log[(1 - w)/I_0 r_1 w] \approx \log c,$$

$$\alpha \approx wc/I_1 r_0(1 - w), \qquad \beta \approx c(1 - w)/I_0 r_1 w, \tag{2.7}$$

$$\mathrm{E}(N|H_0) \approx -\log c/I_0, \qquad \mathrm{E}(N|H_1) \approx -\log c/I_1,$$

$$R_0 \approx -c \log c/I_0, \qquad \text{and} \qquad R_1 \approx -c \log c/I_1. \tag{2.8}$$

Remarks:

1. These results can be verified more rigorously by using Wald's bounds on his approximations when they apply. Our later results will generalize these approximations of R_0 and R_1.
2. The risk corresponding to the optimum strategy is mainly the cost of experimentation.
3. The optimum strategy and its risks depend mainly on c, I_0 and I_1 and are relatively insensitive to the costs r_0 and r_1 of making the wrong decision and to the *a priori* probability w. Note that doubling r_0 and r_1 is equivalent to cutting c in half as far as the strategy is concerned. The consequent change in $\log c$ which determines A and B is relatively small. That is, $\log c$ is changed to $\log c - \log 2$ while $\log 2$ is small compared to $\log c$.

Suppose that the experimenter is given a choice of one of two experiments E_1 and E_2 but that the one chosen must be used exclusively throughout the sequential testing problem. We designate the information numbers by $I_0(E_1)$, $I_1(E_1)$, $I_0(E_2)$ and $I_1(E_2)$. If c is small and $I_0(E_1) > I_0(E_2)$ and $I_1(E_1) > I_1(E_2)$, then it makes sense to select E_1. If, on the other hand, $I_0(E_1) > I_0(E_2)$ and

$I_1(E_1) < I_1(E_2)$, then E_1 would be preferable if H_0 were "true" and E_2 would be preferable if H_1 were. Since the true state of nature is not known, there is not clear cut reason to prefer E_1 to E_2 without resorting to the *a priori* probabilities of H_0 and H_1.

The above rather artificial problem illuminates the more natural one where, after each decision to continue experimentation, one can choose between E_1 and E_2. If the cost of sampling is very small, it may pay to continue sampling even though we are almost convinced about which is the true state of nature. Thus even if we feel quite sure it is H_0, we may still be willing to experiment further, and furthermore it would make sense to select E_1 or E_2 by comparing $I_0(E_1)$ with $I_0(E_2)$.

To be more formal we may select E_1 or E_2 by comparing $I_0(E_1)$ and $I_0(E_2)$ if the maximum likelihood estimate of θ based on the previous observations is θ_0 and by comparing $I_1(E_1)$ and $I_1(E_2)$ if the maximum likelihood estimate is θ_1. Such a procedure may be short of optimal. On the other hand if c is very small, the most damage that could occur would be due to the *nonoptimal* choice of experiment for the first few of the many observations which are expected.

The stopping rule for the single experiment case may be naturally interpreted in terms of *a posteriori* probability as follows: there are two numbers of the order of magnitude of c. Stop experimenting if the *a posteriori* probability of H_0 goes below the first number or if the *a posteriori* probability of H_1 goes below the second number. The expected sample sizes are relatively insensitive to variations in the stopping limits. More specifically, if the *a posteriori* probability limits are any numbers of the order of magnitude of c, α and β are of the order of magnitude of c and $E(N|H_0) \approx -\log c/I_0$ and $E(N|H_1) \approx -\log c/I_1$.

In view of the standard derivations of the Bayes procedures for the one experiment sequential testing problem, it seems natural to stop when the *a posteriori* probability of H_0 or H_1 go below numbers of the order of magnitude of c. An example of such a stopping rule is obtained by selecting A and B equal to $-\log c$ and $\log c$ respectively. Then if $E^{(i)}$ is the experiment selected on the ith trial and x_i is the outcome let $z_i = \log[f_1(x_i, E^{(i)})/f_0(x_i, E^{(i)})]$ where $f_1(x, E^{(i)})$ and $f_0(x, E^{(i)})$ are the densities of the outcome of $E^{(i)}$. Finally, after the nth experiment, continue sampling only if $S_n = \sum_{i=1}^{n} z_i$ lies between B and A.

3. A Special Problem Involving Composite Hypotheses. Comparing Two Probabilities

In the preceding section a method was proposed for the sequential design problem for testing a simple hypothesis vs. a simple alternative. The situation becomes more complicated when the hypotheses are composite. To motivate our procedures for this more complex problem, we shall discuss heuristically the special problem of comparing two probabilities. This problem may be regarded as a prototype of the general sequential design problem for testing hypotheses. We shall devote our main attention to the design aspect and leave the stopping rule in a relatively unrefined state.

It is desired to compare the efficacy of two drugs. The experiments E_1 and E_2 consist of using the first and second drugs respectively. The outcome of these experiments are success or failure, success having probabilities p_1 and p_2 in the two experiments. The two hypotheses are $H_1 : p_1 > p_2$ and $H_2 : p_1 \leqq p_2$.

After n observations consisting of n_1 trials of drug 1 and n_2 trials of drug 2, which led to m_1 and m_2 successes respectively, the maximum-likelihood estimate of $\theta = (p_1, p_2)$ is given by $\hat{\theta} = (\hat{p}_{1n}, \hat{p}_{2n}) = (m_1/n_1, m_2/n_2)$. We shall select our next experiment according to the following idea. Consider the experiment which would be appropriate if we believed that θ were $\hat{\theta}_n$ and we were testing the hypothesis $\theta = \hat{\theta}_n$ vs. the simple alternative $\theta = \tilde{\theta}_n$ which is the "nearest" parameter point under the "alternative hypothesis." To be more specific suppose $m_1/n_1 > m_2/n_2$. Then $\hat{\theta}_n$ is an element of the set of θ for which H_1 is true. The "nearest" element under H_2 is not clearly defined. For the present let us define it as the maximum-likelihood estimate under H_2. Then

$$\tilde{\theta}_n = \left(\frac{m_1 + m_2}{n_1 + n_2}, \frac{m_1 + m_2}{n_1 + n_2}\right)$$
$$= \left[\left(\frac{n_1}{n_1 + n_2}\right)\hat{p}_1 + \left(\frac{n_2}{n_1 + n_2}\right)\hat{p}_2, \left(\frac{n_1}{n_1 + n_2}\right)\hat{p}_1 + \left(\frac{n_2}{n_1 + n_2}\right)\hat{p}_2\right].$$

Note that $\tilde{\theta}_n$ is a weighted average of $(\hat{p}_1, \hat{p}_1)$ and $(\hat{p}_2, \hat{p}_2)$ where the weights are proportional to the frequencies of E_1 and E_2. If we were testing $\theta = \hat{\theta}_n$ vs. $\theta = \tilde{\theta}_n$ and strongly believed in $\theta = \hat{\theta}_n$, we would *select the experiment E for which*

$$I(\hat{\theta}_n, \tilde{\theta}_n, E) = \sum_x \log[f(x, \hat{\theta}_n, E)/f(x, \tilde{\theta}_n, E)]f(x, \hat{\theta}_n, E) \tag{3.1}$$

is as large as possible where $f(x, \theta, E)$ is the probability of the data for experiment E when θ is the value of the parameter.[2] Thus

$$I[(p_1, p_2), (p_1^*, p_2^*), E_1]$$
$$= p_1 \log[p_1/p_1^*] + (1 - p_1) \log[(1 - p_1)/(1 - p_1^*)] \tag{3.2}$$

$$I[(p_1, p_2), (p_1^*, p_2^*), E_2]$$
$$= p_2 \log[p_2/p_2^*] + (1 - p_2) \log[(1 - p_2)/(1 - p_2^*)]. \tag{3.3}$$

It is clear from these expressions and from intuitive considerations that if

[2] We work against the "nearest" alternative under the intuitive assumption that this is the alternative which will make our risk large and which we must guard against.

p_1^* is close to p_1, E_1 is relatively uninformative and if p_2^* is close to p_2, E_2 is relatively uninformative. Thus if n_1 is much larger than n_2, $\hat{\theta}_n$ is close to $(\hat{p}_1, \hat{p}_1)$ and

$$I(\hat{\theta}_n, \tilde{\theta}_n, E_1) < I(\hat{\theta}_n, \tilde{\theta}_n, E_2)$$

and E_2 is called for. Similarly if n_1 is much smaller than n_2, E_1 is called for. For a specified $\hat{\theta}_n$, there is a unique proportion $\lambda(\hat{\theta}_n)$ such that the two informations are equal if $n_2/(n_1 + n_2) = \lambda(\hat{\theta}_n)$. If $n_2/(n_1 + n_2)$ exceeds $\lambda(\hat{\theta}_n)$, E_1 is called for.

In general, the set of (p_1^*, p_2^*) for which

$$I[(p_1, p_2), (p_1^*, p_2^*), E_1)] = I[(p_1, p_2), (p_1^*, p_2^*), E_2]$$

is easy to characterize. See Fig. 1. It seems clear that after many observations $\hat{\theta}_n$ will be close to $\theta = (p_1, p_2)$ and $\tilde{\theta}_n$ will be close to that point $\theta^* = \theta^*(\theta) =$

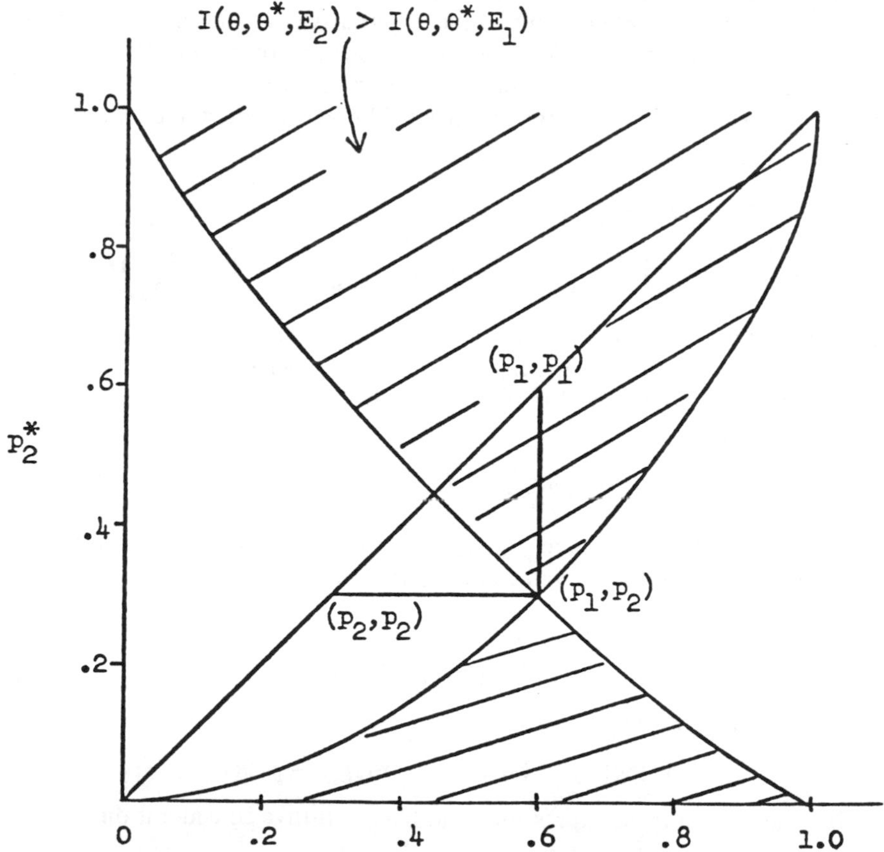

Figure 1. The set of $\theta^* = (p_1^*, p_2^*)$ for which E_2 is preferred to E_1, i.e., $I(\theta, \theta^*, E_2) > I(\theta, \theta^*, E_1)$ for a specified $\theta = (p_1, p_2)$ in the drug testing problem.

Table 1. Tabulation[3] of $p^*(\theta)$, $\lambda(\theta)$, $I(\theta)$ and $e(\theta)$ where $\theta = (p_1, p_2)$ and $\theta^*(\theta) = [p^*(\theta), p^*(\theta)]$.

p_1	p_2	.01	.05	.10	.20	.40
	p^*	.0276				
.05	λ	.560				
	I	.00329				
	e	.988				
	p^*	.0479	.0737			
.10	λ	.579	.526			
	I	.00997	.00200			
	e	.978	.995			
	p^*	.0889	.118	.148		
.20	λ	.585	.547	.520		
	I	.0258	.0120	.00435		
	e	.971	.992	.995		
	p^*	.171	.205	.239	.297	
.40	λ	.588	.557	.537	.525	
	I	.0637	.0429	.0277	.0105	
	e	.973	.988	.992	.999	
	p^*	.260	.297	.333	.394	.500
.60	λ	.576	.551	.534	.515	.500
	I	.111	.0855	.0648	.0375	.00874
	e	.980	.991	.995	.999	1.000
	p^*	.363	.400	.438	.500	
.80	λ	.553	.533	.517	.500	
	I	.174	.145	.120	.0837	
	e	.990	.996	.997	1.000	
	p^*	.425	.463	.500		
.90	λ	.538	.514	.500		
	I	.217	.187	.160		
	e	.998	.999	1.000		

(p_1^*, p_2^*) for which $p_1^* = p_2^*$ and $I(\theta, \theta^*, E_1) = I(\theta, \theta^*, E_2)$. Furthermore the proportion of times that E_1 is applied in the long run is determined by the relation of θ^* to θ. That is to say if $\theta = (p_1, p_2)$ and $\theta^* = [(1 - \lambda)p_1 + \lambda p_2, (1 - \lambda)p_1 + \lambda p_2]$ then $n_2/(n_1 + n_2)$ will tend to be close to $\lambda = \lambda(\theta)$.

The point θ^* and the ratio λ can also be interpreted from another point of view. Essentially θ^* is that point under the alternative hypothesis for which $\max_E I(\theta, \theta^*, E)$ is minimized. (In a sense this property can also be interpreted to say that θ^* is the "nearest" point to θ under the alternative hypotheses.) At θ^*, it doesn't matter which experiment is selected. On the other hand if we regard $I(\theta, \varphi, E)$ as the payoff matrix of a game and an experiment were to be chosen to maximize $\min_{\varphi \varepsilon a} I(\theta, \varphi, E)$ [a is the set corresponding to the alter-

[3] By symmetry we need only consider $p_1 < p_2$ and $p_1 + p_2 \leqq 1$.

native hypothesis], the randomized maximin strategy would give E_1 and E_2 weights $1 - \lambda$ and λ respectively. Thus θ^* and λ correspond to the solutions of a two person zero sum game with payoff $I(\theta, \varphi, E)$ where one player (say nature) selects φ to minimize I and the other player (the experimenter) selects E to maximize I.

It seems clear that the procedure recommended before will not be substantially affected, when c is small, if it is modified so that the $n + 1$st experiment is E_1 with probability $1 - \lambda_n$ and E_2 with probability λ_n where $1 - \lambda_n$ and λ_n correspond to the experimenter's maximin strategy based on the payoff matrix $I(\hat{\theta}_n, \varphi, E)$.

In Table 1, we tabulate as functions of θ,

a) $\theta^*(\theta)$, the minimax choice of nature,
b) $\lambda(\theta)$, the proportion of times E_2 is used in the experimenter's maximin strategy,
c) $I(\theta) = I(\theta, \theta^*(\theta), E_1) = I(\theta, \theta^*(\theta), E_2)$, the value of the game, and
d) $e(\theta) = \min_{\varphi \in \alpha}[I(\theta, \varphi, E_1) + I(\theta, \varphi, E_2)]/2I(\theta, \theta^*(\theta), E_1)$, which represents a measure of the relative efficiency of using each drug half the time to the procedure advocated. Evidently there is no great loss of efficiency in using each drug half the time and this prototype example is mainly useful as an illustrative device.

4. Formal Description of the General Procedures

It is desired to test

$$H_1 : \theta \in w_1$$

vs. the alternative $H_2 : \theta \in w_2$. There is available a set of experiments $\{E\}$ each of which may be replicated. Let $f(x, \theta, E)$ be the density of the outcome x of experiment E with respect to a measure μ_E. Let

$$I(\theta_1, \theta_2, E) = \int \log\left[\frac{f(x, \theta_1, E)}{f(x, \theta_2, E)}\right] f(x, \theta_1, E)\, d\mu_E(x). \tag{4.1}$$

Designate the nth experiment selected by $E^{(n)}$. Although the choice of the $n + 1$st experiment may depend on the past, once it is selected, its outcome is assumed independent of those of the preceding experiments. The maximum likelihood estimate of θ based on the first n observations is designated by $\hat{\theta}_n$. If $\theta \in w_1$, we call H_1 the hypothesis of θ, H_2 the hypothesis alternative to θ, and $a(\theta) = w_2$ the set alternative to θ. If $\theta \in w_2$, the hypothesis of θ is H_2, the alternative hypothesis is H_1 and the alternative set is $a(\theta) = w_1$. Let $\tilde{\theta}_n$ be the maximum likelihood estimate of θ under the hypothesis alternative to $\hat{\theta}_n$. If w_1, w_2, or the union $w_1 \cup w_2$ are not "closed" we may have some difficulty due to the nonexistence of these estimates. We shall assume throughout that suitably closing w_1 and w_2 and, if necessary, taking $\hat{\theta}_n$ and $\tilde{\theta}_n$ on the boundary of w_1 and w_2 eliminates this difficulty. In particular in the drug testing problem $\tilde{\theta}_n$ lies on the boundary separating w_1 and w_2.

Let $E(\theta, \varphi)$ be any experiment which maximizes $I(\theta, \varphi, E)$. We assume that such information maximizing experiments exist. Let us regard $I(\theta, \varphi, E)$ as the payoff matrix of a two-person zero-sum game where one player (seeking to minimize I) selects φ among the elements of the closure of $a(\theta)$, the set alternative to θ, and the other player (seeking to maximize I) selects E as an element of $\{E\}$, the set of available strategies. Now suppose that the experiment E is selected by some random mechanism corresponding to a probability measure λ on $\{E\}$. Then the corresponding information is $\int I(\theta, \varphi, E)\, d\lambda(E)$. Since experiments may be selected in such a fashion we extend the space of available experiments to include this convex set of all randomized experiments. This set is equivalent to the class of all randomized strategies of the second player (experimenter) in the original game. Thus if the original game had randomized or pure solutions, the second player in the extended game has a (not necessarily unique) maximin strategy $E(\theta)$ which is a pure or randomized experiment. The value of the game is given by

$$I(\theta) = \inf_{\varphi \in \alpha(\theta)} I(\theta, \varphi, E(\theta)). \tag{4.2}$$

Let $E^{(n)}$ represent the experiment used for the nth trial, x_n the corresponding outcome,

$$z_n(\theta, \varphi) = \log[f(x_n, \theta, E^{(n)})/f(x_n, \varphi, E^{(n)})], \tag{4.3}$$

$$S_n(\theta, \varphi) = \sum_{i=1}^{n} z_i(\theta, \varphi), \tag{4.4}$$

and

$$S_n = \sum_{i=1}^{n} z_i(\hat{\theta}_n, \tilde{\theta}_n). \tag{4.5}$$

We define our procedure A as follows. *Stop sampling at the nth observation and select the hypothesis of* $\hat{\theta}_n$ *if* $S_n > -\log c$. *If sampling is to be continued let* $E^{(n+1)} = E(\hat{\theta}_n)$ which is to be defined in some unique and measurable way consistent with the above definition of $E(\theta)$.

The stopping rule described above is rather unrefined. It does not require much imagination to see how to go about refining it. On the other hand such refinements will not be necessary for the asymptotic results we want and so we shall not study them here.

5. Asymptotic Characteristics of Procedure A

In this section we shall evaluate the asymptotic characteristics of the procedure A of Section 4 for the case where there are only a finite number of states of nature and a finite number of available (pure) experiments. We shall show that for this procedure the probability of making the wrong decision is $O(c)$ and the expected sample size is asymptotically no larger than $-\log c/I(\theta)$. In the next section we shall show that this is as well as can be done.

The arguments will involve the fact that as $c \to 0$, the required sample size gets large. For large samples, $\hat{\theta}_n$ tends to be equal to θ for all but the first "few" observations. Then, $E^{(n)} = E(\theta)$ and for $\varphi \in a(\theta)$, $\varepsilon\{z_n(\theta, \varphi)\} = I[\theta, \varphi, E(\theta)] \geqq I(\theta)$ and the sample size required for S_n to reach $-\log c$ is approximately $-\log c/I(\theta)$.

For the results in the rest of this paper we make the following assumptions. *There are s possible states of nature which are divided into the two disjoint sets* w_1 *and* w_2. *The set of available (pure) experiments is* $\{E_1, E_2, \ldots, E_k\}$. *The losses due to making the wrong decision are assumed to be positive. That is,* $r(\theta, i)$, *which is equal to the loss due to selecting* H_i *when* θ *is the state of nature, satisfies*

$$\begin{aligned} r(\theta, i) &= 0 \quad \text{if} \quad \theta \in w_i, \qquad i = 1, 2 \\ r(\theta, i) &> 0 \quad \text{if} \quad \theta \notin w_i, \qquad i = 1, 2. \end{aligned} \tag{5.1}$$

If the experiment E yields outcome x and θ *and* φ *are distinct, then*

$$z(\theta, \varphi, E) = \log \frac{f(x, \theta, E)}{f(x, \varphi, E)} \tag{5.2}$$

has finite variance[4] *and*

$$I(\theta, \varphi, E) = \int \log\left[\frac{f(x, \theta, E)}{f(x, \varphi, E)}\right] f(x, \theta, E)\, d\mu_E(x) > 0. \tag{5.3}$$

Hereafter we represent the true state of nature by θ_0 which we assume to be in w_1. Unless clearly specified otherwise, all probabilities and expectations refer to θ_0.

Lemma 1. *If the stopping rule is disregarded and sampling is continued according to any measurable*[5] *procedure,*

$$\hat{\theta}_n \to \theta_0 \qquad \text{w.p.l.} \tag{5.4}$$

In fact there exist K and $b > 0$ *such that*

$$P\{T > n\} \leqq Ke^{-bn} \tag{5.5}$$

where T is defined as the smallest integer such that $\hat{\theta}_n = \theta_0$ *for* $n \geqq T$.

[4] If $z(\theta, \varphi, E_i)$ has finite variance for each of the finite number of pure experiments, it follows easily that the variance of $z(\theta, \varphi, E)$ is bounded for all randomized experiments. Similarly $I(\theta, \varphi, E)$ is bounded away from zero and infinity.

[5] A procedure is considered measurable, if at the nth stage, the experiment selected is a measurable function of the data $x_1, x_2, \ldots, x_n$.

Proof. Assign *a priori* probability $1/s$ to each state of nature. Then $\hat{\theta}_n$ is the value of θ which maximizes the *a posteriori* probability $p_n(\theta)$ after the nth observation. Furthermore,

$$\log\left[\frac{p_n(\theta_0)}{p_n(\theta)}\right] = \sum_{i=1}^{n} \log\left[\frac{f(x_i, \theta_0, E^{(i)})}{f(x_i, \theta, E^{(i)})}\right] = \sum_{i=1}^{i} z_i(\theta_0, \theta) = S_n(\theta_0, \theta) \quad (5.6)$$

and $\hat{\theta}_n = \theta_0$ if $S_n(\theta_0, \theta) > 0$ for all $\theta \neq \theta_0$. It suffices to show that for each $\theta \neq \theta_0$, there is a number b such that $P\{S_n(\theta_0, \theta) \leqq 0\} \leqq e^{-bn}$. But

$$\begin{aligned} &P\{S_n(\theta_0, \theta) \leqq 0\}\mathscr{E}\{e^{tS_n(\theta_0,\theta)} | S_n(\theta_0, \theta) \leqq 0\} \leqq \mathscr{E}\{e^{tS_n(\theta_0,\theta)}\} \\ &P\{S_n(\theta_0, \theta) \leqq 0\} \leqq \mathscr{E}\{e^{tS_n(\theta_0,\theta)}\} \quad \text{for} \quad t \leqq 0. \end{aligned} \quad (5.7)$$

Now $\mathscr{E}\{[f(x, \theta_0, E)/f(x, \theta, E)]^t\}$ is the moment generating function of

$$z(\theta_0, \theta, E) = \log[f(x, \theta_0, E)/f(x, \theta, E)]$$

and is equal to one for $t = -1$ and $t = 0$ and, by convexity, is less than one for $-1 < t < 0$. Thus for a randomized experiment E where E_i is selected with probability p_i,

$$\mathscr{E}\{e^{-z(\theta_0,\theta,E)/2}\} = \sum_{i=1}^{k} p_i \mathscr{E}\{e^{-z(\theta_0,\theta,E_i)/2}\}$$

is bounded below one, and there is a number $b > 0$ such that

$$\max_E \mathscr{E}\{e^{-z(\theta_0,\theta,E)/2}\} = e^{-b} < 1.$$

Thus

$$\mathscr{E}\{e^{-z_i(\theta,\theta)/2} | x_1, x_2, \ldots, x_{i-1}\} \leqq e^{-b}$$

and

$$P\{S_n(\theta_0, \theta) \leqq 0\} \leqq \mathscr{E}\{e^{-S_n(\theta,\theta)/2}\} \leqq e^{-bn}. \qquad \square$$

Lemma 2. *For procedure A, the expected sample size satisfies*

$$\mathscr{E}(N) \leqq -[1 + o(1)] \log c/I(\theta_0). \quad (5.8)$$

Proof. We wish to show that, given any $\varepsilon > 0$, there is a $c^* = c^*(\varepsilon)$ such that

$$\mathscr{E}(N) \leqq -(1 + \varepsilon) \log c/I(\theta_0) \quad \text{for} \quad c < c^*.$$

It is obvious that $N \leqq \max_{\varphi \in w_2} (N_\varphi, T)$ where N_φ is the smallest integer for which $\sum_{i=1}^n z_i(\theta_0, \varphi) > -\log c$ for all $n \geqq N_\varphi$. In view of Lemma 1, it suffices to show that for each $\varphi \in w_2$ and for each $\varepsilon > 0$, there exist $K = K(\varepsilon, \varphi)$ and $b = b(\varepsilon, \varphi) > 0$ such that

$$P\{N_\varphi > n\} \leqq Ke^{-bn} \quad \text{for} \quad n > -(1 + \varepsilon) \log c/I(\theta_0).$$

To show this it suffices to prove that for each $\varphi \in w_2$ and each $\varepsilon > 0$, there exist K and $b > 0$ such that

$$P\left\{\sum_{i=1}^{n} z_i(\theta_0, \varphi) < -\log c\right\} \leqq Ke^{-bn} \quad \text{for} \quad n > -(1 + \varepsilon) \log c/I(\theta_0). \tag{5.9}$$

But

$$\sum_{i=1}^{n} z_i(\theta_0, \varphi) = \sum_{i=1}^{n} [z_i(\theta_0, \varphi) - I(\theta_0, \varphi, E^{(i)})]$$
$$+ \sum_{i=1}^{n} [I(\theta_0, \varphi, E^{(i)}) - I(\theta_0, \varphi, E(\theta_0))] + nI(\theta_0, \varphi, E(\theta_0)).$$

If $\varepsilon_1 > 0$, $z(\theta_0, \varphi, E_i) - I(\theta_0, \varphi, E_i) + \varepsilon_1$ has positive mean and finite moment generating function for $-1 \leqq t \leqq 0$ for each φ and pure experiment E_i. Hence the left-hand derivative of the moment generating function is positive at $t = 0$. Thus there is a $t^* = t^*(\varepsilon) < 0$ and $b_1 = b_1(\varepsilon) > 0$ such that

$$\mathscr{E}\{e^{t^*[z(\theta_0, \varphi, E_i) - I(\theta_0, \varphi, E_i) + \varepsilon_1]}\} \leqq e^{-bn}.$$

Consequently,

$$\mathscr{E}\{e^{t^*[z(\theta_0, \varphi, E) - I(\theta_0, \varphi, E) + \varepsilon_1]}\} \leqq e^{-b_1}$$

for each φ and E. Then, as in our proof of Lemma 1,

$$\mathscr{E}\{e^{t^*[\sum_{i=1}^{n} z_i(\theta_0, \varphi) - I(\theta_0, \varphi, E^{(i)}) + \varepsilon_1]}\} \leqq e^{-b_1 n}$$

and

$$P\left\{\sum_{i=1}^{n} [z_i(\theta_0, \varphi) - I(\theta_0, \varphi, E^{(i)})] < -\varepsilon_1 n\right\} \leqq e^{-b_1 n}. \tag{5.10}$$

Furthermore, it follows from the definition of T that

$$\left|\sum_{i=1}^{n} [I(\theta_0, \varphi, E^{(i)}) - I(\theta_0, \varphi, E(\theta_0))]\right| \leqq K_2 T$$

and hence, applying Lemma 1,

$$P\left\{\sum_{i=1}^{n} [I(\theta_0, \varphi, E^{(i)}) - I(\theta_0, \varphi, E(\theta_0))] < -\varepsilon_2 n\right\} \leqq K_3 e^{-b_3 n}. \tag{5.11}$$

Finally,

$$I(\theta_0, \varphi, E(\theta_0)) \geqq I(\theta_0) \qquad \text{for } \varphi \in \omega_2. \tag{5.12}$$

Then combining (5.10), (5.11), and (5.12) we obtain

$$P\left\{\sum_{i=1}^{n} z_i(\theta_0, \varphi) < n[I(\theta_0) - \varepsilon_3]\right\} \leqq K_4 e^{-b_4 n}$$

from which the desired result (5.9) follows. □

Lemma 3. *For procedure A, the probability of error (rejecting H_1) is $\alpha = O(c)$.*

PROOF. On the set $A_{n\varphi}$ in the sample space for which we reject $H_1 : \theta\varepsilon$ at the nth observation and for which $\hat{\theta}_n = \varphi \in \omega_2$

$$\sum_{i=1}^{n} z_i(\varphi, \theta_0) \geqq \sum_{i=1}^{n} z_i(\hat{\theta}_n, \tilde{\theta}_n) \geqq -\log c$$

and

$$\prod_{i=1}^{n} f(x_i, \theta_0, E^{(i)}) \leqq c \prod_{i=1}^{n} f(x_i, \varphi, E^{(i)}).$$

$$P\{A_{n\varphi}\} = \int_{A_{n\varphi}} \prod_{i=1}^{n} f(x_i, \theta_0, E^{(i)})\, d\mu_{E^{(1)}}(x_1) \cdots d\mu_{E^{(n)}}(x_n)$$

$$\leqq c \int_{A_{n\varphi}} \prod_{i=1}^{n} f(x_i, \varphi, E^{(i)})\, d\mu_{E^{(1)}}(x_1) \cdots d\mu_{E^{(n)}}(x_n).$$

The last integral is the probability of the set $A_{n\varphi}$ when φ is the state of nature. Thus

$$\alpha = \sum_{\varphi \in \omega_2} \sum_{n=1}^{\infty} P\{A_{n\varphi}\} \leqq \sum_{\varphi \in \omega_2} c < sc = O(c).$$

This proof is rather standard and obviously applies to any measurable procedure with the same stopping rule as procedure A. □

Combining Lemmas 2 and 3 we have

Theorem 1. *For procedure A, the risk function $R(\theta)$ satisfies*

$$R(\theta) \leqq -[1 + o(1)]c \log c/I(\theta) \tag{5.13}$$

for all θ.

6. Asymptotic Optimality of Procedure A

We shall state and prove Theorem 2 which together with Theorem 1 will establish the asymptotic optimality of procedure A in the sense described below.

Theorem 2. *Any procedure for which $I(\theta) > 0$ and*

$$R(\theta) = O(-c \log c) \qquad \text{for all } \theta$$

satisfies

$$R(\theta) \geqq -[1 + o(1)]c \log c/I(\theta) \qquad \text{for all } \theta. \tag{6.1}$$

Combining Theorems 1 and 2 we see that for any procedure to do substantially better than A for any θ implies that its risk will be of a *greater order of magnitude* for some θ. In this sense procedure A is asymptotically optimal.

To prove Theorem 2 we shall use two lemmas. The first will show that for the probabilities of error to be small enough,

$$S(\theta_0, \varphi) = \sum_{i=1}^{N} z_i(\theta_0, \varphi) \tag{6.2}$$

must be sufficiently large for all φ in ω_2 with large probability. The second will show that when n is substantially smaller than $-\log c/I(\theta_0)$, it is unlikely that the sums $\sum_{i=1}^{n} z_i(\theta_0, \varphi)$ can be sufficiently large for all $\varphi \in \omega_2$.

Lemma 4. *If* $\varphi \in \omega_2$, $P\{\text{accept } H_1 | \theta = \varphi\} = O(-c \log c)$, $P\{\text{reject } H_1\} = O(-c \log c)$, *and* $0 < \varepsilon < 1$, *then*

$$P\{S(\theta_0, \varphi) < -(1 - \varepsilon) \log c\} = O(-c^{\varepsilon} \log c). \tag{6.3}$$

PROOF. There is a number K such that

$$-Kc \log c \geqq P\{\text{accept } H_1 | \theta = \varphi\} \geqq \sum_{n=1}^{\infty} \int_{A_n} f(x, \varphi)\, d\mu(x)$$

where $f(x, \varphi)$ is the density on the sample space when $\theta = \varphi$ and A_n is the subset of the sample space for which $S(\theta_0, \varphi) < -(1 - \varepsilon) \log c$ and H_1 is accepted at the nth step.

$$\begin{aligned} -Kc \log c &\geqq \sum_{n=1}^{\infty} \int_{A_n} [f(x, \varphi)/f(x, \theta_0)] f(x, \theta_0)\, d\mu(x) \\ &= \sum_{n=1}^{\infty} \int_{A_n} e^{-S(\theta_0, \varphi)} f(x, \theta_0)\, d\mu(x) \geqq c^{1-\varepsilon} \sum_{n=1}^{\infty} P\{A_n\}. \end{aligned}$$

$$\sum_{n=1}^{\infty} P\{A_n\} = O(-c^{\varepsilon} \log c).$$

$$P\{\text{reject } H_1\} = O(-c \log c).$$

$$P\{S(\theta_0, \varphi) < -(1 - \varepsilon) \log c\} \leqq \sum_{n=1}^{\infty} P\{A_n\} + P\{\text{reject } H_1\} = O(-c^{\varepsilon} \log c). \square$$

Lemma 5. *If* $\varepsilon > 0$,

$$P\left\{ \max_{1 \leqq m \leqq n} \min_{\varphi \in \omega_2} \sum_{i=1}^{m} z_i(\theta_0, \varphi) \geqq n[I(\theta_0) + \varepsilon] \right\} \to 0 \quad \text{as} \quad n \to \infty. \tag{6.4}$$

PROOF.

$$\begin{aligned} \sum_{i=1}^{m} z_i(\theta_0, \varphi) &= \sum_{i=1}^{m} [z_i(\theta_0, \varphi) - I(\theta_0, \varphi, E^{(i)})] + \sum_{i=1}^{m} I(\theta_0, \varphi, E^{(i)}) \\ &= A_{1m} + A_{2m}, \end{aligned}$$

where

$$A_{1m} = \sum_{i=1}^{m} [z_i(\theta_0, \varphi) - I(\theta_0, \varphi, E^{(i)})]$$

is a martingale. Now $A_{2m} = \sum_{i=1}^{m} I(\theta_0, \varphi, E^{(i)})$ represents m times the payoff for the game where nature selects φ and the experimenter selects more mixture of his available strategies. Thus $\min_{\varphi \varepsilon \omega_2} A_{2m} \leq mI(\theta_0) \leq nI(\theta_0)$. Thus

$$\min_{\varphi \varepsilon \omega_2}(A_{1m} + A_{2m}) \geqq n[I(\theta_0) + \varepsilon]$$

implies that $A_{1m} \geqq n\varepsilon$ for some φ and

$$P\left\{\max_{1 \leqq m \leqq n} \min_{\varphi \in \omega_2} \sum_{i=1}^{m} z_i(\theta_0, \varphi) \geqq n[I(\theta_0) + \varepsilon]\right\} \leqq \sum_{\varphi \in \omega_2} P\left\{\max_{1 \leqq m \leqq n} A_{1m} \geqq n\varepsilon\right\}.$$

Since A_{1m} is a martingale with mean 0, we may apply Doob's extension of Kolmogoroff's extension of Tchebycheff's inequality ([3], p. 315), to obtain

$$P\left\{\max_{1 \leqq m \leqq n} A_{1m} > n\varepsilon\right\} \leq K/n\varepsilon^2 \quad \text{for each} \quad \varphi \in \omega_2.$$

Lemma 5 follows.

Now we are in position to prove Theorem 2. Let

$$n_c = -(1 - \varepsilon) \log c/[I(\theta_0) + \varepsilon];$$

$$P\{N \leqq n_c\} \leqq P\{N \leqq n_c \text{ and } S(\theta_0, \varphi) \geqq -(1 - \varepsilon) \log c \text{ for all } \varphi \in \omega_2\}$$
$$+ P\{S(\theta_0, \varphi) < -(1 - \varepsilon) \log c \text{ for some } \varphi \in \omega_2\}.$$

By Lemma 5, the first term on the right approaches zero. The condition $R(\theta) = O(-c \log c)$ for all θ permits us to apply Lemma 4 and the second term on the right approaches zero. Hence $\mathscr{E}(N) \geqq -[1 + o(1)] \log c/I(\theta_0)$. Theorem 2 follows. □

7. Miscellaneous Remarks

1. The asymptotic optimality of procedure A may not be especially relevant for the initial stages of experimentation especially if the cost of sampling is high. At first it is desirable to apply experiments which are informative for a broad range of parameter values. Maximizing the Kullback-Leibler information number may give experiments which are efficient only when θ is close to the estimated value.

2. It is clear that the methods and results apply when the cost of sampling varies from experiment to experiment. Here, we are interested in selecting experiments which maximize information per unit cost.

3. The ideas employed in this paper seem equally valid and applicable to problems which involve selecting one of k mutually exclusive hypotheses.

4. A minor modification of the stopping rule would be to continue experimentation as long as $\sum_{i=1}^{n} z_i(\hat{\theta}_i, \tilde{\theta}_i) < -\log c$. This rule which involves

$$\sum_{i=1}^{n} z_i(\hat{\theta}_i, \tilde{\theta}_i)$$

instead of $\sum_{i=1}^{n} z_i(\hat{\theta}_n, \tilde{\theta}_n)$ may be computationally easier to deal with occasionally. It is not difficult to show that Lemma 2 applies for this stopping rule. The author has not proved that Lemma 3 also applies.

(*Editor's note*: Remarks 5–7 have been omitted.)

References

[1] R.N. Bradt, S.M. Johnson, and S.Karlin, "On sequential designs for maximizing the sum of *n* observations," *Ann. Math. Stat.*, Vol. 27 (1956), pp. 1060–1074.
[2] Herman Chernoff, "Large-sample theory; parametric case," *Ann. Math. Stat.*, Vol. 27 (1956), pp. 1–22.
[3] J.L. Doob, *Stochastic Processes*, John Wiley and Sons, Inc., New York, 1953.
[4] S. Kullback and R.A. Leibler, "On information and sufficiency," *Ann. Math. Stat.*, Vol. 22 (1951), pp. 79–86.
[5] H.E. Robbins, "Some aspects of the sequential design of experiments," *Bull. Amer. Math. Soc.*, Vol. 58, No. 5 (1952), pp. 527–535.
[6] Abraham Wald, *Sequential Analysis*, John Wiley and Sons, Inc., New York, 1947.

Introduction to Box and Jenkins (1962) Some Statistical Aspects of Adaptive Optimization and Control

Edward J. Wegman*
George Mason University

1. Introduction

This paper is a long paper full of insights and innovations, many of which have become key elements in modern time series analysis. The paper is a discussion paper and many of the discussants seem to have recognized it as a landmark paper, but perhaps not for the reasons modern statisticians would find it important. While ostensibly on optimization and control, it contains many gems that are precursors to modern statistical and data analytic notions. Indeed, the not unrelated papers by Kalman (1960) and Kalman and Bucy (1961) have proven to have more direct impact on control and linear filtering theory. Nonetheless, this paper surely ranks among the most important statistics papers in the modern era. It represents the first in a series of collaborations between Box and Jenkins that came to full fruition in 1970 with the publication of the now classic text on time domain parametric time series models including autoregressive, moving average, and mixed models. Preliminary discussions of these models appear in this paper. In addition, a description of *technical feedback* is given that anticipates modern exploratory data analysis. The treatment of nonstationarity using difference operators is also noted, as well as the suggestion of modern so-called high-resolution spectral analysis.

This Introduction will proceed as follows. Section 2 contains a short discussion of the notions of empirical and technical feedback with commentary, while Sec. 3 contains a brief description of the optimization model. With

* This work was supported by the Army Research Office under contract DAAL03-87-K-0087, by the Office of Naval Research under contract N00014-89-J-1807, and by the National Science Foundation under grant DMS-90002237.

the benefit of nearly 30 years additional cumulative experience, it is perhaps easier to formulate this model in a slightly clearer way. Section 4 describes the basic Kalman filter model and the connection between the Kalman model and the Box–Jenkins model. Section 5 discusses linear time series modeling and nonstationary time series modeling as found in this paper. And finally in Sec. 6, I discuss spectral density theory a bit and indicate its connection with modern high-resolution spectral analysis.

2. Technical Feedback and Data Analysis

Box and Jenkins (hereafter referred to as BJ) articulate the ideas of empirical and technical feedback. By the former, BJ mean the following rule: "If after testing a modification once, it appears to give an improvement, then include it in the process and start to test for another modification." Thus, empirical feedback includes rather cut-and-dry situations in which the decision rule is completely specified, depending on the outcome of the experiment. In contrast, BJ describe *technical feedback* as occurring "... when information coming from the experiment interacts with technical knowledge contained in the experimenter's mind to lead to some form of action."

The former perspective is very much the concept of classical statistical confirmatory analysis. That is, one models the situation, designs the experiment, carries out the statistical procedure, and draws the appropriate conclusion. The formalism is very strong and there is little left to the choice of the scientist. In contrast, the latter perspective has the distinct flavor of modern exploratory analysis. That is to say, one looks at the results of the experiment in an interactive way, possibly proposing hypotheses, examining their implications, and then reproposing hypotheses, going between theory and experiment in an interactive way until a satisfactory understanding of the experiment is in hand. This interactive, adaptive approach is very much in the spirit of modern data analysis. BJ write, "Inspection and analysis of data will often suggest to the trained scientist explanations which are far from direct." The additional element that BJ bring to bear is the notion that the experimenter may have additional in-depth technical knowledge about the experiment that will guide the direction of the analysis. While no rational data analyst would discard information that may bear on the direction of data analysis, often in contemporary exploratory data analysis the virtue of actually knowing what the experiment is about is somewhat deemphasized. In any case, it is clear that BJ anticipated the attitude and approach to data analysis that we think of as contemporary data analysis.

3. Basic Description of the Model

The goal of this paper is to develop a model for the optimization and control of a system. In most examples, the authors have in mind either a quality-control-oriented manufacturing system or a chemical plant system. In either case, the model is as follows.

We conceive of a discrete system measured discrete times, called *phases* in the Box–Jenkins terminology, the phases or epochs being indexed by the variable p. I slightly prefer the term *epoch* since phase has another technical meaning in time series analysis. A random process ξ_p, which is unmeasurable and uncontrollable by the experimenter, is assumed to affect the operation of the system; ξ_p might be thought of as the state of nature. A variable X is under the experimenter's control, and it is desired to set X either directly or indirectly to optimize (or predict) the system. The (conditional) response of the system is $\eta(X \mid \xi_p)$. The value of X that maximizes the system response $\eta(X \mid \xi_p)$ is given by θ_p when ξ_p is the state of nature. In general, of course, θ_p will be unknown to us. It is clear that $\eta(\theta_p \mid \xi_p) \overset{d}{=} \eta(\theta_p)$ is the maximum of $\eta(X \mid \xi_p)$ in the pth phase or epoch. Notice also that θ_p is a stochastic process even though X may not be. The control variable X will be actually set to X_p during the pth epoch. $\theta_p - X_p \overset{d}{=} \varepsilon_p$ is the deviation of the set point from the optimal.

Box and Jenkins model the system response as a quadratic, that is,

$$\eta(X \mid \xi_p) = \eta(\theta_p) - \tfrac{1}{2}\beta_{11}(X - \theta_p)^2. \tag{1}$$

Taking derivatives of (1) with respect to X, we get

$$\frac{\partial n(X \mid \xi_p)}{\partial X} = \frac{\partial \eta(\theta_p)}{\partial X} - \beta_{11}(X - \theta_p). \tag{2}$$

The first term on the right-hand side of (2) is 0 because $\eta(\theta_p)$ does not depend on X. Thus, we can write

$$(\theta_p - X) = \frac{1}{\beta_{11}}\left\{\frac{\partial \eta(X \mid \xi_p)}{\partial X}\right\}. \tag{3}$$

We want to approximate $X_p - \theta_p$ since this is the deviation of the set point from the optimal value. Equation (3) suggests we can estimate this deviation by calculating an approximation to the derivative. We calculate an approximate derivative by

$$\frac{\partial \eta(X \mid \xi_p)}{\partial X} \cong \frac{\hat{\eta}(X_p + \delta \mid \xi_p) - \hat{\eta}(X_p - \delta \mid \xi_p)}{2\delta}.$$

In the Box–Jenkins paper, $\hat{\eta}$ is given by $\bar{y}$. Thus, our estimate of $\varepsilon_p = \theta_p - X_p$ is given by

$$e_p = \frac{1}{2\beta_{11}\delta}\{\hat{\eta}(X_p + \delta \mid \xi_p) - \hat{\eta}(X_p - \delta \mid \xi_p)\}. \tag{4}$$

BJ assume that $e_p = \varepsilon_p + u_p$, where u_p is the "measurement error." It is a simple matter to see that

$$e_p = \varepsilon_p + u_p = (\theta_p - X_p) + u_p = z_p - X_p.$$

We know the sequence of set points, $X_p, X_{p-1}, X_{p-2}, \ldots$, and using (4), we can calculate the errors, $e_p, e_{p-1}, e_{p-2}, \ldots$. Thus, we can calculate

$$z_p = X_p + e_p. \tag{5}$$

Now we wish to minimize $\eta(\theta_{p+1}) - \eta(X_{p+1} | \xi_{p+1})$, which by (1) is given as

$$\eta(\theta_{p+1}) - \eta(X_{p+1} | \xi_{p+1}) = \tfrac{1}{2}\beta_{11}(X_{p+1} - \theta_{p+1})^2. \tag{6}$$

Thus, we want to minimize

$$E\{(\theta_{p+1} - X_{p+1})^2 | z_p, z_{p-1}, z_{p-2}, \ldots\}.$$

This is done by choosing

$$X_{p+1} = \hat{\theta}_{p+1} = E(\theta_{p+1} | z_p, z_{p-1}, z_{p-2}, \ldots) \stackrel{d}{=} f(z_p, z_{p-1}, z_{p-2}, \ldots).$$

Finally, the choice of f is made by letting f be an infinite moving average, that is,

$$X_{p+1} = \hat{\theta}_{p+1} = \sum_{j=0}^{\infty} \mu_j z_{p-j}. \tag{7}$$

This leap from a general nonlinear function f to a linear infinite moving average is, to be sure, a big leap. Nonetheless, it sets the stage for a reasonable analysis of a rather complex model.

4. Connection to Kalman Filtering

The papers of Kalman (1960) and Kalman–Bucy (1961) introduce the notions of a Kalman filter that is a linear, recursively formulatcd lcast-squares state-space filter. The Kalman filter model can be given briefly as

$$X_p = \phi_p X_{p-1} + \omega_p, \; p = 1, 2, \ldots \text{ (state prediction equation)} \tag{8}$$

and

$$z_p = H_p X_p + v_p, \qquad p = 1, 2, \ldots \text{ (observation equation).} \tag{9}$$

The idea is that the z's are observed values and that the X's are the values to be filtered, predicted, or controlled. In the BJ model, Eq. (5) above is the analog of the observation equation

$$z_p = X_p + e_p, \qquad p = 1, 2, \ldots \tag{10}$$

Equation (7) gives the analog to the state prediction equation, although it is somewhat less obvious in the present form. To see it more clearly, let B be the backward shift operator so that $Bz_p = z_{p-1}$. Then we may write

$$\Theta(B) = \sum_{j=0}^{\infty} \mu_j \mathrm{B}^j.$$

With this, (7) becomes

$$X_{p+1} = \Theta(B) z_p.$$

If $\Theta(B)$ is an invertible operator, then we may write $\Theta^{-1}(B)$ as

$$\Theta^{-1}(B) = \Pi(B) = I - \pi_1 B - \pi_2 B^2 - \cdots - \pi_d B^d$$

so that (7) becomes

$$\Pi(B) X_{p+1} = z_p. \tag{11}$$

Expanding (11), we have

$$X_{p+1} = \sum_{j=1}^{d} \pi_j X_p + z_p, \qquad p = 1, 2, \ldots. \tag{12}$$

It is clear that (12) written in this form is the analog of the state prediction equation. The major difference between the Kalman model and the BJ model is that the Kalman model allows for state-dependent coefficients, whereas the BJ model does not. The BJ model is a much more specialized model for adaptive system control and adaptive optimization. More details of the Kalman filter can be found in Wegman (1983).

5. Connection to Linear Time Series Modeling

A general linear time series model is often given by the infinite moving average model

$$X_t = \psi(B)\varepsilon_t = \sum_{j=0}^{\infty} \psi_j \varepsilon_{t-j}, \qquad t \in T \tag{13}$$

for some index set T. In its inverted form, we have

$$\Pi(B) X_t = X_t - \sum_{j=1}^{\infty} \pi_j X_{t-j} = \varepsilon_t, \qquad t \in T. \tag{14}$$

In practice, these infinite series are often truncated, giving rise, respectively, to the moving average (MA), the autoregressive (AR), or the mixed autoregressive-moving average (ARMA) process. As indicated above, Eq. (7) is in the form of the infinite moving average. In Sec. 2.4 of BJ, there is a quite clear discussion of these general linear models and Table 2 on p. 408 illustrates alternate forms of the "controller weight," including exponentially weighted, first-order autoregressive, and first-order moving average. Section 3.4 of BJ (1962) discusses the nature of stochastic processes for which the predictor is optimal. The section begins with the representation in Eq. (14) above. Moreover, in Sec. 3.3 and 4.1 of BJ, it is clear that the fitting and prediction techniques that were to become those aspects of linear time series modeling which proved to be so powerful in Box and Jenkins (1970) were explained already in 1962.

One quite interesting aspect of the discussion is that BJ recognized the need to account for nonstationary time series. This is done in BJ (1970) with differencing and that discussion of differencing is found here also, although the notion of the rather powerful integrated models, the ARIMA models, does not appear. One amusing insight into the character of these always practical authors is the manipulation of infinite sums as typified in their discussion on p. 414 of BJ (1962). In this discussion they manipulate divergent infinite series with little concern for convergence issues. The same is true of BJ (1970), and although the results can be rigorously justified, the practical BJ show little regard for mathematical subtleties. In this, they share a heritage with Euler who was also given to manipulating divergent series to good effect. See, for example, Struik (1948).

6. High-Resolution Spectral Analysis

During the 1950s and 1960s, there was considerable interest in the estimation of spectral densities connected with the emerging space technology. Space technology is also much of the reason for the interest in control theory and the explanation of the powerful reaction to the Kalman filter. Typical of this era was the smoothed periodogram approach to spectral density estimation. Much of the smoothed periodogram approach is codified in Blackman and Tukey (1959). The periodogram is the square modulus of the discrete Fourier transform of the raw time series. It is known to be an inconsistent estimator of the spectral density, but with appropriate smoothing, it can be made consistent. Much effort in determining smoothing kernels and kernel bandwidths was expended in this era.

In the late 1970s and through much of the 1980s, there was much discussion of the so-called maximum entropy or high-resolution spectral analysis. Basically, one can conceive of the smoothed periodogram as a nonparametric approach to spectral estimation. Because the approach used smoothing heavily, it was weak in identifying narrow bands or lines in the spectrum. The maximum entropy or high-resolution techniques popular more recently are based on fitting a linear model, autoregressive, moving average, or ARMA, and then using the parametric form of the spectral density that can be fitted using the parameters of the linear model. In particular, if the set of roots of the auxiliary equation for the autoregressive part of the model contain a complex conjugate pair near the unit circle, then the covariance function has a nearly sinusoidal component. Since the spectral density is the inverse Fourier transform of the covariance, the spectral density will have a narrow band or line component. Thus, whereas the nonparametric smoothed periodogram approach is weak in identifying line components to the spectrum, the parametric linear time series model approach is quite good at it. The impact of this is hard to overestimate. The idea is usually attributed to John Burg dating around 1975, but Eq. (5.2.3) on p. 395 of BJ (1962) clearly indicates that BJ were cognisant of the form of the linear model spectral density, and their general discussion in Sec. 5 of the 1962 paper clearly suggests that they were aware of the connection of linear models to spectral density estimation.

8. About the Authors

G.E.P. Box was born in Gravesend, Kent, England in 1919. He obtained his first degree in chemistry. While engaged in war work, he met R.A. Fisher at Cambridge in 1944 [see Box (1978, p. 245)] and received much useful advice that greatly influenced his subsequent career. After the war, he entered the department of statistics at University College, London, obtaining his Ph.D. in 1952. He later received a D. Sc. degree, also from London University, in 1960.

Following experience in industry with Imperial Chemical Industries and serving as the director of the Statistical Techniques Research Group at Princeton University, he has been a professor at the University of Wisconsin since 1960. In this position, he was a moving spirit in developing the department of statistics (of which he was the founding chairman) and fostering its work in industrial applications, leading *inter alia* to the formation of the Center for Quality and Productivity, now a prominent focus for work on statistical methods in quality assurance.

G.M. Jenkins obtained a Ph.D. at University College, London in 1956. From 1957–65, he was on the faculty at Imperial College. London. During this period, he visited Stanford University and Princeton University as a visiting professor in 1959–60, and the University of Wisconsin in 1964–65. From 1959, he began collaborating with G.E.P. Box—one product of this association is the present paper—a later one was the book, *Time Series Analysis Forecasting and Control* [Box and Jenkins (1970)].

To summarize his later career, in 1965, Jenkins became professor of systems engineering at the University of Lancaster and there emphasized the practical use of statistical techniques and their coordination with other features of applied problems. He left the university in 1974 to form a company specializing in such applications and also to take up an appointment as a visiting professor at the London Business School. He died in 1982, at the young age of 49.

References

Blackman, R.B., and Tukey, J.W. (1959). *The Measurement of Power Spectra from the Point of View of Communications Engineering*. Dover, New York.

Box, G.E.P., and Jenkins, G.M. (1970). *Time Series Analysis, Forecasting and Control*. Holden-Day, San Francisco, Calif.

Box, J.F. (1978). *R.A. Fisher: The Life of a Scientist*. Wiley, New York.

Kalman, R.E. (1960). "A new approach to linear filtering and prediction problems" *Trans. ASME: J. Basic Engineer.*, **82D**, 35–45.

Kalman, R.E., and Bucy, R.S. (1961). "New results in linear filtering and prediction theory" *Trans. ASME: J. Basic Engineer.*, **83D**, 95–108.

Struik, D.J. (1948). *A Concise History of Mathematics*. Dover, New York, pp. 167–179.

Wegman, E.J. (1983). Kalman filtering, in *Encyclopedia of Statistical Sciences*, S. Kotz, N.L. Johnson, and C.B. Read, eds.), Vol. 4. pp. 342–346. J. Wiley & Sons: New York.

Some Statistical Aspects of Adaptive Optimization and Control

G.E.P. Box
University of Wisconsin, Madison

G.M. Jenkins
Imperial College, London

Summary

It is often necessary to adjust some variable X, such as the concentration of consecutive batches of a product, to keep X close to a specified target value. A second more complicated problem occurs when the independent variables X in a response function $\eta(X)$ are to be adjusted so that the derivatives $\partial\eta/\partial X$ are kept close to a target value zero, thus maximizing or minimizing the response. These are shown to be problems of prediction, essentially, and the paper is devoted mainly to the estimation from past data of the "best" adjustments to be applied in the first problem.

1. Introduction

1.1. Experimentation as an Adaptive Process

Our interest in adaptive systems arises principally because we believe that experimentation itself is an adaptive system. This iterative or adaptive nature of experimentation, as characterized by the closed loop, *conjecture-design-experiment-analysis* has been discussed previously by Box (1957).

In some circumstances, the probabilities associated with particular occurrences which are part of this iterative process may be of little direct interest, as may be the rightness or wrongness of particular "decisions" which are made along the way. For instance, on one admittedly idealized model of an evolutionary operation programme applied during the finite life of a chemical process, Box has shown in an unpublished report that if a virtually limitless number of modifications are available for trial, then to obtain maximum profit, each modification should be tested once and once only no matter what the standard deviation of the test procedure is.

Results of this kind are important but for the most part are not generally applicable to the experimental process itself. To see why this is so, we must distinguish between empirical and technical feedback.

1.2. Empirical and Technical Feedback

Empirical feedback occurs when we can give a simple rule which describes unequivocally what action should be taken and what new experiments should be done in every conceivable situation. In the above example the rule is: If after testing a modification once, it appears to give an improvement, then include it in the process and start to test another modification. If it fails to do this, leave it out and test another modification.

Technical feedback occurs when the information coming from the experiment interacts with technical knowledge contained in the experimenter's mind to lead to some form of action. The reasoning here is of a more complex kind not predictable by a simple rule. Inspection and analysis of data will often suggest to the trained scientist explanations which are far from direct. These will provide leads which he follows up in further work. To allow technical feedback to occur efficiently, the scientist must have a reasonably sure foundation on which to build his conjectures. To provide this, the effects about which he has to reason will need to be known with fairly good and stated precision and this may required considerable replication, hidden or direct. This is one reason why the $n = 1$ rule is not recommended for most applications of evolutionary operation, or for any other procedure which benefits heavily from technical as well as from empirical feedback.

However, when the feedback is essentially empirical, the use of conservative rules appropriate for technical feedback can lead to considerable inefficiency. Two problems which we feel should be considered from the empirical feedback point of view are those of adaptive optimization and adaptive quality control. It is hoped that the discussion of these problems here may provide a starting point for a study of the much wider adaptive aspects of the experimental method itself.

1.3. Adaptive Quality Control

Quality control undoubtedly has many different motivations and aims. Quality control charts are used to spotlight abnormal variations, to indicate when something went wrong and, perhaps most important of all, to give a pictorial and readily understandable representation of the behaviour of the process in a form which is continually brought to the notice of those responsible for running it. These are all examples of the technical feedback aspects of quality control, the importance of which cannot be easily overrated but which will not be considered here.

Quality control charts are, however, also used as part of an adaptive loop. With Shewhart charts, for example, some rule will be instituted that when a charted point goes outside a certain limit, or a succession of points fall between some less extreme limits, then a specified action is taken to adjust the mean value. In this circumstance we are clearly dealing with empirical rather than with technical feedback and the problem should be so treated.

A paper which pointed out the advantages of a new type of quality control chart using cumulative sums was recently read before this Society by Barnard (1959). In it the important feedback aspect of the quality control problem was clearly brought out and we owe much to this work.

1.4. Adaptive Optimization

Our interest in adaptive optimization was stimulated in the first place by experiences with evolutionary operation. By evolutionary operation we mean a simple technique used by plant operators by means of which over the life of the plant many modifications are tried and, where they prove to be useful, are included in the process. In practice, evolutionary operation, like quality control, is used for a number of rather different purposes. Its main purpose seems to be to act as a permanent incentive to ideas and as a device to screen these ideas as they are born, either spontaneously or out of previous work. As with quality control, many important advances are made here by the use of technical rather than empirical feedback.

However, evolutionary operation has occasionally also been used as a kind of manual process of adaptive optimization. It sometimes happens that there exist some unmeasurable and largely unpredictable variables, ξ, such as catalyst activity, whose levels change with time. When this happens, evolutionary operation is sometimes used to indicate continuously how the levels of some other controllable variables X should be moved so as to ensure, as nearly as possible, that the current maximum value in some *objective function*, or *response*, η is achieved at any given time t.

Other examples of variables which affect efficiency and, like catalyst activity, are not easily measured when the chemical process is running, and for which adaptive compensation in other variables may be desirable, are: degree of fouling of heat exchangers, permeability of packed columns, abrasion of particles in fluidized beds and, in some cases, quality of feed stock. When such uncontrollable and unmeasurable variables ξ occur as a permanent and inevitable feature of chemical processes, it may become worth while to install special equipment to provide automatic adaptive compensation of the controllable and measurable variables X (Box, 1960). It also becomes worth while to consider the theory of such systems so that the adaptive optimization may be made as efficient as possible.

1.5. Scope of the Present Investigation

In section 2 we consider a simplified discrete-time model for the adaptive optimization problem. We find that we are led to problems of adaptive control which lead in turn to problems of prediction and smoothing of a time-series. Our results are then applicable to the problem of adaptive statistical quality control, which is given special attention in sections 2 and 3. These sections incorporate ideas well known to control engineers. In section 4, the methods are illustrated by means of examples and in section 5 the results are expressed in terms of power spectra.

The problem of maximizing or minimizing a response will be considered in a later paper.

2. A Discrete-Time Model

2.1. Discrete Adaptive Optimization Model

We shall first consider a simplified model in which data are available only at discrete and equal intervals of time each of which we call a *phase*. We suppose further that the situation remains constant during a phase but may change from phase to phase. The model is thus directly appropriate to a batch-type chemical process and provides a discrete approximation to the operation of a continuous process from which discrete data are taken at equal intervals of time.

Suppose the uncontrollable and immeasurable variables have levels ξ_p during the pth phase and these levels change from one phase to the next. Then the conditional response function $\eta(X|\xi_p)$ will change correspondingly as in Fig. 1. Suppose that in the pth phase it may be approximated by the quadratic equation

$$\eta_p = \eta(X|\xi_p) = \eta(\theta_p) - \tfrac{1}{2}\beta_{11}(X - \theta_p)^2, \tag{2.1.1}$$

where $\theta_p = \{X_{\max}|\xi_p\}$ is the conditional maximal setting during the pth phase and β_{11} is known from prior calibration and does not change appreciably with ξ_p. We suppose finally that because of fluctuation in ξ_p, the conditional maximal value θ_p follows some stochastic process usually of a non-stationary kind. Let X_p be the *set-point* at which the controllable variable X is held in the pth phase (Fig. 1). Then the standardized slope of the response function

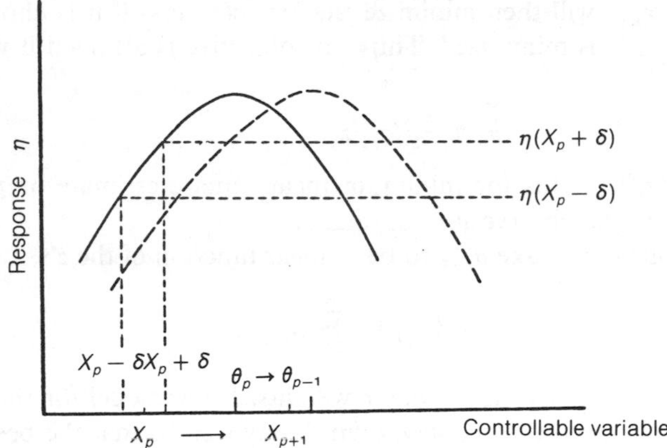

Figure 1. Simple adaptive optimization.

at $X = X_p$,

$$\frac{1}{\beta_{11}}\left(\frac{d\eta_p}{dX}\right)_{X=X_p} = \theta_p - X_p = \varepsilon_p,$$

measures the extent to which X_p deviates from θ_p. If experiments are performed at $X_p + \delta$ and at $X_p - \delta$ and the average response observed at these levels is $\bar{y}(X_p + \delta)$ and $\bar{y}(X_p - \delta)$, then an estimate e_p of $\varepsilon_p = \theta_p - X_p$ is given by

$$e_p = \tfrac{1}{2}\{\bar{y}(X_p + \delta) - \bar{y}(X_p - \delta)\}/(\delta\beta_{11}) = \varepsilon_p + u_p, \tag{2.1.2}$$

where u_p is called the measurement error, although it will be produced partly by uncontrolled fluctuations in the process. Now

$$e_p = \varepsilon_p + u_p = \theta_p - X_p + u_p = z_p - X_p, \tag{2.1.3}$$

where $z_p = \theta_p + u_p$ is an estimate of the position of the optimal setting θ_p during the pth phase.

If a series of adjustments has actually been made on some basis or other, we have a record of a sequence of set points $X_p, X_{p-1}, X_{p-2}, \ldots$ and of a sequence of deviations $e_p, e_{p-1}, e_{p-2}, \ldots$. From these it will be possible to reconstruct the sequence $z_p, z_{p-1}, z_{p-2}, \ldots$, of estimated positions of the maxima in phases $p, p-1, p-2, \ldots$. From such data we wish to make an adjustment x_{p+1} to the set point X_p so that the adjusted set point $X_{p+1} = X_p + x_{p+1}$ which will be maintained during the *coming* $(p+1)$th phase will be in some sense "best" in relation to the coming and unknown value of θ_{p+1}.

Now we suppose that the objective function η is such that the loss sustained during the $(p+1)$th phase is realistically measured by

$$\eta(\theta_{p+1}) - \eta(X_{p+1}|\xi_{p+1}) = \tfrac{1}{2}\beta_{11}(X_{p+1} - \theta_{p+1})^2, \tag{2.1.4}$$

the amount by which η falls short of the value theoretically attainable. The adjustment x_{p+1} will then minimize the expected loss if it is chosen so that $E(\theta_{p+1} - X_{p+1})^2$ is minimized. Thus our objective is attained if we set X_{p+1} equal to

$$\hat{\theta}_{p+1} = f(z_p, z_{p-1}, z_{p-2}, \ldots),$$

where $f(z_p, z_{p-1}, \ldots)$ is the minimum mean square estimate or *predictor* of θ_{p+1}, based on the observations $z_p, z_{p-1}, \ldots$.

For simplicity we take $\hat{\theta}_{p+1}$ to be a linear function of the z's,

$$\hat{\theta}_{p+1} = \sum_{j=0}^{\infty} \mu_j z_{p-j}, \tag{2.1.5}$$

and refer to the μ_j's as the *predictor* weights. If the model for the stochastic process $\{\theta_p\}$ is specified precisely, then it may well be that the best predictor is a non-linear function of the previous z's, as in the model used by Barnard (1959). In general, however, there is no empirical evidence for making such precise assumptions for the model so that the simplicity of the linear predictor is well worth maintaining.

At the beginning of the $(p + 1)$th phase we should therefore apply to the previous setting X_p an adjustment $x_{p+1} = \hat{\theta}_{p+1} - \hat{\theta}_p$. Now in practice we do not observe the z's directly but only the e's. This means that we must calculate the adjustment from the estimated slopes $e_p, e_{p-1}, \ldots$, in the form

$$x_{p+1} = X_{p+1} - X_p = \hat{\theta}_{p+1} - \hat{\theta}_p = \sum_{j=0}^{\infty} w_j e_{p-j}, \tag{2.1.6}$$

where we call the w_j the *controller weights.* In section 2.3 we show that (2.1.5) and (2.1.6) imply that a definite relationship exists between the predictor weights μ_j and the controller weights w_j. The optimal adjustment is then obtained by choosing the w's, or equivalently the μ's, to minimize $E(\varepsilon_{p+1}^2) = E(\theta_{p+1} - \hat{\theta}_{p+1})^2$.

Now $\hat{\theta}_{p+1} = f(z_p, z_{p-1}, \ldots)$ and $z_{p+1} = \theta_{p+1} + u_{p+1}$, so that if the "measurement error" u_{p+1} is distributed about zero with variance σ_u^2 *independently* of $u_p, u_{p-1}, \ldots$, and of $\theta_p, \theta_{p-1}, \ldots$, then

$$E(e_{p+1}^2) = E(z_{p+1} - \hat{\theta}_{p+1})^2 = E(\varepsilon_{p+1}^2) + \sigma_u^2. \tag{2.1.7}$$

The loss $E(\varepsilon_{p+1}^2)$ is then minimal when $E(z_{p+1} - \hat{\theta}_{p+1})^2 = E(e_{p+1})^2$ is minimized and $\hat{z}_{p+1}$, the best predictor of z_{p+1}, supplies in this case the best predictor $\hat{\theta}_{p+1}$ of θ_{p+1}. In general, σ_u^2 will depend on the number n of experiments performed in estimating the slope and also on δ, the magnitude of the perturbations. For any fixed values of n and δ, however, the best way of tracking θ_{p+1} is always to make an adjustment

$$x_{p+1} = \hat{\theta}_{p+1} - \hat{\theta}_p = \sum_{j=0}^{\infty} w_j e_{p-j}.$$

In those cases where the measurement errors u are independent of the θ's and of each other than $\hat{\theta}_{p+1} = \hat{z}_{p+1}$.

2.2. Dynamics of Adjustment

On our assumptions if, using data acquired during the pth and previous phases of operation, we wish to change the set point X_p of a controlled variable, such as temperature, to a level X_{p+1} so as to minimize the loss during the $(p + 1)$th phase, we must set X_{p+1} equal to $\hat{\theta}_{p+1}$, the optimal value to be expected in the $(p + 1)$th phase from the predictions of past data. In practice it would often not be possible to change a controlled variable X such as temperature directly. This would have to be done indirectly by adjusting some other variable X^* such as steam pressure. In general, we refer to the variable X^* which we can directly adjust as the manipulated variable. Now frequently, a change in the manipulated variable X^* (steam pressure) would not be immediately felt in the controlled variable X (temperature). The effect would build up over a considerable period of time.

It is convenient to suppose that the manipulated variable has been calibrated in units of its ultimate effect on the controlled variable. Then if we turn the steam valve forwards by "5 units", an increase of 5 units in temperature will ultimately result. Suppose that from data available in the pth phase an adjustment x^*_{p+1} can be made in the manipulated variable at a time such that proportions $v_0, v_1, v_2, \ldots,$ of this change are experienced in the controlled variable in phases $p + 1, p + 2, p + 3, \ldots,$ where $v_0 + v_1 + \cdots = 1$. Then we call $v_0, v_1, v_2, \ldots$ the *dynamic weights.*

It is shown in section 5.4 that we can induce the controlled variable X to undergo the optimal adjustment

$$x_{p+1} = \sum_{j=0}^{\infty} w_j e_{p-j}$$

if we adjust the manipulated variable in accordance with

$$x^*_{p+1} = \sum_{j=0}^{\infty} w_j^* e_{p-j}, \tag{2.2.1}$$

where the w_j^*'s are such that the w_j's are the convolution of the controller and dynamic weights,

$$w_j = v_0 w_j^* + v_1 w_{j-1}^* + v_2 w_{j-2}^* + \cdots. \tag{2.2.2}$$

It follows that if we know the optimal w_j's and the dynamic weights v_i, then provided $v_0 \neq 0$ we can compute the w_j^*'s which tell us how to change the manipulated variable X^*.

2.3. Discrete Adaptive Quality Control

In this section we retain the previous symbols but change the context in which they are used. Suppose we are now concerned with some "quality characteristic", such as the concentration of consecutive batches, and that if no steps were taken to control its behaviour, its observed value at the pth phase would be

$$z_p = \theta_p + u_p. \tag{2.3.1}$$

As before, we assume that θ_p follows some stochastic process, in general non-stationary, and u_p is a measurement error. Suppose that our object is to hold θ as closely as possible to some *target value* which, without loss of generality, we can take to be zero. To achieve this, suppose a method is available whereby the mean value of the stochastic process z can be adjusted up or down at will. Suppose that at the pth stage a *total correction* of $-X_p$ is being applied to the mean. Then the quantity actually observed is the apparent deviation from target

$$z_p - X_p = \theta_p - X_p + u_p = \varepsilon_p + u_p = e_p.$$

We wish now to calculate a further adjustment x_{p+1} computed from data in the pth and previous phases so that when the total correction $-X_{p+1} = -(X_p + x_{p+1})$ is applied during the pth phase the actual deviation from target $\varepsilon_{p+1} = \theta_{p+1} - X_{p+1}$ will be small.

We now assume a quadratic loss function, that is we suppose that the loss involved through θ being off-target by an amount ε_p is proportional to ε_p^2. On this assumption we must choose x_{p+1} as before so that $E(\varepsilon_{p+1})^2 = E(\theta_{p+1} - X_{p+1})^2$ is minimized. Once more than, this requires that

$$x_{p+1} = \hat{\theta}_{p+1} - \hat{\theta}_p = \sum_{j=0}^{\infty} w_j e_{p-j},$$

where the w_j's are chosen so that $\hat{\theta}_{p+1}$ is the minimum mean square error estimate of θ_{p+1}. As before, if the measurement errors are uncorrelated with each other and with the θ's, then $\hat{z}_{p+1} = \hat{\theta}_{p+1}$ and we can then take

$$x_{p+1} = \hat{z}_{p+1} - \hat{z}_p = \sum_{j=0}^{\infty} w_j e_{p-j},$$

where now we choose the w's so that $\hat{z}_{p+1} = \hat{\theta}_{p+1}$ is the minimum mean square error estimate of z_{p+1}.

Finally, suppose the mean value could only be controlled indirectly via a manipulated variable. Then, as before, with

$$x_{p+1} = \sum_{j=0}^{\infty} w_j e_{p-j}$$

denoting the adjustment induced into the controlled variable by a correction

$$x_{p+1}^* = \sum_{j=0}^{\infty} w_j^* e_{p-j}$$

directly applied to the manipulated variable, we should obtain optimal control by choosing the w^*'s so that the weights

$$w_j = \sum_{i=0}^{\infty} v_i w_{j-1}^*$$

were those required for optimal prediction of θ_{p+1}.

2.4. Relation Between Optimization, Control and Prediction Problems

We have seen that with the assumptions and simplifications introduced, the adaptive optimization and adaptive control problems are identical. Furthermore, in both cases optimal action on data accumulated during the pth and previous phases is taken when X_p is adjusted either directly or indirectly to a value

$$X_{p+1} = X_p + x_{p+1} = \hat{\theta}_{p+1},$$

where $\hat{\theta}_{p+1}$ is the mean square error predictor of θ_{p+1}.

In Table 1, the roles which the various elements play in the three problems are summarized and in Fig. 2 the simultaneous behaviour of the three series z_p, X_p, ε_p is displayed for a typical situation. By reference to Table 1 we can interpret Fig. 2 for each problem.

In the prediction problem the z's are directly observed and the predictor $\hat{\theta}_{p+1}$ can be conveniently calculated from

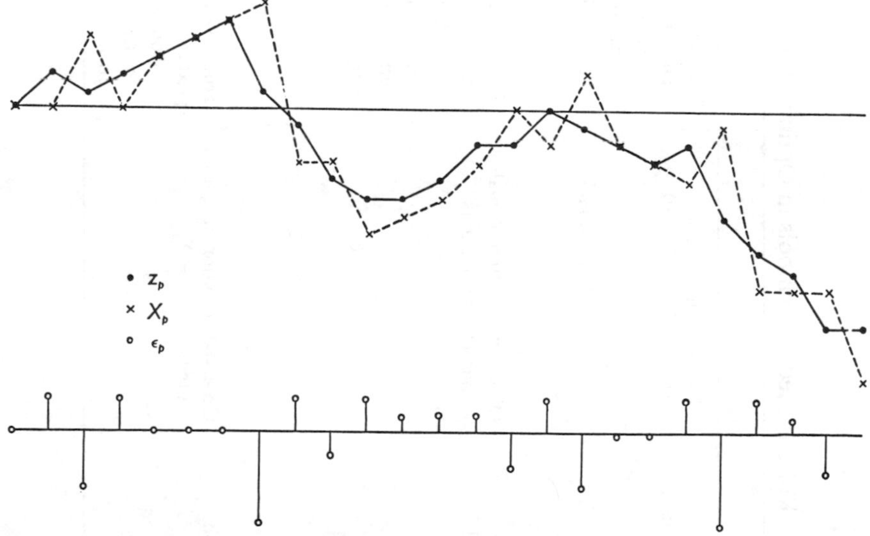

Figure 2. Typical control and associated prediction problem.

Table 1. Roles Played by Symbols in Optimization, Control and Prediction Problems.

	e_p	$\varepsilon_p = e_p - u_p$	z_p	$\theta_p = z_p - u_p$	X_p	x_{p+1}
Adaptive Optimization	Observed slope $\dfrac{\bar{y}(X_p + \delta) - \bar{y}(X_p - \delta)}{2\delta\beta_{11}}$ measures $\theta_p - X_p + u_p$, distance of X_p from max.; subject to error u_p	True slope, uncontaminated by measurement error u_p, measures $\theta_p - X_p$ distance of X_p from max.	Inferred position of max, can be calculated from $X_p + e_p$	Hypothetical true position of max, uncontaminated by error u_p	Set point of controlled variable	Calculated adjustment to be applied to set point X_p throughout pth phase
Adaptive control	Observed deviation of quality characteristic from its target	True deviation from target uncontaminated by measurement error u_p	Inferred value of quality characteristic if no control had been applied; can be calculated from $X_p + e_p$.	True value of quality characteristic if no control had been applied	Total correction applied negatively to mean	Calculated correction to be applied negatively to mean throughout the $(p + 1)$th phase
Prediction	Observed deviation of z_p from prediction $\hat{\theta}_p = X_p$	True deviation from prediction uncontaminated by measurement error u_p	Directly observed characteristic	True value of characteristic uncontaminated by measurement error	Predicted value $\hat{\theta}_p$	Difference in predicted values $\hat{\theta}_{p+1} - \hat{\theta}_p$

$$\hat{\theta}_{p+1} = \sum_{j=0}^{\infty} \mu_j z_{p-j}, \tag{2.4.1}$$

where the μ's are constants suitably chosen to minimize $E(\varepsilon_{p+1}^2) = E(\theta_{p+1} - \hat{\theta}_{p+1})^2$.

In the optimization and control problems where optimal action is taken by setting X_{p+1} equal to $\hat{\theta}_{p+1}$ we do not observe the z's directly but only the e's. Now since $z_p = \hat{\theta}_p + e_p$ we could, as each new observation came to hand, reconstruct the z's and so use (2.4.1) to calculate the value $\hat{\theta}_{p+1}$. In practice, it is simpler to calculate the optimal adjustments $x_{p+1} = X_{p+1} - X_p = \hat{\theta}_{p+1} - \hat{\theta}_p$ from the directly observed e's according to

$$\hat{\theta}_{p+1} - \hat{\theta}_p = \sum_{j=0}^{\infty} w_j e_{p-j}, \tag{2.4.2}$$

where the weights w_j are functions of the μ's given by the recurrence relations

$$w_j = \mu_j - \mu_{j-1} + \sum_{k=0}^{j-1} \mu_k w_{j-1-k}. \tag{2.4.3}$$

To derive these relations we introduce the backward shift operator B such that $B^j z_p = z_{p-j}$. Then (2.4.1) may be written

$$\hat{\theta}_{p+1} = \sum_{j=0}^{\infty} \mu_j B^j z_p = \mathscr{L}_\mu(z_p), \tag{2.4.4}$$

where $\mathscr{L}_\mu$ is the linear operator on the z_p which produces $\hat{\theta}_{p+1}$. Similarly we may rewrite (2.4.2) in the form

$$(1 - B)\hat{\theta}_{p+1} = \sum_{j=0}^{\infty} w_j B^j (z_p - \hat{\theta}_p),$$

so that

$$\left(1 - B + B \sum_{j=0}^{\infty} w_j B^j\right) \hat{\theta}_{p+1} = \sum_{j=0}^{\infty} w_j B^j z_p. \tag{2.4.5}$$

Denoting the linear operator

$$\sum_{j=0}^{\infty} w_j B^j$$

by $\mathscr{L}_w$, we obtain from (2.4.4) and (2.4.5) that

$$\mathscr{L}_w = (1 - B + B\mathscr{L}_w)\mathscr{L}_\mu. \tag{2.4.6}$$

Since the linear operators are infinite series in powers of B, it follows that this equality implies equality of every term in these series. If we identify coefficients of B^j on both sides of (2.4.5) we are led to the recurrence relationship (2.4.3) between the controller weights w_j and the predictor weights μ_j. Some examples of this relationship are given in Table 2.

Table 2. Some Examples of Relationships Between Predictor and Controller Weights.

Nature of weights	j	0	1	2	···	j
Exponentially weighted mean (proportional control)	μ_j	λ	$\lambda(1-\lambda)$	$\lambda(1-\lambda)^2$		$\lambda(1-\lambda)^j$
	w_j	λ	0	0		0
First order autoregressive	μ_j	a_1	0	0		0
	w_j	a_1	$-a_1(1-a_1)$	$-a_1^2(1-a_1)$		$-a_1^j(1-a_1)$
First order moving average (first difference control)	μ_j	β_1	$-\beta_1^2$	β_1^3		$(-1)^{j+1}\beta_1$
	w_j	β_1	$-\beta_1$	0		0

Example 1 corresponds to applying a weight λ to the current deviation and ignoring all the others and may be seen to correspond to the use of predictor weights which are exponentially weighted moving averages. These have been used with considerable success for the prediction of sales and production figures by Holt *et al.* (1960), Brown (1959), Muth (1960) and Cox (1961). The second and third examples take their names from the fact that they are best in a sense to be described in section 3 when the z_p process is of the type specified.

3. Choice of Weights

3.1. Choice of Predictor

In this section we shall consider various specifications of the series z_p and how to choose the corresponding predictor weights with desirable properties. The associated controller weights obtained from the recurrence relation (2.4.3) will, of course, have the same desirable properties. The controller weights may be obtained directly but we adopt the present development via prediction partly because the prediction problem is of considerable interest in its own right.

As before, we suppose that the $(p-j)$th observation may be represented by

$$z_{p-j} = \theta_{p-j} + u_{p-j}, \tag{3.1.1}$$

where u_{p-j} is a measurement error and our object is to predict θ_{p+1} from previous values of z.

If our theory is to have practical value it would be unrealistic to assume that θ_{p-j} followed a stationary process. Instead we suppose that θ_{p-j} has a mean $E(\theta_{p-j}) = m_{p-j}$ which may vary. Our model can then be written

$$z_{p-j} = m_{p-j} + (\theta_{p-j} - m_{p-j}) + u_{p-j}. \tag{3.1.2}$$

The series $(\theta_{p-j} - m_{p-j})$ is assumed to be stationary and the $h \times h$ covariance matrix $\sigma_\theta^2 \mathbf{R}_h$ for h successive observations from this series is a Toeplitz matrix with elements $\{\rho_{|i-j|}\}\sigma_\theta^2$, where ρ_s is the autocorrelation of the θ's of lag s. The corresponding $h \times h$ covariance matrix for the u's, which we also assume stationary with mean zero, is denoted by $\sigma_u^2 \mathbf{S}_h$. The covariance matrix for the z's is thus $\sigma_\theta^2 \mathbf{T}_h$, where

$$\mathbf{T}_h = \mathbf{R}_h + \frac{\sigma_u^2}{\sigma_\theta^2} \mathbf{S}_h.$$

If the mean value $E(\theta_{p-j}) = m_{p-j}$ were known for every j then we should employ for our predictor of θ_{p+1} the quantity

$$m_{p+1} + \sum_{j=0}^{\infty} \mu_j \{z_{p-j} - m_{p-j}\}$$

and our prediction error would be

$$\varepsilon_{p+1} = (\theta_{p+1} - m_{p+1}) - \sum_{j=0}^{\infty} \mu_j (z_{p-j} - m_{p-j}). \tag{3.1.3}$$

Now in most practical circumstances the predictor will place negligible weight on values of the series remote from the present time. Suppose all values of μ_j can be supposed zero for all $j \geqslant h$; then the predictor becomes

$$m_{p+1} + \sum_{j=0}^{h-1} \mu_j \{z_{p-j} - m_{p-j}\}.$$

The mean square error $E(\varepsilon_{p+1})^2 = \mathrm{var}(\varepsilon_{p+1})$ may then be written in matrix form as

$$\mathrm{var}(\varepsilon_{p+1}) = \sigma_\theta^2 \{1 - 2\boldsymbol{\mu}'\boldsymbol{\rho} + \boldsymbol{\mu}'\mathbf{T}\boldsymbol{\mu}\}, \tag{3.1.4}$$

where

$$\boldsymbol{\mu}_h' = (\mu_0, \mu_1, \ldots, \mu_{h-1})$$

and

$$\boldsymbol{\rho}_h' = (\rho_1, \rho_2, \ldots, \rho_h).$$

Differentiating with respect to the vector $\boldsymbol{\mu}$, we find that the mean square error is minimized when

$$\boldsymbol{\mu} = \mathbf{T}^{-1}\boldsymbol{\rho}. \tag{3.1.5}$$

The optimal predictor of θ_{p+1} is thus

$$\hat{\theta}_{p+1} = \mathbf{z}_p' \mathbf{T}^{-1} \boldsymbol{\rho}, \tag{3.1.6}$$

where

$$\mathbf{z}'_p = (z_p, z_{p-1}, \ldots, z_{p-h+1})$$

and this has variance $\boldsymbol{\rho}'\mathbf{T}^{-1}\boldsymbol{\rho}\sigma_\theta^2$. If in particular there were no measurement error and the z process, and hence the θ process, followed a kth order autoregressive scheme,

$$z_{p+1} = a_1 z_p + a_2 z_{p-1} + \cdots + a_{k+1} z_{p-k} \qquad (k \leqslant h)$$

then we obtain the well-known result that $\boldsymbol{\mu} = \mathbf{a}$ where $\mathbf{a}' = (a_0, a_1, \ldots, a_{k+1})$.

3.2. Constrained Predictors

In most practical situations the mean m_{p-j} would not be known. If we employed the quantity $\mu_0 z_p + \mu_1 z_{p-1} + \ldots$ as a predictor of θ_{p+1} then the error of prediction would be

$$\varepsilon_{p+1} = (\theta_{p+1} - m_{p+1}) - \sum_{j=0}^{h-1} \mu_j(z_{p-j} - m_{p-j}) + \left(m_{p+1} - \sum_{j=0}^{h-1} \mu_j m_{p-j}\right) \qquad (3.2.1)$$

and the mean square error,

$$E(\varepsilon_{p+1}^2) = \left(m_{p+1} - \sum_{j=0}^{h-1} \mu_j m_{p-j}\right)^2 + \operatorname{var}(\varepsilon_{p+1}), \qquad (3.2.2)$$

would in general be inflated by an amount which depended on the behaviour of the mean m_{p-j}. One way in which this difficulty might be overcome would be to constrain the predictor weights μ_j so that the bias term

$$m_{p+1} - \sum_{j=0}^{h-1} \mu_j m_{p-j}$$

was eliminated or at least rendered small.

For example, if at the pth observation it was assumed that the mean, although unknown, was constant over the last h observations and would remain constant for one further observation, so that $m_{p+1} = m_p = m_{p-1} = \cdots = m_{p-h+1} = m$, then the bias term would vanish if the μ's were chosen subject to the restriction

$$\sum_{j=0}^{h-1} \mu_j = 1. \qquad (3.2.3)$$

As observed by Cox (1961), by introducing this constraint we endow the predictor with the property of adjusting to the mean of the series. This is a property which is, of course, not possessed in general by the unrestricted Wiener predictors. For example, for predicting an autoregressive series the optimal weights $\boldsymbol{\mu} = \mathbf{a}$ will not in general sum to unity. This is because the existence of a known fixed mean is implicit in the assumption of an autoregressive series. For reasons which will be clear later, we would call the predictor constrained by (3.2.3) a *mean-projecting* predictor. More generally, if we can represent the behaviour of the $h + 1$ values

$$m_{p+1}, m_p, m_{p-1}, \ldots, m_{p-h+1}$$

exactly by a polynomial of degree $0 \leqslant d \leqslant h - 1$ then the bias term would vanish if

$$\sum_{j=0}^{h-1} j^k \mu_j = (-1)^k \qquad (k = 0, 1, 2, \ldots, d). \tag{3.2.4}$$

Constrained predictors so obtained may be referred to as linear trend projecting, quadratic trend projecting, and so on.

Suppose then that the $h + 1$ values $m_{p+1}, \ldots, m_{p-h+1}$ are exactly represented by a polynomial of degree $d < h - 1$. Then since the bias term is now zero, the mean square error will be minimized by minimizing $\text{var}(\varepsilon_{p+1})$ subject to the constraints (3.2.4).

These constraints may be written

$$\mathbf{A}'\boldsymbol{\mu} = \mathbf{c}, \tag{3.2.5}$$

where $\mathbf{A}$ is an $h \times (d + 1)$ matrix $\{a_{jk}\} = \{j^k\}$ with $j = 0, 1, \ldots, h - 1; k = 0, 1, \ldots, d$, and $\mathbf{c}$ is a $(d + 1) \times 1$ vector $\{(-1)^k\}$, $k = 0, 1, \ldots, d$. Introducing a vector of Lagrange multipliers $\boldsymbol{\lambda}' = (\lambda_0, \lambda_1, \ldots, \lambda_d)$, we have that the minimum mean square error is given by the unconditional minimum of

$$\{1 + 2\boldsymbol{\mu}'\boldsymbol{\rho} + \boldsymbol{\mu}'\mathbf{T}\boldsymbol{\mu}\} - 2(\boldsymbol{\mu}'\mathbf{A} - \mathbf{c})\boldsymbol{\lambda}. \tag{3.2.6}$$

On equating derivatives to zero, we obtain the simultaneous equations

$$\boldsymbol{\mu} = \mathbf{T}^{-1}\boldsymbol{\rho} + \mathbf{T}^{-1}\mathbf{A}\boldsymbol{\lambda}, \qquad \mathbf{A}'\boldsymbol{\mu} = \mathbf{c}. \tag{3.2.7}$$

On multiplying the first equation by $\mathbf{A}'$ and substituting the result in the second, we obtain

$$\boldsymbol{\lambda} = (\mathbf{A}'\mathbf{T}\mathbf{A})^{-1}\{\mathbf{c} - \mathbf{A}'\mathbf{T}^{-1}\boldsymbol{\rho}\}; \tag{3.2.8}$$

hence finally

$$\boldsymbol{\mu}_c = \mathbf{T}^{-1}\boldsymbol{\rho} + \mathbf{T}^{-1}\mathbf{A}(\mathbf{A}'\mathbf{T}^{-1}\mathbf{A})^{-1}(\mathbf{c} - \mathbf{A}'\mathbf{T}^{-1}\boldsymbol{\rho}) \tag{3.2.9}$$

is the vector of optimal constrained predictor weights.

The corresponding predictor $\boldsymbol{\mu}_c'\mathbf{z}$ has variance

$$\{\boldsymbol{\rho}'\mathbf{T}^{-1}\boldsymbol{\rho} + (\mathbf{c} - \mathbf{A}'\mathbf{T}^{-1}\boldsymbol{\rho})'(\mathbf{A}'\mathbf{T}\mathbf{A})^{-1}(\mathbf{c} - \mathbf{A}'\mathbf{T}^{-1}\boldsymbol{\rho})\}\sigma_\theta^2, \tag{3.2.10}$$

where the second term in the curly brackets represents the inflation of the variance which occurs due to the constraints.

Given that the mean m_{p-j} follows a polynomial of the type assumed, this predictor $\boldsymbol{\mu}_c$ provides in fact the minimum variance linear unbiased predictor of θ_{p+1} as has been shown in an unpublished report by E. Parzen. We refer also to some earlier work on constrained predictors by Zadeh and Ragazzini (1950).

The predictor $\boldsymbol{\mu}_c$ has an alternative interpretation. Using it, the predicted value of θ_{p+1} may be written

$$\boldsymbol{\mu}_c'\mathbf{z} = \mathbf{c}'(\mathbf{A}'\mathbf{T}^{-1}\mathbf{A})^{-1}\mathbf{A}'\mathbf{T}^{-1}\mathbf{z} + \boldsymbol{\rho}\mathbf{T}^{-1}\{\mathbf{z} - \mathbf{A}(\mathbf{A}'\mathbf{T}^{-1}\mathbf{A})^{-1}\mathbf{A}'\mathbf{T}^{-1}\mathbf{z}\}. \quad (3.2.11)$$

Now we have seen that if the mean was known we should use as the predictor for θ_{p+1} the quantity

$$m_{p+1} + \boldsymbol{\mu}'(\mathbf{z} - \mathbf{m}) \qquad \text{with } \boldsymbol{\mu}' = \boldsymbol{\rho}'\mathbf{T}^{-1}, \quad (3.2.12)$$

where $\mathbf{m}' = (m_p, m_{p-1}, \ldots, m_{p-h+1})$.

If in (3.2.12) we replace $m_{p+1} = \mathbf{c}'\boldsymbol{\beta}$ and $\mathbf{m} = \mathbf{A}'\boldsymbol{\beta}$ by the corresponding quantities where $\hat{\beta}$ is replaced by its best linear unbiased estimate

$$\hat{\boldsymbol{\beta}} = (\mathbf{A}'\mathbf{T}^{-1}\mathbf{A})^{-1}\mathbf{A}'\mathbf{T}^{-1}\mathbf{z},$$

then this is precisely the constrained predictor $\hat{\theta}_{p+1}$ given by (3.2.11).

Suppose, for example, we assume that over the last h observations z can be represented by a mean m subject to a linear trend, plus a stochastic variable $\theta - m$ for which the variance-covariance matrix $\mathbf{R}\sigma_\theta^2$ is known, plus a measurement error for which the variance-covariance matrix $\mathbf{S}\sigma_u^2$ is known.

Then the prediction of θ_{p+1} using the constrained predictor $\boldsymbol{\mu}_c'\mathbf{z}$ would be equivalent to the following:

(i) estimate m_{p+1} by projecting the line which best fits the last h z's through one further observation;
(ii) add to this an estimate of $\theta_{p+1} - m_{p+1}$, the deviation from the projected line at the next observation.

The latter estimate would be a weighted sum of the residuals from the line using the *unconstrained* predictor $\boldsymbol{\mu} = \mathbf{T}^{-1}\boldsymbol{\rho}$.

3.3. A More Practical Predictor

The above approach would be extremely difficult to apply in practice. To use it we would have to choose d and h, the degree of the polynomial and the length of series to which it was to be fitted. We should also need to know the covariance matrices $\mathbf{R}\sigma_\theta^2$ and $\mathbf{S}\sigma_u^2$. We are therefore led to seek an alternative approach which can be applied more readily whilst retaining so far as possible the advantages of the previous system.

We shall suppose in what follows either that the measurement errors u are uncorrelated one with another and with the θ's, so that $\hat{z}_{p+1} = \hat{\theta}_{p+1}$ or that the u's are negligible, in which case $z_{p+1} = \theta_{p+1}$ and again $\hat{z}_{p+1} = \hat{\theta}_{p+1}$.

Table 3 shows the optimal μ and w weights obtained for various series of interest. In each case the series is assumed to be represented over the last h observations by a polynomial of degree d, and the variation about the polynomial is assumed to follow the process indicated. In some instances the errors

Table 3. Some Constrained Predictors and Their Associated Controller Weights.

Description	j	0	1	2	3	4	5	6	7	8	9
1 $d = 0$ $h = 1$	μ_j	1	.	.	.	.	.	.	.	.	.
Errors independent	w_j	1	.	.	.	.	.	.	.	.	.
2 $d = 1$ $h = 2$	μ_j	2	−1	.	.	.	.	.	.	.	.
Errors independent	w_j	2	1	1	1	1	1	1	1	1	1
3 $d = 2$ $h = 3$	μ_j	3	−3	1	.	.	.	.	.	.	.
Errors independent	w_j	3	3	4	5	6	7	8	9	10	11
4 $d = 0$ $h = \infty$	μ_j	γ_0	$\gamma_0(1-\gamma_0)$	$\gamma_0(1-\gamma_0)^2$	$\gamma_0(1-\gamma_0)^3$	.	.	.	.	.	.
Errors independent	w_j	γ_0	0	0	0	0	0	0	0	0	0
5 $d = 1$ $h = 4$	μ_j	1.00	.50	.00	−.50	.00	.00	.00	.00	.00	.00
Errors independent	w_j	1.00	.50	.50	.25	.38	.38	.44	.38	.38	.38
6 $d = 1$ $h = 4$	μ_j	1.14	.27	.04	−.45	.00	.00	.00	.00	.00	.00
First order A. R.											
($a_1 = 0.3$)	w_j	1.14	.43	.57	.32	.47	.45	.40	.45	—	—
7 $d = 1$ $h = 4$	μ_j	1.22	.15	.04	−.41	.00	.00	.00	.00	.00	.00
First order A. R.											
($a_1 = 0.5$)	w_j	1.22	.42	.58	.37	.47	.48	.43	.47	—	—
8 $d = 1$ $h = 4$	μ_j	1.33	.01	.00	−.34	.00	.00	.00	.00	.00	.00
First order A. R.											
($a_1 = 0.9$)	w_j	1.33	.45	.60	.46	.51	.53	.51	.52	—	—
9 $d = 1$ $h = 6$	μ_j	0.67	.47	.27	.07	−.13	−.33	.00	.00	.00	.00
Errors independent	w_j	0.67	.24	.27	.27	.22	.08	.32	.21	.19	.20

are supposed independent and in others that they follow a first order autoregressive series with coefficient a.

The first three series in the table are all examples of what may be called fully saturated predictors. We fit a polynomial of degree d to the last $h = d + 1$ observations. The weight μ_j is then given by minus the coefficient of t^{j+1} in $(1 - t)^{d+1}$. The form of the weights w_j is of special interest. In general, the change in the predictor $\hat{z}$ at the $(p + 1)$th stage may be written as in (2.4.2) in the form

$$\Delta\hat{z}_{p+1} = \sum_{j=0}^{\infty} w_j e_{p-j}$$

and for the first three series in Table 3 this becomes

series 1, $\Delta\hat{z}_{p+1} = e_p$;

series 2, $\Delta\hat{z}_{p+1} = e_p + S^1 e_p$;

series 3, $\Delta\hat{z}_{p+1} = e_p + S^1 e_p + S^2 e_p$,

where

$$S^1 e_p = \sum_{j=0}^{\infty} e_{p-j}, \qquad S^2 e_p = \sum_{j=0}^{\infty} \sum_{k=0}^{\infty} e_{p-j-k},$$

and in general $S^j e_p$ denotes the jth multiple sum over the past history of the e's. For a dth degree polynomial fitted to the last $h = d + 1$ observations we would have

$$\Delta\hat{z}_{p+1} = e_p + S^1 e_p + S^2 e_p + \cdots + S^d e_p.$$

We now recall that a mean projecting predictor which has proved to be of great practical value is the exponentially weighted mean, shown here as series 4. Technically, $h = \infty$ for this predictor, but in practice the exponential weights become negligible after a moderate number of observations so that local changes in mean are satisfactorily followed. For this series

$$\Delta\hat{z}_{p+1} = \gamma_0 e_p.$$

This important mean projecting series does not fit naturally into our previous formulation but it is now seen to occur as a simple generalization of series 1.

In series 5, 6, 7, 8 and 9 we have tabulated sets of weights for various linear trend projectors. Bearing in mind that the properties of these series are not very sensitive to moderate changes in the weights, we see that all of them could be approximated by an obvious modification of series 2, namely

$$\Delta\hat{z}_{p+1} = \gamma_0 e_p + \gamma_1 S^1 e_p.$$

Finally, therefore, as a natural generalization, we consider the model where $\Delta\hat{z}_{p+1}$, the change to be applied at the $(p + 1)$th stage to the previous predictor $\hat{z}_p$, is given by

$$\Delta\hat{z}_{p+1} = (\gamma_{-l}S^{-l} + \cdots + \gamma_{-2}S^{-2} + \gamma_{-1}S^{-1} + \gamma_0 + \gamma_1 S^1 + \gamma_2 S^2 + \cdots + \gamma_m S^m)e_p, \tag{3.3.1}$$

where $S^{-j}e_p = \Delta^j e_p$.

3.4. The Nature of the Stochastic Process for Which the Predictor Is Optimal

We now ask the question: what stochastic process would z need to follow for such a predictor to be optimal? Consider a process generated by

$$z_{p+1} = \sum_{j=0}^{\infty} \eta_j z_{p-j} + \alpha_{p+1} \tag{3.4.1}$$

which is, in general, non-stationary and where $\alpha_{p+1}, \alpha_p, \alpha_{p-1}, \ldots$ are uncorrelated identically distributed random variables, with mean zero. Suppose we predict the series using

$$\hat{z}_{p+1} = \sum_{j=0}^{\infty} \mu_j z_{p-j}. \tag{3.4.2}$$

Then since

$$E(e_{p+1})^2 = E(z_{p+1} - \hat{z}_{p+1})^2 = E\left\{\sum_{j=0}^{\infty} (\eta_j - \mu_j) z_{p-j}\right\}^2 + E(\alpha_{p+1})^2, \tag{3.4.3}$$

prediction is optimal if $\mu_j = \eta_j$ ($j = 0, 1, 2, \ldots$). It follows that the equivalent predictor

$$\Delta\hat{z}_{p+1} = \sum_{j=0}^{\infty} w_j e_{p-j}$$

is optimal for the equivalent stochastic process

$$\Delta z_{p+1} = \Delta\alpha_{p+1} + \sum_{j=0}^{\infty} w_j \alpha_{p-j}$$

and that when this optimal predictor is used the e's become α's and are uncorrelated.

In the present instance then our predictor is optimal for a series generated by

$$\Delta z_{p+1} = \Delta\alpha_{p+1} + \sum_{j=-l}^{m} \gamma_j S^j \alpha_p. \tag{3.4.4}$$

Differencing m times we have

$$\Delta^{m+1} z_{p+1} = \Delta^{m+1}\alpha_{p+1} + \sum_{j=0}^{l+m} \gamma_j \Delta^{l+m-j}\alpha_p.$$

On rearranging this expression, we find that our predictor would be optimal for any stochastic variable z whose $(m + 1)$st difference may be represented by a moving average process of order $l + m + 1$,

$$\Delta^{m+1} z_{p+1} = \alpha_{p+1} + \sum_{j=0}^{l+m} \delta_j \alpha_{p-j}, \tag{3.4.5}$$

so that all serial covariances after that of lag $l + m + 1$ are zero. Thus we have a result which is of considerable practical value. If, after differencing our series z, which in general will be non-stationary, m times, we could render it stationary and if the population serial covariances of lag greater than some value $l + m + 1$ were then zero, a predictor of the type (3.3.1) would be optimal.

The widely applied predictor obtained by taking an exponentially weighted mean

$$\hat{z}_{p+1} = \gamma_0 \sum_{j=0}^{\infty} (1 - \gamma_0)^j z_{p-j}$$

corresponds simply to taking the single central term in our general series namely $\Delta\hat{z}_{p+1} = \gamma_0 e_p$. It is optimal for the stochastic process

$$\Delta z_{p+1} = \Delta\alpha_{p+1} + \gamma_0 \alpha_p = \alpha_{p+1} - (1 - \gamma_0)\alpha_p,$$

i.e.

$$z_{p+1} = m + \alpha_{p+1} + \gamma_0 S^1 \alpha_p,$$

for which the first difference is a first order moving average. The addition of further terms can be thought of, therefore, as an appropriate generalization of this exponential predictor. In particular, for series which are highly non-stationary and exhibit marked trends, the additional term in $S^1 e_p$ will be of particular value since this will allow the predictor to adjust to changes in linear trend as well as to changes in mean.

Bearing in mind the great success of the exponential predictor, it might be expected that the simple generalization

$$\Delta\hat{z}_{p+1} = \gamma_{-1}\Delta e_p + \gamma_0 e_p + \gamma_1 S e_p \tag{3.4.6}$$

might be adequate for many practical purposes. Experience of two kinds indicates that this is so. Our own somewhat limited experience in applying this theory to industrial series has shown that for those series so far tried, this generalization has been adequate. In fact, so far as *prediction* is concerned, the term in Δe_p has not so far been needed. A further vast fund of experience in this area is possessed by control engineers. We have seen already that if there were no dynamics then the adjustment x_{p+1} of the control set point should be made equal to $\Delta\hat{z}_{p+1}$ the predicted change. A form of automatic control commonly used in industrial plants in continuous time makes a correction proportional to a linear combination of (i) the first derivative of the current deviation, (ii) the deviation itself, (iii) the integral of the deviations over all past history. If, therefore, we were using our predictor (3.4.6) for control purposes, we would employ a discrete time analogue of what control engineers have been using on automatic equipment for many years. The success of their efforts suggests that stochastic processes for which (3.4.6) is optimal adequately describe many industrial series.

The types of continuous control mentioned above are called *derivative*, *proportional* and *integral* respectively. For the discrete process we shall refer to the corresponding terms in (3.4.6) as *first difference*, *proportional* and *cumulative* terms and when we use the prediction equation for control purposes we shall talk of first difference control, proportional control and cumulative control. We shall adopt the three-term model in what follows bearing in mind that it is readily elaborated if need be.

(*Editor's note*: Section 3.5 has been omitted.)

3.6. Representation of Dynamics by Exponential Stages

Suppose the control dynamics can be represneted by an exponential transfer function so that $v_i = (1 - \phi)\phi^i$ $(i = 0, 1, 2, \ldots)$.

From (2.2.2) it may be shown that to induce action in the controlled variable according to

$$x_{p+1} = \gamma_{-1}\Delta e_p + \gamma_0 e_p + \gamma_1 S^1 e_p \qquad (3.6.1)$$

an adjustment

$$x_{p+1}^* = \gamma_{-2}^*\Delta^2 e_p + \gamma_{-1}^*\Delta e_p + \gamma_0^* e_p + \gamma_1^* S^1 e_p \qquad (3.6.2)$$

must be applied to the manipulated variable, where $\gamma_{-2}^* = k\gamma_{-1}$, $\gamma_{-1}^* = \gamma_{-1} + k\gamma_0$, $\gamma_0^* = \gamma_0 + k\gamma_1$, $\gamma_1^* = \gamma_1$ and $k = \phi/(1 - \phi)$. In other words the matrix $\mathbf{T}$, which transforms the γ's into γ^*'s when exponential dynamics apply, is given by the matrix relation

$$\boldsymbol{\gamma}^* = \mathbf{T}\boldsymbol{\gamma} \qquad (3.6.3)$$

with

$$\mathbf{T} = \begin{bmatrix} k & \cdot & \cdot \\ 1 & k & \cdot \\ \cdot & 1 & k \\ \cdot & \cdot & 1 \end{bmatrix}. \qquad (3.6.4)$$

The corresponding matrix $\mathbf{T}$ appropriate for the more general control model (3.3.1) is of the same form. We see that to compensate for exponential dynamics we must add difference control to the manipulated variable of one order higher than we require to be applied to the controlled variable. Furthermore, if we know

(i) the γ's required for optimal prediction,

(ii) the constant k of the exponential transfer, then the γ^*'s, which determine the appropriate combination of the various kinds of control to be applied to the manipulated variable, are very easily obtained.

It frequently happens in practice that only the two terms involving γ_0 and γ_1 which produce proportional and cumulative action are needed for near-optimal prediction (and hence, in the absence of dynamics, for near-optimal control). Yet as soon as we must compensate for simple exponential dynamics our third term γ_{-1} will be equal to $k\gamma_0$ and so difference action is needed.

In some cases the dynamics are such that more than one exponential stage is involved. If the action x_{p+1} which it was desired to induce at the controlled variable was represented by an expression involving c terms and the dynamic behaviour was as if s exponential stages were involved with constants $k_1, k_2, \ldots, k_s$, then control involving differences of s orders higher would be required at the manipulated variable. If $\mathbf{T}_i$ is the $(c + i) \times (c + i - 1)$ transfer matrix for the ith stage of the same form as (3.6.2) but containing the constant k_i, then the $(c + s) \times c$ matrix $\mathbf{T}$ representing the overall transfer matrix is given by $\mathbf{T} = \mathbf{T}_1\mathbf{T}_2\ldots\mathbf{T}_s$. For example, if an exponentially weighted mean with constant γ_0 adequately predicts the next observation, so that simple proportional control according to $x_{p+1} = \gamma_0 e_p$ is required to be applied to the controlled variable, and if the dynamics can be represented by two exponential stages with constants k_1 and k_2, then $\mathbf{T}' = [k_1 k_2, k_1 + k_2, 1]$ and we control the manipulated variable in acccordance with

$$x_{p+1}^* = k_1 k_2 \gamma_0 \Delta^2 e_p + (k_1 + k_2)\gamma_0 \Delta e_p + \gamma_0 e_p.$$

4. Examples

4.1. A Study of Three Series

In Fig. 3 are shown two or three hundred observations from each of three series associated with typical chemical operations. The series show values respectively of concentration, temperature and viscosity when no control was applied. Each series consists of discrete observations although the plotted points are joined in this figure by lines. Series II was obtained by temporarily disconnecting an automatic temperature controller, series I and III were obtained by reconstructing the uncontrolled series. In each of these latter cases some manual control had been employed. From (i) the values for the controlled series, (ii) the control action taken, and, where appropriate, (iii) the approximate system dynamics, the course which the series would have taken had no control been exerted could be calculated to a sufficient degree of approximation. The concentration of active ingredient in successive batches shown in series I was controlled by the addition of the appropriate amount of pure chemical on a batch-to-batch basis. In this example, therefore, no

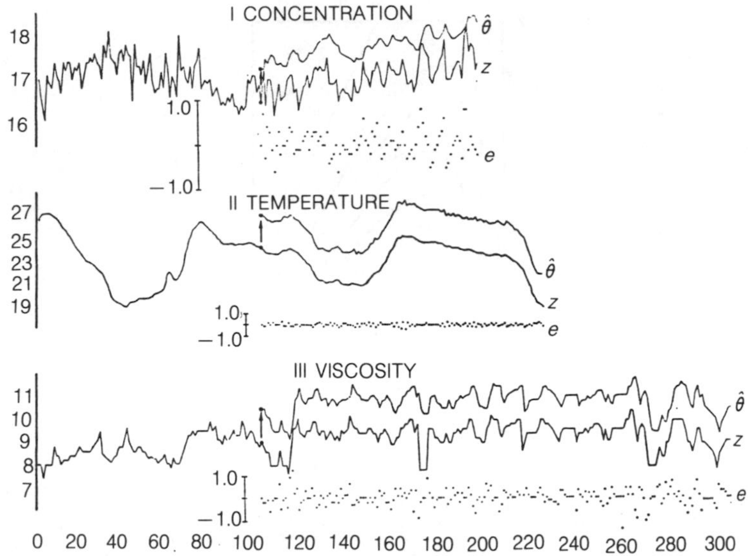

Figure 3. Three uncontrolled industrial series, predicted series θ_D, and error e_D shown for latter sections.

dynamics had to be considered in reconstructing the series. The dynamics appropriate to series III are discussed later. It was found in each case that the serial covariances after the third are small, so that we feel justified in proceeding with our analysis based on the use of difference, proportional and cumculative control.

In Fig. 4 contour plots of the sums of squares surfaces are shown for different values of γ_0 and γ_1 when $\gamma_{-1} = 0$.

These diagrams show that:

(i) The choice of the coefficients is not very critical. A sum of squares very little greater than the smallest value can be obtained over a fairly wide area of the $(\gamma_{-1}, \gamma_0, \gamma_1)$ space.

(ii) For none of the series is there any strong indication that a difference term is necessary. Formal interpolation among the calculated sums of squares gives a minimum in each case in which γ_{-1} is slightly different from zero. However, the increase produced by setting $\gamma_{-1} = 0$ is entirely negligible.

(iii) For series I the simple proportional predictor

$$x_{p+1} = 0.3e_p$$

is near optimal. This corresponds to the prediction of the next value in the series by an exponentially weighted mean of previous observations. The small value $\gamma_0 = 0.30$ ensures that the exponential μ weights (0.30, 0.21, 0.15, 0.10, 0.07, etc.) die out rather slowly. This is to be expected because of the rather noisy nature of the series. In general, the choice of

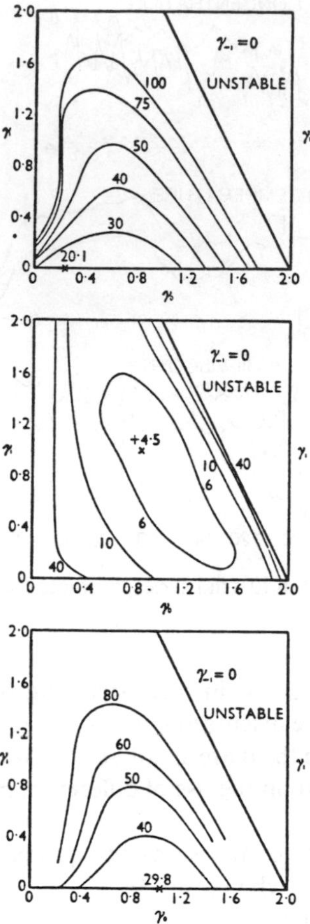

Figure 4. Sums of squares contours for series I–III.

the constant in a proportional (i.e. exponentially weighted) predictor is a compromise between the desirability of emphasis on "up-to-date values", obtained by choosing γ_0 close to unity and the desirability of "smoothing out the noise" obtained by taking a smaller value of γ_0.

(iv) For series II the necessity for cumulative control in addition to proportional control is clearly seen, the near-optimal predictor being

$$x_{p+1} = 1.1e_p + 0.8 \sum_{j=0}^{\infty} e_{p-j}.$$

Almost equally good prediction would be obtained by setting both γ_0 and γ_1 equal to unity. The corresponding μ weights would then be $\mu_0 = 2$, $\mu_1 = -1$ with all other values zero. The predictor $X_{p+1} = 2z_p - z_{p-1}$ is that obtained by projecting a straight line drawn through the last two observations and would be optimal when the second differences were independent random deviates. This type of prediction would be expected to be good for the present series (Fig. 3) in which linear trend extrapolation is clearly desirable and where the superimposed noise is small. We notice as expected that for this series the serial covariances of the second differences are *all* small in magnitude.

(v) For series III a near optimal predictor is $x_{p+1} = e_p$. This means that $X_{p+1} = z_p$ so that the previous observation is used to predict the next. Since an important characteristic of series III is the predominance of step changes, this is clearly sensible.

4.2. Predicting the Series

To illustrate the extent to which series of the various kinds can be predicted, we have applied predictors with optimal weights obtained from the first hundred values of each series to forecast values in the remainder. The predicted values are plotted above the second part of each series in Fig. 3. To avoid confusion between the original values and the predicted values, the latter have been moved upwards by the amount indicated and plotted above the original series. The errors of prediction are indicated by the series of dots plotted below each series. Table 4 shows for the first and the second part of each series, (i) the serial correlations of second differences, (ii) the optimal weights, (iii) the reduction in variance achieved by control of the second part of each series using weights calculated from the first part. Although no formal tests have been made there seems reason to doubt strict homogeneity particularly in the first series. The optimal weights for the first and second parts of each series agree remarkably well, however.

(*Editor's note*: The remainder of Section 4 has been omitted.)

5. Spectral Approach in Discrete Time

5.1. Basic Spectral Formulae

In this section we shall present some of the previous results in spectral form and thus prepare the way for a more realistic approach to the adaptive optimization problem to be considered in a later paper.

We have shown that the error ε_p could be expressed as a linear function over past history of the random variables θ_p and u_p. We represent such a linear function by

$$T_l = \sum_{j=0}^{\infty} \phi_j^{(l)} \theta_{p-j}, \tag{5.1.1}$$

Table 4. Serial Correlations ρ_s and Optimal Control Weights for Series I-III.

	Series I			Series II			Series III		
	1st 100 obs.	2nd 97 obs.	Total 197 obs.	1st 100 obs.	2nd 126 obs.	Total 226 obs.	1st 100 obs.	2nd 210 obs.	Total 310 obs.
Var(Δz^2)	.57	.22	.39	.032	.012	.020	17.7	22.2	20.7
ρ_1	−.69	−.53	−.67	.09	−.39	−.09	−.64	−.46	−.50
ρ_2	.22	.07	.22	−.18	.11	−.07	.25	−.07	.00
ρ_3	−.03	−.13	−.09	−.15	−.07	−.13	−.26	.08	.00
ρ_4	−.01	.22	.05	−.05	−.08	−.05	.30	−.13	−.01
ρ_5	.01	−.17	−.03	−.03	.10	.02	−.29	.06	−.02
ρ_6	−.06	−.02	−.06	.02	−.05	−.01	.25	−.01	.04
ρ_7	.12	.21	.15	.01	.17	.06	−.26	.06	−.02
ρ_8	−.05	−.24	−.12	−.07	−.05	−.06	.18	−.06	.01
ρ_9	.00	.14	.05	−.09	−.17	−.13	−.02	.03	.01
ρ_{10}	.03	.00	.03	.10	.14	.12	−.05	−.04	−.04
γ_0†	.30	.50	.30	1.20	1.00	1.10	.80	1.00	1.00
γ_1†	.00	.00	.00	.80	.60	.80	.00	.00	.00
Variance‡ before control		.139			2.313			29.0	
Variance‡ after control		.089			.013			11.0	

† Optimal values of γ_0 and γ_1, when $\gamma_{-1} = 0$.
‡ Variances refer to second part of series.

and a simple calculation shows that the covariance between two such linear functions is

$$\text{cov}(T_l, T_m) = \sum_{p=0}^{\infty} \sum_{q=0}^{\infty} \phi_p^{(l)} \phi_q^{(m)} \zeta_{|p-q|}, \tag{5.1.2}$$

where $\zeta_s = \sigma_\theta^2 \rho_s$ is the autocovariance of the θ's of lag s. If the θ's are observed at unit time intervals, then we may write

$$\zeta_s = \sigma_\theta^2 \int_{-\pi}^{+\pi} g_\theta^*(\omega) e^{i\omega s}\, d\omega, \tag{5.1.3}$$

where $g_\theta^*(\omega)$ is the spectral density function, defined in the frequency range $[-\pi, \pi]$ for operational convenience, and is equal to $\frac{1}{2} g_\theta(\omega)$, where $g_\theta(\omega)$ is defined in $[0, \pi]$. If we now substitute for (5.1.3) in (5.1.2) we obtain

$$\text{cov}(T_l, T_m) = \sigma_\theta^2 \int_{-\pi}^{+\pi} g_\theta^*(\omega) H_l(\omega) \overline{H}_m(\omega)\, d\omega, \tag{5.1.4}$$

where

$$H_l(\omega) = \sum_{j=0}^{\infty} \phi_j^{(l)} e^{-ij\omega}$$

is the discrete Fourier transform of the weights $\phi_j^{(l)}$ and the bar denotes a complex conjugate. In the special case where $l = m$, (5.1.4) reduces to

$$\text{var}(T_l) = \sigma_\theta^2 \int_{-\pi}^{+\pi} g_\theta^*(\omega) |H_l(\omega)|^2\, d\omega. \tag{5.1.5}$$

More generally, we may write the variance-covariance matrix of several such linear combinations in the form

$$\sigma_\theta^2 \mathbf{R} = \sigma_\theta^2 \int_{-\pi}^{+\pi} \mathbf{H}(\boldsymbol{\omega}) \overline{\mathbf{H}}(\boldsymbol{\omega})' g_\theta^*(\omega)\, d\omega, \tag{5.1.6}$$

where $\mathbf{H}(\boldsymbol{\omega})' = \{H_1(\omega), H_2(\omega), \ldots, H_k(\omega)\}$ and the matrix $\mathbf{H}(\boldsymbol{\omega})\overline{\mathbf{H}}(\boldsymbol{\omega})'$ is Hermitian.

Equation (5.1.5) says that the variance of any linear combination may be expressed as a *spectral average* with weight function or kernel $|H_l(\omega)|^2$. The generating function $H_l(\omega)$ for the weights $\phi_j^{(l)}$ is important and is called the *frequency response function* of the linear operation performed by (5.1.1), or else the *transfer function* when expressed as a function of $p = i\omega$. In engineering language, (5.1.1) is called a *linear filter* and is characterized by the important property that if the input θ_p is periodic of the form $Ae^{ip\omega}$, then the output is given by $AH(\omega)e^{i\omega p}$, i.e. a periodic disturbance with the same frequency. Since $H(\omega) = G(\omega)e^{i\phi(\omega)}$, it follows that the output is multiplied by a factor $G(\omega)$, the *gain*, and phase-shifted by an amount $\phi(\omega)$. We may now interpret (5.1.5) heuristically as follows; the random variable θ_p may be decomposed into contributions $\theta_p(\omega)$ at each frequency with variance $\sigma_\theta^2 g_\theta^*(\omega)$

which when passed through the linear filter (5.1.1) will be multiplied by the factor $|H(\omega)|$, so that its variance is $\sigma_\theta^2|H(\omega)|^2 g_\theta(\omega)$. The total variance is then obtained by adding up the contributions at each frequency.

The importance of these ideas to statistical estimation becomes apparent when it is realized that any linear estimate may be regarded as a filtering operation performed on the observations. We illustrate this by considering the estimation of a mean

$$\bar{\theta} = \frac{1}{2n+1} \sum_{p=-n}^{n} \theta_p \tag{5.1.7}$$

and a harmonic regression coefficient in the form

$$b = \frac{2}{2n+1} \sum_{p=-n}^{+n} \theta_p \cos(\omega_0 p). \tag{5.1.8}$$

For convenience we have arranged that the number of observations in the sample $N = 2n + 1$ is odd and that the time origin occurs at the $(n + 1)$st observation. Regarding (5.1.7) and (5.1.8) as filtering operations, it is readily seen that their frequency response functions $H_1(\omega)$ and $H_2(\omega)$ are respectively given by

$$H_1(\omega) = \sin(\tfrac{1}{2}N\omega)/\sin(\tfrac{1}{2}\omega) \tag{5.1.9}$$

and

$$H_2(\omega) = \frac{\sin\{\frac{1}{2}N(\omega - \omega_0)\}}{\sin\{\frac{1}{2}(\omega - \omega_0)\}} + \frac{\sin\{\frac{1}{2}N(\omega + \omega_0)\}}{\sin\{\frac{1}{2}(\omega + \omega_0)\}}. \tag{5.1.10}$$

By choice of time origin we have arranged that the phase-shift is zero and hence the frequency response function is a gain only. Also, these functions differ only in that (5.1.9) is centred about zero frequency and (5.1.10) about the frequency ω_0. The function (5.1.10) is exactly the same as the spectral window associated with the truncated periodogram estimate of the spectral density and has been plotted along with other forms of $H(\omega)$ in a paper recently published by one of us (Jenkins, 1961). It has the property that it is highly peaked about the frequency ω_0 and in fact behaves like a Dirac δ function as $n, N \to \infty$. It is not surprising, therefore, that it may be shown rigorously that

$$\lim_{N\to\infty} N \operatorname{var}(\bar{\theta}) = \pi\sigma_\theta^2 g_\theta(0) \tag{5.1.11}$$

and

$$\lim_{N\to\infty} N \operatorname{var}(b) = 2\pi\sigma_\theta^2 g_\theta(\omega_0), \tag{5.1.12}$$

so that the variances for large samples may be approximated respectively by $\pi\sigma_\theta^2 g_\theta(0)/N$ and $2\pi\sigma_\theta^2 g_\theta(\omega_0)/N$. However, this approximation may be seen to be a good one even for low values of N by observing that the width of the spectral window (5.1.10) will be narrow in general. Thus if the spectral density remains fairly constant over the width of the windows associated with the two estimates T_l and T_m, then

$$\operatorname{cov}(T_l, T_m) \sim \sigma_\theta^2 g_\theta(\omega_0) \int_{-\pi}^{+\pi} H_l(\omega)\overline{H}_m(\omega)\, d\omega = 2\pi\sigma_\theta^2 g_\theta(\omega_0) \sum_{j=0}^{\infty} \phi_j^{(l)}\phi_j^{(m)}. \tag{5.1.13}$$

In the case of a variance, if $g_\theta(\omega_0)$ remains fairly constant over the bandwidth of the window, then

$$\operatorname{var}(T_l) \sim 2\pi\sigma_\theta^2 g_\theta(\omega_0) \sum_{j=0}^{\infty} (\phi_j^{(l)})^2 \tag{5.1.14}$$

from which (5.1.11) and (5.1.12) are readily obtained. From (5.1.13) it may be seen that T_l and T_m are approximately uncorrelated provided the weights $\phi_j^{(l)}$ and $\phi_j^{(m)}$ are orthogonal, which is true for the estimators $\bar{\theta}$ and b.

5.2. Prediction in Discrete Time

Suppose that we can observe the mean process θ_p exactly without any measurement error u_p, so that $z_p = \theta_p$. Then the problem of pure prediction one step ahead as defined by (2.4.1) is to minimize the prediction error ε_p given by

$$\varepsilon_{p+1} = \theta_{p+1} - \sum_{j=0}^{\infty} \mu_j \theta_{p-j}. \tag{5.2.1}$$

With

$$\mu(\omega) = \sum_{j=0}^{\infty} \mu_j e^{i\omega j},$$

the frequency response function associated with the predictor weights, it is readily seen that the $H(\omega)$ associated with the prediction error (5.2.1) is $1 - e^{-i\omega}\,\mu(\omega)$ and hence that

$$\operatorname{var}(\varepsilon_{p+1}) = \sigma_\theta^2 \int_{-\pi}^{+\pi} |1 - e^{-i\omega}\mu(\omega)|^2 g_\theta^*(\omega)\, d\omega. \tag{5.2.2}$$

If it is now assumed that the spectral density may be written in the form

$$g_\theta(\omega) = \frac{\sigma_\varepsilon^2}{2\pi\sigma_\theta^2} |\psi(\omega)|^{-2} \qquad (-\pi \leqslant \omega \leqslant \pi), \tag{5.2.3}$$

then, using fairly standard techniques in the calculus of variations, it may be shown that the loss (5.2.2) is minimized with respect to variations in $\mu(\omega)$ when

$$1 - e^{-i\omega}\mu(\omega) = \psi(\omega). \tag{5.2.4}$$

Two special cases are of interest:

(i) If the θ_p series represents an autoregressive process, then $\psi(\omega)$ is a polynomial of the form

$$\psi(\omega) = 1 - \sum_{j=1}^{k} a_j e^{-i\omega j}.$$

It follows that $\mu(\omega) = a_1 + a_2 e^{-i\omega} + \cdots + a_k e^{-i(k-1)\omega}$ so that the predictor weights μ_j are equal to a_j, as was shown in section 3.

(ii) If the θ_p series constitutes a moving average process

$$\theta_p = \varepsilon_p + \beta_1 \varepsilon_{p-1} + \cdots + \beta_m \varepsilon_{p-m}, \tag{5.2.5}$$

then

$$\psi^{-1}(\omega) = 1 + \beta_1 e^{-i\omega} + \cdots + \beta_m e^{-im\omega} \tag{5.2.6}$$

and the solution for the optimum weights as given by (5.2.4) becomes extremely complicated. An example when $m = 1$ has been given previously in Table 2.

5.3. Control of the Mean when the Added Noise Is Absent

In this situation, $\varepsilon_p = e_p$ and the *loop equation* (2.4.2) may be written

$$\Delta\varepsilon_{p+1} = \Delta\theta_{p+1} - \sum_{j=0}^{\infty} w_j \varepsilon_{p-j} \tag{5.3.1}$$

which may be expressed symbolically as

$$\varepsilon_{p+1} = \left(1 - \frac{B\mathscr{L}_w}{1 - B + B\mathscr{L}_w}\right)\theta_{p+1} = (1 - \mathscr{L}_\mu)\theta_{p+1}, \tag{5.3.2}$$

where $\mathscr{L}_w = \Sigma w_j B^j$ and B is the backward shift or delay operator. If the operator $\mathscr{L}_\mu$ is expanded in an infinite series in powers of B, then the terms of this series represent the weights given to previous θ's in obtaining a linear prediction of θ_{p+1}. The transform of these weights will be the frequency response function of the filter associated with (5.3.2) and this is obtained simply by replacing B by $e^{-i\omega}$, i.e.

$$H(\omega) = 1 - \frac{e^{-i\omega}W(\omega)}{1 - e^{-i\omega} + e^{-i\omega}W(\omega)}, \tag{5.3.3}$$

where $W(\omega) = \Sigma w_j e^{-i\omega j}$ is the frequency response function of the controller weights. Since $H(\omega) = 1 - \mu(\omega)e^{-i\omega}$, it follows that the relation between the predictor and controller weights may be expressed in transforms by

$$\mu(\omega) = \frac{W(\omega)}{1 - e^{-i\omega} + e^{-i\omega}W(\omega)} \tag{5.3.4}$$

yielding the inverse relation

$$W(\omega) = \frac{\mu(\omega)(1 - e^{-i\omega})}{1 - e^{-i\omega}\mu(\omega)}. \tag{5.3.5}$$

It follows from (5.2.4) that the optimum controller weights for the spectrum

(5.2.3) are given by

$$W(\omega) = (e^{i\omega} - 1)\{\psi^{-1}(\omega) - 1\}. \tag{5.3.6}$$

In the case of an autoregressive scheme, $\psi^{-1}(\omega)$ is rather complicated, but for the moving average process with $\psi^{-1}(\omega)$ given by (5.2.6), the best controller weights are

$$W(\omega) = (e^{i\omega} - 1) \sum_{j=1}^{m} \beta_j e^{-i\omega j}, \tag{5.3.7}$$

leading to various forms of difference control depending on the value of m. When $m = 1$, the weights correspond to first difference control. It is interesting to note the duality which exists between the autoregressive and moving average processes in so far as the optimum choice of predictor and controller weights is concerned.

If we use (5.3.4) and (5.3.5), it is possible to draw some general conclusions about the conditions which have to be satisfied in order for the predictor and controller weights to satisfy some of the constraints discussed in section 3, for example that they sum to unity. Since the sum of any set of weights is given by the value of the frequency response function at zero frequency, it follows from (5.3.5) that if the predictor weights do not sum to unity, the controller weights will always sum to zero. The disadvantages of the predictors which are not mean correcting are thus seen to carry over to the controller weights. If the latter sum to zero then in a situation where, for example, the last few deviations ε_p were all equal to some quantity δ, the controller would recommend no control action whereas a correction $-\delta$ is required for effective control.

When the predictor weights do add to unity, it is necessary to apply L'Hôpital's rule to evaluate the frequency response functions at zero frequency since both numerator and denominator vanish in (5.3.4) and (5.3.5). A simple calculation gives

$$\sum_{j=0}^{\infty} \mu_j = \sum_{j=0}^{\infty} jw_j \Big/ \left\{1 + \sum_{j=0}^{\infty} jw_j\right\}, \tag{5.3.8}$$

$$\sum_{j=0}^{\infty} w_j = \sum_{j=0}^{\infty} \mu_j \Big/ \left\{\sum_{j=0}^{\infty} \mu_j + \sum_{j=0}^{\infty} j\mu_j\right\}. \tag{5.3.9}$$

It may be seen, therefore, that for a mean and trend correcting predictor, the controller weights must have an infinite sum and so we are led to some form of integral control.

5.4. Control with Dynamics and Added Noise

In discrete time there is a delay of one time unit before correction is made, so that even when all the correction becomes effective at the next instant, there are dynamic considerations present. For the moment we consider the dynamic aspects as referring to the extent to which the current correction becomes fully effective or not at the next instant, i.e. we ignore the initial delay of one time unit.

The loop equation may then be written, as before, in the form

$$\Delta\theta_{p+1} = \Delta\varepsilon_{p+1} + x_{p+1}, \tag{5.4.1}$$

but now x_{p+1} is made up of two operations acting independently. The controller produces, from the past deviations ε_p, a correction whose negative value is $x^*_{p+1} = \mathscr{L}^*_w \varepsilon_p$, where $\mathscr{L}^*_w$ is the linear operator associated with the controller. This is acted upon in turn by the dynamic characteristics of the process, characterized by a linear operator $\mathscr{L}_v$, finally producing $x_{p+1} = \mathscr{L}_v \mathscr{L}^*_w \varepsilon_p$. From (5.4.1), it is seen that the linear operator connecting ε_{p+1} and the previous θ's is given by

$$\varepsilon_{p+1} = \left(1 - \frac{B\mathscr{L}_v\mathscr{L}^*_w}{1 - B + B\mathscr{L}_v\mathscr{L}^*_w}\right)\theta_{p+1}, \tag{5.4.2}$$

so that the loss may be written as

$$\operatorname{var}(\varepsilon_{p+1}) = \int_{-\pi}^{+\pi} \left| 1 - \frac{e^{-i\omega}V(\omega)W^*(\omega)}{1 - e^{-i\omega} + e^{-i\omega}V(\omega)W^*(\omega)} \right|^2 g^*_\theta(\omega)\, d\omega. \tag{5.4.3}$$

Hence the effect of the dynamics is to replace the controller frequency response function $W(\omega)$ in (5.3.4) by the product of the controller transfer function and the process or valve transfer function in (5.4.3).

By writing down explicitly what is meant by the operation $x_{p+1} = \mathscr{L}_v \mathscr{L}^*_w \varepsilon_p$, it is possible to determine the weights which are actually applied to the errors ε_p. We may write this operation in the form

$$x_{p+1} = \sum_{r=0}^{\infty} v_r x^*_{p-r}$$

which in turn may be written as

$$x_{p+1} = \sum_{r=0}^{\infty} v_r \sum_{s=0}^{\infty} w^*_s \varepsilon_{p-r-s} = \sum_{l=0}^{\infty} w_l \varepsilon_{p-l},$$

where

$$w_l = \sum_{r=0}^{l} v_r w^*_{l-r}. \tag{5.4.4}$$

It follows that the effective weights which are applied to the errors ε_p are the convolution of the controller weights w^*_j and the weights v_j characterizing the dynamic response of the process.

In order to consider the effect of the added noise, it is necessary to derive the modified loop equation in this case. We must now distinguish between ε_{p+1} and e_{p+1} because if the u_p process is not white noise, then it is no longer true that $E(\varepsilon^2_{p+1})$ is minimized when $E(e^2_{p+1})$ is minimized. Ignoring dynamics, the loop equation becomes

$$\varepsilon_{p+1} = \varepsilon_p + (\theta_{p+1} - \theta_p) - x_{p+1} + u_{p+1}. \tag{5.4.5}$$

This differs from the previous loop equation in that the added error u_{p+1} now inflates the loss. The linear operators connecting ε_{p+1} and the past history of the series and the u series may be obtained by rearranging (5.4.5) in the form

$$(1 - B + B\mathscr{L}_w)\varepsilon_{p+1} = (1 - B)\theta_{p+1} + u_{p+1},$$

so that the loss may be written as

$$\operatorname{var}(\varepsilon_{p+1}) = \sigma_\theta^2 \int_{-\pi}^{+\pi} \left| 1 - \frac{e^{-i\omega}W(\omega)}{1 - e^{-i\omega} + e^{-i\omega}W(\omega)} \right|^2 g_\theta^*(\omega)\, d\omega$$
$$+ \sigma_u^2 \int_{-\pi}^{+\pi} \left| \frac{W(\omega)}{1 - e^{-i\omega} + e^{-i\omega}W(\omega)} \right|^2 g_u^*(\omega)\, d\omega. \qquad (5.4.6)$$

This formula shows that a compromise must now be set up between direct control of the θ_p series and smoothing out the effect of the u_p series, since a choice of $W(\omega)$ which makes the first integral small would tend to make the second integral large and vice versa.

(*Editor's note*: The remainder of Section 5 has been omitted.)

Acknowledgement

This research was supported by the National Science Foundation.

References

Barnard, G.A. (1959), "Control charts and stochastic processes", *J. R. statist. Soc.* B, **21**, 239–272.

——— Jenkins, G.M. and Winsten, C.B. (1962), "Likelihood inference and time-series", *J. R. statist. Soc.* A, **125**, 321–372.

Box, G.E.P. (1957), "Integration of techniques in process development", *Trans. of the 11th Annual Convention of the Amer. Soc. Qual. Control*, 687.

——— (1960), "Some general considerations in process optimisation", *Trans. Am. Soc. mech. Engrs*, Ser. D (*Basic Eng.*), **82**, 113–119.

——— and Chanmugam, J.R. (1962), "Adaptive optimisation of continuous processes", *Industrial and Engineering Chemistry* (*Fundamentals*), **1**, 2–16.

Brown, R.G. (1959), *Statistical Forecasting for Inventory Control.* New York: McGraw-Hill.

Cox, D.R. (1961), "Prediction by exponentially weighted moving averages and related methods", *J. R. statist. Soc.* B, **23**, 414–422.

Holt, C.C., Modigliani, F., Muth, T.F. and Simon, H.A. (1960), *Planning Production, Inventories and Work Force.* London: Prentice-Hall.

Jenkins, G.M. (1961), "General considerations in the analysis of spectra", *Technometrics*, **3**, 133–166.

Muth, J.F. (1960), "Optimal properties of exponentially weighted forecasts", *J. Amer. statist. Soc.*, **55**, 299–306.

Whittle, P. (1950), "The simultaneous estimation of a time-series harmonic components and covariance structure", *Trabajos de Estadistica*, **3**, 43–57.

Zadeh, L.A. and Ragazzini, J.R. (1950), "An extension of Wiener's theory of prediction", *J. appl. Phys.*, **21**, 645–655.

Introduction to Tukey (1962) The Future of Data Analysis

Lyle V. Jones
University of North Carolina at Chapel Hill

Among articles from the *Annals of Mathematical Statistics* (*A*MS), "The Future of Data Analysis" is unusual in many ways.

The manuscript is reported to have been received on July 1, 1961, the first day of the term of a new journal editor (J.L. Hodges, Jr.). The paper appeared as the lead article of the first issue of 1962, while numerous papers that had been processed and accepted by the previous editor (William Kruskal) remained unpublished even at the end of 1962 [Hodges (1962)].

The length of the paper is 67 pages. The distribution of length for the 147 articles published in *AMS* in 1962 displays a median of 8 pages, with first and third quartiles of 6 and 14.5 pages. Except for this paper, the longest article is 29 pages. (An addendum on p. 812 of Vol. 33 consists of nine successive references inadvertedly omitted from the published article, representing perhaps a misplaced page of the manuscript.)*

Unlike other articles from the *AMS*, this paper presents no derivations, proves no theorems, espouses no optimality conditions. Instead, it argues that alternatives to optimal procedures often are required for the statistical analysis of real-world data and that it is the business of statisticians to attend to such alternatives. While the development of optimal solutions is judged to be natural and desirable for mathematicians, Tukey deems such developments neither natural nor desirable for statisticians in their contributions to data analysis, because overdependence on optimal properties of procedures is bound to stultify progress, producing "a dried-up, encysted field."

Tukey argues against viewing statistics "as a monolithic, authoritarian structure designed to produce the 'official' results." He enthusiastically supports a flexible attitude for data analysis and the use of alternative approaches

* *Editor's note*: These references have been incorporated into the version reproduced in this volume.

and techniques for attacking a given problem. The competing techniques then may be compared, not solely with respect to one criterion, but with respect to alternative criteria as well. We are urged to "face up to the fact that *data analysis is intrinsically an empirical science*;" therefore, data analysts should adopt the attitudes of experimental science, giving up "the vain hope that data analysis can be founded upon a logico-deductive system."

In perspective, this paper presented a clear challenge to established standards of academic statistics, and it outlined the framework for the future development of exploratory data analysis. In earlier papers, Tukey (e.g., 1954, 1960a) had hinted at the need for reformation. He had argued for statistics as a science rather than a branch of mathematics [Tukey (1953, 1961a)], had urged statisticians to study real-world problems for which normality is a myth [Tukey (1960a, 1960b)], and had emphasized the importance of distinguishing decision problems and decision procedures from conclusion problems and conclusion procedures [Tukey (1960c)].

Many of Tukey's ideas about data analysis had been communicated in a seminar that he presented in 1957–58 at the Center for Advanced Study in the Behavioral Sciences, and a revised version of the seminar notes appeared later in mimeographed form [Tukey, (1961b)]. Mosteller relates that some of Tukey's exploratory data analysis had appeared as memoranda long before, in the early 1940s, and that "many in the invisible college were aware of that work" [Mosteller (1984)]. However, the impetus for including exploratory analysis as a part of statistics seems attributable more to Tukey's initiative, dedication, and persistence than to the efforts of any other individual. Indeed, this 1962 paper provides considerable support for this view.

Tukey proposes that work in mathematical statistics should fall in either of two categories: (1) work motivated by a desire to contribute to the practice of data analysis, and (2) work not so motivated, which then should be judged by the standards of pure mathematics. Work justified by neither data analysis nor pure mathematics is "doomed to sink out of sight." He pleads for novelty in data analysis, as illustrated by "wholly new questions to be answered," by "more realistic frameworks" for familiar problems, and by developing "unfamiliar summaries" of data that are likely to be "unexpectedly revealing."

Tukey's thesis is motivated in part by the need for procedures to analyze "spotty data": data from long-tailed distributions, data from a near-Gaussian distribution except for occasional "stragglers," and data from a mixture of distributions (perhaps nearly Gaussian) that display different levels of variability. If stragglers or outliers are sprinkled at random in a sample of data, the analysis of variance is known to be robust because the various mean squares are expected to be affected equally, ensuring against too many false positive results. Note, however, that the cost of this advantage is a loss of power, a possibly substantial loss. To ameliorate the loss of power or efficiency, Tukey commends procedures for trimming or Winsorizing sample data, each tailored to features of the particular sample. For complex data, Tukey proposes and illustrates automated procedures for detecting outliers, based on a full normal plot (FUNOP), and for modifying their values (FUNOM).

These procedures often are usefully followed by "vacuum cleaning," also illustrated in the paper, a regression procedure for "cleaning" residuals by removing from the (modified) data the systematic main effects and interaction effects.

Earlier, in a study of contaminated distributions, Tukey had presented some striking results that "nearly imperceptible nonnormalities may make conventional relative efficiencies of estimates of scale and location entirely useless" [Tukey (1960b)]. It is of interest to note that Hampel (1991) cites this 1960 Tukey paper as having provided the impetus for Huber's work on robust estimation [e.g., Huber (1964)].

Many of the ideas presented in "The Future of Data Analysis" have been developed further by Tukey and others. Among notable contributions are *Exploratory Data Analysis* [Tukey (1977a)], *Data Analysis and Regression* [Mosteller and Tukey (1978)], *Robust Statistics* [Huber (1981)], and *Robust Statistics: The Approach Based on Influence Functions* [Hampel et al. (1986)]. The analysis of data from long-tailed distributions is addressed in Dixon and Tukey (1968), Hoaglin et al. (1981, 1986), Rogers and Tukey (1972), and Tukey (1979). Issues of multiplicity and multiple comparisons are discussed by Braun and Tukey (1983) and Tukey (1977b), among many others. The direct assessment of standard errors using jackknife procedures is illustrated in Mosteller and Tukey (1968). Some recent examples of contributions that build on Tukey's exploratory data analysis appear in Hoaglin, et al. (1983, 1985). "The Future of Data Analysis" anticipated the general acceptance of exploratory analysis within statistics, as reflected by a wealth of literature on exploratory topics as well as the inclusion in textbooks, courses, and software systems of many procedures that Tukey introduced.

John Wilder Tukey (1915–) earned a master's degree in chemistry at Brown University in 1937. (His formal education prior to college had been limited to some high-school courses in Providence, R.I. His earlier education had been supervised by his parents at home.) He was awarded a Ph.D. in mathematics at Princeton in 1939, and he remained at Princeton as an instructor in mathematics. In 1942, he joined the staff of the Fire Control Research Office in Princeton. Among those with whom he associated during World War II, all working for one or another Army activity in Princeton, were George W. Brown, W.G. Cochran, Wilfred Dixon, Paul Dwyer, Merrill Flood, Philip McCarthy, A.M. Mood, Frederick Mosteller, S.S. Wilks, and Charles P. Winsor. In 1945, at Wilks' initiative, Tukey rejoined the faculty of the Princeton mathematics department, and he became a charter faculty member of the Princeton statistics department. In 1945, he also became a member of the research staff at Bell Laboratories. Tukey occupied both of these positions until his retirement in 1985. He has continued to maintain a high level of scientific activity since then.

Over a span of more than 50 years, Tukey is credited with some 800 publications. Also, he has participated in a myriad of national research activities. He has contributed to methodology in the physical sciences and engineering, environmental sciences, medical sciences, social sciences, and educational research, and he has influenced public policy in many of these areas, as a member of the president's Science Advisory Committee and numerous other policy committees. Not infrequently, in fulfilling his responsibilities on a committee, Tukey was inspired to invent new procedures that later were found to have wide applicability, e.g., Gentleman et al. (1969) and Tukey (1969).

Some of Tukey's published articles, along with a number of unpublished papers, have been selected to appear in *The Collected Works of John W. Tukey*. Six volumes of *Collected Works* were published between 1984 and 1990: two on time series, two on the principles and philosophy of data analysis, one on graphics, and one on more mathematical topics. Additional volumes are scheduled to cover Tukey's contributions to applied problems (especially environmental), and within statistics to factorial design, multiple comparisons, additive fits, transformation of variables, research design and sampling, and robust/resistant procedures.

References

Braun, H.I., and Tukey, J.W. (1983). Multiple comparisons through orderly partitions: The maximum subrange procedure, in *Principals of Modern Psychological Measurement: A Festschrift for Frederic M. Lord* (W. Wainer, and S. Messick, eds.). Laurence Erlbaum Associates. Hillsdale, N.J. 55–65.

Dixon, W.J., and Tukey, J.W. (1968). Approximate behavior of the distribution of Winsorized *t*, *Technometrics*, **10**, 83–98.

Gentleman, W.M., Gilbert, J.P., and Tukey, J.S. (1969). The smear-and-sweep analysis, in *The National Halothane Study*. (J.P. Bunker, W.H. Forrest, Jr., F. Mosteller, and L.D. Vandam, eds.). U.S. Government Printing Office, Washington, D.C. 287–315.

Hampel, F. (1991). Introduction to P.J. Huber (1964): Robust estimation of a location parameter, In *Breakthroughs in Statistics*, Vol. II (S. Kotz, and N.L. Johnson, eds.). Springer-Verlag, New York.

Hampel, F.R., Ronchetti, E.M., Rousseeuw, P.J., and Stahel, W.A. (1986). *Robust Statistics: The Approach Based on Influence Functions*. Wiley. New York.

Hoaglin, D.C., Iglewicz, B., and Tukey, J.W. (1981). Small-sample performance of a resistant rule for outlier detection, in *1980 Proceedings of the Statistical Computing Section*. American Statistical Association, Washington, D.C.

Hoaglin, D.C., Iglewicz, B., and Tukey, J.W. (1986). Performance of some resistant rules for outlier labeling, *J. Amer. Statist. Assoc.* **81**, 991–999.

Hoaglin, D.C., Mosteller, F., and Tukey, J.W. (eds.) (1983). *Understanding Robust and Exploratory Data Analysis*. Wiley, New York.

Hoaglin, D.C., Mosteller, F., and Tukey, J.W. (eds.) (1985). *Exploring Data Tables, Trends, and Shapes*. Wiley, New York.

Hodges, J.L., Jr. (1962). Report of the editor for 1962, *Ann. Math. Statist.*, **33**, 517–518.

Huber, P.J. (1964). Robust estimation of a location parameter. *Ann. Math. Statist.*, **35**, 73–101.

Huber, P.J. (1981). *Robust Statistics*. Wiley, New York.

Mosteller, F. (1984). Biography of John W. Tukey, in *The Collected Works of John W. Tukey, Vol. I: Time Series, 1949–1964* (D.R. Brillinger, ed.) Wadsworth, Monterey, Calif.

Mosteller, F., and Tukey, J.W. (1968). Data analysis, including statistics, in *Handbook of Social Psychology*, 2nd ed. Vol. 2. (G. Lindzey, and E. Aronson, eds.). Addison-Wesley, Reading, Mass. [Also in *The Collected Works of John W. Tukey, Vol. IV: Philosophy and Principles of Data Analysis, 1965–1986* (L.V. Jones, ed.) 1986, Wadsworth, Monterey, Calif.], 601–720.

Mosteller, F., and Tukey, J.W. (1978). *Data Analysis and Regression: A Second Course in Statistics.* Addison-Wesley, Reading, Mass.

Rogers, W.H., and Tukey, J.W. (1972). Understanding some long-tailed symmetrical distributions, *Statistica Neerlandica*, **26**, 211–226.

Tukey, J.W. (1953). The growth of experimental design in a research laboratory, in *Research Operations in Industry.* King's Crown Press, New York [also in *The Collected Works of John W. Tukey, Vol. III: Philosophy and Principles of Data Analysis, 1949–1964* (L.V. Jones, ed.), 1986, Wadsworth, Monterey, Calif., 65–76.

Tukey, J.W. (1954). Unsolved problems of experimental statistics, *J. Amer. Statist Assoc.*, **55**, 80–93. [also in *The Collected Works of John W. Tukey, Vol. III: Philosophy and Principles of Data Analysis, 1949–1964* (L.V. Jones, ed.), 1986, Wadsworth, Monterey, Calif.].

Tukey, J.W. (1960a). Where do we go from here? *J. Amer. Statist. Assoc.*, **55**, 80–93 [also in *The Collected Work of John W. Tukey, Vol. III: Philosophy and Principles of Data Analysis, 1949–1964* (L.V. Jones, ed.), 1986, Wadsworth, Monterey, Calif.].

Tukey, J.W. (1960b). A survey of sampling from contaminated distributions, in *Contributions to Probability and Statistics: Essays in Honor of Harold Hotelling* (I. Olkin, S.G. Ghurye, W. Hoeffding, W.G. Madow, and H.B. Mann, eds.). Stanford University Press.

Tukey, J.W. (1960c). Conclusions vs. decisions, *Technometrics*, **2**, 423–433 [also in, *The Collected Works of John W. Tukey, Vol. III: Philosophy and Principles of Data Analysis, 1949–1964* (L.V. Jones, ed.), 1986, Wadsworth, Monterey, Calif.].

Tukey, J.W. (1961a). Statistical and quantitative methodology, in *Trends in Social Sciences* (D.P. Ray, ed.). Philosophical Library, New York [also in *The Collected Works of John W. Tukey, Vol. III: Philosophy and Principles of Data Analysis, 1949–1964* (L.V. Jones, ed.), 1986, Wadsworth, Monterey, Calif.].

Tukey, J.W. (1961b). Data analysis and behavioral science or learning to shun the quantitative man's burden by shunning badmandments (mimeo) [also in *The Collected Works of John W. Tukey, Vol. III: Philosophy and Principles of Data Analysis, 1949–1964* (L.V. Jones, ed.), 1986, Wadsworth, Monterey, Calif.].

Tukey, J.W. (1969). A further inquiry into institutional differences by means of superstandardization (a regression adjustment beyond standardization), in *The National Halothane Study* (J.P. Bunker, W.H. Forrest, Jr., F. Mosteller, and L.D. Vandam, eds.). U.S. Government Printing Office, Washington, D.C.

Tukey, J.W. (1977a). *Exploratory Data Analysis.* Addison-Wesley, Reading, mass.

Tukey, J.W. (1977b). Some thoughts on clinical trials, especially problems of multiplicity, *Sci.* **198**, 679–684.

Tukey, J.W. (1979). Study of robustness by simulation: Particularly improvement by adjustment and combination, in *Robustness in Statistics* (R.L. Launer, and G.N. Wilkinson, eds.). Academic Press, New York.

The Future of Data Analysis[1]

John W. Tukey
Princeton University and
Bell Telephone Laboratories

I. General Considerations

1. Introduction

For a long time I have thought I was a statistician, interested in inferences from the particular to the general. But as I have watched mathematical statistics evolve, I have had cause to wonder and to doubt. And when I have pondered about why such techniques as the spectrum analysis of time series have proved so useful, it has become clear that their "dealing with fluctuations" aspects are, in many circumstances, of lesser importance than the aspects that would already have been required to deal effectively with the simpler case of very extensive data, where fluctuations would no longer be a problem. All in all, I have come to feel that my central interest is in *data analysis*, which I take to include, among other things: procedures for analyzing data, techniques for interpreting the results of such procedures, ways of planning the gathering of data to make its analysis easier, more precise or more accurate, and all the machinery and results of (mathematical) statistics which apply to analyzing data.

Large parts of data analysis are inferential in the sample-to-population sense, but these are only parts, not the whole. Large parts of data analysis are

Received July 1, 1961.

[1] Prepared in part in connection with research sponsored by the Army Research Office through Contract DA36-034-ORD-2297 with Princeton University. Reproduction in whole or part is permitted for any purpose of the United States Government.

incisive, laying bare indications which we could not perceive by simple and direct examination of the raw data, but these too are only parts, not the whole. Some parts of data analysis, as the term is here stretched beyond its philology, are allocation, in the sense that they guide us in the distribution of effort and other valuable considerations in observation, experimentation, or analysis. Data analysis is a larger and more varied field than inference, or incisive procedures, or allocation.

Statistics has contributed much to data analysis. In the future it can, and in my view should, contribute much more. For such contributions to exist, and be valuable, it is not necessary that they be direct. They need not provide new techniques, or better tables for old techniques, in order to influence the practice of data analysis. Consider three *examples*:

(1) The work of Mann and Wald (1942) on the asymptotic power of chi-square goodness-of-fit tests has influenced practice, even though the results they obtained were for impractically large samples.
(2) The development, under Wald's leadership, of a general theory of decision functions has influenced, and will continue to influence data analysis in many ways. (Little of this influence, in my judgment, will come from the use of specific decision procedures; much will come from the forced recognition of the necessity of considering "complete classes" of procedures among which selection must be made by judgment, *perhaps* codified á la Bayes.)
(3) The development of a more effective procedure for determining properties of samples from non-normal distributions by experimental sampling is likely, if the procedure be used wisely and widely, to contribute much to the practice of data analysis.

To the extent that pieces of *mathematical statistics* fail to contribute, or are not intended to contribute, even by a long and tortuous chain, to the practice of data analysis, they must be judged as pieces of *pure* mathematics, and criticized according to its purest standards. Individual parts of mathematical statistics must look for their justification toward either data analysis or pure mathematics. Work which obeys neither master, and there are those who deny the rule of both for their own work, cannot fail to be transient, to be doomed to sink out of sight. And we must be careful that, in its sinking, it does not take with it work of continuing value. Most present techniques of data analysis, statistical or not, have a respectable antiquity. Least squares goes back more than a century and a half (e.g., Gauss, 1803). The comparison of a sum of squares with the value otherwise anticipated goes back more than 80 years (e.g., cp., Bortkiewicz, 1901). The use of higher moments to describe observations and the use of chi-square to assess goodness of fit are both more than 60 years old (Pearson, 1895, 1900). While the last century has seen great developments of regression techniques, and of the comparison of sums of squares, comparison with the development of other sciences suggests that novelty has entered data analysis at a slow, plodding pace.

By and large, the great innovations in statistics have not had correspondingly great effects upon data analysis. The extensive calculation of "sums of squares" is the outstanding exception. (Iterative maximum likelihood, as in the use of working probits, probably comes next, but is not widely enough used to represent a great effect.)

Is it not time to seek out novelty in data analysis?

2. Special Growth Areas

Can we identify *some* of the areas of data analysis which today offer unusual challenges, unusual possibilities of growth?

The treatment of "spotty data" is an ancient problem, and one about which there has been much discussion, but the development of organized techniques of dealing with "outliers", "wild shots", "blunders", or "large deviations" has lagged far behind both the needs, and the possibilities presently open to us for meeting these needs.

The analysis of multiple-response data has similarly been much discussed, but, with the exception of psychological uses of factor analysis, we see few analyses of multiple-response data today which make essential use of its multiple-response character.

Data of the sort today thought of as generated by a stochastic process is a current challenge which both deserves more attention, and different sorts of attention, than it is now receiving.

Problems of selection often involve multi-stage operations, in which later-stage actions are guided by earlier-stage results. The essential role of data analysis, even when narrowly defined, is clear. (To many of us the whole process is a part of data analysis.) Here we have learned a little, and have far to go.

We have sought out more and more subtle ways of assessing error. The half-normal plot typifies the latest, which will have more extensive repercussions than most of us have dreamed of.

Data which is heterogeneous in precision, or in character of variation, and data that is very incomplete, offer challenges that we have just begun to meet. Beyond this we need to stress flexibility of attack, willingness to iterate, and willingness to study things as they are, rather than as they, hopefully, should be.

Again let me emphasize that these are only *some* of the growth areas, and that their selection and relative emphasis has been affected by both personal idiosyncrasies and the importance of making certain general points.

3. How Can New Data Analysis Be Initiated?

How is novelty most likely to begin and grow? Not through work on familiar problems, in terms of familiar frameworks, and starting with the results of applying familiar processes to the observations. Some or all of these familiar constraints must be given up in each piece of work which may contribute novelty.

We should seek out wholly new questions to be answered. This is likely to require a concern with more complexly organized data, though there will be exceptions, as when we are insightful enough to ask new, useful kinds of questions about familiar sorts of data.

We need to tackle old problems in more realistic frameworks. The study of data analysis in the face of fluctuations whose distribution is rather reasonable, but unlikely to be normal, provides many important instances of this. So-called non-parametric methods, valuable though they are as first steps toward more realistic frameworks, are neither typical or ideal examples of where to stop. Their *ultimate* use in data analysis is likely to be concentrated (i) upon situations where relative ranks are really all the data available, and (ii) upon situations where unusually quick or portable procedures are needed. In other situations it will be possible, and often desirable, to analyze the data more thoroughly and effectively by other methods. Thus while nonparametric analysis of two-way tables leading to confidence statements for differences of typical values for rows and columns is quite possible, it is computationally difficult enough to keep me from believing that such techniques will ever be widely used. (The situation for three- and more-way tables is even worse.) As means for defending the results of an analysis against statistical attack on distributional grounds, nonparametric methods will retain a pre-eminent place. And as an outstanding first step in eliminating unacceptable dependence upon normality assumptions they have been of the greatest importance. But these facts do not suffice to make them the methods of continuing choice.

We should seek out unfamiliar summaries of observational material, and establish their useful properties.

But these are not the only ways to break out into the open. The comparison, under suitable or unsuitable assumptions, of different ways of analyzing the same data for the same purpose has been a unifying concept in statistics. Such comparisons have brought great benefits to data analysis. But the habit of making them has given many of us a specific mental rigidity. Many seem to find it essential to begin with a probability model containing a parameter, and then to ask for a good estimate for this parameter (too often, unfortunately, for one that is optimum). Many have forgotten that data analysis can, sometimes quite appropriately, precede probability models, that progress can come from asking what a specified indicator (= a specified function of the data) may reasonably be regarded as estimating. Escape from this constraint can do much to promote novelty.

And still more novelty can come from finding, and evading, still deeper lying constraints.

It can help, throughout this process, to admit that our first concern is "data analysis". I once suggested in discussion at a statistical meeting that it might be well if statisticians looked to see how data was actually analyzed by many sorts of people. A very eminent and senior statistician rose at once to say that this was a novel idea, that it might have merit, but that young statisticians should be careful not to indulge in it too much, since it might distort their ideas. *The ideas of data analysis ought to survive a look at how data is analyzed.* Those who try may even find new techniques evolving, as my colleague Martin Wilk suggests, from studies of the nature of "intuitive generalization".

4. Sciences, Mathematics, and the Arts

The extreme cases of science and art are clearly distinguished, but, as the case of the student who was eligible for Phi Beta Kappa because mathematics was humanistic and for Sigma Xi because it was scientific shows, the place of mathematics is often far from clear. There should be little surprise that many find the places of statistics and data analysis still less clear.

There are diverse views as to what makes a science, but three constituents will be judged essential by most, viz:

(a1) intellectual content,
(a2) organization into an understandable form,
(a3) reliance upon the test of experience as the ultimate standard of validity.

By these tests, mathematics is not a science, since its ultimate standard of validity is an agreed-upon sort of logical consistency and provability.

As I see it, data analysis passes all three tests, and I would regard it as a science, one defined by a ubiquitous problem rather than by a concrete subject. (Where "statistics" stands is up to "statisticians" and to which ultimate standard—analysis of data or pure mathematics—they adhere. A useful mixed status may be gained by adhering sometimes to one, and sometimes to the other, although it is impossible to adopt both *simultaneously as ultimate* standards.)

Data analysis, and the parts of statistics which adhere to it, must then take on the characteristics of a science rather than those of mathematics, specifically:

(b1) Data analysis must seek for scope and usefulness rather than security.
(b2) Data analysis must be willing to err moderately often in order that inadequate evidence shall more often *suggest* the right answer.
(b3) Data analysis must use mathematical argument and mathematical results as bases for judgment rather than as bases for proof or stamps of validity.

These points are meant to be taken seriously. Thus, for example, (b1) does not mean merely "give up security by acting as if 99.9% confidence were certainty", much more importantly it means "give general advice about the use of techniques as soon as there is reasonable ground to think the advice sound; be prepared for a reasonable fraction (not too large) of cases of such advice to be *generally wrong*"!

All sciences have much of art in their makeup. (It is sad that the phrase "the useful and mechanic arts" no longer reminds us of this frequently.) As well as teaching facts and well-established structures, all sciences must teach their apprentices how to think about things in the manner of that particular science, and what are its current beliefs and practices. Data analysis must do the same. Inevitably its task will be harder than that of most sciences. "Physicists" have usually undergone a long and concentrated exposure to those who are already masters of the field. "Data analysts", even if professional statisticians, will have had far less exposure to professional data analysts during their training.

Three reasons for this hold today and can at best be altered slowly:

(c1) Statistics tends to be taught as part of mathematics.
(c2) In learning statistics *per se* there has been limited attention to data analysis.
(c3) The number of years of intimate and vigorous contact with professionals is far less for statistics Ph.D.'s than for physics (or mathematics) Ph.D.'s.

Thus data analysis, and adhering statistics, faces an unusually difficult problem of communicating certain of its essentials, one which cannot presumably be met as well as in most fields by indirect discourse and working side-by-side.

For the present, there is no substitute for making opinions more obvious, for being willing to express opinions and understandings in print (knowing that they may be wrong), for arguing by analogy, for reporting broad conclusions supported upon a variety of moderately tenuous pieces of evidence (and then going on, later, to identify and make critical tests of these conclusions), for emphasing the importance of judgment and illustrating its use, (as Cox (1957) has done with the use of judgment indexes as competitors for covariance), not merely for admitting that a statistician needs to use it. And for, while doing all this, continuing to use statistical techniques for the confirmatory appraisal of observations through such conclusion procedures as confidence statements and tests of significance.

Likewise, we must recognize that, as Martin Wilk has put it, "The hallmark of good science is that it uses models and 'theory' but never believes them."

5. Dangers of Optimization

What is needed is progress, and the unlocking of certain of rigidities (ossifications?) which tend to characterize statistics today. Whether we look back over this century, or look into our own crystal ball, there is but one natural chain of growth in dealing with a specific problem of data analysis, viz:

(a1′) recognition of problem,
(a1″) one technique used,
(a2) competing techniques used,
(a3) rough comparisons of efficacy,

(a4) comparison in terms of a precise (and thereby inadequate) criterion,
(a5′) optimization in terms of a precise, and similarly inadequate criterion,
(a5″) comparison in terms of several criteria.

(Number of primes does not indicate relative order.)

If we are to be effective in introducing novelty, we must heed two main commandments in connection with new problems:

(A) Praise and use work which reaches stage (a3), or only stage (a2), or even stage (a1″).
(B) Urge the extension of work from each stage to the next, with special emphasis on the earlier stages.

One of the clear signs of the lassitude of the present cycle of data analysis is the emphasis of many statisticians upon certain of the later stages to the exclusion of the earlier ones. Some, indeed, seem to equate stage (a5′) to statistics—an attitude which if widely adopted is guaranteed to produce a dried-up, encysted field with little chance of real growth.

I must once more quote George Kimball's words of wisdom (1958).

> There is a further difficulty with the finding of "best" solutions. All too frequently when a "best" solution to a problem has been found, someone comes along and finds a still better solution simply by pointing out the existence of a hitherto unsuspected variable. In my experience when a moderately good solution to a problem has been found, it is seldom worth while to spend much time trying to convert this into the "best" solution. The time is much better spent in real research ....

As Kimball says so forcefully, optimizing a simple or easy problem is not as worthwhile as meliorizing a more complex or difficult one. In data analysis we have no difficulty in complicating problems in useful ways. It would, for example, often help to introduce a number of criteria in the place of a single one. It would almost always help to introduce weaker assumptions, such as more flexibility for parent distributions. And so on.

It is true, as in the case of other ossifications, that attacking this ossification is almost sure to reduce the apparent neatness of our subject. But neatness does not accompany rapid growth. Euclidean plane geometry is neat, but it is still Euclidean after two millenia. (Clyde Coombs points out that this neatness made it easier to think of non-Euclidean geometry. If it were generally understood that the great virtue of neatness was that it made it easier to make things complex again, there would be little to say against a desire for neatness.)

6. Why Optimization?

Why this emphasis on optimization? It is natural, and desirable, for mathematicians to optimize; it focusses attention on a small subset of all possibilities, it often leads to general principles, it encourages sharpening of concepts, particularly when intuitively unsound optima are regarded as reasons for reexamining concepts and criteria (e.g., Cox, (1958, p. 361) and the criterion of power). Danger only comes from mathematical optimizing when the results are taken too seriously. In elementary calculus we all optimize surface-to-volume relations, but no one complains when milk bottles turn out to be neither spherical or cylindrical. It is understood there that such optimum problems are unrealistically oversimplified, that they offer *guidance*, not the *answer*. (Treated similarly, the optimum results of mathematical statistics can be most valuable.)

There is a second reason for emphasis upon the optimum, one based more upon the historical circumstances than upon today's conditions. Let others speak:

> It is a pity, therefore, that the authors have confined their attention to the relatively simple problem of determining the approximate distribution of arbitrary criteria and have failed to produce any sort of justification for the tests they propose. In addition to those functions studied there are an infinity of others, and unless some principle of selection is introduced we have nothing to look forward to but an infinity of test criteria and an infinity of papers in which they are described. (Box, 1956, p. 29.)

> More generally still, one has the feeling that the statistics we are in the habit of calculating from our time series tend to be unduly stereotyped. We are, in a way, in the reverse situation to that which obtained when Fisher wrote about how easy it was to invent a great multiplicity of statistics, and how the problem was to select the good statistics from the bad ones. With time series we could surely benefit from the exploration of the properties of many more statistics than we are in the habit of calculating. We are more sober than in the days of Fechner, Galton, and "K.P."; perhaps we are too sober. (Barnard, 1959a, p. 257.)

The first quotation denies that (a2) above should precede (a3) to (a5′), while the second affirms that (a2) and (a3) should be pushed, especially in fields that have not been well explored. (The writer would not like to worry about an infinity of methods for attacking a question until he has at least four such which have *not* been shown to have distinguishable behavior.)

7. The Absence of Judgment

The view that "statistics is optimization" is perhaps but a reflection of the view that "data analysis should not *appear to* be a matter of judgment". Here "appear to" is in italics because many who hold this view would like to suppress these words, even though, when pressed, they would agree that the optimum *does* depend upon the assumptions and criteria, whose selection may, perhaps, even be admitted to involve judgment. It is very helpful to replace the use of judgment by the use of knowledge, but only if the result is the use of knowledge with judgment.

Pure mathematics differs from most human endeavor in that assumptions are not criticized because of their relation to something outside, though they are, of course, often criticized as unaesthetic or as unnecessarily strong. This cleavage between pure mathematics and other human activities has been deepening since the introduction of non-Euclidean geometries by Gauss, Bolyai, and Lobachevski about a century and a half ago. Criticism of assumptions on the basis of aesthetics and strength, without regard for external correspondence has proved its value for the development of mathematics. But we dare not use such a wholly internal standard anywhere except in pure mathematics. (For a discussion of its dangers in pure mathematics, see the closing pages of von Neumann, 1947.)

In data analysis we must look to a very heavy emphasis on judgment. At least three different sorts or sources of judgment are likely to be involved in almost every instance:

(a1) judgment based upon the experience of the particular field of subject matter from which the data come,
(a2) judgment based upon a broad experience with how particular techniques of data analysis have worked out in a variety of fields of application,
(a3) judgment based upon abstract results about the properties of particular techniques, whether obtained by mathematical proofs or empirical sampling.

Notice especially the form of (a3). It is consistent with actual practice in every field of science with a theoretical branch. A scientist's actions are *guided*, not determined, by what has been derived from theory or established by experiment, *as is his advice to others.* The judgment with which isolated results are put together to guide action or advice in the usual situation, which is too complex for guidance to be *deduced* from available knowledge, will often be a mixture of individual and collective judgments, but judgement will play a crucial role. Scientists know that they will sometimes be wrong; they try not to err too often, but they accept some insecurity as the price of wider scope. Data analysts must do the same.

One can describe some of the most important steps in the development of mathematical statistics as attempts to save smaller and smaller scraps of certainty (ending with giving up the certainty of using an optimal technique for the certainty of using an admissible one, one that at least cannot be unequivocally shown to be non-optimal). Such attempts must, in large part at least, be attempts to maintain the mathematical character of statistics at the expense of its data-analytical character.

If data analysis is to be well done, much of it must be a matter of judgment, and "theory", whether statistical or non-statistical, will have to guide, not command.

8. The Reflection of Judgment upon Theory

The wise exercise of judgment can hardly help but to stimulate new theory. While the occurrence of this phenomenon may not be in question, its appropriateness is regarded quite differently. Three quotations from the discussion of Kiefer's recent paper before the Research Section of the Royal Statistical Society* point up the issue:

> Obviously, if a scientist asks my advice about a complex problem for which I cannot compute a good procedure in the near future, I am not going to tell him to cease his work until such a procedure is found. But when I give him the best that my intuition and reason can now produce, I am not going to be satisfied with it, no matter how clever a procedure it may appear on the surface. The aim of the subject is not the construction of nice looking procedures with intuitive appeal, but the determination of procedures which are proved to be good. (Kiefer (1959, p. 317) in reply to discussion.)

> A major part of Dr. Kiefer's contribution is that he is forcing us to consider very carefully what we want from a design. But it seems to me to be no more reprehensible to start with an intuitively attractive design and then to search for optimality criteria which it satisfies, than to follow the approach of the present paper, starting from (if I may call it so) an intuitively attractive criterion, and then to search for designs which satisfy it. Either way one is liable to be surprised by what comes out; but the two methods are complementary. (Mallows, 1959, p. 307.)

> The essential point of the divergence is concisely stated at the end of section A: "The rational approach is to state the problem and the optimality criterion and then to find the appropriate design, and not alter the statements of the problem and criterion just to justify the use of the design to which we are wedded by our prejudiced intuition." I would agree with this statement if the word "deductive" were inserted in place of the word "rational". As a rationalist I feel that the word "rational" is one which indicates a high element of desirability, and I think it is much broader in its meaning than "deductive". In fact what appears to me to be the rational approach is to take designs which are in use already, to see what is achieved by these designs by consideration of the general aims to evaluate such designs, in a provisional way, and then to seek to find designs which improve on existing practice. Having found such designs the cycle should be repeated again. The important thing about this approach is that we are always able to adjust our optimality criteria to designs as well as adjusting our designs to our optimality criteria (Barnard, 1959b, p. 312).

9. Teaching Data Analysis

The problems of teaching data analysis have undoubtedly had much to with these unfortunate rigidities. Teaching data analysis is not easy, and the time allowed is always far from sufficient. But these difficulties have been enhanced by certain views which have been widely adopted, such as those caricatured in:

(a1) "avoidance of cookbookery and growth of understanding come only by mathematical treatment, with emphasis upon proofs".

(a2) "It is really quite intolerable for the teacher then to have to reply, 'I don't know'." (An actual quotation from Stevens (1950, p. 129).)

(a3) "whatever the facts may be, we must keep things simple so that we can teach students more easily".

* *Editor's note*: J.C. Kiefer's (1959) paper is reproduced in Volume I of this series.

(a4) "even if we do not know how to treat this problem so as to be either good data analysis or good mathematics, we should treat it somehow, because we must teach the students something".

It would be well for statistics if these points of view were not true to life, were overdrawn, but it is not hard to find them adopted in practice, even among one's friends.

The problem of cookbookery is not peculiar to data analysis. But the solution of concentrating upon mathematics and proof is. The field of biochemistry today contains much more detailed knowledge than does the field of data analysis. The over-all teaching problem is more difficult. Yet the text books strive to tell the facts in as much detail as they can find room for. (Biochemistry was selected for this example because there is a clear and readable account by a coauthor of a leading text of how such a text is revised (Azimov, 1955).)

A teacher of biochemistry does not find it intolerable to say "I don't know". Nor does a physicist. Each spends a fair amount of time explaining what his science does not know, and, consequently, what are some of the challenges it offers the new student. Why should not both data analysts and statisticians do the same?

Surely the simplest problems of data analysis are those of

(b1) location based upon a single set of data,
(b2) relative location based upon two sets of data.

There are various procedures for dealing with each of these problems. Our knowledge about their relative merits under various circumstances is far from negligible. Some of it is a matter of proof, much of the rest can be learned by collecting the results when each member of any class draws a few samples from each of several parent distributions and applies the techniques. Take the one-sample problem as an example. What text, and which teachers, teach the following simple facts about one-sample tests?

(c1) for symmetrical distributions toward the rectangular, the mid-range offers high-efficiency of point-estimation, while the normally-calibrated range-midrange procedure (Walsh, 1949a) offers conservative but still efficient confidence intervals,
(c2) for symmetrical distributions near normality, the mean offers good point estimates and Student's t offers good confidence intervals, but signed-rank (Walsh, 1949b, 1959) confidence intervals are about as good for small and moderate sample sizes,
(c3) for symmetrical distributions with slightly longer tails, like the logistic perhaps, a (trimmed) mean omitting a prechosen number of the smallest and an equal number of the largest observations offers good point estimates, while signed-rank procedures offer good interval estimates,
(c4) for symmetric distributions with really long tails (like the central 99.8% of the Cauchy distribution, perhaps) the median offers good point estimates, and the sign test offers good interval estimates.

(c5) the behavior of the one-sample t-test has been studied by various authors (cp., Gayen, 1949, and references cited by him) with results for asymmetric distributions which can be digested and expressed in understandable form.

These facts are specific, and would not merit the expenditure of a large fraction of a course in data analysis. But they can be said briefly. And, while time spent showing that Student's t is optimum for exactly normal samples may well, on balance, have a negative value, time spent on these points would have a positive one.

These facts are a little complex, and may not prove infinitely easy to teach, but any class can check almost any one of them by doing its own experimental sampling. Is it any better to teach everyone, amateurs and professionals alike, about only a single one-sample procedure (or *as is perhaps worse* about the comparative merits of various procedures in sampling from but a *single* shape of parent population) than it would be to teach laymen and doctors alike that the only pill to give is aspirin (or to discuss the merits and demerits of various pills for but a single ailment, mild headache)?

Some might think point (a4) above to be excessively stretched. Let us turn again to the Stevens article referred to in (a2), in which he introduced randomized confidence intervals for binomial populations. In an addendum, Stevens notes an independent proposal and discussion of this technique, saying: "It was there dismissed rather briefly as being unsatisfactory. This may be granted but since ... some solution is necessary [because the teacher should not say "I don't know" (J.W.T.)], it seems that this one deserves to be studied and to be used by teachers of statistics until a better one can be found." The solution is admittedly unsatisfactory, and not just the best we have to date, yet it is to be taught, and used!

Not only must data analysis admit that it uses judgment, it must cease to hide its lack of knowledge by teaching answers better left unused. The useful must be separated from the unuseful or antiuseful.

As Egon Pearson pointed out in a Statistical Techniques Research Group discussion where this point was raised, there is a real place in *discussing* randomized confidence limits in advanced classes for statisticians; not because they are useful, not because of aesthetic beauty, but rather because they may stimulate critical thought. As Martin Wilk puts it:

> We dare not confine ourselves to emphasizing properties (such as efficiency or robustness) which, although sometimes useful as guides, only hold under classes of assumptions all of which may be wholly unrealistic; we must teach an understanding of *why* certain sorts of techniques (e.g., confidence intervals) are indeed useful.

10. Practicing Data Analysis

If data analysis is to be helpful and useful, it must be practice There are many ways in which it can be used, some good and some evil. Laying aside unethical practices, one of the most dangerous (as I have argued elsewhere (Tukey, 1961b)) is the use of formal data-analytical procedures for sanctification, for the preservation of conclusions from all criticism, for the granting of an *imprimatur*. While statisticians have contributed to this misuse, their share has been small. There is a corresponding danger for data analysis, particularly in its statistical aspects. This is the view that all statisticians *should* treat a given set of data in the same way, just as all British admirals, in the days of sail, maneuvered in accord with the same principles. The admirals could not communicate with one another, and a single basic doctrine was essential to coordinated and effective action. Today, statisticians can communicate with one another, and have more to gain by using special knowledge (subject-matter or methodological) and flexibility of attack than they have to lose by not all behaving alike.

In general, the best account of current statistical thinking and practice is to be found in the printed discussions in the *Journal of the Royal Statistical Society*. While reviewing some of these lately, I was surprised, and a little shocked to find the following:

> I should like to give a word of warning concerning the approach to tests of significance adopted in this paper. It is very easy to devise different tests which, on the average, have similar properties, i.e., they behave satisfactorily when the null hypothesis is true and have approximately the same power of detecting departures from that hypothesis. Two such tests may, however, give very different results when applied to a given set of data. This situation leads to a good deal of contention amongst statisticians and much discredit of the science of statistics. The appalling position can easily arise in which one can get any answer one wants if only one goes around to a large enough number of statisticians. (Yates, 1955, p. 31).

To my mind this quotation, if taken very much more seriously than I presume it to have been meant, nearly typifies a picture of statistics as a monolithic, authoritarian structure designed to produce the "official" results. While the possibility of development in this direction is a real danger to data analysis, I find it hard to believe that this danger is as great as that posed by over-emphasis on optimization.

11. Facing Uncertainty

The most important maxim for data analysis to heed, and one which many statisticians seem to have shunned, is this: "Far better an approximate answer to the *right* question, which is often vague, than an *exact* answer to the wrong question, which can always be made precise." Data analysis must progress by approximate answers, at best, since its knowledge of what the problem really is will at best be approximate. It would be a mistake not to face up to this fact, for by denying it, we would deny ourselves the use of a great body of approximate knowledge, as well as failing to maintain alertness to the possible importance in each particular instance of particular ways in which our knowledge is incomplete.

II. Spotty Data

12. What Is It?

The area which is presently most obviously promising as a site of vigorous new growth in data analysis has a long history. The surveyor recognizes a clear distinction between "errors" with which he must live, and "blunders", whose effects he must avoid. This distinction is partly a matter of size of deviation, but more a matter of difference in the character or assignability of causes. Early in the history of formal data analysis this recognition led to work on the "rejection of observations". Until quite recently matters rested there.

One main problem is the excision of the effects of occasional potent causes. The gain from such excision should not be undervalued. Paul Olmstead, for example, who has had extensive experience with such data, maintains that engineering data typically involves 10% of "wild shots" or "stragglers". A ratio of 3 between "wild shot" and "normal" standard deviations is far too low (individual "wild shots" can then hardly be detected). Yet 10% wild shots with standard deviation 3σ contribute a variance equal to that of the remaining 90% of the cases with standard deviation 1σ. Wild shots can easily double, triple, or quadruple a variance, so that really large increases in precision can result from cutting out their effects.

We are proud, and rightly so, of the "robustness" of the analysis of variance. A few "wild shots" sprinkled at random into data taken in a conventional symmetric pattern will, on the average, affect each mean square equally. True, but we usually forget that this provides only "robustness of validity", ensuring that we will not be led by "wild shots" to too many false positives, or to too great security about the precision of our estimates. Conventional analysis of variance procedures offer little "robustness of efficiency", little tendency for the high efficiency provided for normal distributions of fluctuations-and-errors to be preserved for non-normal distributions. A few "wild shots", either spread widely or concentrated in a single cell, can increase all mean squares substantially. (Other spreadings may have different results.) From a hypothesis-testing point of view this decreases our chances of detecting real effects, and increases the number of false negatives, perhaps greatly. From an estimation point of view it increases our variance of estimate and decreases our efficiency, perhaps greatly. Today we are far from adopting an adequately sensitive technique of analysis, even in such simple situations as randomized blocks.

Now one cannot look at a single body of data alone and be sure which are the "wild shots". One can usually find a statistician to argue that some particular observation is not unusual in cause, but is rather "merely a large deviation". When there are many such, he will have to admit that the distribution of deviations (of errors, of fluctuations) has much longer tails than a Gaussian (normal) distribution. And there will be instances where he will be correct in such a view.

It would be unfortunate if the proper treatment of data was seriously different when errors-and-fluctuations have a long-tailed distribution, as compared to the case where occasional causes throw in "wild shots". Fortunately, this appears not to be the case; cures or palliatives for the one seem to be effective against the other. A simple indication that this is likely to be so is furnished by the probability element

$$[(1-\theta)(2\pi)^{-1/2}e^{-1/2y^2}+\theta[h^{-1}(2\pi)^{-1/2}]e^{-y^2/2h^2}]\,dy,$$

which can be construed in at least three ways:

(a1) as a unified long-tailed distribution which is conveniently manipulable in certain ways,

(a2) as representing a situation in which there is probability θ that an occasional-cause system, which contributes additional variability when it acts, will indeed act,

(a3) as representing a situation in which variability is irregularly nonhomogeneous.

It is convenient to classify together most instances of long-tailed fluctuation-and-error distributions, most instances of occasional causes with large effects, and a substantial number of cases with irregularly non-constant variability; the term "spotty data" is surely appropriate.

13. An Appropriate Step Forward

Clearly the treatment of spotty data is going to force us to abandon normality. And clearly we can go far by studying cases of sampling from long-tailed distributions. How far to go? Which long-tailed distributions to face up to?

To seek complete answers to these questions would be foolish. But a little reflection takes us a surprising distance. We do not wish to take on any more new problems than we must. Accordingly it will be well for us to begin with long-tailed distributions which offer the minimum of doubt as to what should be taken as the "true value". If we stick to symmetrical distributions we can avoid all difficulties of this sort. The center of symmetry is the median, the mean (if this exists), and the $\alpha\%$ trimmed mean for all α. (The $\alpha\%$ trimmed mean is the mean of the part of the distribution between the lower $\alpha\%$ point and the upper $\alpha\%$ point.) No other point on a symmetrical distribution has a particular claim to be considered the "true value". Thus we will do well to *begin* by restricting ourselves to symmetric distributions.

Should we consider all symmetric distributions? This would be a counsel of perfection, and dangerous, since it would offer us far more flexibility than we know how to handle. We can surely manage, even in the beginning, to face a single one-parameter family of shapes of distribution. Indeed we may be able to face a few such families. Which ones will be most effective? Experience to date suggests that we should be interested both in families of shapes in which the tails behave more or less as one would expect from the remainder of the distribution (conforming tails) and in families in which the tails are longer than the remainder suggests (surprising tails). The latter are of course peculiarly applicable to situations involving occasional causes of moderately rare occurrence.

What one-parameter families of distribution shapes with conforming tails should be considered? And will the choice matter much? We may look for candidates in two directions, symmetric distributions proposed for graduation, and symmetric distributions which are easy to manipulate. The leading classical candidate consists of:

(a1) the symmetric Pearson distributions, namely the normal-theory distributions of Student's t and Pearson's r.

The only similar system worth particular mention is:

(a2) N.L. Johnson's [1949] distributions, which, in the symmetric case, are the distributions of tanh N/δ or sinh N/δ, where N follows a unit normal distribution, and δ is an appropriate constant.

While the symmetrical Johnson curves are moderately easy to manipulate in various circumstances, even more facility is frequently offered by what may be called the lambda-distributions, viz:

(a3) the distributions of $P^{\lambda} - (1 - P)^{\lambda}$ where P is uniformly distributed on $(0, 1)$.

(It is possible that it would be desirable to introduce a further system of symmetrical distributions obtained from the logistic distribution by simple transformation.)

While the analytic descriptions of these three conforming systems are quite different, there is, fortunately, little need to be careful in choosing among them, since they are remarkably similar (personal communication from E.S. Pearson for (a2) vs. (a1) and Tukey, 1962 for (a3) vs. (a1)).

The extreme cases of all of these symmetrical families will fail to have certain moments; some will fail to have variances, others will even fail to have means. It is easy to forget that these failures are associated with the last ε of cumulative probability in each tail, no matter how small $\varepsilon > 0$ may be. If we clip a tail of probability 10^{-40} off each end of a Cauchy distribution (this requires clipping in the vicinity of $x = \pm 10^{20}$), and replace the removed 2.10^{-40} probability at any bounded set of values, the resulting distribution will have finite moments of all orders. But the behavior of samples of sizes less than, say, 10^{30} from the two distributions will be practically indistinguishable.

The finiteness of the moments does not matter directly; the extendedness of the tails does. For distributions given in simple analytic form, infiniteness of moments *often* warns of practical extendedness of tails, but, as the infinite moments of

$$dF = [(1 - 10^{-50})(2\pi)^{-1/2}e^{-1/2x^2} + 10^{-50}(4/\pi)(1 + x^2)^{-1}]\,dx$$

show in the other direction, there is no necessary connection. A sure way to drive ourselves away from attaching undue significance to infinite moments, and back to a realistic view is to study some one-parameter families with finite moments which converge to infinite-moment cases. These are easy to find, and include

(b1) the distribution of tan X, where $(2/\pi)X$ is uniform on $(-\theta, +\theta)$ for $0 < \theta < 1$ (as $\theta \to 1$, this goes to the Cauchy distribution)
(b2) the distribution of $(1 - P)^{-1} - P^{-1}$, where P is uniform on $(\varepsilon, 1 - \varepsilon)$ for $0 < \varepsilon < 0.5$ (as $\varepsilon \to 0$, this goes to a rough approximation to a Cauchy distribution).

(Note that all distributions satisfying (b1) and (b2) have all moments finite.)

Beyond this we shall want in due course to consider some symmetrical distributions with surprising tails (i.e., some which have longer tails than would be expected from the rest of the distribution). Here we might consider, in particular:

(c1) contaminated distributions at scale 3, with probability element

$$\{(1 - \theta)(2\pi)^{-1/2}e^{-y^2/2} + \theta[3^{-1}(2\pi)^{-1/2}]e^{-y^2/18}\}\,dy,$$

(c2) contaminated lambda distributions such as

$$(1 - \theta)[P^{0.2} - (1 - P)^{0.2}] + \theta[\log P - \log(1 - P)]$$

and

$$(1 - \theta)[\log P - \log(1 - P)] + \theta[(1/P) - (1 - P)^{-1}].$$

There is no scarcity of one-shape-parameter families of symmetrical distributions which are reasonably easily manipulated and whose behavior can offer valuable guidance.

Once we have a reasonable command of the symmetrical case, at least in a few problems, it will be time to plan our attack upon the unsymmetrical case.

14. Trimming and Winsorizing Samples

In the simplest problems, those involving only the location of one or more simple random samples, it is natural to attempt to eliminate the effects of "wild shots" by *trimming* each sample, by removing equal numbers of the lowest and highest observations, and then proceding as if the trimmed sample were a complete sample. As in all procedures intended to guard against "wild shots" or long-tailed distributions, we must expect, in comparison with procedures tailored to exactly normal distributions:

(a1) some loss in efficiency when the samples do come from a normal distribution.
(a2) increased efficiency when the samples come from a long-tailed distribution.

An adequate study of the effects of trimming naturally proceeds step by step. It may well start with comparisons of variances of trimmed means with those of untrimmed means, first for normal samples, and then for long-tailed samples. Order-statistic moments for samples from long-tailed distributions are thus required. The next step is to consider alternative modifications of the classical t-statistic with a trimmed mean in the numerator and various denominators. Order statistic moments, first for the normal and rectangular, and then for suitable long-tailed distributions are now used to determine the ratios

$$\frac{\text{variance of numerator}}{\text{average squared denominator}}$$

for various modified t-statistics and various underlying distributions. This enables the choice of a denominator which will make this ratio quite closely constant. (It is, of course, exactly constant for the classical t-statistic.) Detailed behavior for both normal and non-normal distributions is now to be obtained; sophicated experimental sampling is almost certainly necessary and sufficient for this. Close agreement of critical values can now replace close agreement of moment ratios as a basis for selection, and critical values can then be supplemented with both normal and non-normal power functions. At that point we will know rather more about the symmetrical-distribution behavior of such modified t-statistics based upon trimmed samples than we presently do about Student's t itself. (This program is actively under way at the Statistical Techniques Research Group of Princeton University.)

Both this discussion, and the work at S.T.R.G., began with the case of trimmed samples because it is the simplest to think about. But it is not likely to be the most effective for moderately or sub-extremely long-tailed distribution. When I first met Charles P. Winsor in 1941 he had already developed a clear and individual philosophy about the proper treatment of apparent "wild shots". When he found an "outlier" in a sample he did not simply reject it. Rather he changed its value, replacing its original value by the nearest value of an observation not seriously suspect. His philosophy for doing this, which applied to "wild shots", can be supplemented by a philosophy appropriate to long-tailed distributions which leads to the same actions (cp., Anscombe and Tukey, 1962). It seems only appropriate, then, to attach his name to the process of replacing the values of certain of the most extreme observations in a sample by the nearest unaffected values, to speak of Winsorizing or Winsorization.

For normal samples, Winsorized means are more stable than trimmed means (Dixon, 1960, 1907). Consequently there is need to examine the advantages of modified t-statistics based upon Winsorized samples.

The needs and possibilities will not come to an end here. So far we have discussed only the case where a fixed number of observations are trimmed from, or Winsorized at, each end of a sample. But intelligent rejection of observations has always been guided by the configuration of the particular sample considered, more observations being discarded from some samples than from others. Tailoring is required.

Tailored trimming and tailored Winsorizing, respectively, may thus be expected to give better results than fixed trimming or fixed Winsorizing, and will require separate study. (Given a prescribed way of choosing the number to be removed or modified at each end, there will, in particular, be a common factor (for given sample size) by which the length of the corresponding fixed-mode confidence limits can be multiplied in order to obtain tailored-mode confidence limits with the indicated confidence. This is of course only one way to allow for tailoring each sample separately.)

All these procedures need to be studied first for the single sample situation, whose results should provide considerable guidance for those multiple-sample problems where conventional approaches would involve separate estimates of variance. Situations leading classically to pooled variances are well-known to be wisely dealt with by randomization theory so far as robustness of validity is concerned (cp., Pitman, 1937; Welch, 1938). While randomization of trimmed samples could be studied, it would seem to involve unnecessary complications, especially since the pure-randomization part of the randomization theory of Winsorized samples is inevitably the same as for unmodified samples. In particular, robustness of validity is certain to be preserved by Winsorization. And there is every reason to expect robustness of efficiency to be considerably improved in such problems by judicious Winsorization in advance of the application of randomization thcory.

Before we leave this general topic we owe the reader a few numbers. Consider the means of a sample of 11, of that sample with 2 observations trimmed from each end, and of that sample with 2 observations Winsorized on each end. The resulting variances, for a unit normal parent, and for a Cauchy parent, are as follows (cp., Dixon, 1960; Sarhan and Greenberg, 1958 for normal variances of symmetrically Winsorized means):

	Normal parent	Cauchy parent
plain mean	0.091	infinite
trimmed mean	0.102	finite
Winsorized mean	0.095	finite*

* Presumably larger than for the corresponding trimmed mean.

The small loss due to Winsorization or trimming in the face of a normal parent contrasts interestingly with the large gain in the face of a Cauchy parent.

15. How Soon Should Such Techniques Be Put into Service?

The question of what we need to know about a new procedure before we recommend its use, or begin to use it, should not be a difficult one. Yet when one friendly private comment about a new non-parametric procedure (that in Siegel and Tukey, 1960) was that it should not appear in print, and I presume should *a fortiori* not be put into service, even in those places where such a non-parametric procedure is appropriate, until its power function was given, there are differences of opinion of some magnitude. (Especially when the sparsity of examples of non-parametric power functions which offer useful guidance to those who might use the procedure is considered.)

Let us cast our thoughts back to Student's t. How much did we know when it was introduced into service? Not even the distribution for a normal parent was known! Some empirical sampling, a few moments, and some Pearson curves produced the tables of critical values. But, since it filled an aching void, it went directly into use.

As time went on, the position of the *single-sample t* procedure became first better, as its distribution, and its optimality, under normal assumptions was established, and then poorer, as more was learned about its weaknesses for nonnormal samples, whose frequent occurrence had been stressed by Student himself (1927). (For references to the history of the discovery of the weaknesses see, e.g., Geary, 1947; and Gayen, 1949. For references which tend to support Geary's (1949, p. 241) proposal that all textbooks state that "*Normality is a myth; there never has, and never will be, a normal distribution.*" see Tukey, 1960.)

The case of modified t's based upon trimmed or Winsorized means is somewhat different. There are already well-established procedures with known advantages, viz:

(a1) Student's t, optimum for normality
(a2) the signed-rank procedure, robust of validity within symmetry
(a3) the sign-test procedure, robust of validity generally.

How much do we need to know about a competitor before we introduce it to service? For my own part, I expect to begin recommending the first product of the program sketched above, a modified t based upon a trimmed mean with fixed degree of trimming, as a standard procedure for most single-sample location problems, as soon as I know:

(b1) that the variance of the numerator for normality is suitably small.
(b2) that the ratio of ave (denominator)^2 to var (numerator) is constant to within a few % for a variety of symmetrical parent distributions.

(b3) estimates based upon empirical sampling of the critical values (% points) appropriate for normality.

In bringing such a procedure forward it would be essential to emphasize that it was not the final answer; that newer and better procedures were to be expected, perhaps shortly. (Indeed it would be wrong to feel that we are ever going to completely and finally solve any of the problems of data analysis, even the simple ones.)

Surely the suggested amount of knowledge is not enough for anyone to guarantee either

(c1) that the chance of error, when the procedure is applied to real data, corresponds precisely to the nominal levels of significance or confidence, or

(c2) that the procedure, when applied to real data, will be optimal in any one specific sense.

But we have never been able to make either of these statements about Student's *t*. Should we therefore never have used Student's t?

A judgment of what procedures to introduce, to recommend, to use, must always be a judgment, a judgment of whether the gain is relatively certain, or perhaps only likely, to outweigh the loss. And this judgment can be based upon a combination of the three points above with general information of what longer-tailed distributions do to trimmed and untrimmed means.

III. Spotty Data in More Complex Situations

16. Modified Normal Plotting

Spotty data are not confined to problems of simple structure. Good procedures for dealing with spotty data are not always going to be as forthright and direct as those just discussed for simple samples, where one may make a single direct calculation, and where even tailored trimming or Winsorizing need only require carrying out a number of direct calculations and selecting the most favorable. Once we proceed to situations where our "criticism" of an apparent deviation must be based upon apparent deviations corresponding to different conditions, we shall find iterative methods of calculation convenient and almost essential. Iteration will, in fact, take place in "loops within loops" since the details of, e.g., making a particular fit may involve several iterative cycles (as in non-linear least squares, e.g., cp., Deming, 1943, or in probit analysis; e.g., cp., Finney, 1947, 1952), while human examination of the results of a particular fit may lead to refitting all over again, with a different model or different weights. Even such simple problems as linear regression, multiple regression, and cross-classified analyses of variance will require iteration.

The two-way table offers simple instance which is both illuminating and useful. A class of procedures consistent with the philosophies noted in Section 14 [cp., Anscombe and Tukey, 1962] operate in the following way:

(a1) if a particular deviation is much "too large" in comparison with the bulk of the other deviations, its effect upon the final estimates is to be made very small
(a2) if a particular deviation is only moderately "too large", its effect is to be decreased, but not made negligible.

The first task in developing such a procedure must be the setting up of a sub-procedure which identifies the "too muchness", if any, of the extreme apparent deviations. In the first approximation, this procedure may take no account of the fact that we are concerned with residuals, and may proceed as if it were analyzing a possibly normal sample.

The natural way to attack this problem graphically would be to fit row and column means, calculate residuals (=apparent deviations), order them, plot them against typical values for normal order statistics, draw a straight line through the result, and assess "too muchness" of extreme deviations in terms of their tendency to fall off the line. Such an approach would be moderately effective. Attempts to routinize or automate the procedure would find difficulty in describing just how to fit the straight line, since little weight should be given to apparently discrepant points.

The display can be made more sensitive, and automation easier, if a different plot is used where ordinates correspond to secant slopes on the older plot. More precisely, if we make a *conventional* probability plot on *probability graph paper*, we plot the observed ith smallest value in a set of n over (against) a corresponding fraction on the probability scale. (Different choices are conventionally made for this fraction, such $(i - \frac{1}{2})/n$ or $i/(n + 1)$.) The same plot could be made on ordinary graph paper if we were to plot y_i, the ith smallest, against $a_{i|n}$, the standard normal deviate corresponding to the selected probability.

With this meaning for the plotting constants $a_{i|n}$, denote the median of the values to be examined by $\dot{y}$ (read "y split"). Then $(0, \dot{y})$ and $(a_{i|n}, y_i)$ are points in the old plot, and the slope of the line segment (secant) joining them is

$$z_i = \frac{y_i - \dot{y}}{a_{i|n}}.$$

A plot of z_i against i is a more revealing plot, and one to which we should like to fit a horizontal line. We have then only to select a typical z, and can avoid great weight on aberrant values by selecting a median of a suitable set of z's.

The choice of $a_{i|n}$ has been discussed at some length (cp., Blom, 1958, pp. 144–146 and references there to Weibull, Gumbel, Bliss, Ipsen and Jerne; cp. also, Chernoff and Lieberman, 1954). The differences between the various choices are probably not very important. The choice $a_{i|n} = \text{Gau}^{-1}[(3i - 1)/(3n + 1)]$, where $P = \text{Gau}(y)$ is the normal cumulative, is simple and surely an adequate approximation to what is claimed to be optimum (also cp., Blom, 1958, pp. 70–71).

17. Automated Examination

Some would say that one should not automate such procedures of examination, that one should encourage the study of the data. (Which is somehow discouraged by automation?) To this view there are at least three strong counter-arguments:

(1) Most data analysis is going to be done by people who are not sophisticated data analysts and who have very limited time; if you do not provide them tools the data will be even less studied. Properly automated tools are the easiest to use for a man with a computer.
(2) If sophisticated data analysts are to gain in depth and power, they must have both the time and the stimulation to try out *new* procedures of analysis; hence the *known* procedures must be made easy for them to apply as possible. Again automation is called for.
(3) If we are to study and intercompare procedures, it will be much easier if the procedures have been fully specified, as must happen if the process of being made routine and automatizable.

I find these counterarguments conclusive, and I look forward to the automation of as many standardizable statistical procedures as possible. When these are available, we can teach the man who will have access to them the "why" and the "which", and let the "how" follow along.

(*Editor's note*: Subsections 18–20 have been omitted.)

IV. Multiple-Response Data

21. Where Are We, and Why?

Multivariate analysis has been the subject of many pages of papers, and not a few pages of books. Yet when one looks around at the practical analysis of data today one sees few instances where multiple responses are formally analyzed in any way which makes essential use of their multiplicity of response. The most noticeable exception is factor analysis, about which we shall make some more specific remarks below. There is, to be sure, a fairly large amount of data-analysis by multiple regression, some quite sophisticated and incisive, much quite bookish and bumbling. (Cochran's remark of a number of years ago that "regression is the worst taught part of statistics" has lost no validity.) But ordinary multiple regression is overtly not multiple-*response*, no matter how multivariate it may be.

Why is it that so much formalized and useful analysis of single-response data, and of single-response aspects of multiple-response data, is accompanied by so little truly multiple-response analysis? One interesting and reasonable suggestion is that it is because multiple-response procedures have been modeled upon how early single-response procedures were supposed to have been used, rather than upon how they were in fact used.

Single-response techniques started as significance procedures designed in principle to answer the question: "is the evidence strong enough to contradict a hypothesis of no difference?", or perhaps, as became clearer later, the more meaningful question: "is the evidence strong enough to support a belief that the observed difference has the correct sign?" But in early use by the naive these techniques, while sometimes used for the second purpose, were often used merely to answer "should I believe the observed difference?". While this last form is a misinterpretation to a statistician, its wide use and the accompanying growth of the use of such techniques, suggests that it was useful to the practitioner. (Perhaps the spread of confidence procedures will go far to replace it by a combination of technique and interpretation that statisticians can be happy with. Cp., e.g., Roy and Gnanadesikan, 1957, 1958.)

But there was one essential to this process that is often overlooked. Single-response differences are (relatively) simple, and the practitioner found it moderately easy to think, talk, and write about them, either as to sign, or as to sign and amount. Even when fluctuations have negligible influence, multiple-response differences are not simple, are usually not easy to think about, are usually not easy to describe. Better ways of description and understanding of multiple-response differences, and of multiple-response variabilities, may be essential if we are to have wider and deeper use of truly multiple-response techniques of data analysis. While it can be argued that the provision of such ways of description is not a problem of statistics, if the latter be narrowly enough defined, it is surely a central problem of data analysis.

Let me turn to a problem dear to the heart of my friend and mentor, Edgar Anderson, the shape of leaves. It is, I believe, fair to say that we do not at present have a satisfactory way of describing to the *mind* of another person either:

(a1) the nature of the difference in typical pattern (shape and size) of two populations of leaves, or
(a2) the nature of the variability of dimensions for a single population of leaves.

Photographs, or tracings, of shapes and sizes of random samples may convey the information to his *eyes* fairly well, but we can hardly, as data analysts, regard this as satisfactory summarization—better must be possible, but how?

In view of this difficulty of description, it is not surprising that we do not have a good collection of ideal, or prototype multivariate problems and solutions, indeed it is doubtful if we have even one (where many are needed). A better grasp of just what we want from a multivariate situation, and why, could perhaps come without the aid of better description, but only with painful slowness.

We shall treat only one specific instance of the multiple-response aspects of analysis of variance. (This instance, in Section 25 below, combines the boiling-down and factor-identification techniques of factor analysis with the patterned-observation and additive-decomposition techniques of analysis of variance.) A variety of schemes for the analysis of variance of multiple-response data have been put forward (cp., e.g., Bartlett, 1947; Tukey, 1949a; Rao, 1952; Roy, 1958; Rao, 1959), but little data has been analyzed with their aid. A comparative study of various approaches *from the point of view of data analysis* could be very valuable. However, such a study probably requires, as a foundation, a better understanding of what we really want to do with multiple-response data.

It seems highly probable that as such better ways of description are developed and recognized there will be a great clarification in what multiple-response analysis needs in the way of inferential techniques.

22. The Case of Two Samples

It would not be fair to give the impression that there is no use of truly multiple-response techniques except in factor analysis. There are a few other instances, mainly significance testing for two groups and simple discriminant functions.

Hotelling's T^2 serves the two-sample problem well in many cases.* Only the need for lengthy arithmetic and the inversion of a matrix stands in its way. The requirement for lengthy arithmetic, both in accumulating sums of products and squares, and in inverting the resulting matrix, which is probably inevitable for a procedure that is affine-invariant (i.e., one whose results are unchanged by the introduction of new responses that are non-singular linear combinations of the given responses), used to stand in the way of its use. More and more of those who have multivariate data to analyze now have access to modern computers, and find the cost of arithmetic low. Those who are not in so favorable a state may benefit substantially from the use of validity-conserving "quick" methods, specifically from applying to more simply obtainable "relatively highly apparent" comparisons, the critical values which would be appropriate to affine-invariant "most apparent" comparison, and which are therefore generally applicable. Such procedures would still be able to establish significance in a large proportion of those cases in which it is warranted. Such "quick" multivariate techniques would not be thorough in establishing lack of significance, but would be far more thorough than would application, at an appropriate error-rate, of a single-response procedure to each component of the multiple response.

* *Editor's note*: Hotelling's (1948) paper on T^2 is reproduced in Volume I of this series.

Let us sketch how two or three such "quick" multiple-response procedures might be constructed. First, consider calculating means and variances for each component, applying Penrose's method (1947) to obtain an approximate discriminant function, computing values of this discriminant for every observation, calculating Student's t for these values and then referring t^2 to the critical values of Hotelling's T^2 to test the result. A simpler version, under some circumstances, might be to use sample ranges in place of sample variances. A more stringent version would calculate a Penrose discriminant, regress each component (each original response) on this to obtain means and variances of residuals, extract a second Penrose discriminant from the residuals, and combine the two discriminants before testing. Each of these procedures is feasible for high multiplicity of response, since their labor is proportional to only the first power of the number of components, whereas Hotelling's T^2 requires labor in collection of SS and SP proportional to the square of this number, and labor in the matrix inversion proportional to its cube.

For very large numbers of components, as compared to the number of observations, it is necessary to give up affine invariance. Dempster's method (1958, 1960), which involves judgment selection of a measure of size, is applicable in such cases and may well also prove useful in many cases where Hotelling's T^2 would be feasible. There is always a cost to providing flexibility in a method. Affine-invariance implies an ability of the analysis to twist to meet any apparent ellipsoid of variability. Unless the number of cases is large compared to the number of responses, this flexibility requires loose critical values. For moderate numbers of cases, judgment "sizes" may give greater power than would affine invariance. (The observations are still being asked to provide their own error term, thus validity can be essentially the same for the two approaches.)

Many statisticians grant the observer or experimenter the right to use judgment in selecting the response to be studied in a single-response analysis. (Judgment is badly needed in this choice.) It is no far cry to the use of judgment, perhaps by the two parties together, in selecting a "size" for analysis in a multiple-response situation. In this connection, the promising work of Wilk and Gnanadesikan (1961) in using multiple-response sizes to generalize the half-normal-plot analysis for single-response data must be pointed out.

There are many ways to use flexibility in opening up the practice of the analysis of multiple-response data. Few, if any, involve matrices "six feet high and four feet wide", or the solution of very complex maximum likelihood equations.

23. Factor Analysis: The Two Parts

Factor analysis is a procedure which has received a moderate amount of attention from statisticians, and rather more from psychometricians. Issues of principle, of sampling fluctuation, and of computational ease have been confounded. By and large statisticians have been unsatisfied with the result.

Any reasonable account of factor analysis from the data analyst's point of view must separate the process into two quite distinct parts. Every type of factor analysis includes a "boiling-down" operation in which the dimensionality of the data is reduced by introducing new coordinates. In most cases this process is followed, or accompanied, by a "rotation" process in which a set of new coordinates are located which might be believed to approximate a set with some intrinsic meaning. The problems connected with the second phase are rather special, and tend to involve such questions, upon which we should not enter here, as whether simple structure is a hypothesis about the tested, or about the test-makers.

The "boiling-down" process is something else again. If we strip it of such extraneous considerations as:

(a1) a requirement to extract as much useful information as possible with as *few* coordinates as possible,
(a2) a requirement to try to go far enough but not one bit further,

we find that it is an essentially simple problem of summarization, one with which most, if not all, statisticians can be content (so long as not too much is claimed for it). If we start with 40 responses, and really need only 8 coordinates to express the 'meat' of the matter, then finding 10, 12, or even 15 coordinates which are almost sure to contain what is most meaningful is a very useful process, and one not hard to carry out. (The early stages of almost any method of factor analysis will serve, provided we do not stop them too soon.) And, before we go onward beyond such rough summarization, it may be very much worth while to investigate how changes in mode of expression, either of the original coordinate or the new ones, may lead to a further reduction in dimensionality. Adequate packaging, and broad understanding, of effective methods of "boiling down, but not too far" which are either simple or computationally convenient could contribute much to the analysis of highly multiple-response data.

24. Factor Analysis: Regression

Whatever the conclusion as to the subject-matter relevance and importance of the "rotation" part of the process, we may, as statisticians, be sure of its demand for more and more data. For it is essentially based, at best, upon the detailed behavior of multivariate variance components. It is commonplace that variance-component techniques applied to single responses will require large samples. When they are applied to multiple responses, their demands for large samples cannot help being more extensive. Consequently there must be a constant attempt to recognize situations and ways in which techniques basically dependent upon regression rather than variance components can be introduced into a problem, since regression techniques always offer hopes of learning more from less data than do variance-component techniques.

When we have repeated measurements, we can, if we wish, make a multivariate analysis of variance, and factor-analyze a variance component for individuals (or tests) rather than factor-analyzing the corresponding mean square (or sum of squares). It has already been pointed out that it is better to turn to the variance component (Tukey, 1951) in such instances.

A somewhat related situation arises when the multiple responses are accompanied by a (small) number of descriptive variables. We clearly can replace the original responses by residuals after regression on descriptive variables before proceeding to factor analysis. Should we?

If this question has a general answer, it must be that we should. For a description of the variation of the multiple responses in terms of a mixture of

(a1) regression upon descriptive variables, and
(a2) variation of residual factors,

will convey much more meaning than the results of a simple factor analysis, if only because of the lower sensitivity of the regression coefficients to fluctuations. And the residual factor analysis will often be far more directly meaningful than the raw factor analysis. Consider a study of detailed aspects of children's personalities, as revealed in their use of words. Suppose the sexes of the individual children known. Elimination of the additive effect of sex would almost surely lead to more meaningful "factors", and elimination of reading speed as well, if available for elimination, might lead us to even closer grips with essentials.

(*Editor's note*: Subsection 25 has been omitted.)

26. Taxonomy; Classification; Incomplete Data

Problems which can be fitted under the broad heading of taxonomy (or, if you must, nosology) are truly multiple response problems, whether plants, retrievable information, or states of ill-health are to be classified. It is already clear that there are a variety of types of problems here. In plants, for example, where the separation of species does not usually call for formal data analysis, the work of Edgar Anderson (e.g., 1949, 1957) has shown how even simple multiple response techniques (e.g., Anderson, 1957) can be used to attack such more complex questions as the presence and nature of inter-specific hybridization.

The rise of the large, fast, and (on a per-operation basis) cheap electronic computer has opened the way to wholly mechanized methods of species-finding. Here biologists have had a prime role in attacking the problem (cp.,

Sneath, 1957a, 1957b; Michener and Sokal, 1957; Sneath and Cowan, 1958; Rogers and Tanimoto, 1960).

Formal and semiformal techniques for identifying "species" from multiple-response data are certain to prove important, both for what they will help us learn, and for the stimulus their development will give to data analysis.

Once "species" are identified, the problem of assigning individuals to them is present. There is a reasonable body of work on the classical approaches to discrimination, but much remains to be done. The possibilities of using the Penrose technique repeatedly (see Section 22 above) have neither been investigated or exploited. And the assessment, as an end product, for each individual of its probabilities of belonging to each "species", in place of the forcible assignment of each individual to a single species, has, to my knowledge, only appeared in connection with so-called *probability weather forecasts.* (For an example of assessment as an expression of the state of the evidence, see Mosteller and Wallace, 1962.)

When we realize that classification includes medical diagnosis, and we recognize the spread of affect, from naive optimism to well-informed hope for slow gains as a result of extensive and coordinated effort, with which the application of electronic computer to medical diagnosis is presently being regarded by those who are beginning to work in this field, we cannot find the classification problem simple or adequately treated.

Once either taxonomy or classification is an issue, the problem of incomplete data arises. This is particularly true in medical diagnosis, where no patient may have had all the tests. There is a small body of literature on the analysis of incomplete data. Unfortunately it is mostly directed to the problems of using incomplete data to estimate population parameters. Such results and techniques can hardly be used in either taxonomy or classification. Great progress will be required of us here also.

V. Some Other Promising Areas

27. Stochastic-Process Data

The number of papers about the probability theory of stochastic processes has grown until it is substantial, the number of papers about statistical inference in stochastic processes is seeking to follow the same pattern, yet, with a few outstanding exceptions, there does not seem to be anything like a comparable amount of attention to *data analysis* for stochastic processes. About the cause of this there can be quite varied views. A tempting possibility is that we see the ill effects of having the empirical analysis of data wait upon theory, rather than leading theory a merry chase.

Techniques related to circular frequencies, cosines, or complex exponentials, and to linear, time-invariant black boxes are one of the outstanding exceptions. The estimation of power spectra has proved a very useful tool in the hands of many who work in diverse fields. Its relation to variance components had been discussed quite recently (Tukey, 1961a). Its more powerful brother, the analysis of cross-spectra, which possesses the strength of methods based upon regression, is on its way to what are likely to be greater successes. All these techniques involve quadratic or bilinear expressions, and their variability involves fourth-degree expressions. Their use is now beginning to be supplemented by quite promising techniques associated with (individual, pairs of, or complex-valued) linear expressions, such as (approximate) Hilbert transforms and the results of complex demodulation.

The history of this whole group of techniques is quite illuminating as to a plausible sequence of development which may appear in other fields:

(a1) the basic ideas, in an imprecise form, of distribution of energy over frequency became a commonplace of physics and engineering,
(a2) a belief in the possibility of exact cycles, of the concentration of energy into lines, led to the development of such techniques as Schuster's periodograms and the various techniques for testing significance in harmonic analysis (e.g., Fisher, 1929),
(a3) an abstract theory of generalized harmonic analysis of considerable mathematical subtlety was developed by Wiener (e.g., 1930).
(a4) various pieces of data were analyzed in various ways, mostly variously unsatisfactory,
(a5) a Bell Telephone Laboratories engineer wanted to show a slide with a curve which represented the rough dependence of energy upon frequency for certain radar, and had some data analyzed,
(a6) examination showed a curious alternation of the estimates above and below a reasonable curve, and R.W. Hamming suggested (0.25, 0.50, 0.25) smoothing. (The striking success of this smoothing seized the writer's attention and led to his involvement in the succeeding steps.)
(a7) so the 4th degree theory of the variability and covariability of the estimates was worked out for the Gaussian case, to considerable profit,
(a8) gradually the simpler 2nd degree theory for the average values of spectral estimates, which does not involve distributional assumptions, came to be recognized as of greater and greater use,
(a9) more recently we have learned that still simpler 1st degree theory, especially of complex demodulation (cp., Tukey, 1961a, pp. 200–201), offers new promise,
(a10) and the next step will call for investigation of the theory, roughly of 2nd degree, of the variability of the results of such techniques.

In this history note that:

(b1) decades were lost because over-simple probability models in which there was a hope of estimating everything were introduced and taken seriously (in step a2),

(b2) the demands of actual data analysis have driven theoretical understanding rather than vice versa (e.g., cp., steps a6, a7, a9, a10),
(b3) by and large, the most useful simpler theory was a consequence of more complex theory, although it could have been more easily found separately (e.g., cp., steps a7 and a8).

There seems to be no reason why we should not expect (b1) (b2) and (b3) to hold in other areas of stochastic-process data analysis.

If I were actively concerned with the analysis of data from stochastic processes (other than as related to spectra), I believe that I should try to seek out techniques of data processing which were not too closely tied to individual models, which might be likely to be unexpectedly revealing, and which were being pushed by the needs of actual data analysis.

28. Selection and Screening Problems

Data analysis, as used here, includes planning for the acquisition of data as well as working with data already obtained. Both aspects are combined in multistage selection and screening problems, where a pool of candidates are tested to differing extents, and the basic questions are those of policy. How many stages of testing shall there be? How much effort shall be spent on each? How is the number of candidates passed on from one stage to another to be determined?

This subject has been studied, but the results so far obtained, though quite helpful, leave many questions open. (Cp., Dunnett, 1960, for some aspects; Falconer, 1960, and Cochran, l9ol, for others.) This is in part due directly to the analytical difficulty of the problerns, many of whose solutions are going to require either wholly new methods, or experimental simulation. An indirect effect of analytical difficulty is that available solutions refer to criteria, such as "mean advance" which do not fit all applications.

Not only do problems of this class occur in widely different fields— selecting scholarship candidates, breeding new plant varieties, screening molds for antibiotic production are but three—but the proper criterion varies within a field of application. The criterion for selecting a single national scholarship winner should be different from that used to select 10,000 scholarship holders; the criterion for an antibiotic screen of earth samples (in search of new species of antibiotic producers) should be different from that for an antibiotic screen of radiation-induced mutants (in search of a step toward more efficient producers); and so on.

Practical selection (cp., e.g., page 223 in Lerner, 1954) has long made use of selection for the next stage, not of a fixed number, nor of a fixed fraction, but of those whose indicated quality is above some apparent gap in indicated quality. It is reasonable to believe (as I for one do) that such flexible selection is more efficient than any fixed % methods. But, to my knowledge, we have no evidence from either analytic techniques or empirical simulation as to whether this is indeed the case. This is but one of many open questions.

29. External, Internal, and Confounded Estimates of Error

The distinction between external and internal estimates of error is a tradition in physical measurement, where external estimates may come from comparisons between the work of different investigators, or may even be regarded as requiring comparisons of measurements of the same quantity by different methods. A similar distinction is of course involved in the extensive discussions of "the proper error term" in modern experimental statistics. No one can consider these questions to be of minor importance.

But there is another scale of kinds of error estimate whose importance, at least in a narrower field, is at least as great; a scale which can be regarded as a scale of subtlety or a scale of confusion. The first substantial step on this scale may well have been the use of "hidden replication" (cp., Fisher, 1935) in a factorial experiment as a basis of assessing variability appropriate as a measure of error. This can be regarded, from one standpoint, as merely the use of residuals from an additive fit to assess the stability of the fitted coefficients. As such it would not be essentially different from the use of residuals from a fitted straight line. But if, in this latter case, the fluctuations vary about a crooked line, we know of no sense in which the residuals, which include both variability and crookedness, are a specifically appropriate error term. In the factorial case, however, the presence of arbitrary interactions does not affect the correctness of the error term, provided the versions of each factor are randomly sampled from a population of versions (e.g., Cornfield and Tukey, 1956) whose size is allowed for. Thus "hidden replication" may reasonably be regarded as a substantial step in subtlety.

The recovery of inter-block information in incomplete block experiments (including those in lattices) is another stage, which has been extended to the recovery of inter-variety information (Bucher, 1957).

But the latest step promises much. Until it was taken, we all felt that error should be assessed in one place, effects in another, even if different places were merely different *pre-chosen* linear combinations of the observations. The introduction of the half-normal plot (cp., Daniel, 1959) has shown us that this need not be the case, that, under the usual circumstance where many effects are small, while a few are large enough to be detected, we may confound *effects* with *error* and still obtain reasonable analyses of the result.

(*Editor's note*: Subsections 30–31 have been omitted.)

32. Two Samples with Unequal Variability

Why is the treatment of heterogeneous variability in such poor shape? In part, perhaps, because we are stuck at the early stage of the Behrens-Fisher problem. Analytical difficulties are undoubtedly a contributing cause, but it is doubtful that they would have been allowed to hold us up if we had reached a clear conclusion as to what do with the two-sample problem when the variances are to be separately estimated with appreciable uncertainty.

While it has been the fashion to act as though we should solve this problem in terms of high principle, either a high principle of "making use of all the information" or a high principle of "making exact probability statements", it now seems likely that we may, in due course, decide to settle this question in a far more realistic way. Both the highly-principled approaches before us require *precise* normality of the two parent distributions. We know this condition is not met in practice, but we have not asked how much the case of non-normal populations can tell us about what technique can reasonably be applied, or whether either of the well-known proposals is reasonably robust. We should ask, and think about the answer.

Martin Wilk is vigorous in pointing out that solving the two-sample *location* problem will not solve the problem of comparing two samples whose variances are likely to differ. He considers that the crux of the issue lies beyond, where both difference in variance and difference in mean are equally interesting to describe and test. Discussion of this problem, like that of so many left untouched here, would open up further interesting and rewarding issues for which we have no space.

While no clear general line has yet been identified, along which major progress in meeting heterogeneity of variability is likely to be made, there are a number of clear starting points.

VI. Flexibility of Attack

33. Choice of Modes of Expression

Clearly our whole discussion speaks for greater flexibility of attack in data analysis. Much of what has been said could be used to provide detailed examples. But there are a few broader issues which deserve specific mention.

Shall the signal strength be measured in volts, in volts^2 (often equivalent to "watts"), in $\sqrt{\text{volts}}$, or in log volts (often described as "in decibels")? As a question about statement of final results this would not be for data analysis and statisticians to answer alone, though they would be able to assist in its answering. But, as a question about how an analysis is to be conducted, it is their responsibility, though they may receive considerable help from subject-matter experts.

This is an instance of a choice of a mode of expression for a single response, a question about which we now know much more than we did once upon a time. Clarification of what data characteristics make standard techniques of analysis more effective, and of the nature of measurement, as it has been developed in science, has made it possible to set down goals, and to arrange them in order of usual importance. Moreover, there are reasonably effective techniques for asking sufficiently large bodies of data about the mode in which their analysis should go forward. Today, our first need here is to assemble and purvey available knowledge.

The related question of expressing multiple-response data in a desirable way is almost unattacked. It is substantially more complex than the single-response case, where the order of the different responses is usually prescribed, so that all that remains is to assign numerical values to these responses in a reasonable way.

First and last, a lot is said about changes of coordinates, both in pure mathematics, and in many fields of mathematical application. But little space is spent emphasizing what "a coordinate" is. From our present point of view, a coordinate is the combination of two things:

(a1) a set of level surfaces, namely a classification of the points, objects, responses or response-lists into sets such that any two members of a set are equivalent so far as that particular coordinate is concerned (as when a 200 lb. woman aged 53 has the same height as a 110 lb. man aged 22), and

(a2) an assignment of numerical values, one to each of these level surfaces or equivalence classes.

In dealing with modes of expression for single-response data, part (a1) is usually wholly settled, and we have to deal only with part (a2).

In dealing with multiple-response data, the least we are likely to face is a need to deal with part (a2) for each response, at least individually but often somewhat jointly. Actually, this simplest case, where the identification of the qualitative, (a1), aspects of all coordinates is not at all at the analyst's disposal, is rather rare. A much more frequent situation is one in which coupled changes in (a1) and (a2) are to be contemplated, as when any linearly independent set of linear combinations of initially given and quantitatively expressed coordinates make up an equally plausible system of coordinates for analysis. In fact, because of its limited but real simplicity, we often assume this situation to hold with sometimes quite limited regard for whether this simplicity, and the consequent emphasis on affine invariance of techniques, is appropriate in the actual instance at hand.

In many instances of multiple-response data we can appropriately consider almost any system of coordinates, giving at most limited attention to the qualitative and quantitative relations of the coordinates for analysis to the initial coordinates. Our knowledge of how to proceed in such circumstances is limited, and poorly organized. We need to learn much more; but we also need to make better use of what we do know. All too often we approach problems in terms of conventional coordinates although we know that the effect to be studied is heavily influenced by an *approximately-known com-*

bination of conventional coordinates. Time and temperature in the stability testing of foods, chemicals, or drugs is one instance. (The combination of time and temperature corresponding to 30 kcal/mole activation energy can be profitably substituted for time in most problems. A little information will allow us to do even better.) Temperature and voltage in the reliability testing of electronic equipment is another.

As in so many of the instances we have described above, *the main need in this field* is for techniques of indication, for ways to allow the data to express their *apparent* character. The need for significance and confidence procedures will only begin to arise as respectable indication procedures come into steady use.

34. Sizes, Nomination, Budgeting

The choice of qualitative and quantitative aspects of coordinates is not the only way in which to approximately exercise judgment in approaching the analysis of multiple response data. The work of Dempster (1958, 1960) and of Wilk and Gnanadesikan (1962) points the way toward what seems likely to prove an extensive use of judgment-selected measures of "size" for differences of multiple response. The considerations which should be involved in such choices have not yet been carefully identified, discussed, and compared. Still, it is, I believe, clear that one should not limit oneself to information and judgment about the actual variability, individual and joint, of the several responses (or of more useful coordinates introduced to describe these responses). It will also be wise and proper to give attention to what sorts (= what directions) of real effects seem more likely to occur, and to what sorts of effects, if real, it is more likely to be important to detect or assess.

The problems which arise in trying to guide the wise choice of "size" are new, but not wholly isolated. The practice at East Malling, where experiments take a major fraction of a scientific lifetime, of *nominating* (cp., Pearce, 1953) certain comparisons, appears to have been the first step toward what seems to be an inevitable end, the budgeting of error rates in complex experiments. We consider it appropriate to combine subject-matter wisdom with statistical knowledge in planning what factors shall enter a complex experiment, at how many versions each shall appear, and which these versions shall be. This granted, how can there be objection to using this same combination of wisdom and knowledge to determine, in advance of the data, what level of significance shall be used at each of the lines of the *initial* analysis of variance. If wisdom and knowledge suffice to determine whether or not a line is to *appear* in the initial analysis, surely they suffice to determine whether 5%, 1%, or 0.1% is to be the basis for immediate attention. Yet budgeting of error rate does not seem to have yet been done to any substantial extent.

35. A Caveat About Indications

It may be that the central problem of complex experimentation may come to be recognized as a psychological one, as the problem of becoming used to a separation between indication and conclusion. The physical sciences are used to "praying over" their data, examining the same data from a variety of points of view. This process has been very rewarding, and has led to many extremely valuable insights. Without this sort of flexibility, progress in physical science would have been much slower. Flexibility in analysis is often only to be had honestly at the price of a willingness not to demand that what has *already* been observed shall establish, or prove, what analysis *suggests*. In physical science generally, the results of praying over the data are thought of as something to be put to further test in another experiment, as indications rather than conclusions.

If complex experiment is to serve us well, we shall need to import this freedom to reexamine, rearrange, and reanalyze a body of data into all fields of application. But we shall need to bring it in *alongside*, and not in place of, preplanned analyses where we can assign known significance or confidence to conclusions about preselected questions. We must be prepared to have indications go far beyond conclusions, or even to have them suggest that what was concluded about was better not considered. The development of adequate psychological flexibility may not be easy, but without we shall slow down our progress.

This particular warning about the place of indication besides conclusion applies in many places and situations. It appears at this particular point because sizes, nomination, and error-rate budgeting are all directed toward the attainment of conclusions with valid error-rates of reasonable overall size. It would have been dangerous to have left the impression that suggestive behavior which did not appear conclusive because of the wise use of such somewhat conservative techniques should, consequently, not be taken as an indication, should not even be considered as a candidate for following up in a later study. Indications are not to be judged as if they were conclusions.

(*Editor's note*: Subsections 36–37 have been omitted.)

VII. A Specific Sort of Flexibility

38. The Vacuum Cleaner

In connection with stochastic process data we noted the advantages of techniques which were revealing in terms of many different models. This is a topic which deserves special attention.

If one technique of data analysis were to be exalted above all others for its ability to be revealing to the mind in connection with each of many different models, there is little doubt which one would be chosen. The simple graph has brought more information to the data analyst's mind than any other device. It specializes in providing indications of unexpected phenomena. So long as we have to deal with the relation of a single response to a single stimulus, we can express almost everything qualitative, and much that is quantitative, by a graph. We may have to plot the differences between observed response and a function of the stimulus against another function of stimulus; we may have to re-express the response, but the meat of the matter can usually be set out in a graph.

So long as we think of direct graphing of stimulus against response, we tend to think of graphing as a way to avoid computation. But when we consider the nature and value of indirect graphs, such as those mentioned above, we come to realize that a graph is often the way in which the results of a substantial computational effort is made manifest.

We need to seek out other tools of data analysis showing high flexibility of effectiveness in other situations. Like the simple graph they will offer us much. We should not expect them to be free of a substantial foundation of computation, and we should not expect their results to be necessarily graphical. Our aim should be flexibility.

In the area of time-series-like phenomena, as we have already noted, spectrum analysis and, more particularly, cross-spectrum analysis (and their extensions) offer such tools.

For data set out in a two-way table, exhibiting the combined relation of two factors to a response, we have long had a moderately flexible approach: the fitting of row, column, and grand means, *and the calculation of residuals.* We have had much profit from the fitting of row, column, and grand means, though not as much as if we had usually gone on to calculate individual residuals, rather than stopping with calculation of the sum of their squares (of the "sum of squares for interaction" or "for discrepancy" etc. in the corresponding analysis of variance). But it is easy to write down two-way tables of quite distinct structure where the fitting of row, column, and grand means fail to exhaust the bulk of this structure. We need to have general techniques that go farther than any just mentioned.

A first step was practiced under such titles as "the linear-by-linear interaction", and was later formalized as "one degree of freedom for non-additivity" (Tukey, 1949b). By isolating a further single (numerical) aspect of the tabulated values, it became possible to ask the data just one more question, and to retrieve just one more number as an answer.

How does one ask still more questions of a two-way table in such a way as to detect as much orderly behavior as possible? The answer here must depend upon the nature of the two factors. If one or both is quantitative, or naturally ordered, we will have access to techniques not otherwise available.

Let us be Spartan, and suppose that neither factor has natural quantitative expression or natural order.

If we are to ask reasonably specific questions we must plan to be guided by the table itself in choosing which ones we ask. (This is true to the extent that general prescriptions can be given. Subject-matter knowledge and insight can, and of course should, guide us in specific instances.) If the start is to be routine, prechosen, there is little chance that the fitting of row, column, and grand means can be replaced by some other first step that is both equally simple and better. And the question becomes, once these have been fitted, and residuals formed, how are we to be guided to a next step? Only the fitted means offer us new guidance.

The table of coefficients whose entries are products of "row mean minus grand mean" with "column mean minus grand mean" designates the one degree of freedom for non-additivity. To go further we should perhaps make use of the individual factors rather than their product. When we seek for ways to apply "row mean minus grand mean", for example, we see that we can apply these entries separately in each column, obtaining one regression coefficient per column, a whole row of regression coefficients in all; and so on. This procedure generates what, since its initial use was to produce "vacuum cleaned" residuals (residuals more completely free of systematic effect than those obtained by mean fitting), is conveniently called the basic *vacuum cleaner*. (Those who feel that FUNORFUNOM really removes the dirt, may wish to adopt Denis Farlie's suggestion that the present procedure be called FILLET, (i) because it is like taking the bones out of a fish (a herring, perhaps), (ii) from the acronym "Freeing Interaction Line from the Error Term".)

(*Editor's note*: Subsections 39–42 have been omitted.)

VIII. How Shall We Proceed?

43. What Are the Necessary Tools?

If we are to make progress in data analysis, as it is important that we should, we need to pay attention to our tools and our attitudes. If these are adequate, our goals will take care of themselves.

We dare not neglect any of the tools that have proved useful in the past. But equally we dare not find ourselves confined to their use. If algebra and analysis cannot help us, we must press on just the same, making as good use of intuition and originality as we know how.

In particular we must give very much more attention to what specific techniques and procedures do when the hypotheses on which they are customarily developed do not hold. And in doing this we must take a positive attitude, not a negative one. It is not sufficient to start with what it is supposed to be desired to estimate, and to study how well an estimator succeeds in doing this. We must give even more attention to starting with an estimator and discovering what is a reasonable estimate, to discovering what is it reasonable to think of the estimator as estimating. To those who hold the (ossified) view that "statistics is optimization" such a study is hindside before, but to those who believe that "the purpose of data analysis is to analyze data better" it is clearly wise to learn what a procedure really seems to be telling us about. It would be hard to overemphasize the importance of this approach as a tool in clarifying situations.

Procedures of diagnosis, and procedures to extract indications rather than conclusions, will have to play a large part in the future of data analyses. Graphical techniques offer great possibilities in both areas, and deserve far more extensive discussion than they were given here. Graphs will certainly be increasingly "drawn" by the computer without being touched by hands. More and more, too, as better procedures of diagnosis and indication are automated, graphs, and other forms of expository output, will, in many instances, cease to be the vehicle through which a man diagnoses or seeks indications, becoming, instead, the vehicle through which the man supervises, and approves or disapproves, the diagnoses and indications already found by the machine.

44. The Role of Empirical Sampling

Numerical answers about the absolute or comparative performance of data analysis procedures will continue to be of importance. Approximate answers will almost always serve as well as exact ones, provided the quality of the approximation is matched to the problem. There will, in my judgment, be no escape from a very much more extensive use of experimental sampling (empirical sampling, Monte Carlo, etc.) in establishing these approximate answers. And while a little of this experimental sampling can be of a naive sort, where samples are drawn directly from the situation of concern, the great majority of it will have to be of a more sophisticated nature, truly deserving the name of Monte Carlo. (Cp., Kahn, 1966, for an introduction.)

It is, incidentally, both surprising and unfortunate that those concerned with statistical theory and statistical mathematics have had so little contact with the recent developments of sophisticated procedures of empirical sampling. The basic techniques and insights are fully interchangeable with those of survey sampling, the only difference being that many more "handles" are easily available for treating a problem of statistical theory than are generally available for treating a problem about a human population or about an aggregation of business establishments. (cp., Tukey, 1957, for an instance of interchangeability.)

As one comes to make use of all that he knows, both in replacing the original problem by one with an equivalent answer, and in being more subtle in analysis of results, one finds that no more than a few hundred samples suffice to answer most questions with adequate accuracy. (The same modifications tend to reduce the demands for extreme high quality of the underlying "random numbers".) And with fast electronic machines such numbers of samples do not represent great expenditures of time or money. (Programming time is likely to be the bottleneck.)

45. What Are the Necessary Attitudes?

Almost all the most vital attitudes can be described in a type form: *willingness to face up to X*. Granted that facing up can be uncomfortable, history suggests it is possible.

We need to face up to *more realistic problems*. The fact that normal theory, for instance, may offer the only framework in which some problem can be tackled simply or algebraically may be a very good reason for *starting* with the normal case, but never can be a good reason for **stopping** there. We must expect to tackle more realistic problems than our teachers did, and expect our successors to tackle problems which are more realistic than those we ourselves dared to take on.

We need to face up to the *necessarily approximate nature of useful results in data analysis*. Our formal hypotheses and assumptions will never be broad enough to encompass actual situations. Even results that pretend to be precise in derivation will be approximate in application. Consequently we are likely to find that results which are approximate in derivation or calculation will prove no more approximate in application than those that *pretend* to be precise, and even that some admittedly approximate results will prove to be *closer* to fact in application than some supposedly exact results.

We need to face up to the need for *collecting the results of actual experience with specific data-analytic techniques*. Mathematical or empirical-sampling studies of the behavior of techniques in idealized situations have very great value, but they cannot replace experience with the behaviour of techniques in real situations.

We need to face up to the *need for iterative procedures in data analysis*. It is nice to plan to make but a single analysis, to avoid finding that the results of one analysis have led to a requirement for making a different one. It is also nice to be able to carry out an individual analysis in a single straightforward step, to avoid iteration and repeated computation. But it is not realistic to believe that good data analysis is consistent with either of these niceties. As we learn how to do better data analysis, computation will get more extensive, rather than simpler, and reanalysis will become much more nearly the custom.

We need to face up to the need for *both indication and conclusion in the same analysis.* Appearances which are not established as of definite sign, for example, are not all of a muchness. Some are so weak as to be better forgotten, others approach the borders of establishment so closely as warrant immediate and active following up. And the gap between what is required for an interesting indication and for a conclusion widens as the structure of the data becomes more complex.

We need to face up to the need for *a free use of* ad hoc *and informal procedures in seeking indications.* At those times when our purpose is to ask the data what it suggests or indicates it would be foolish to be bound by formalities, or by any rules or principles beyond those shown by empirical experience to be helpful in such situations.

We need to face up to the fact that, as we enter into new fields or study new kinds of procedures, *it is natural for indication procedures to grow up before the corresponding conclusion procedures do so.* In breaking new ground (new from the point of view of data analysis), then, we must plan to learn to ask first of the data what it suggests, leaving for later consideration the question of what it establishes. This means that almost all considerations which explicitly involve probability will enter at the later stage.

We must face up to the need for a *double standard in dealing with error rates, whether significance levels or lacks of confidence.* As *students* and *developers* of data analysis, we may find it worth while to be concerned about small difference among error rates, perhaps with the fact that a nominal 5% is really 4% or 6%, or even with so trivial a difference as from 5% to 4.5% or 5.5%. But as *practitioners* of data analysis we must take a much coarser attitude toward error rates, one which may sometimes have difficulty distinguishing 1% from 5%, one which is hardly ever able to distinguish more than one intermediate value between these conventional levels. To be useful, a conclusion procedure need not be precise. As working data analysts we need to recognize that this is so.

We must face up to the fact that, in any experimental science, *our certainty about what will happen in a particular situation does not usually come from directly applicable experiments or theory*, but rather comes mainly through analogy between situations which are *not known* to behave similarly. Data analysis has, of necessity, to be an experimental science, and needs therefore to adopt the attitudes of experimental science. As a consequence our choices of analytical approach will usually be guided by what is known about simpler or similar situations, rather than by what is known about the situation at hand.

Finally, we need to give up the vain hope that data analysis can be founded upon a logico-deductive system like Euclidean plane geometry (or some form of the propositional calculus) and to face up to the fact that *data analysis is intrinsically an empirical science.* Some may feel let down by this, may feel that if data analysis cannot be a logico-deductive system, it inevitably falls to the state of a crass technology. With them I cannot agree. It will still be true that there will be aspects of data analysis well called technology, but there will also be the hallmarks of stimulating science: intellectual adventure, demanding calls upon insight, and a need to find out "how things really are" by investigation and the confrontation of insights with experience.

46. How Might Data Analysis Be Taught?

If we carry the point of view set forth here to its logical conclusion, we would teach data analysis in a very different way from any that I know to have been tried. We would teach it like biochemistry, with emphasis on what we have learned, with some class discussion of how such things were learned perhaps, but with relegation of all question of detailed methods to the "laboratory work". If we carried through the analogy to the end, all study of detailed proofs, as well as all trials of empirical sampling or comparisons of ways of presentation would belong in "the laboratory" rather than "in class". Moreover, practice in the use of data analysis techniques would be left to other courses in which problems arose, just as applications of biochemistry are left to other courses.

It seems likely, but not certain, that this would prove to be too great a switch to consider putting into immediate effect. Even if it is too much for one step, what about taking it in two or three steps?

I can hear the war cry "cookbookery" being raised against such a proposal. If raised it would fail, because the proposal is really to go in the opposite direction from cookbookery; to teach not "what to do", nor "how we learned what to do", but rather "what we have learned". This last is at the opposite pole from "cookbookery", goes beyond "the conduct of taste-testing panels", and is concerned with "the art of cookery". Dare we adventure?

(*Editor's note*: Subsection 47 has been omitted.)

48. What of the Future?

The future of data analysis can involve great progress, the overcoming of real difficulties, and the provision of a great service to all fields of science and technology. Will it? That remains to us, to our willingness to take up the rocky road of real problems in preference to the smooth road of unreal assumptions, arbitrary criteria, and abstract results without real attachments. Who is for the challenge?

References

Anderson, Edgar (1949). *Introgressive Hybridization.* Wiley, New York.

Anderson, Edgar (1957). A semigraphical method for the analysis of complex problems. *Proc. Nat. Acad. Sci. USA* **13** 923–927. (Reprinted with appended note: *Technometrics* **2** 387–391).

Anscombe, F.J. and Tukey, J.W. (1962) The examination and analysis of residuals, *Technometrics.*

Azimov, Isaac (1955). The sound of panting. *Astounding Science Fiction* **55** June 104–113.

Barnard, G.A. (1959a). Control charts and stochastic processes. *J. Rov. Statist. Soc.* Ser. B. **21** 239–257 and reply to discussion, 269–271.

Barnard, G.A. (1959b). (Discussion of Kiefer, 1959). *J. Roy. Statist. Soc.* Ser. B. **21** 311–312.

Bartlett, M.S. (1947). Multivariate analysis. *J. Roy. Statist. Soc.* Suppl. **9** 176–197.

Beale, E.M.L. and Mallows, C.L. (1958). On the analysis of screening experiments. Princeton Univ. Stat. Tech. Res. Grp. Tech. Rpt. 20.

Benard, A. and Bos-Levenbach, E.C. (1953). The plotting of observations on probability paper. (Dutch) *Statistica* (Rijkswijk) **7** 163–173.

Blom, Gunnar (1958). *Statistical Estimates and Transformed Beta-Variables.* Wiley, New York.

Bortkiewicz, Ladislaus von (1901). Andwendungen der Wahrscheinlichkeitsrechnung auf Statistik. *Encyklopadie der Math. Wissenschaften* (Teil 2, 1900–1904) 822–851. Leipzig, Teubner. (Especially p. 831).

Box, G.E.P. (1956). (Discussion) *J. Roy. Statist. Soc.* Ser. B. **18** 28–29.

Bucher, Bradley D. (1957). The recovery of intervariety information in incomplete block designs. Ph.D. Thesis, Princeton, 108 pp.

Chernoff, Herman and Lieberman, Gerald J. (1954). Use of normal probability paper. *J. Amer. Statist. Assoc.* **49** 778–785.

Chernoff, Herman and Lieberman, Gerald J. (1956). The use of generalized probability paper for continuous distributions. *Ann. Math. Statist.* **27** 806–818.

Cochran, W.G. (1951). Improvement by means of selection. *Proc. Second Berkeley Symp. Math. Stat. Prob.* 449–470. Univ. of California Press.

Cornfield, Jerome and Tukey, John W. (1956). Average values of mean squares in factorials. *Ann. Math. Statist.* **27** 907–949.

Cox, D.R. (1957). The use of a concomitant variable in selecting an experimental design. *Biometrika* **44** 150–158 and 534.

Cox, D.R. (1958). Some problems connected with statistical inference. *Ann. Math. Statist.* **29** 357–372.

Creasy, M.A. (1957). Analysis of variance as an alternative to factor analysis. *J. Roy. Statist. Soc. Ser. B* **19** 318–325.

Daniel, C. (1959). Use of half-normal plots in interpreting factorial two-level experiments. *Technometrics* **1** 311–341.

Deming, W.E. (1943). *Statistical Adjustment of Data.* Wiley, New York.

Dempster, A.P. (1958). A high dimensional two sample significance test. *Ann. Math. Statist.* **28** 995–1010.

Dempster, A.P. (1960). A significance test for the separation of two highly multivariate small samples. *Biometrics* **16** 41–50.

Dixon, W.J. (1957). Estimates of the mean and standard deviation of a normal population. *Ann. Math. Statist.* **28** 806–809

Dixon, W.J. (1960). Simplified estimation from censored normal samples. *Ann. Math. Statist.* **31** 385–391.

Dunnett, C.W. (1960). On selecting the largest of k normal population means. *J. Roy. Statist. Soc. Ser. B* **22** 1–30 and reply to discussion 38–40.

Eysenck, H.J. (1950). Criterion analysis. An application of the hypotheticodeductive method to factor analysis. *Psychol. Bull.* **57** 38–53.

Eysenck, H.J. (1952). *The Scientific Study of Personality.* Routledge and Kegan, London

Finney, David J. (1947), (1952). *Probit Analysis; a Statistical Treatment of the Sigmoid Response Curve.* Cambridge Univ. Press.

Fisher, R.A. (1929). Tests of significance in harmonic analysis. *Proc. Roy. Soc. London* Ser. A. **125** 54–59.

Fisher, R.A. (1935), (1937, 1942, 1947, ...). *The Design of Experiments.* Oliver and Boyd, Edinburgh and London. (especially Sec. 41).

Gauss, Carl Friedrich (1803). Disquisitiones de elementis ellipticis pallidis. *Werke* **6** 20–24 (French translation by J. Bertrand, 1855. English translation by Hale F. Trotter. Princeton Univ. Stat. Tech. Res. Grp. Tech. Rpt. 5.)

Gayen, A.K. (1949). The distribution of 'Student's' *t* in random samples of any size drawn from non-normal universes. *Biometrika* **36** 353–369 (and references therein.)

Geary, R.C. (1947). Testing for normality. *Biometrika* **34** 209–242.

Hannan, E.J. (1958). The estimation of spectral density after trend removal. *J. Roy. Statist. Soc.* Ser. B. **20** 323–333.

Johnson, N.L. (1949). Systems of frequency curves generated by methods of translation. *Biometrika* **36** 149–176.

Kahn, Herman (1956). Use of different Monte Carlo sampling techniques. *Symposium on Monte Carlo Methods* 146–190 (edited by Herbert A. Meyer). Wiley, New York.

Kiefer, J. (1959). Optimum experimental designs. *J. Roy. Statist. Soc.* Ser. B. **21** 272–304 and reply to discussion 315–319.

Kimball, George E. (1958). A critique of operations research. *J. Wash. Acad. Sci.* **48** 33–37 (especially p. 35).

Lerner, I. Michael (1950). *Population Genetics and Animal Improvement.* Cambridge Univ. Press.

Mallows, C.L. (1959). (Discussion of Kiefer, 1959). *J. Roy. Statist. Soc.* Ser. B. **21** 307–308.

Mann, H.B. and Wald, A. (1942). On the choice of the number of intervals in application of the chi square test. *Ann. Math. Statist.* **13** 306–317.

Michener, C.D. and Sokal, R.R. (1957). A quantitative approach to a problem in classification. *Evolution* **11** 130–162.

Mosteller, Frederick and Wallace, David L. (1962). Notes on an authorship problem. *Proc. Harvard Symp. Digital Computers and their Applications.* Harvard Univ. Press.

Pearce, S.C. (1953). Field experimentation with fruit trees and other perennial plants. Commonwealth Bur. Horticulture and Plantation Crops. Tech. Comm. 23.

Pearson, Karl (1895). Contributions to the mathematical theory of evolution. II. Skew variation in homogeneous material. *Philos. Trans. Roy. Soc. London* Ser. A. **186** 343–414.

Pearson, Karl (1900). On the criterion that a given system of deviations from the probable in the case of a correlated system of variables is such that it can be reasonably supposed to have arisen from random sampling. *Phil. Mag.* (5) **50** 157–175.

Penrose, L.S. (1947). Some notes on discrimination. *Ann. Eugenics* **13**, 228–237.

Pitman, E.J.G. (1938). Significance tests which may be applied to samples from any populations. III. The analysis of variance test. *Biometrika* **29** 322–335.

Rao C. Radhakrishna (1952). *Advanced Statistical Methods in Biometric Research.* Wiley, New York.

Rao, C. Radhakrishna (1959). Some problems involving linear hypotheses in multivariate analysis. *Biometrika* **46** 49–58.

Rogers, David J. and Tanimoto, Taffee T. (1960). A computer-program for classifying plants. *Science* **132** 1115–1118.

Roy, J. (1958). Step-down procedure in multivariate analysis. *Ann. Math. Statist.* **29** 1177–1187.

Roy, S.N. and Gnanadesikan, R. (1957). Further contributions to multivariate confidence bounds. *Biometrika* **44** 399–410.

Roy, S.N. and Gnanadesikan, R. (1958). A note on "Further contributions to multivariate confidence bounds". *Biometrika* **45** 581.

Sarhan, Ahmed E. and Greenberg, Bernard G. (1958). Estimation of location and scale parameters by order statistics from singly and doubly censored samples. *Ann. Math. Statist.* **29** 79–105.

Siegel, Sidney and Tukey, John W. (1960). A non-parametric sum of ranks procedure for relative spread in unpaired samples. *J. Amer. Statist. Assoc.* **55** 429–445.

Sneath P.H.A. (1957a). Some thoughts on bacterial classifications. *J. Gen. Microbiol.* **17** 184–200.

Sneath, P.H.A. (1967b) The application of computers to taxonomy. *J. Gen. Microbiol.* **17** 201–226.

Sneath, P.H.A. and Cowan, S.T. (1958). An electrotaxonomic survey of bacteria. *J. Gen. Microbiol.* **19** 551–565.

Stevens, W.L. (1950). Fiducial limits of the parameter of a discontinuous distribution. *Biometrika* **37** 117–129.

Student (William Sealy Gosset) (1927). Errors of routine analysis. *Biometrika* **19** 151–164. (Reprinted as pages 135–149 of "Student's" *Collected Papers* (edited by E.S. Pearson and John Wishart) Biometrika Office, Univ. Coll. London 1942 (1947).

Tucker, Ledyard R. (1958). An interbattery method of factor analysis. *Psychometrika* **23** 111–136.

Tukey, John W. (1949a). Dyadic anova—An analysis of variance for vectors. *Human Biology* **21** 65–110.

Tukey, John W. (1949b). One degree of freedom for non-additivity. *Biometrics* **5** 232–242.

Tukey, John W. (1951). Components in regression. *Biometrics* **7** 33–69. (especially pp. 61–64).

Tukey, John W. (1957). Antithesis or regression. *Proc. Camb. Philos. Soc.* **53** 923–924.

Tukey, John W. (1960). A survey of sampling from contaminated distributions. Paper 39 (pp. 448–485) in *Contributions to Probability and Statistics* (edited by I. Olkin *et al*). Stanford Univ. Press.

Tukey, John W. (1961a). Discussion, emphasizing the connection between analysis of variance and spectrum analysis. *Technometrics* **3** 191–219.

Tukey, John W. (1961b). Statistical and quantitative methodology. In *Trends in Social Sciences*, (edited by Donald P. Ray) Philosophical Library, New York.

Tukey John W. (1962). The symmetric λ-distributions.

Various Authors (1959). Various titles. *Technometrics* **1** 111–209.

Von Neumann, John (1947). The mathematician, pp. 180–196 of the *Works of the Mind.* (edited by R.B. Heywood). Chicago Univ. Press. (Reprinted as pp. 2053–2069 of the *World of Mathematics* (edited by James R. Newman). Simon and Schuster, New York, 1956).

Walsh, John E. (1949a). On the range-midrange test and some tests with bounded significance levels. *Ann. Math. Statist.* **20** 257–267.

Walsh, John E. (1949b). Some significance tests for the median which are valid under very general conditions. *Ann. Math. Statist.* **20** 64–81.

Walsh, John E. (1959). Comments on "The simplest signed-rank tests". *J. Amer. Statist. Assoc.* **54** 213–224.

Welch, B.L. (1937). On the *z*-test in randomized blocks and latin squares. *Biometrika* **29** 21–52.

Wiener, Norbert (1930). Generalized harmonic analysis. *Acta Math.* **55** 118–258.

Wilk, M.B. and Gnanadesikan, R. (1961). Graphical analysis of multiple response experimental data using ordered distances. *Proc. Nat. Acad. Sci. USA* **47** 1209–1212.

Yates, F. (1955). (Discussion) *J. Roy. Statist. Soc.* Ser. B. **17** 31.

Introduction to Birch (1963) Maximum Likelihood in Three-Way Contingency Tables

Stephen E. Fienberg
York University

General Background

During the 1960s, there was an extremely rapid expansion of the literature on categorical data analysis, especially related to the use of loglinear models for multidimensional contingency tables, culminating in a series of papers by Goodman [e.g., see those collected in Goodman (1978)] and the monographs by Bishop et al. (1975) and Haberman (1974). The problems addressed by this literature paralleled for discrete/categorical data those addressed by multiple regression, analysis of variance, and the general linear model for continuous data. Loglinear model methods have had a major impact on the current statistical methodology in use today for the analysis of categorical data, and a crucial paper by Martin Birch that appeared in 1963 provides the underpinnings of the general statistical theory involved.

Yule (1900) had introduced some of the notions of independence and partial association for contingency tables, but without the formal methodology to accompany them. Thirty-five years later, Bartlett (1935) discussed the concept of no second-order interaction in $2 \times 2 \times 2$ tables and derived maximum likelihood estimates for the cell probabilities under this model. Although many authors attempted to extend these ideas, Roy and Kastenbaum's (1956) presentation of no second-order interaction for $r \times c \times t$ tables illustrates one of the problems that faced those working in the area, the computational difficulties associated with the practical application of the methods developed.

Part of the impetus for the burgeoning literature of the 1960s was the availability of high-speed computers that allowed researchers to carry out calculations that were deemed unreasonably complex only a decade earlier. Another part came from several large-scale medical and epidemiological studies in the United States that involved multiple categorical outcome variables and large numbers of explanatory variables, both continuous and discrete. For example, the National Halothane Study [see Bunker et al. (1969)] involved the analysis of contingency tables with seven or more variables and thousands of cells assessing the impact of various anesthetics on postoperative outcomes, and the Framingham study [see Truett et al. (1967)] also involved multiple categorical outcome variables with numerous explanatory variables in a multivariate analysis of the risk of coronary heart disease in the Massachusetts town of Framingham.

The final component of impetus came from the development of a framework for understanding categorical data models and related estimation and testing procedures. Key pieces of this framework can be traced to the early 1960s and a trio of papers by Darroch (1962), Good (1963), and Birch (1963), each of which identified key theoretical and practical issues of interest for contingency table analysis. [There were also related papers by Goodman (1963) and Plackett (1962), but they are less closely linked to the Birch paper than those by Darroch and Good.] Clearly, the ideas for extensions and a framework were in the air at the time, but it was Birch's paper that seemed to cast these ideas in their clearest and broadest form, and as a result, it played a breakthrough role, despite its overlap with the other two papers.

In the following sections, I first provide some personal background on Birch and then highlight the contributions and insights of Birch's paper itself. Following these sections, I indicate in somewhat greater detail the related contributions of Darroch and Good. Finally, I mention something about the extensive literature that can be traced back directly to Birch's paper.

Personal Background

Martin William Birch was born on June 6, 1939 in England. He received a B.A. in mathematics from Cambridge University in 1960. Subsequently, while working on his diploma course in mathematical statistics at Cambridge, he was introduced to applications of contingency table analysis in an applied project on genetics taught by W.F. Bodmer and P.A. Parsons. At the time, Birch was already thinking creatively about maximum likelihood ideas for categorical models, and he attended a lecture on the topic of contingency table analysis given by Robin Plackett, who was the external examiner for the diploma course in 1961 (McLaren, personal communication).

Following the completion of his diploma in 1961, Birch went first to the University of Liverpool for two years as a research assistant, where he was elected a Fellow of the Royal Statistical Society in 1962, and then to the University of Glasgow, where he was assistant in mathematical statistics for approximately three years. During this period, he completed a Ph.D. thesis on the analysis of contingency tables that observers are convinced would have been accepted but for personal reasons he chose not to submit it. In 1966, Birch gave up his post at Glasgow and joined Government Communications Headquarters, the British intelligence agency. He died tragically at the age of 30 on July 31, 1969.

During Birch's short academic career, he published five statistical papers. Three of these on loglinear models appeared in *Series B* of the *Journal of the Royal Statistical Society* [Birch (1963, 1964b, 1965)], and another on the asymptotic distribution of the Pearson chi-square statistic appeared in the *Annals of Mathematical Statistics* [Birch (1964a)]. The fifth paper, on errors in the variables in regression analysis, appeared in the *Journal of the American Statistical Association* [Birch (1964c)]. As mentioned above, Birch had already begun to think about maximum likelihood methods for categorical data models while at Cambridge, and he clearly was influenced by Plackett who was at Liverpool when Birch arrived there in 1961. As far as I have been able to ascertain, he had little or no link to others working in the field.

Although Birch's four categorical data papers are widely cited, by far the most influential one was his 1963 paper on maximum likelihood methods for log-linear models.

Birch's 1963 Paper

Birch organizes his 1963 paper around a series of problems in the analysis of three-way contingency tables, but in the process he introduces a number of more general results and an analysis-of-variance-like notation that facilitates a unified approach to maximum likelihood estimation for independence like models, as well as the no second-order (three-factor) interaction model.

Section 2 presents a simple yet powerful result on the equivalence of maximum likelihood estimates (MLE's) for what we now know as Poisson sampling (independent Poisson variates for each cell in a tabular array) and various product-multinomial sampling schemes (independent multinomials for several nonoverlapping sets of cells in a tabular array). Here Birch gives a necessary and sufficient condition for the equivalence that the MLE's for Poisson sampling satisfy the marginal constraints implied by each of the multinomials in a given product-multinomial sampling scheme. Equivalence of MLE's for two-way and multiway tables for various models of independence had been noted earlier, e.g., by Roy and Kastenbaum, but no one had set forth the general underlying result or used the device of the Poisson sampling model, which turns out later in the paper to facilitate various proofs.

In Sec. 3, Birch begins with the no second-order interaction model of Bartlett and of Roy and Kastenbaum and then recasts it in terms of contrasts of the logarithms of the expected cell values, which can be rewritten in the form of the loglinear model

$$\ln m_{ijk} = u + u_{1i} + u_{2j} + u_{3k} + u_{12ij} + u_{13ik} + u_{23jk},$$

where m_{ijk} is the expected count in the (i, j, k) cell corresponding to the observed count, n_{ijk}, and the sums of the u terms over any subscript are zero, i.e.,

$$u_{1.} = u_{2.} = u_{3.} = 0; \qquad u_{13i.} = u_{12i.} = 0 \qquad \text{for each } i;$$

$$u_{23j.} = u_{12.j} = 0 \qquad \text{for each } j; \qquad u_{23.k} = u_{13.k} = 0 \qquad \text{for each } k.$$

Note that Birch's double subscript notation facilitates generalization to higher dimensions. For the Poisson sampling scheme, Birch then proves that the two-way margins of the observed counts are sufficient statistics for this model, that the marginal totals are equal to their maximum likelihood estimates, i.e.,

$$\hat{m}_{ij.} = n_{ij} \qquad \text{for all } i, j,$$

$$\hat{m}_{i.k} = n_{i.k} \qquad \text{for all } i, k,$$

$$\hat{m}_{.jk} = n_{.jk} \qquad \text{for all } j, k,$$

and that, if all $n_{ijk} > 0$, the MLEs exist and are unique.

At this point, Birch notes, using the results of Sec. 2, that the MLEs for the Poisson sampling scheme are the same as those for the multinomial scheme that fixes only the table total, as well as for any product-multinomial scheme that fixes a single set of one- or two-way marginal totals. He then refers the reader to previous papers, including Darroch's, for iterative methods to solve the likelihood equations.

In Sec. 4, Birch writes out the general log-linear model for q-way tables where $q \geq 3$ and then describes hierarchical models, in which setting a given interaction term equal to zero implies that all of its higher-order relatives are equal to zero. Birch then states the generalization of the results from Sec. 3 for hierarchical models. (His use of the word symmetrical is a bit misleading here; it simply means that the u terms in a particular group, e.g., $\{u_{12ij}\}$, are either all zero or all arbitrary.) He also notes that for nonhierarchical models there are similar results, but that the sufficient statistics are no longer only marginal totals (this then prevents thc application of the results of Sec. 2). Toward the end of this section, he notes that the the form of the sufficient statistics for all these models is a consequence of the many-parameter exponential family distribution with which we are working.

Sections 5 and 6 of the paper revisit the various models of independence and conditional independence for three-way tables, reexpressing them as log-linear models, exploring relationships among them, and demonstrating that their MLEs have closed-form estimates and thus iteration is not required for their solution. (The exception is the model of marginal independence, which is not expressible as a loglinear model for the full table.) Once again, Birch notes the applicability of the equivalence result from Sec. 2.

The paper covers much ground in an innovative fashion, unifying results already available in the literature, as well as suggesting the form of general results not actually discussed. It is chock full of results, many of which are surprisingly elegant. Because of the many contributions that we see here, the failure of Birch to explore certain topics is worth noting. These include conditions for the existence of MLEs in the presence of some zero cell counts, general forms for closed-form estimates in four- and higher-way tables, efficient algorithms for the solution of the likelihood equations in higher-way tables, other exponential family generalizations (e.g., for most of the models in the GLIM family).

Related Contributions in Darroch (1962) and Good (1963)

Darroch's paper also notes the analogy between interaction models in contingency tables and the analysis of variance for continuous responses, and although he works with the no second-order interaction model in the multiplicative form, he also takes logarithms of the cell probabilities and discusses contrasts. Actually, a good chunk of the Darroch paper focuses on alternatives to the Bartlett–Roy–Kastenbaum approach, but in the end he returns to a multiplicative formulation in terms of the equality of odds ratios for three-way tables. For the no second-order interaction model, he conjectures the uniqueness of the MLEs when they exist and notes that he has existence and uniqueness results only for the $2 \times 2 \times 2$ and $3 \times 2 \times 2$ table cases. For the former, he gives the now well-known condition that the sums of counts in adjacent cells and in the diagonally opposite cells [i.e., the four pairs of cells: (1, 1, 1), and (2, 2, 2); (1, 2, 1) and (2, 1, 2); (2, 1, 1) and (1, 2, 2); and (2, 2, 1) and (1, 2, 2)] must be positive. This is an important component of Darroch's contribution to the literature that is missing from Birch's work; recall that Birch takes $n_{ijk} > 0$ for all (i, j, k) in order to avoid nonexistence problems.

Darroch also spends a considerable portion of his paper discussing perfect contingency tables in which

$$\frac{\sum_i p_{ij.}p_{i.k}}{p_{i..}} = p_{.j.}p_{..k} \qquad \text{for all } j, k,$$

$$\frac{\sum_j p_{.jk}p_{ij.}}{p_{.j.}} = p_{i..}p_{..k} \qquad \text{for all } i, k,$$

$$\frac{\sum_k p_{i.k}p_{.jk}}{p_{..k}} = p_{i..}p_{.j.} \qquad \text{for all } i, j.$$

Perfect tables satisfy the no second-order interaction model but are also collapsible. (Collapsibility as an issue was to be the focus of several papers in the 1970s and 1980s.) Darroch then links a model for perfect tables back to the Roy–Kastenbaum model writing out the likelihood as a linear function of the logarithms of the multiplicative parameters and stopping just short of a full log-linear formulation.

Finally, Darroch considers the likelihood equations for the no second-order interaction model and suggests an iterative procedure for their solution without a proof or claim for convergence. We can now recognize the algorithm as a variant of iterative proportional fitting, but Darroch gives no references to any other literature on the algorithm.

Darroch's interest in this problem was stimulated by a talk given by Robin Plackett on $2 \times 2 \times t$ tables at Manchester in December, 1960 (perhaps a lecture similar to the one attended by Birch in Cambridge). The paper was submitted for publication in the fall of 1961, and it appears unlikely that Birch would have had any knowledge of Darroch's results until the paper appeared in June 1962. Darroch does not recall having ever met Birch even though they worked only 50 miles apart.

Good's paper has a very different form, notation, and orientation from Darroch's and Birch's, and it contains several related results. The paper is a pathbreaking one in many senses. In it, Good introduces the principle of maximum conditional entropy and applies it to the generation models for m-way contingency tables with all rth-order subtables assigned. This yields a model that can be rewritten in hierarchical loglinear form, which Good proceeds to do but in a somewhat difficult-to-follow notation. Note that Good does not quite give the most general form of hierarchical loglinear model since the maximal interactions terms in his models are all of the same order. One of Good's results notes that the vanishing of all rth-order and higher-order interactions in m-way tables is equivalent to the vanishing of all rth-order interactions in all $(r + 1)$-dimensional subtables. (This result appears to be related to the Darroch notion of perfect tables, but it is somewhat different and thus applies to all symmetric log-linear models with matching *highest-order* terms.)

Up to this point, the paper deals only with model formulation, but then Good turns to maximum likelihood estimation for the cell probabilities subject to the vanishing of symmetrically chosen interactions. This yields a result similar to that found in Birch's paper for this class of loglinear models. Good gives the likelihood equations, and he proves that the MLEs are equivalent to maximum entropy estimates subject to the corresponding marginal constraints on the observed table, provided that the maximum of the likelihood is achieved at a stationary value. Good also proposes the method of iterative proportional fitting to solve these equations and gives reference for the method in the literature on information and control. He provides no proof of convergence for the algorithm, noting that "The matter requires further thought and numerical calculation, possibly rather extensive." A bit later, Good describes the approach for two-way tables in Deming and Stephan's (1940) paper, noting how it differs from maximum likelihood, but he fails to recognize that their proposed algorithm is, in effect, equivalent to the one he suggests using. He does comment that zero cell values create possible problems.

Good specifically references the Darroch paper, as well as Plackett's 1962 paper (referenced by both Darroch and Birch), but notes that these papers appeared after his was submitted for publication. Good's paper contained many of the same or at least results related to those of Birch, yet because of the notation, the form of the proofs, and the focus on model generation, it had less influence on the developments in the literature in the next several years. Later, other authors began to recognize the many innovative ideas that appear in the paper and that Good too had seen the role that loglinear models might play. By this time, however, many of the foundations of the log-linear model literature had already been set.

Thus, as we noted in the opening section of this introduction, all three of the papers by Birch, Darroch, and Good had common, overlapping elements that could be thought of as providing the foundation for many of the subsequent methodological developments.

The Subsequent Loglinear Literature

While working as part of a major medical project known as the National Halothane Study during 1966–67, Yvonne Bishop picked up on the ideas in Birch's paper and adapted the Deming–Stephan version of iterative proportional fitting to solve the likelihood equations for various loglinear and logit models applied to large sparse multiway arrays. She recognized that Birch's restriction to table with positive cell counts was unnecessary, although several years passed before others were to provide necessary and sufficient conditions for the existence of MLE's. The application of these loglinear methods using high-speed computers to very large problems was a major contribution to the Halothane study [Bunker et al. (1969)], and much of the subsequent work of authors such as Fienberg, Goodman, Haberman, and others can be traced to Birch (1963) through the work in her 1967 dissertation and in Bishop (1969). Her implementation of the Deming–Stephan algorithm still forms the basis of several popular computer programs for multiway contingency table analysis.

At about the same time, a paper by Ireland and Kullback (1968) revived the contributions of Good and linked them more clearly with the earlier work of Kullback and his students on minimum discrimination information methods for a contingency table and the use of the Deming-Stephan algorithm. Later this strand of literature became intertwined with that on the use of maximum likelihood estimation because of the equivalence result pointed out by Good.

Yet another strand of work on loglinear models is also traceable to the seminal work of Roy and Kastenbaum through Darroch and Good (but not Birch). Bhapkar and Koch (1968), both of whom worked with S.N. Roy, attempted to develop an alternative approach to estimation using the method of minimum modified chisquare and a form of weighted least squares. This line of work developed in parallel with that on maximum likelihood estimation.

Beginning in the late 1960s, a number of papers appeared carrying the extensions of the loglinear model ideas to other related problems, such as the analysis of incomplete contingency tables, clarifying a number of the key issues required to fill out the underpinnings of the methodology such as those left unresolved by Birch, and more practical issues of model selection that facilitated the widespread application of the techniques during the 1970s. Nelder and Wedderburn (1972), in their exposition of the methodology for the analysis of generalized linear models (GLIM), pay special attention to the special cases of log-linear models for Poisson sampling and logit models for binomial sampling, and they specifically reference both Birch and Bishop. The books by Haberman (1974) and Bishop et al. (1975) captured the theoretical and methodological developments of the decade following the publication of Birch's paper. Subsequently, the number of papers developing the use and application of loglinear model methods numbered in the thousands.

At the time of the preparation of this introduction, there are at least two dozen statistics books available that focus primarily or in large part on maximum likelihood methods for loglinear models applied to categorical data [e.g, see Agresti (1990), Andersen (1990); Fienberg (1980); Goodman (1984); Haberman (1978, 1979); McCullagh and Nelder (1989); Read and Cressie (1988); Santner and Duffy (1989); and Whittaker (1990)] and several others are in preparation. Not an issue of a major statistical journal goes by without at least one new paper on the topic. The theory in Birch's 1963 paper underpins much of this current literature and the paper is still widely cited.

Acknowledgments

I wish to thank the Hebrew University where I served as Berman Visiting Professor on leave from Carnegie Mellon University during the preparation of this introduction. John Darroch and David McLaren provided me with invaluable recollections.

References

Agresti, A. (1990). *Categorical Data Analysis*. Wiley, New York.

Andersen, E.B. (1990). *The Statistical Analysis of Categorical Data*. Springer-Verlag, Heidelberg.

Bartlett, M.S. (1935). Contingency table interactions, *Suppl. J. Roy. Statist. Soc.*, **2**, 248–252.

Bhapkar, V.P., and Koch, G.G. (1968). On the hypothesis of "no interaction" in contingency tables, *Biometrics*, **24**, 567–594.

Birch, M.W. (1963). Maximum likelihood in three-way contingency tables, *J. Roy. Statist. Soc., Ser. B.*, **25**, 220–233.

Birch, M.W. (1964a). A new proof of the Pearson-Fisher theorem, *Ann. Math. Statist.*, **35**, 718–824.

Birch, M.W. (1964b). The detection of partial association, I: the 2 × 2 case, *J. Roy. Statist. Soc., Ser. B.*, **26**, 313–324.
Birch, M.W. (1964c). A note on the maximum likelihood estimation of a linear structural relationship, *J. Amer. Statist. Assoc.*, **59**, 1175–1178.
Birch, M.W. (1965).The detection of partial association, II: the general case, *J. Roy. Statist. Soc., Ser. B.*, **27**, 11–124.
Bishop, Y.M.M. (1967). Multi-dimensional contingency tables: Cell estimates. Ph.D. Dissertation, Department of Statistics, Harvard University, Cambridge, Mass.
Bishop, Y.M.M. (1969). Full contingency tables, logits, and split contingency tables, *Biometrics*, **25**, 383–399.
Bishop, Y.M.M., Fienberg, S.E., and Holland, P.W. (1975). *Discrete Multivariate Analysis: Theory and Practice*. MIT Press, Cambridge, Mass.
Bunker, J.P., Forrest, W.H.Jr., Mosteller, F., and Vandam, L. (1969). *The National Halothane Study*. U. S. Government Printing Office, Washington, D.C.
Darroch, J.N. (1962). Interactions in multi-factor contingency tables, *J. Roy. Statist. Soc., Ser. B*, **24**, 251–263.
Deming, W.E., and Stephan, F.F. (1940). On a least squares adjustment of a sampled frequency table when the expected marginal totals are known, *Ann. Math. Statist.*, **11**, 427–444.
Fienberg, S.E. (1980). *The Analysis of Cross-Classified Categorical Data*, 2nd ed. MIT Press, Cambridge, Mass.
Good, I.J. (1963). Maximum entropy for hypothesis formulation, especially for multidimensional contingency tables, *Ann. Math. Statist.*, **34**, 911–934.
Goodman, L.A. (1963). On methods for comparing contingency tables, *J. Roy. Statist. Soc., Ser. A*, **26**, 94–108.
Goodman, L.A. (1978). *Analyzing Qualitative/Categorical Data.* Abt Books, Cambridge, Mass.
Goodman, L.A. (1984). *Analysis of Cross-Classified Data Having Ordered Categories.* Harvard University Press, Cambridge, Mass.
Haberman, S.J. (1974). *The Analysis of Frequency Data.* University of Chicago Press, Chicago, Il.
Haberman, S.J. (1978). *The Analysis of Qualitative Data, Vol. 1, Introductory Topics.* Academic Press, New York.
Haberman, S.J. (1979). *The Analysis of Qualitative Data, Vol. 2, New Developments.* Academic Press, New York.
Ireland, C.T., and Kullback, S. (1968). Minimum discrimination information estimation. *Biometrika*, **55**, 707–713.
McCullagh, P., and Nelder, J. A. (1989). *Generalized Linear Models*, 2nd ed. Chapman and Hall, London.
Nelder, J.A., and Wedderburn, R.W.M. (1972). Generalized linear models, *J. Roy. Statist. Soc., Ser. A*, **135**, 370–384.
Plackett, R.L. (1962). A note on interactions in contingency tables. *J. Roy. Statist. Soc., Ser. B*, **24**, 162–166.
Read, T.R.C., and Cressie, N.A.C. (1988). *Goodness-of-Fit Statistics for Discrete Multivariate Data.* Springer-Verlag, New York.
Roy, S.N., and Kastenbaum, M.A. (1956). "On the hypothesis of no "interaction" in a multiway contingency table, *Ann. Math. Statist.*, **27**, 749–757.
Santner, T.J., and Duffy, D.E. (1989). *The Statistical Analysis of Discrete Data.* Springer-Verlag: New York.
Truett, J., Cornfield, J., and Kannel, W. (1967). A multivariate analysis of the risk of caronary heart disease in Framingham, *J. Chron. Dis.*, **20**, 511–524.
Yule, G.U. (1900). On the association of attributes in statistics, *Philos. Trans. Roy. Soc. London., Ser. A*, **194**, 257–319.
Whittaker, J. (1990). *Graphical Models in Applied Multivariate Statistics.* Wiley, New York.

Maximum Likelihood in Three-Way Contingency Tables

Martin W. Birch
University of Liverpool

Summary

Interactions in three-way and many-way contingency tables are defined as certain linear combinations of the logarithms of the expected frequencies. Maximum-likelihood estimation is discussed for many-way tables and the solutions given for three-way tables in the cases of greatest interest.

1. Introduction

Three-way contingency tables are a straightfoward generalization of two-way contingency tables. They can, of course, be imagined as rectangular parallelepipeds built out of cubical boxes each containing a number. In writing out an $r \times s \times t$ table we will usually write out in order each component $r \times s$ table with marginal totals at the bottom and at the right, and finally the sum of all the $r \times s$ tables.

In everything that is to follow we will ignore any logical order in the groups of each classification. If the groups are at equally spaced levels according to some scale, it should be possible to split interactions up into linear, quadratic, ..., components, but we shall not attempt this here.

Let m_{ijk} be the expected number of individuals which are classified as i by factor I, j by factor II and k by factor III in an $r \times s \times t$ table. Let

$$m_{ij.} = \sum_{k=1}^{t} m_{ijk}, \quad m_{i..} = \sum_{j=1}^{s} \sum_{k=1}^{t} m_{ijk},$$

and so on. In the two-way contingency table there is only one null hypothesis that is commonly tested, namely that the two criteria are independent or, equivalently, that there is no (two-factor) interaction. For the three-way table there are at least five ways of generalizing the independence-no-interaction hypothesis of the two-way table.

H_a: I, II and III are mutually independent,

i.e. $$m_{ijk} = m_{i..}m_{.j.}m_{..k}/m_{...}^2.$$

H_{b3}: III is independent of I and II together,

i.e. $$m_{ijk} = m_{ij.}m_{..k}/m_{...}.$$

H_{c3}: I and II are independent (marginally),

i.e. $$m_{ij.} = m_{i..}m_{.j.}/m_{...}.$$

H_{d3}: I and II are independent in each layer of III,

i.e. $$m_{ijk} = m_{i.k}m_{.jk}/m_{..k}.$$

H_e: There is no three-factor interaction between I, II and III. We shall define three-factor interaction in Section 3.

Hypotheses H_{b1}, H_{b2}, H_{c1}, H_{c2}, H_{d1}, H_{d2} are defined in the obvious way.

Data which can be set out in a three-way contingency table may have arisen in a variety of different ways. We may have taken counts on several Poisson processes for a preassigned period of time, or we may have sorted a fixed number of randomly selected objects into different groups. Or, for each i, we may have sorted a fixed number of randomly selected objects, all in category i according to factor I, into different groups. It might be thought that each method of sampling should lead to a totally different estimation problem. But it is shown in Section 2 that one may have the same maximum-likelihood estimates for a whole family of different sampling methods. This result is used later on in this paper.

Section 3 is concerned with Bartlett's (1935) condition for no three-factor interaction as generalized by Roy and Kastenbaum (1956). This has been criticized on the grounds that the maximum-likelihood estimates under this definition are not implicit functions of the frequencies and are not very easily computed. It is shown in Section 3 of this paper that the maximum-likelihood estimates do have some very pleasant properties nevertheless. First, the likelihood equations have a unique solution at which the likelihood is maximized. Secondly, the maximum-likelihood estimates are the same under a wide variety of sampling conditions. Thirdly, they are sufficient statistics. It is found convenient to regard the logarithms of the expected frequencies as linear functions of unknown parameters.

In Section 4 the results of Section 3 are generalized to many-way contingency tables. It is found that the pleasant properties of the estimates result from the logarithms of the expected frequencies being linear functions of the unknown parameters.

Section 5 is in some ways a sequel to Simpson's (1951) paper. Two further pathological contingency tables are given and the hypotheses H_{c3}, $H_{c3} \cap H_e$ and H_{d3} are expressed in parametric terms.

In Section 6 the maximum-likelihood estimates under the hypotheses $H_{c3} \cap H_e$ and H_{c3} are obtained. Estimation under the other hypotheses (apart from H_e, which has been dealt with already) does not cause any trouble.

2. Sampling Procedures

Let us compare two experiments.

Experiment I. We have σ independent Poisson distributions with expected frequencies $m_i(\boldsymbol{\theta})$, $i = 1, \ldots, \sigma$ and corresponding observed frequencies n_i, $i = 1, \ldots, \sigma$.

Experiment II. We have $\rho - 1$ independent multinomial distributions. For the αth we have taken a sample of fixed size

$$N_\alpha = \sum_{i=R(\alpha-1)+1}^{R(\alpha)} n_i,$$

and we have observed numbers $n_{R(\alpha-1)+1}, \ldots, n_{R(\alpha)}$ respectively in the 1st, ..., $r(\alpha)$th categories, where

$$R(\alpha) = \begin{cases} \sum_{\beta=1}^{\alpha} r(\beta) & (\alpha \geqslant 1), \\ 0 & (\alpha = 0). \end{cases}$$

The corresponding expected frequencies are $m_{R(\alpha-1)+1}(\boldsymbol{\theta}), \ldots, m_{R(\alpha)}(\boldsymbol{\theta})$. Thus there is a probability $m_{R(\alpha-1)+\gamma}/N_\alpha$ of the random variable falling in the γth category. Suppose we also have $r(\rho)$ Poisson distributions with expected frequencies $m_{R(\rho-1)+1}(\boldsymbol{\theta}), \ldots, m_\sigma(\boldsymbol{\theta})$ and corresponding observed frequencies $n_{R(\rho-1)+1}(\boldsymbol{\theta}), \ldots, n_\sigma(\boldsymbol{\theta})$, where

$$\sigma = R(\rho) = \sum_{\beta=1}^{\rho} r(\beta).$$

$m_i(\boldsymbol{\theta})$ is the same function of $\boldsymbol{\theta}$ in both experiments for $i = 1, \ldots, \sigma$.

Let us denote the maximum-likelihood estimate of $\boldsymbol{\theta}$ in experiment I by $\hat{\boldsymbol{\theta}}$, and the maximum-likelihood estimate of $\boldsymbol{\theta}$ in experiment II by $\hat{\boldsymbol{\theta}}^*$. Let us write $\hat{m}_i$ for $m_i(\hat{\boldsymbol{\theta}})$ and $\hat{m}_i^*$ for $m_i(\hat{\boldsymbol{\theta}}^*)$, $i = 1, \ldots, \sigma$. Then, provided

$$\sum_{i=R(\alpha-1)+1}^{R(\alpha)} \hat{m}_i = N_\alpha \quad (\alpha = 1, \ldots, \rho - 1), \tag{2.1}$$

$\hat{\boldsymbol{\theta}} = \hat{\boldsymbol{\theta}}^*$ and $\hat{m}_i = \hat{m}_i^*$, for $i = 1, \ldots, \sigma$, i.e. the maximum-likelihood estimates are the same in both cases.

It should be noted that for experiment II we have, for all $\boldsymbol{\theta}$,

$$\sum_{i=R(\alpha-1)+1}^{R(\alpha)} m_i(\boldsymbol{\theta}) = N_\alpha \quad (\alpha = 1, \ldots, \rho - 1), \tag{2.2}$$

and therefore

$$\sum_{i=R(\alpha-1)+1}^{R(\alpha)} \hat{m}_i^* = N_\alpha \quad (\alpha = 1, \ldots, \rho - 1). \tag{2.3}$$

Thus (2.1) is both sufficient and necessary for the maximum-likelihood estimates to be the same.

To prove the result, consider the likelihood function. For experiment I, it is

$$\prod_{i=1}^{\sigma} \frac{m_i^{n_i}}{n_i!} e^{-m_i}.$$

For experiment II, it is

$$\sum_{i=R(\rho-1)+1}^{R(\rho)} \frac{m_i^{n_i}}{n_i!} e^{-m_i} \prod_{\alpha=1}^{\rho-1} \left\{ \sum_{i=R(\alpha-1)+1}^{R(\alpha)} \frac{m_i^{n_i}}{n_i!} \right\} N_\alpha! = \prod_{i=1}^{\sigma} \frac{m_i^{n_i}}{n_i!} e^{-m_i} \prod_{\alpha=1}^{\rho-1} N_\alpha! e^{N_\alpha}$$

by (2.2). But $N_1, \ldots, N_{\rho-1}$ do not depend on $\boldsymbol{\theta}$, and therefore the two likelihoods are constant multiples of one another. The only difference between experiments I and II in the problem of maximizing the likelihood is that for experiment II we have $\rho - 1$ additional constraints on $\boldsymbol{\theta}$. If (2.1) is true, these constraints are automatically satisfied and we get the same maximum-likelihood estimates for both experiments.

The importance of this result lies in its applications. It often happens in contingency tables that we get the same maximum-likelihood estimates under a wide variety of sampling conditions.

It is convenient to describe experiment I as unrestricted random sampling and experiment II as random sampling with

$$\sum_{i=R(\alpha-1)+1}^{R(\alpha)} n_i$$

fixed for $\alpha = 1, \ldots, \rho - 1$.

3. The Hypothesis of No Three-Factor Interaction

The condition for no three-factor interaction is given by Bartlett (1935) for the $2 \times 2 \times 2$ table as

$$p_{111}p_{122}p_{212}p_{221} = p_{222}p_{211}p_{121}p_{112} \tag{3.1}$$

where $p_{ijk} = m_{ijk}/m_{\ldots}$.

Roy and Kastenbaum (1956) generalize this to the $r \times s \times t$ table. They define the condition for no three-factor interaction to be that

$$\frac{p_{rst}p_{ijt}}{p_{ist}p_{rjt}} = \frac{p_{rsk}p_{ijk}}{p_{isk}p_{rjk}} \quad (1 \leqslant i \leqslant r-1; 1 \leqslant j \leqslant s-1; 1 \leqslant k \leqslant t-1). \quad (3.2)$$

(It is tacitly assumed that all the expected frequencies are non-zero. We will always assume this henceforth since anomalies occur if zero m's are allowed.) Suppose now we have unrestricted sampling conditions, Roy and Kastenbaum's condition may be rewritten

$$\ln m_{ijk} = \ln m_{rjk} + \ln m_{isk} + \ln m_{ijt} - \ln m_{ist} - \ln m_{rjt} - \ln m_{rsk} + \ln m_{rst}$$
$$(1 \leqslant i \leqslant r-1; 1 \leqslant j \leqslant s-1; 1 \leqslant k \leqslant t-1).$$

This condition is satified if, and only if, $\ln m_{ijk}$ can be written in the form

$$\ln m_{ijk} = u + u_{1i} + u_{2j} + u_{3k} + u_{12ij} + u_{13ik} + u_{23jk}$$
$$(1 \leqslant i \leqslant r; 1 \leqslant j \leqslant s; 1 \leqslant k \leqslant t), \quad (3.3)$$

where

$$u_{1.} = u_{2.} = u_{3.} = 0;$$

$$u_{13i.} = u_{12i.} = 0, \text{ each } i; u_{23j.} = u_{12 \cdot j} = 0, \text{ each } j; u_{23.k} = u_{13.k} = 0, \text{ each } k;$$

but otherwise $u, u_{1i}, \ldots, u_{23st}$ are completely arbitrary.

In general, we can write $\ln m_{ijk}$ in the form

$$\ln m_{ijk} = u + u_{1i} + u_{2j} + u_{3k} + u_{12ij} + u_{13ik} + u_{23jk} + u_{123ijk} \quad (3.4)$$

where $u_{123ij.} = 0$, each (i, j); $u_{123i.k} = 0$, each (i, k); and $u_{123.jk} = 0$, each (j, k). The u_{123}'s are then the three-factor interactions.

We shall consider the maximum-likelihood estimation of the m_{ijk}'s on the hypothesis (3.3) of no three-factor interaction. The m_{ijk}'s are functions of the u's. We shall take

$$u, u_{1i} \, (2 \leqslant i \leqslant r), u_{2j} \, (2 \leqslant j \leqslant s), u_{3k} \, (2 \leqslant k \leqslant t), u_{12ij} \, (2 \leqslant i \leqslant r, 2 \leqslant j \leqslant s),$$
$$u_{13ik} \, (2 \leqslant i \leqslant r, 2 \leqslant k \leqslant t), \text{ and } u_{23jk} \, (2 \leqslant j \leqslant s, 2 \leqslant k \leqslant t)$$

as our independent parameters. We denote maximum-likelihood estimates of parameters by circumflex accents. Thus $\hat{u}$ is the maximum-likelihood estimate of u and $\hat{m}_{ijk}$ the maximum-likelihood estimate of m_{ijk}. We assume that none of the observed frequencies are zero. (The probability that this is so tends to 1 if $m_{\ldots} \to \infty$ while the expected frequencies remain in the same ratio.)

Then the following statements hold.

(3.5) The marginal totals $n_{.jk}, n_{i.k}, n_{ij.}$ are sufficient statistics.

(3.6) The marginal totals are equal to the maximum-likelihood estimates of their expectations, i.e. $n_{ij.} = \hat{m}_{ij.}$.

(3.7) There is one set of u's (or for that matter of m's), and one only, for which the likelihood function is stationary.

(3.8) The maximum-likelihood estimates are determined uniquely by the marginal totals being equal to the maximum-likelihood estimates of their expectations.

Proof of (3.5). Let us write L for the logarithm of the likelihood. Then

$$\begin{aligned} L &\equiv \sum_i \sum_j \sum_k \{n_{ijk} \ln m_{ijk} - m_{ijk} - \ln(n_{ijk}!)\} \\ &\equiv n_{...}u + \sum_i n_{i..}u_{1i} + \sum_j n_{.j.}u_{2j} + \sum_k n_{..k}u_{3k} + \sum_i \sum_j n_{ij.}u_{12ij} + \sum_i \sum_k n_{ik}u_{13ik} \\ &\quad + \sum_j \sum_k m_{.jk}u_{23jk} - m_{...} - \sum_i \sum_j \sum_k \ln(n_{ijk}!). \end{aligned}$$

(3.5) follows immediately. □

Proof of (3.6). If θ is any parameter,

$$\begin{aligned} \frac{\partial L}{\partial \theta} &= \sum_i \sum_j \sum_k \left\{ n_{ijk} \frac{\partial}{\partial \theta}(\ln m_{ijk}) - \frac{\partial}{\partial \theta}(m_{ijk}) \right\} \\ &= \sum_i \sum_j \sum_k (n_{ijk} - m_{ijk}) \frac{\partial}{\partial \theta}(\ln m_{ijk}). \end{aligned} \tag{3.9}$$

(3.9) is a useful formula for our purpose since $\ln m_{ijk}$ is a linear function of the u's.

$$\frac{\partial L}{\partial u} = n_{...} - m_{...},$$

so that

$$\hat{m}_{...} = n_{...}, \tag{3.10}$$

$$\frac{\partial L}{\partial u_{1i}} = (n_{i..} - m_{i..}) - (n_{1..} - m_{1..}) \quad (2 \leqslant i \leqslant r),$$

giving

$$\hat{m}_{i..} - n_{i..} = \hat{m}_{1..} - n_{1ii} \quad (1 \leqslant i \leqslant r),$$

and so

$$\hat{m}_{i..} - n_{i..} = \frac{1}{r}(\hat{m}_{...} - n_{...}) \quad (1 \leqslant i \leqslant r). \tag{3.11}$$

By using (3.9), we get

$$\hat{m}_{i..} = n_{i..} \quad (1 \leqslant i \leqslant r). \tag{3.12}$$

$$\frac{\partial L}{\partial u_{12ij}} = (n_{ij.} - m_{ij.}) - (n_{i1.} - m_{i1.}) - (n_{1j.} - m_{1j.}) + (n_{11.} - m_{11.})$$

$$(2 \leqslant i \leqslant r; 2 \leqslant j \leqslant s).$$

Therefore

$$\hat{m}_{ij.} - n_{ij.} = (\hat{m}_{i1.} - n_{i1.}) + (\hat{m}_{1j.} - n_{1j.}) - (\hat{m}_{11.} - n_{11.})$$
$$(1 \leqslant i \leqslant r; 1 \leqslant j \leqslant s),$$

giving

$$\hat{m}_{ij.} - n_{ij.} = \frac{1}{s}(\hat{m}_{i..} - n_{i..}) + \frac{1}{r}(\hat{m}_{.j.} - n_{.j.}) - \frac{1}{rs}(\hat{m}_{...} - n_{...})$$
$$(1 \leqslant i \leqslant r; 1 \leqslant j \leqslant s). \tag{3.13}$$

Using (3.10), (3.11) and an analogue of (3.11) gives us

$$\hat{m}_{ij.} = n_{ij.} \quad (1 \leqslant i \leqslant r; 1 \leqslant j \leqslant s). \tag{3.14}$$

Similarly we can show that $\hat{m}_{i.k} = n_{i.k}$ and that $\hat{m}_{.jk} = n_{.jk}$. We have proved (3.6). □

PROOFS OF (3.7) AND (3.8). The proviso that all the n's are non-zero ensures that L attains its supremum for finite values of the u's. For then

$$n_{ijk} \ln m_{ijk} - m_{ijk} - \ln(n_{ijk}!)$$

for fixed n_{ijk} and variable m_{ijk} is bounded above and tends to $-\infty$ as $\ln m_{ijk} \to \pm\infty$. Therefore $L \to -\infty$ as the u's tend to infinity in any way and is bounded above.

Since L is twice differentiable, L has a maximum for finite values of the u's which is also a stationary point. The uniqueness of this point follows if it can be shown that the matrix $\{-\partial^2 L/\partial\theta_h \partial\theta_l\}$ is positive definite. It is easy to see that the matrix is non-negative definite since a typical element may be written in the form

$$\sum_i \sum_j \sum_k m_{ijk} \frac{\partial}{\partial\theta_h}(\ln m_{ijk}) \frac{\partial}{\partial\theta_l}(\ln m_{ijk}). \tag{3.15}$$

Therefore the quadratic form

$$\sum_h \sum_l -\frac{\partial^2 L}{\partial\theta_h \partial\theta_l} x_h x_l$$

may be written as the sum of squares

$$\sum_i \sum_j \sum_k m_{ijk} \left\{\sum \frac{\partial}{\partial\theta_h}(\ln m_{ijk}) x_h\right\}^2.$$

To prove that $\{-\partial^2 L/\partial\theta_h \partial\theta_l\}$ is non-singular it is sufficient to show that

$$\sum_i \sum_j \sum_k m_{ijk} \left\{\sum \frac{\partial}{\partial\theta_h}(\ln m_{ijk}) x_h\right\}^2 = 0 \tag{3.16}$$

implies $\mathbf{x} = \mathbf{0}$. Now (3.16) implies

$$\sum_h \frac{\partial}{\partial \theta_h}(\ln m_{ijk})x_h = 0$$

for each i, j, k since $m_{ijk} > 0$. Denote the x_h corresponding to $\theta_h = u_{1i}$ $(2 \leqslant i \leqslant r)$ by v_{1i} and put $v_{11} = -v_{12} - \cdots - v_{1r}$; similarly for the other interaction groups. Then

$$\sum \frac{\partial}{\partial \theta_h}(\ln m_{ijk})x_h = v + v_{1i} + v_{2j} + v_{3k} + v_{12ij} + v_{13ik} + v_{23jk}$$

$$(1 \leqslant i \leqslant r;\ 1 \leqslant j \leqslant s;\ 1 \leqslant k \leqslant t).$$

Therefore,

$$\sum \frac{\partial}{\partial \theta_h}(\ln m_{ijk})x_h = 0$$

for each (i, j, k) implies that

$$0 = v = v_{1i} = v_{2j} = v_{3k} = v_{12ij} = v_{13ik} = v_{23jk}$$

$$(1 \leqslant i \leqslant r;\ 1 \leqslant j \leqslant s;\ 1 \leqslant k \leqslant t),$$

i.e. $\mathbf{x} = \mathbf{0}$.

This proves (3.7). (3.8) follows from (3.7) and the fact that the conditions for stationarity, viz. equations (3.10), (3.11), (3.13) and analogous equations, are satisfied if and only if the observed marginal totals agree with the maximum-likelihood estimates of their expectations. □

It should be noted that (3.8) is a purely analytic statement, namely that, given any set of positive integers, n_{ijk}, $1 \leqslant i \leqslant r$, $1 \leqslant j \leqslant s$, $1 \leqslant k \leqslant t$, there is one and only one set of positive numbers n^*_{ijk} satisfying the condition (3.2) and also the equations

$$n^*_{ij.} = n_{ij.}, \quad n^*_{i.k} = n_{i.k}, \quad n^*_{.jk} = n_{.jk}.$$

A purely analytic proof of (3.8) can be abstracted from the proof given, moreover, one in which we do not need to know that the n's are integers, only that they are positive. A corollary of this is that the expected frequencies in a three-way contingency table are given uniquely by the marginal totals if the expected frequencies are known to be positive and to satisfy H_e, a result conjectured by Darroch (1962).

(3.6) is well known already but it has not been remarked upon. It is important because it means that we have the same maximum-likelihood estimates under sampling conditions other than unrestricted sampling. Since $\hat{m}_{...} = n_{...}$ we have the same maximum-likehood estimates in the case of a single multinomial. Since $\hat{m}_{i..} = n_{i..}$ for each i, we have the same maximum-likelihood estimates in the case of r multinomials when the total for each category according to factor I is fixed. Since $\hat{m}_{ij.} = n_{ij.}$ for each (i, j) we have the same maximum-likelihood estimates in the case of rs multinomials when the total for each category of (I, II) is fixed.

We have said nothing about the numerical solution of equations (3.6). Methods have been given by Norton (1945), Kastenbaum and Lamphiear (1959) and Darroch (1962). These are all iterative methods.

4. Generalization

The results (3.5) to (3.8) may be generalized to a q-way $r_1 \times r_2 \times \cdots \times r_q$ contingency table. Suppose we have $r_1 r_2 \ldots r_q$ independent Poisson radom variables $n_{i_1 i_2 \ldots i_q}$ with means $m_{i_1 i_2 \ldots i_q}$ where

$$\ln m_{i_1 i_2 \ldots i_q} = u + u_{1i_1} + \cdots + u_{qi_q} + u_{12i_1i_2} + u_{13i_1i_3} + \cdots + u_{q-1qi_{q-1}i_q} + u_{123i_1i_2i_3} + \cdots + u_{12\ldots qi_1i_2\ldots i_q}, \tag{4.1}$$

where

$$u_{1.} = 0, \ldots, u_{q.} = 0, u_{12i_1.} = u_{12.i_2} = 0, \ldots, u_{12\ldots qi_1\ldots i_{q-1}.} = \cdots = u_{12\ldots q.i_2\ldots i_q} = 0.$$

Let us call the group of interactions

$$u_{ab\ldots z11\ldots 1}, u_{ab\ldots z21\ldots 1}, \ldots, u_{ab\ldots zr_a1\ldots 1}, u_{a1b\ldots z121\ldots 1}, \ldots, u_{ab\ldots zr_ar_b\ldots r_z}$$

the interaction group $u_{ab\ldots z}$. We will say $u_{ab\ldots z} = 0$ if each interaction of the group is zero. Let us say that the interaction group $u_{a'b'\ldots z'}$ implies the interaction group $u_{a''b''\ldots z''}$ if, and only if, $(a'', b'', \ldots, z'')$ is a selection from $a', b', \ldots, z'$, e.g. u_{123} implies u_{13} but neither u_{13} nor u_{1245} implies u_{123}. We define a symmetrical estimation problem as one in which all interactions of certain groups are given, and all interactions of other groups are completely arbitrary, except for the conditions of (4.1), and are to be estimated. We further define a hierarchical symmetrical estimation problem as a symmetrical estimation problem in which all groups of interactions implied by arbitrary groups are themselves arbitrary.

Then for a symmetrical estimation problem we can generalize (3.5) and (3.7) respectively as follows.

(4.2) The marginal totals corresponding to unknown interaction groups are sufficient statistics.

(4.3) If the observed frequencies are all non-zero the likelihood function has a unique stationary point. At this the likelihood is maximized.

Further, for a hierarchical symmetrical estimation problem we can generalize (3.6) and (3.8) respectively as follows.

(4.4) The marginal totals corresponding to unknown interaction groups are equal to the maximum-likelihood estimates of their expectations.

(4.5) The maximum-likelihood estimates are given uniquely by (4.4).

For non-hierarchical symmetrical estimation problems analogues of (3.11) and (3.13) still hold, although (4.4) in general does not. And (4.4) can be generalized to say that in symmetrical estimation problems marginal totals corresponding to interaction groups, such that all groups implied by them are zero, are equal to the maximum-likelihood estimates of their expectations. In particular, if u is arbitrary, the grand total is equal to the maximum-likelihood estimate of its expectation.

Results (4.2) to (4.5) have their analogues in the analysis of factorial experiments where the experimental errors are assumed to be normally and identically distributed. But here it is the logarithm of the expectation which is a linear function of the interactions. As a result, the maximum-likelihood equations are non-linear and, as we shall see, we often fail to have orthogonality, even asymptotically.

To prove (4.2), (4.3), (4.4) and (4.5) we first consider a more general problem similar to the general logistic problem formulated by Dyke and Patterson (1952). Suppose we have σ independent Poisson distributions with means m_i and observed frequencies n_i. Suppose $\ln m_i = \sum_k a_{ik}\theta_k$ where $\{a_{ik}\}$ is a $\sigma \times \tau$ matrix of rank τ and the θ's are unknown parameters. Assume that the n_i's are all positive. Then:

(4.6) The totals $\sum_i a_{ik} n_i$, $k = 1, \ldots, \tau$ are sufficient statistics.
(4.7) $\sum_i a_{ik} \hat{m}_i = \sum_i a_{ik} n_i$, $k = 1, \ldots, \tau$.
(4.8) The likelihood function has a unique stationary point. At this the likelihood is maximized.
(4.9) The maximum-likelihood estimates are given uniquely by the equations of (4.7).

(*Editor's note*: Proofs have been omitted.)

Considerations of space and elegance forbid telling in detail how (4.2) ... (4.5) follow from (4.6) ... (4.9). But what we do is to regard

$$u_{ab\ldots z i_a i_b \ldots i_z} \quad (2 \leqslant i_a \leqslant r_a, 2 \leqslant i_b \leqslant r_b, \ldots, 2 \leqslant i_z \leqslant r_z)$$

as independent variables for each arbitrary interaction group. We then apply (4.6–4.9). The equations given by (4.7) are then shown to be equivalent to those given by (4.2). The methods used are adaptations of methods used in Section 3.

For a mixture of independent multinomial and Poisson distributions we get the same maximum-likelihood estimates under certain circumstances as for the case of no three-factor interactions in the three-way contingency table. In particular, if u is arbitrary, we get the same maximum-likelihood estimates in the case of single multinomial distribution.

For $r_1 r_2 \ldots r_q$ independent Poisson distributions the estimate of the information matrix takes the simple form

$$\sum_{i_1} \cdots \sum_{i_q} m_{i_1 \ldots i_q} \frac{\partial}{\partial \theta_h} (\ln m_{i_1 \ldots i_q}) \frac{\partial}{\partial \theta_l} (\ln m_{i_1 \ldots i_q}), \tag{4.11}$$

which may be rewritten in the form $\mathbf{V}'\mathbf{W}\mathbf{V}$, where $\mathbf{W}$ is a positive definite diagonal $R \times R$ matrix $\mathbf{V}$ is an $R \times r$ matrix, where $R = r_1 r_2 \ldots r_q$ and $r \leqslant R$. $r = R$ when all the interactions are arbitrary. The information matrix is then readily inverted as $\mathbf{V}^{-1}\mathbf{W}^{-1}(\mathbf{V}^{-1})'$. When $r < R$ the algebraic inversion is more difficult.

5. Independence of Two Attributes

Simpson (1951) has thrown much light on ideas of independence between two attributes in a $2 \times 2 \times 2$ contingency table. An example of his shows that H_{c3} and H_e may both be true while H_{d3} is not. Positive association in each layer of I coexists with zero marginal association. Examples can be given (Simpson does not give one) in which H_{d3} is true while H_{c3} is not. Independence in each layer coexists with positive association in the margin, e.g.

$$
\begin{array}{cc|c}
4 & 2 & 6 \\
2 & 1 & 3 \\
\hline
6 & 3 & 9
\end{array}
\qquad
\begin{array}{cc|c}
1 & 2 & 3 \\
2 & 4 & 6 \\
\hline
3 & 6 & 9
\end{array}
\qquad
\begin{array}{cc|c}
5 & 4 & 9 \\
4 & 5 & 9 \\
\hline
9 & 9 & 18.
\end{array}
\tag{5.1}
$$

Simpson shows further that, in a $2 \times 2 \times 2$ table, if H_{c3} and H_{d3} are both true then one of the hypotheses H_{b1}, H_{b2} is true. Thus independence in both layers and marginally implies that I is independent of II and III together or that II is independent of I and III together. This result can be generalized to $r \times s \times 2$ tables with arbitrary r and s. If H_{d3} holds then we can put

$$m_{ij1} = \alpha\beta_i\gamma_j, \quad \text{where} \quad \beta_. = \gamma_. = 0,$$

and

$$m_{ij2} = \alpha'\beta_i'\gamma_j', \quad \text{where} \quad \beta_.' = \gamma_.' = 0. \tag{5.2}$$

If H_{c3} holds, then

$$m_{ij.}m_{...} = m_{i..}m_{.j.},$$

$$(\alpha\beta_i\gamma_j + \alpha'\beta_i'\gamma_j')(\alpha + \alpha') = (\alpha\beta_i + \alpha'\beta_i')(\alpha\gamma_j + \alpha'\gamma_j'),$$

$$\alpha\alpha'(\beta_i - \beta_i')(\gamma_j - \gamma_j') = 0. \tag{5.3}$$

If H_{b1} is not true, then $\gamma_j \neq \gamma_j'$ for some j. Since (5.3) is true for all (i, j) and $\alpha, \alpha' > 0$, $\beta_i = \beta_i'$ for all i. Thus, for $r \times s \times 2$ tables H_{d3} and H_{c3} cannot both hold without at least one of the two stronger hypotheses, H_{b1}, H_{b2}, holding. On the other hand, we can construct $2 \times 2 \times 3$ tables like

$$
\begin{array}{cc|c}
4 & 2 & 6 \\
2 & 1 & 3 \\
\hline
6 & 3 & 9
\end{array}
\qquad
\begin{array}{cc|c}
2 & 1 & 3 \\
4 & 2 & 6 \\
\hline
6 & 3 & 9
\end{array}
\qquad
\begin{array}{cc|c}
1 & 4 & 5 \\
1 & 4 & 5 \\
\hline
2 & 8 & 10
\end{array}
\qquad
\begin{array}{cc|c}
7 & 7 & 14 \\
7 & 7 & 14 \\
\hline
14 & 14 & 28
\end{array}
\tag{5.4}
$$

in which H_{c3} and H_{d3} are both true but neither H_{b2} nor H_{b3} is.

In the paragraphs above we have discussed the hypotheses H_{d3} and H_e as if they were distinct. In fact H_{d3} implies H_e. For H_e may be interpreted as meaning that the association in each layer of III is the same from layer to layer while H_{d3} may be interpreted as meaning that the association is zero in each layer.

H_{d3} may be expressed concisely in terms of the u's as defined in (3.4):

$$H_{d3} \Leftrightarrow u_{12} = 0, \quad u_{123} = 0.$$

(*Editor's note*: Proof has been omitted.)

In similar ways it can be shown that $H_{b3} \Leftrightarrow u_{13} = 0, u_{23} = 0$ and $u_{123} = 0$ and that $H_a \Leftrightarrow u_{12} = 0, u_{13} = 0, u_{23} = 0$ and $u_{123} = 0$. Therefore, in the terminology of Section 3, when H_a, H_{b3}, H_{d3} or H_e are known to be true we have hierarchical symmetrical estimation problems.

Unfortunately, H_{c3} cannot be expressed conveniently on terms of the u's. We might hope H_{c3} would correspond to $u_{12} = 0$. But then $H_{c3} \cap H_e$ would be equivalent to H_{d3}. Instead we write m_{ijk} as $m_{ij.}m''_{ijk}$ where $m''_{ijk} = m_{ijk}/m_{ij.}$. Then $\ln m_{ij.}$ may be written as $u' + u'_{1i} + u'_{2j}$ where $u'_{1.} = u'_{2.} = 0$ and the m'''s are arbitrary except for the conditions $m''_{ij.} = 0$. We may further express m''_{ijk} as

$$\exp(u'' + u''_{1i} + u''_{2j} + u''_{12ij} + u''_{13ik} + u''_{23jk} + u''_{123ijk}),$$

where $u''_{1.} = 0$, $u''_{2.} = 0$, and so on. The conditions $m''_{ij.} = 1$ reduce to

$$u'' + u''_{1i} + u''_{2j} + u''_{12ij} = -\ln \textstyle\sum_k \exp(u''_{3k} + u''_{13ik} + u''_{23jk} + u''_{123ijk}). \quad (5.5)$$

Thus, when H_{c3} holds, the expected frequencies can be made to depend on

$$rst - (s-1)(t-1)$$

parameters each varying from $-\infty$ to $+\infty$ independently of each other.

In general we can write

$$\ln m_{ijk} = u' + u'_{1i} + u'_{2j} + u'_{12ij} + u'' + u''_{1i} + u''_{2j} + u''_{12ij} + u''_{3k} + u''_{23ik} + u''_{13jk} + u''_{123ijk} \quad (5.6)$$

where relations like $u'_{1.} = 0$, $u''_{13i.} = 0$ hold and equations (5.5) define u'', u''_{1i}, u''_{2j}, u''_{12ij}. H_{c3} is then equivalent to $u'_{12} = 0$, and H_e to $u''_{123} = 0$. Imposing H_{c3} removes $(r-1)(s-1)$ degrees of freedom, imposing H_e removes $(r-1)(s-1)(t-1)$ degrees of freedom and imposing both removes $(r-1)(s-1)t$ degrees of freedom.

Simpson also shows that in one situaton H_{c3} may be a reasonable hypothesis while H_{d3} is irrelevant and vice versa. One has to know where the data have come from before one can analyse a three-way contingency table properly. It is doubtful whether there is any practical situation in which H_{c3} and H_{d3} might both be expected to hold as in example (5.4).

In the absence of three-factor interactions, $H_{c2} \cap H_{c3}$ implies H_{b1} and therefore H_{d2} and H_{d3}. Roy and Kastenbaum (1956) state this result for the $r \times s \times t$ table but fail to give a convincing proof. Darroch (1962) gives a satisfactory proof, a shortened version of which is given below.

By (3.8), if we have a set of numbers satisfying the equations

$$m'_{ij.} = m_{ij.}, \quad m'_{i.k} = m_{i.k}, \quad m'_{.jk} = m_{.jk}$$

and the m's and m''s both satisfy H_e, the condition for no three factor interaction, then $m'_{ijk} = m_{ijk}$ for each i, j, k. Now suppose we have m_{ijk}'s satisfying H_e, and H_{c2} and H_{c3} as well. Put $m'_{ijk} = m_{i..}m_{.jk}/m_{...}$. Then the m''s satisfy H_e. Also

$$m'_{ij.} = m_{i..}m_{.j.}/m_{...} = m_{ij.} \quad (\text{by } (H_{c3}),$$

$$m'_{i.k} = m_{i..}m_{..k}/m_{...} = m_{i.k} \quad (\text{by } H_{c2}),$$

and

$$m'_{.jk} = m_{.jk}.$$

Therefore, by (3.8), $m_{ijk} = m_{i..}m_{.jk}/m_{...}$, i.e. m_{ijk} satisfies H_{b1}. It can be seen further that

$$H_{c2} \cap H_{c3} \cap H_e \Leftrightarrow H_{d2} \cap H_{c3} \Leftrightarrow H_{c2} \cap H_{d3} \Leftrightarrow H_{d2} \cap H_{d3} \Leftrightarrow H_{b1}.$$

6. Maximum-Likelihood Estimates

In Section 3 we consider the maximum-likehood estimates of the m's when constrained to satisfy H_e. We will now consider the maximum-likelihood estimates of the m's when constrained to satisfy each of the hypotheses H_a, H_{b3}, H_{c3}, H_{d3}, $H_{c3} \cap H_e$ and $\mathscr{H}$. Here $H_a, \ldots, H_e$ are as defined in the Introduction while $\mathscr{H}$ is the universal hypothesis (completely arbitrary m's).

Let us start by assuming unrestricted random sampling. The maximum-likelihood estimates under the hypotheses H_a, H_{b3}, H_{c3}, and $\mathscr{H}$ are easily obtained either from (4.4) or first principles.

(i) H_a *constrained to hold*:

$$\hat{m}_{ijk} = \frac{n_{i..}n_{.j.}n_{..k}}{n_{...}^2}; \tag{6.1}$$

(ii) H_{b3} *constrained to hold*:

$$\hat{m}_{ijk} = \frac{n_{ij.}n_{..k}}{n_{...}}; \tag{6.2}$$

(iii) H_{d3} *constrained to hold*:

$$\hat{m}_{ijk} = \frac{n_{i.k}n_{.jk}}{n_{..k}}; \tag{6.3}$$

(iv) $\mathscr{H}$ *constrained to hold* (*i.e. no constraint*):

$$\hat{m}_{ijk} = n_{ijk}. \tag{6.4}$$

The maximum-likelihood estimates under H_{c3} and $H_{c3} \cap H_e$ are less easily obtained and we will consider these separately.

(v) H_{c3} *constrained to hold.*

We may put

$$m_{ijk} = \exp(u' + u'_{1i} + u'_{2j})m''_{ijk}$$

where we have rs constraints

$$m''_{ij.} = 1 \quad (1 \leqslant i \leqslant r; 1 \leqslant j \leqslant s)$$

on the m'''s as well as

$$u'_{1.} = u'_{2.} = 0 \quad (m''_{ijk} = m_{ijk}/m_{ij.}).$$

The maximum-likelihood equations are

$$n_{..} - \hat{m}_{...} = 0; \tag{6.5}$$

$$(n_{i..} - \hat{m}_{i..}) = (n_{r..} - \hat{m}_{r..}) = 0 \quad (2 \leqslant i \leqslant r); \tag{6.6}$$

$$(n_{.j.} - \hat{m}_{.j.}) - (n_{.s.} - \hat{m}_{.s.}) = 0 \quad (2 \leqslant j \leqslant s); \tag{6.7}$$

$$(n_{ijk} - \hat{m}_{ijk})/\hat{m}''_{ijk} + \lambda_{ij} = 0 \quad (1 \leqslant i \leqslant r; 1 \leqslant j \leqslant s; 1 \leqslant k \leqslant t). \tag{6.8}$$

The λ's are Lagrange multipliers. (6.5), (6.6) and (6.7) together with H_{c3} give

$$\hat{m}_{ij.} = \frac{n_{i..}n_{.j.}}{n_{...}}. \tag{6.9}$$

(6.8) gives

$$\begin{aligned} n_{ijk} &= \hat{m}_{ijk} - \lambda_{ij}\hat{m}''_{ijk}, \\ &= \hat{m}_{ijk}(1 - \lambda_{ij}/\hat{m}_{ij.}). \end{aligned} \tag{6.10}$$

Summing (6.10) over k and using (6.9) gives

$$n_{ij.} = \frac{n_{i..}n_{.j.}}{n_{...}}(1 - \lambda_{ij}/\hat{m}_{ij.}). \tag{6.11}$$

(6.10) and (6.11) then give

$$\hat{m}_{ijk} = \frac{n_{i..}n_{.j.}n_{..k}}{n_{...}n_{ij.}}. \tag{6.12}$$

(vi) H_{c3} *and* H_e *both constrained to hold.*

We may put

$$m_{ijk} = \exp(u' + u'_{1i} + u'_{2j})m''_{ijk}$$

where we have rs constraints

$$m''_{ij.} = 1 \quad (1 \leqslant i \leqslant r; 1 \leqslant j \leqslant s)$$

on the m'''s and m''_{ijk} is expressible in the form

$$u'' + u''_{1i} + u''_{2j} + u''_{3k} + u''_{12ij} + u''_{13ik} + u''_{23jk}$$

since it satisfies H_e. If we differentiate with respect to the u''s and the u'''s the expression

$$L + \sum_{i=1}^{r} \sum_{j=1}^{s} \sum_{k=1}^{t} \lambda_{ij}(m''_{ijk} - 1)$$

and equate the partial derivations to zero, we find that

$$\hat{m}_{ijk} = \frac{n_{i..}n_{.j.}n^*_{ijk}}{n_{...}n_{ij.}}, \tag{6.13}$$

where n^*_{ijk} is the maximum-likelihood estimate of m_{ijk} under the hypothesis H_e.

We have evaluated the estimates in the case of unrestricted random sampling. If, however, we have ordinary multinomial sampling with $n_{...}$ alone fixed we get the same maximum-likelihood estimates since $n_{...} = \hat{m}_{...}$ for all cases considered ((i)...(vi) and the H_e case). The same thing is true if we have r multinomials with $n_{i..}$ fixed for each i except in the case of H_{c1} and $H_{c1} \cap H_e$. However, if we have rs multinomials with $n_{ij.}$ fixed for each (i, j) we get the same maximum-likelihood estimates only in the cases of H_{b3}, H_{d1}, H_{d2}, H_e and $\mathscr{H}$ (doubtfully H_{c3} and $H_{c3} \cap H_e$). In the cases of H_a, H_{b1}, H_{b2}, H_{d3}, H_{c1}, H_{c2}, $H_{c1} \cap H_e$, $H_{c2} \cap H_e$ we get different estimates in general since we do not have $\hat{m}_{ij.} = n_{ij.}$ for unrestricted sampling. H_{c3} and $H_{c3} \cap H_e$ are queer hypotheses when $n_{ij.}$ is fixed for each i and j since H_{c3} reduces to conditions on the $n_{ij.}$'s.

The maximum-likelihood estimate of m_{ijk} when H_{c3} is constrained to hold is of special interest. The χ^2 goodness-of-fit statistic for testing H_{c3} (against $\mathscr{H}$) is

$$\sum_i \sum_j \sum_k (\hat{m}_{ijk} - n_{ijk})^2/\hat{m}_{ijk},$$

where $\hat{m}_{ijk}$ is the maximum-likelihood estimate of m_{ijk} when H_{c3} is constrained to hold. Because of (6.12), the statistic is

$$\sum_i \sum_j \sum_k n_{ijk}\left(\frac{n_{i..}n_{.j.}}{n_{...}n_{ij.}} - 1\right)^2 \Bigg/ \frac{n_{i..}n_{.j.}}{n_{...}n_{ij.}} = \sum_i \sum_j \left(\frac{n_{i..}n_{.j.}}{n_{...}} - n_{ij.}\right)^2 \Bigg/ \frac{n_{i..}n_{.j.}}{n_{...}}, \tag{6.14}$$

which ignores factor III. The use of (6.14) has been advocated, apparently on the grounds that factor III is irrelevant. But it is reassuring to find that it is yielded by the general method of constructing goodness-of-fit statistics.

7. Applications

The author hopes, in a further paper, to apply the results of this paper to hypothesis testing in three-way tables.

Acknowledgements

I would like to thank Professor R.L. Plackett and a referee for helpful criticisms of earlier versions of this paper.

References

Bartlett, M.S. (1935), "Contingency table interactions", *Suppl. J.R. statist. Soc.*, **2**, 248–252.

Darroch, J.N. (1962), "Interactions in multifactor contingency tables", *J.R. statist. Soc.* B, **24**, 251–263.

Dyke, G.V. and Patterson, H.D. (1952), "Analysis of factorial arrangements when the data are proportions", *Biometrics*, **8**, 1–12.

Kastenbaum, M.A. and Lamphiear, D.E. (1959), "Calculation of chi-square to test the no three-factor interaction hypothesis", *Biometrics*, **15**, 107–115.

Norton, H.W. (1945), "Calculation of chi-square from complex contingency tables", *J. Amer. statist. Ass.*, **40**, 251–258.

Plackett, R.L. (1962), "A note on interactions in contingency table", *J.R. statist. Soc.* B, **24**, 162–166.

Roy, S.N. and Kastenbaum, M.A. (1956), "On the hypothesis of no 'interaction' in a multiway contingency table", *Ann math. Statist.*, **27**, 749–757.

Simpson, E.H. (1951), "The interpretation of interaction in contingency tables", *J.R. statist. Soc.* B, **13**, 238–241.

Introduction to Huber (1964) Robust Estimation of a Location Parameter

Frank R. Hampel
ETH Zürich

1. General Remarks on the Paper

1. Importance of the Paper

Huber's first paper on robust statistics is outstanding in several respects. It contains the first encompassing mathematical definitions of the "*approximate* validity of a parametric model" and thus became the founding paper of the "stability theory of statistical procedures" that by a historical accident was called "robust statistics." Since parametric models, like the model of normality, are almost always at best only approximately valid, the paper for the first time fills a conceptual gap left open in Fisher's theory of exact parametric models [Fisher (1922), see in particular p. 314]. The practical importance of this gap was convincingly demonstrated by E.S. Pearson (1931) and others for tests and by Tukey (1960) for estimation. But even apart from the practical aspect, Huber's paper is important for the logical chain of statistical reasoning.

Every reasonable and independently thinking user of statistics wonders where, for example, the normal distribution comes from which he or she is told to use. Even if he agrees that it might seem a reasonable approximation to his data, the classical statistical theory tells him nothing about the effects of his unknown deviation from the model. He will be even more appalled when he encounters the widespread pseudologic such as that used in: "If your test of normality comes out nonsignificant, you can act as if/can believe that/have proven that you have an exactly normal distribution" [see Hampel et al. (1986, p. 55)]. The point is that we can never prove (only disprove statistically) that our true distribution belongs to an exact parametric model, but that we *can prove statistically* that it belongs to some known *neighborhood* of it (in a suitable, namely, the weak topology). Huber's paper is the first one that works out the consequences of such a more realistic assumption.

In doing this, Huber introduced several important concepts and used general and abstract mathematical theory for deriving practically useful tools and results that became standards in robustness theory and its applications. (For more details, see Sec. 1.4 and 3 below.) Such a successful blend is rare in statistics. Huber also worked out and checked a numerical algorithm for his main estimator, and he even gave tables full of numbers—of great importance for the applicability and applications of his methods, although tables were generally deemed rather unimportant in mathematical statistics at that time. Huber clearly saw—and solved—the problem as a whole, and not just some isolated mathematical aspect of it.

2. Some Comments for Applied Statisticians

Although the paper contains much forbidding mathematics, this mathematics leads to inviting procedures that are both accessible and important for applied statisticians. Most M-estimators, and in particular, the Huber estimators for location (regression) and scale, are nothing but maximum likelihood estimators, although for rather unfamiliar parametric models. Those who think they do not understand maximum likelihood estimation may just use a computer subroutine. Huber's estimators provide an important way for dealing with outliers (among other things), different from and in many ways better than the customary rejection of outliers. Like trimmed (or truncated) and Winsorized means and variances, they are yet another (and, in fact, globally better) formalization of the idea Tukey (1962, p. 18) was taught by Winsor in 1941, namely, that outliers should not be rejected, but rather moved close to the remaining sample in a suitable way.

Huber's estimators (although not his theory) can be easily (and partly have been) generalized to most parametric models (not only regression and covariance matrices, for which even his theory works). This is mainly a task for development work. As with likelihood procedures in general, there is no exact finite sample theory (for exact although generally invalid models), but there are only asymptotic standard errors, tests, and confidence intervals; these, however, have been proved to be reliable also if the model is only approximately true. Practical problems still exist with very small samples and/or many parameters, when the asymptotic results may become rather inaccurate. For computer programs, see Sec. 3.11 of this paper.

Some fairly easy introductory readings are Hampel (1974, 1980, 1985) and Hampel et al. (1986, Chap. 1, 8.1, and 8.2). They also contain the synthesis of Winsor's idea with full rejection of distant outliers in the form of "smooth rejection" by redescending M-estimators.

3. Prehistory of the Paper

The general robustness problem ("What happens if our ideal modeling assumptions are more or less violated? How are they violated, and what can we do about this?") is probably as old as statistics, and a number of eminent statisticians were clearly aware of it [see, e.g., Newcomb (1886), K. Pearson (1902), "Student" (Gosset) (1927), and Jeffreys (1939)]. Newcomb (1886) worked already with mixtures of several normal distributions [a generalization of Tukey's (1960) mixture model], and Jeffreys (1939, Chap. 5.7) proposed a redescending M-estimator (maximum likelihood t_7, as a compromise based on valuable empirical evidence) and gave a table for its computation; in retrospect, Newcomb even can be seen to have considered a kind of one-step Huber estimator [Stigler (1973)]. However, their isolated attempts, based on much common sense and experience but lacking a general robustness theory to back them up, were simply ignored by the mainstream of statistics. In this connection, also the exchange of letters between Gosset and Fisher on the robustness problem is quite interesting. Meanwhile, the users of statistics tried to get along with the subjective or "objective" rejection of outliers, the latter largely ill-understood, ill-justified, sometimes ill-working, but usually preventing the worst [see Hampel (1985) or Hampel et al. (1986, Chap. 1.4) for the relationship to robustness theory].

It was only the drastic nonrobustness of the levels of the χ^2- and F-tests for variances that led to wider awareness and more extensive work on the robustness problem—mostly on tests and mostly only on the level of tests [refer to E.S. Pearson (1931) and others; also see Box (1953) and Box and Andersen (1955) where the term "robust" was introduced]. Also the flourishing research on "nonparametric statistics," especially rank methods, around the 1950s can, in part, be seen as an overreaction to the dangers of using strict parametric models. (Despite very close historical, personal, and other ties, *conceptually* robust statistics, being the statistics of *approximate parametric models*, has nothing to do with nonparametric statistics.) Apart from the favorable atmospheric background in Berkeley, it was Tukey's (1960) paper on robust *estimation*, summarizing earlier highly original work by Tukey and his co-workers, that influenced Huber the most.

4. Main Points of the Paper

In his paper, Huber introduced the "gross error model" (as a strong generalization of Tukey's and Newcomb's contamination model), where a fraction ε of the distribution of the data is allowed to be completely arbitrary (the "contamination" or "gross errors"), whereas the other part strictly follows the parametric model. The gross error model can be viewed as "half" of a topological neighborhood (namely, that given by the total variation distance).

Huber also considered the Kolmogorov distance neighborhood that contains weak topology (and total variation) neighborhoods and leads to similar results. It thus takes care of the types of deviations from exact parametric models occurring in reality, whereas the gross error model provides simplified and still relevant results [see also Hampel (1968)].

Huber then defined (M)-estimators (maximum likelihood type estimators, now called M-estimators) and investigated their properties in depth refer also to Huber (1967)]. These estimators, with all their nice properties and far-reaching generality, play a central role in robustness theory. (Every "usual," i.e., asymptotically normal, estimator is locally equivalent to an M-estimator. They were and are also used by other statisticians under different names, such as "estimating equations" or "estimating functions.")

In a fictitious game against "Nature" (see Sec. 3.5 below), Huber derived the optimal robust M-estimator for rather general location models and, in particular, what later were called the "Huber estimator" for the normal location model (a prototype robust estimator) and "Huber's scale estimator" for the normal scale model. He also investigated his "Proposal 2" for the joint estimation of location and scale.

These estimators are essentially proven to be never much worse for *approximately* normal data than the maximum likelihood (and least-squares) estimators for *strictly* normal data, whereas the latter estimators can become arbitrarily bad in *any* realistic neighborhood of the normal model (e.g., by a single unchecked outlier).

5. Further Developments

Huber's paper laid the basis for the development of a general robustness theory. It contains the first of three major (interconnected) approaches to such a theory, the others being described in Huber (1965, 1968), Huber and Strassen (1973), and in Hampel (1968). The former is probably mathematically deepest and the latter most generally applicable (to *arbitrary* "reasonable" parametric models). The relationship between the latter and Huber's first approach is extensively discussed in Hampel et al. (1986, Chap. 1.3 and 2.7, especially 1.3e, which also contains a synthesis of the two approaches). The theory has been extended and applied to regression, multivariate statistics, time series analysis, and some more specialized models; it has been and is being used in real applications; several related large computer programs exist. Also, the robustness problem of the violation of the assumption of independence has more recently found the attention it deserves. These lines of development have preliminarily culminated in the books by Huber (1981) and Hampel et al. (1986). Some related books with special or different emphasis are Andrews et al. (1972), Mosteller and Tukey (1977), Hoaglin et al. (1983), and Rousseeuw and Leroy (1987). There are also several other books and monographs, and robustness research is still going on. A sign of the freshness and importance of Huber's original paper is that the Science Citation Index has still contained about 10 to 20 citations of it every year over the last few years.

If we take the term "robust" not in a technical sense [see Hampel (1968)] but more loosely, we find a bewildering amount of literature using it, including "Bayesian robustness" [for some brief remarks on the latter from the above line of thought, refer to Hampel (1973, p. 95), Huber (1981, p. 263); Hampel et al. (1986, p. 52f); Huber 1987)]. Some literature discusses only mathematical problems without considering applications, some tries pragmatic and ad hoc applications without a solid theoretical background (although often with good intuition); some writers just pay lip service to "robustness" or use the term because it became fashionable.

6. Misunderstandings

Huber's paper, as well as the following robustness work, has also been the subject of much criticism, misunderstandings, lack of understanding, and prejudice. There is not the space here to repeat all the refutations and clarifications that can be found, for example, in Hampel (1973), the discussion of Hampel (1975), Hampel (1980, pp. 17–19; 1989), and Hampel et al. (1986, Chap. 1 and 8.2); Chap. 8.2a (pp. 397–403) of the latter book devotes about five pages to a discussion of specifically 10 common objections against Huber's 1964 paper. A central point of debate, the question of asymmetry, is very briefly discussed in Sec. 3.1 below.

7. Summary

Huber has found a new paradigm that, where applicable, supersedes Fisher's paradigm of strict parametric models. [See also the perfectly general paradigm resulting from merging Huber's first and Hampel's paradigm in Hampel et al. (1986, Chap. 1.3e.)] Huber has made the "dirty" problem area of model specification, deviations from a model, gross errors etc. mathematically precise in a tractable way; he thereby made this area mathematically respectable and built a mathematical theory for it leading to very reasonable and applicable results. The practical applications of this theory are becoming more and more important as more and larger data sets are processed on a computer.

It seems very fitting to quote the end of Sec. 16 from Tukey (1960, p. 473):

> The time has come to study problems where the specification is neither so precise as a Pearson curve nor so diffuse as all distributions with a continuous density function. The precise formulation of such specifications as "the underlying distribution is nearly normal" in such a way as to be *precise*, *manageable*, and *relevant to practice* [italics added] is now both important and overdue. It is time to return to the problem of specification.

This is just what Huber did.

2. Personal Background of the Paper

Peter J. Huber was born in 1934 in Wohlen (Aargau, Switzerland). After finishing the Kantonsschule Aarau and one semester at the University of Zürich, he studied mathematics and physics at the ETH (Eidgenössische Technische Hochschule, Swiss Federal Institute of Technology) Zürich, beginning in the fall of 1954. Here he obtained his "Mathematikdiplom" (master's degree in mathematics), and in 1961, he obtained his Ph.D. in mathematics with Beno Eckmann. The title of his dissertation was "Homotopy Theory in General Categories."

He thus seemed headed toward full abstraction in pure mathematics. However, he felt the need for starting anew with something more concrete. Stiefel and Saxer, who were then working in applied mathematics at ETH and noticed a gap in statistics, encouraged him to go to the University of California in Berkeley in order to study statistics.

During his studies at ETH, Huber had only occasionally used some statistics in the context of his work on ancient Babylonian astronomy. In this he worked together with B.L. van der Waerden, of algebra fame, then (and still) at the University of Zürich. Although Huber used van der Waerden's (1957) statistics book, they hardly, if ever, discussed statistics then.

Huber first learned some statistics from the books by van der Waerden (1957) and Wald (1950). He then studied for two years at Berkeley (1961–63), with a Swiss National Science Foundation and a Miller fellowship. There he learned mainly from Lehmann, LeCam, and (in the area of probability theory) Loève. After having already written several smaller publications on Babylonian astronomy, algebraic topology, and statistics, he wrote there his first and most famous paper on robust statistics, before spending another year at Cornell University and returning as professor to ETH Zürich.

Huber (like van der Waerden by the way) was thus one of those newcomers to statistics who, equipped with a broad and deep general mathematical knowledge and unspoiled by unreflected statistical traditions, could take a fresh look at the field. Right from the beginning, he was bothered by the fact that all the nice mathematical assumptions from which the statistical theory is derived are only approximations to reality. And he felt extremely uncomfortable after reading Tukey's (1960) demonstration of the drastic effects even of tiny deviations from idealized mathematical assumptions.

If we compare Huber and Tukey, who also had a strong background in pure mathematics, it is interesting to see how Tukey (1962, p. 2) tries to improve statistics from outside mathematical statistics, so to speak (but still using its results), whereas Huber (1975) tries to do it from within mathematical statistics (but maintaining a broader statistical and scientific scope).

The time and atmosphere in Berkeley were just right for Huber's task. The work on rank tests there, the study of the Hodges–Lehmann estimator, but also the van der Waerden X-test, Wald's and Wolfowitz's previous work on maximum likelihood estimators and minimax strategies, and LeCam's asymptotic work provided stimulating ingredients and, in part, useful tools for making the jump from Tukey's (1960) contamination model to the general gross error model.

For some interesting personal recollections on the genesis of the paper, mentioning the detours he first tried (nonparametric maximum likelihood and Bayesian approaches), see Huber (1987).

3. Contents of the Paper

1. Introduction and Summary

The concisely written and highly interesting *introduction and summary* contains a preview of the contents of the paper, introducing in particular the gross error model (and Kolmogorov distance model), M-estimators, the basic idea of robust regression using a general ρ-function instead of least squares [worked out in Huber (1973)], and what later became known as the Huber estimator, Huberizing (or metrically Winsorizing, see below), and Huber's least favorable distribution with its surprisingly short tails. (In retrospect, for someone who has built up intuition for M-estimators, its memoryless exponential tails are not surprising anymore.) There are brief historical remarks going back to Gauss [see Huber (1972, p. 1042ff) for more details] and touching on Tukey (1960) [refer to Huber (1981, p. 1ff) for more details]. Huber sensibly argues for the use of the asymptotic variance instead of the actual variance even for moderate n [as is indirectly also corroborated by the results of the large Monte Carlo study in Andrews et al. (1972); see, e.g., Sec. 6F].

The problem of "what to estimate" is also discussed. Since the "unavoidable indeterminacy" still seems to cause some confusion [see Hampel et al. (1986, Chap. 8.2a, p. 401f, and Chap. 8.2d, pp. 409–416)], it should be pointed out that at the end of Part 10 (scale estimation), Huber eventually settled on Fisher consistency at the ideal model, in full agreement with Tukey (1960, p. 452)

> The basic problem is to study the behavior *on the alternative pencil* of methods of estimation which are *correctly calibrated on the standard reference pencil* (in our instances the normal one).

and with the conceptual framework of Hampel (1968; 1974, Sec. 6). Furthermore, replacing the "exact asymptotic variances" (themselves only approximations for finite n) by extrapolations from the ideal model yields (typically) almost exactly the same numerical results, dispels the ambiguity between model distribution and least favorable distribution, allows one to enter a bias into an "extrapolated asymptotic MSE" in a natural way (leading to a gross error model with truly arbitrary contamination), and clarifies even quantitatively the interplay between variance and bias for varying n [Hampel et al. (1986, Sec. 1.3e)].

But it may be considered a basic weakness of Huber's first approach in the narrow sense that he first has to replace the ideal (e.g., normal) parametric model by an auxiliary one of least favorable distributions (which has not always been possible), while eventually he would like to obtain parameter estimates for the ideal model. In general (cases with trivial symmetries excepted), the appropriate estimator for one model is not consistent for the other model, and a choice must be made. These problems do not exist in the approach of Hampel (1968).

There is a slight slip in the introduction [and also in Part 4, Ex. (iii) and (iv), and Part 11], very natural at that time, but still often misunderstood recently: Huberizing (or metrically Winsorizing, namely, the "bringing in" toward the rest the Huber estimator does in effect to outlying data, with the random transformation determined by the ψ-function) is *not* "asymptotically equivalent to Winsorizing" (but rather, in first order, to *trimming*, which, in turn, is *not* a rejection procedure). See Hampel (1968; 1974, Sec. 3.2, 3.3, 4, 5.1) for clarification by means of the influence curve. The reason is that the places of the two point masses depend on the whole sample and not only on two order statistics. (The other remarks on Winsorizing are correct.)

2. Asymptotic Normality of (M)-Estimators for Convex ρ

Part 2 proves almost sure convergence and asymptotic normality for M-estimators of location with convex ρ and includes the asymptotic information inequality for M-estimators.

3. The Case of Non-Convex ρ

Part 3 briefly discusses nonconvex ρ, or equivalently, nonmonotone (e.g., redescending) ψ-functions and influence functions. There are indeed global consistency problems. A presently recommended practice for "redescenders" is to iterate fully with a bounded monotone ψ-function and then to do one to three iterations with the redescender [see, e.g., Hampel (1980, p. 16) or Huber (1981, p. 192)]. A thorough analysis of convergence regions is given by Clarke (1986).

4. Examples

Part 4 contains as examples the mean, median, Huber estimator, and what later [Hampel (1974, p. 392f)] was called "Huber-type skipped mean" (to avoid confusion with trimming), together with their asymptotic variances.

The last example provided the main motivation for Huber (1967), where the conjectured asymptotic variance formula was rigorously proven. A simple "one-step" version of this last example turned out to be a very successful rejection rule [Hampel (1985)].

5. Minimax Question; A Special Case

Part 5 contains a main result of the paper, namely, the explicit minimax solution of the 2-person 0-sum game of the statistician (selecting an M-estimator of location) against Nature (selecting a distribution from a gross error model, when the basic ideal model distribution has concave log density), with the loss for the statistician being the resulting asymptotic variance. The saddlepoint consists of a least favorable distribution with an unchanged center (apart from norming) and exponential tails, and its maximum likelihood estimator with bounded and monotone ψ [thus merely bounding the score function of the MLE for the ideal model distribution; see the very similar estimators resulting from other approaches in Hampel et al. (1986)]. Huber's result is applicable for many more models than the normal one, including asymmetric ones (though not, e.g., for the Cauchy location model); in Part 10, he uses it for the normal scale model.

6. The Contaminated Normal Distribution

The corollary in Part 6 gives the first and most famous optimality property of the Huber estimator and describes Huber's least favorable distribution. The numbers in Table I are very important for practical applications of the theory: They show that in this case, the minimax theory is *not* too pessimistic, and that the (usually unknown) fraction of contamination (or gross errors) hardly matters for Huber estimators with k between 1 and 2. In fact, for all k between 1.2 and 1.8, and for all fractions of arbitrary contamination (or outliers) ε between 0 and 0.1 (the range usually relevant in practice), the asymptotic variance is never more than 7% higher than the minimal maximum variance that could be guaranteed if ε (but not the distribution of outliers) were known. In particular, none of the estimators for this range of k loses more than 7% efficiency under the strict model of normality, whereas the arithmetic mean can have efficiency zero under *any* positive fraction of outliers [and even *after* rejection of outliers, the "mean" typically loses about 15 to 20% efficiency under more realistic situations such as t_3; see Table 1 in Hampel (1985)]. For $k \approx 1.57$, the asymptotic variance exceeds the minimax variance by at most about 3% for all ε between 0 and 10% (the best possible result for this range of ε). Practically the whole table can now be reproduced very accurately by means of simple linear extrapolation from the normal model, using a Taylor expansion [Hampel et al. (1986, pp. 50 and 173)].

7. The Question of Bias

The short Part 7 on bias has long been neglected, although bias, being of 1st order, is theoretically more important than asymptotic variance (the actual variance going to 0 for $n \to \infty$). Only recently has this part been taken up again [see Martin and Zamar (1989) and subsequent work]. Huber showed that for very large n, the simple median is the best estimator (in accordance with applied statistical folklore, and contrary to the spirit of adaptive estimation and shrinking neighborhood models), but he also stressed that for small n, the variance is more important. He came rather close to combining bias and variance [see Hampel et al. (1986, Sec. 1.3e, and Sec. 3.1 above)]. A slight error in the least favorable distribution against bias is corrected in Huber (1981, p. 74f; see also the sketch there): Its density is not constant in the middle, but piecewise normal up to the center of symmetry, and hence bimodal.

8. Minimax Theory

Part 8 contains a general abstract minimax theory, including a generalized technical definition of the Fisher information. In part, it may be viewed as a concession to the spirit of the time. (Huber repeatedly remarked ironically that many statisticians believe only in proving optimality results.) However, Huber himself used the results several times: for Part 9, his unpublished version of the *tanh* estimators [see Collins (1976)], and for Huber (1974). The proofs in Huber (1981, Chap. 4, in particular, Sec. 4.4) are shorter and more elegant than the original proofs. Theorem 4 needs for the uniqueness some additional condition like the convexity of the (open) carrier; refer to Huber (loc. cit., p. 81).

9. The Minimax Solution for ε-Normal Distributions

Part 9 provides the conceptually important solution for the Kolmogorov distance model (which is a proper topological neighborhood, as opposed to the gross error model). The result is very similar to that of Part 6, only more complicated, and thus shows that the gross error model already captures the most essential features of deviations from the ideal parametric model and is esthetically more satisfying. Other types of topological neighborhoods (for different robustness problems) are discussed in Huber (1965, 1968), Hampel (1968), and Huber and Strassen (1973).

10. Estimation of a Scale Parameter

Part 10 derives the important Huber scale estimator by taking logarithms and reducing the problem to an (asymmetric!) location problem. Actually Huber discusses only one part of the range of solutions (namely, that for large c); otherwise, he would have obtained the median (absolute) deviation as limiting most robust scale estimator (Hampel 1968, 1974). After some vacillating, he decided to make the estimator Fisher-consistent at the normal model (see Sec. 3.1 above).

11. Estimation of a Location Parameter, if Scale and ε are Unknown

Part 11 contains three proposals for the simultaneous estimation of location and scale, including the famous Proposal 2, which generalizes nicely to linear models, including a convergence proof, but has only a moderately high breakdown point [refer to Andrews et al. (1972) for its finite sample behavior, also that of several other closely related estimators]. Proposal 2 (and its generalization to linear models) became one of the most often used robust estimators, whereas the other two proposals, especially Proposal 1, are hardly ever mentioned later. The Monte Carlo study in Andrews et al. (1972) also contains some estimators roughly in the spirit of Proposal 3, notably LJS and JAE (and other adaptive trimmed means), which were run overall with moderate success. For Proposal 2, Huber then proves the existence and uniqueness of the solution for the defining implicit equations and describes a highly specific algorithm, whose properties he checks by an exploratory Monte Carlo study [see Huber (1981), Sec. 6.7, 7.8, and 8.11 and the references therein for this and other algorithms, and the references in Hampel et al. (1986), Sec. 6.4c for the major robust computer programs and the algorithms used by them]. Table IV gives the suprema of the asymptotic variances of the location part under symmetric contamination; compare Exhibit 6.6.1 (p. 144) in Huber (1981) for the symmetric and asymmetric asymptotic breakdown points ("symmetric" breakdown occurs just outside, general breakdown already within Table IV; the lower finite sample gross error breakdown points, i.e., the fractions of outliers still tolerated, are even noticeably lower in small samples).

12. A Generalization

The last part of Huber's paper contains a generalization of M-estimators in the spirit of U-statistics. This new class does not seem to have been used later, except that Huber shows one of its simplest members to be identical to the well-known Hodges–Lehmann estimator, usually treated as an R-estimator. Huber gives numerical comparisons between this estimator and the Huber estimators, indicating how similarly they behave. This similarity turned out to also be striking in several cooperative Monte Carlo studies by Tukey and others analyzed in Andrews et al. (1972, p. 155) and Hampel (1983, Sec. 2.4, 3.3, 3.9).

Acknowledgments.

Much of the more personal information was given by Peter Huber during a long telephone conversation. Some useful suggestions were made by the editors and a referee.

References

Andrews, D.F., Bickel, P.J., Hampel, F.R., Huber, P.J., Rogers, W.H., and Tukey, J.W. (1972). *Robust Estimates of Location: Survey and Advances*. Princeton University Press, Princeton, N.J.

Box, G.E.P. (1953). Non-normality and tests on variances, *Biometrika*, **40**, 318–335.

Box, G.E.P., and Andersen, S.L. (1955). Permutation theory in the derivation of robust criteria and the study of departures from assumption, *J. Roy. Statist. Soc., Ser. B*, **17**, 1–34.

Clarke, B.R. (1986). Asymptotic theory for description of regions in which Newton-Raphson iterations converge to location M-estimators, *J. Statist. Plann. Infer.* **15**, 71–85.

Collins, J.R. (1976). Robust estimation of a location parameter in the presence of asymmetry, *Ann. Statist.*, **4**, 68–85.

Fisher, R.A. (1922). On the mathematical foundations of theoretical statistics, *Philos. Trans. Roy. Soc. Lond. Ser. A*, **222**, 309–368.

Hampel, F.R. (1968). Contributions to the theory of robust estimation. Ph.D. thesis, University of California, Berkeley.

Hampel, F.R. (1973). Robust estimation: A condensed partial survey, *Z. Wahrsch. verw. Geb.*, **27**, 87–104.

Hampel, F.R. (1974). The influence curve and its role in robust estimation, *J. Amer. Statist. Assoc.*, **69**, 383–393.

Hampel, F.R. (1975). Beyond location parameters: Robust concepts and methods, *Proc. 40th Session ISI*, **XLVI**, Book 1, 375–391 (with discussion).

Hampel, F.R. (1980). Robuste Schätzungen: Ein anwendungsorientierter Ueberblick, *Biometrical J.*, **22**, 3–21.

Hampel, F.R. (1983). The robustness of some nonparametric procedures, in *A Festschrift for Erich L. Lehmann* P.J. Bickel, K.A. Doksum, and J.L. Hodges Jr., eds. Wadsworth, Belmont, Calif., pp. 209–238.

Hampel, F.R. (1985). The breakdown points of the mean combined with some rejection rules, *Technometrics*, **27**, 95–107.

Hampel, F. (1989). Nonlinear estimation in linear models. In *Statistical Data Analysis and Inference* Y. Dodge, (ed.). Elsevier Science (North-Holland), Amsterdam, pp. 19–32.

Hampel, F.R., Ronchetti, E.M., Rousseeuw, P.J., and Stahel, W.A. (1986). *Robust Statistics: The Approach Based on Influence Functions*. Wiley, New York.

Hoaglin, D.C., Mosteller, F., and Tukey, J.W. (eds.) (1983). *Understanding Robust and Exploratory Data Analysis*. Wiley, New York.

Huber, P.J. (1965). A robust version of the probability ratio test, *Ann. Math. Statist.*, **36**, 1753–1758.

Huber, P.J. (1967). The behavior of maximum likelihood estimates under nonstandard conditions, in *Proceedings of 5th Berkeley Symposium on Mathematical Statistics and Probability*, Vol. 1, pp. 221–233, University of California Press, Berkeley.

Huber, P.J. (1968). Robust confidence limits, *Z. Wahrsch. verw. Geb.*, **10**, 269–278.

Huber, P.J. (1972). Robust statistics: A review, *Ann. Math. Statist.*, **43**, 1041–1067.
Huber, P.J. (1973). Robust regression: Asymptotics, conjectures, and Monte Carlo, *Ann. Statist.*, **1**, 799–821.
Huber, P.J. (1974). Fisher information and spline interpolation, *Ann. Statist.*, **2**, 1029–1033.
Huber, P.J. (1975). Application vs. abstraction: The selling out of mathematical statistics? *Suppl. Adv. Appl. Prob.* **7**, 84–89.
Huber, P.J. (1981). *Robust Statistics.* Wiley, New York.
Huber, P.J. (1987). Comment (pp. 279–281) on James S. Hodges: Uncertainty, policy analysis and statistics, *Statist. Sci.* **2**, No. 3, 259–291.
Huber, P.J., and Strassen, V. (1973). Minimax tests and the Neyman–Pearson lemma for capacities, *Ann. Statist.*, **1**, 251–263; **2**, 223–224.
Jeffreys, H. (1939, 1948, 1961, 1983). *Theory of Probability.* Clarendon Press, Oxford.
Martin, R.D., and Zamar, R.H. (1989). Asymptotically min-max robust M-estimates of scale for positive random variables, *J. Amer. Statist. Assoc.* **84**, 494–501.
Mosteller, F., and Tukey, J.W. (1977). *Data Analysis and Regression.* Addison-Wesley, Reading, Mass.
Newcomb, S. (1886). A generalized theory of the combination of observations so as to obtain the best result, *Amer. J. Math.*, **8**, 343–366.
Pearson, E.S. (1931). The analysis of variance in cases of non-normal variation, *Biometrika*, **23**, 114–133.
Pearson, K. (1902). On the mathematical theory of errors of judgement, with special reference to the personal equation, *Philos. Trans. Roy. Soc., Ser. A*, **198**, 235–299.
Rousseeuw, P.J., and Leroy, A. (1987). *Robust Regression and Outlier Detection.* Wiley, New York.
Stigler, S.M. (1973). Simon Newcomb, Percy Daniell, and the history of robust estimation 1885–1920. *J. Amer. Statist. Assoc*, **68**, 872–879.
Student (1927). Errors of routine analysis, *Biometrika*, **19**, 151–164.
Tukey, J.W. (1960). A survey of sampling from contaminated distributions, *Contributions to Probability and Statistics* I. Olkin, S.G. Ghurye, W. Hoeffding, W.G. Madow, and H.B. Mann. (eds). Stanford University Press, pp. 448–485.
Tukey, J.W. (1962). The future of data analysis, *Ann. Math. Statist*, **33**, 1–67.
van der Waerden, B.L. (1957). *Mathematische Statistik.* Springer-Verlag, Berlin. (English translation, *Mathematical Statistics*, 1969, Springer-Verlag, New York.)
Wald, A. (1950). *Statistical Decision Functions.* Wiley, New York.

Robust Estimation of a Location Parameter[1]

Peter J. Huber[2]
University of California, Berkeley

1. Introduction and Summary

This paper contains a new approach toward a theory of robust estimation; it treats in detail the asymptotic theory of estimating a location parameter for contaminated normal distributions, and exhibits estimators—intermediaries between sample mean and sample median—that are asymptotically most robust (in a sense to be specified) among all translation invariant estimators. For the general background, see Tukey (1960) (p. 448 ff.)

Let $x_1, \ldots, x_n$ be independent random variables with common distribution function $F(t - \xi)$. The problem is to estimate the location parameter ξ, but with the complication that the prototype distribution $F(t)$ is only approximately known. I shall primarily be concerned with the model of indeterminacy $F = (1 - \varepsilon)\Phi + \varepsilon H$, where $0 \leqq \varepsilon < 1$ is a known number, $\Phi(t) = (2\pi)^{-1/2} \int_{-\infty}^{t} \exp(-\frac{1}{2}s^2)\, ds$ is the standard normal cumulative and H is an unknown contaminating distribution. This model arises for instance if the observations are assumed to be normal with variance 1, but a fraction ε of them is affected by gross errors. Later on, I shall also consider other models of indeterminacy, e.g., $\sup_t |F(t) - \Phi(t)| \leqq \varepsilon$.

Some inconvenience is caused by the fact that location and scale parameters are not uniquely determined: in general, for fixed ε, there will be several values of ξ and σ such that $\sup_t |F(t) - \Phi((t - \xi)/\sigma)| \leqq \varepsilon$, and similarly for the contaminated case. Although this inherent and unavoidable indeterminacy is

Received 4 June 1963.

[1] This research was performed while the author held an Adolph C. and Mary Sprague Miller Fellowship.

[2] Now at Cornell University.

small if ε is small and is rather irrelevant for practical purposes, it poses awkward problems for the theory, especially for optimality questions. To remove this difficulty, one may either (i) restrict attention to symmetric distributions, and estimate the location of the center of symmetry (this works for ξ but not for σ); (ii) one may define the parameter to be estimated in terms of the estimator itself, namely by its asymptotic value for sample size $n \to \infty$; or (iii) one may define the parameters by arbitrarily chosen functionals of the distribution (e.g., by the expectation, or the median of F). All three possibilities have unsatisfactory aspects, and I shall usually choose the variant which is mathematically most convenient.

It is interesting to look back to the very origin of the theory of estimation, namely to Gauss and his theory of least squares. Gauss was fully aware that his main reason for assuming an underlying normal distribution and a quadratic loss function was mathematical, i.e., computational, convenience. In later times, this was often forgotten, partly because of the central limit theorem. However, if one wants to be honest, the central limit theorem can at most explain why many distributions occurring in practice are approximately normal. The stress is on the word "approximately."

This raises a question which could have been asked already by Gauss, but which was, as far as I know, only raised a few years ago (notably by Tukey): What happens if the true distribution deviates slightly from the assumed normal one? As is now well known, the sample mean then may have a catastrophically bad performance: seemingly quite mild deviations may already explode its variance. Tukey and others proposed several more robust substitutes—trimmed means, Winsorized means, etc.—and explored their performance for a few typical violations of normality. A general theory of robust estimation is still lacking; it is hoped that the present paper will furnish the first few steps toward such a theory.

At the core of the method of least squares lies the idea to minimize the sum of the squared "errors," that is, to adjust the unknown parameters such that the sum of the squares of the differences between observed and computed values is minimized. In the simplest case, with which we are concerned here, namely the estimation of a location parameter, one has to minimize the expression $\sum_i (x_i - T)^2$; this is of course achieved by the sample mean $T = \sum_i x_i/n$. I should like to emphasize that no loss function is involved here; I am only describing how the least squares estimator is defined, and neither the underlying family of distributions nor the true value of the parameter to be estimated enters so far.

It is quite natural to ask whether one can obtain more robustness by minimizing another function of the errors than the sum of their squares. We shall therefore concentrate our attention to estimators that can be defined by a minimum principle of the form (for a location parameter):

$$T = T_n(x_1, \ldots, x_n) \textit{ minimizes } \textstyle\sum_i \rho(x_i - T), \quad \textit{where } \rho \textit{ is a non-constant function.} \qquad (M)$$

Of course, this definition generalizes at once to more general least squares type problems, where several parameters have to be determined.

This class of estimators contains in particular (i) the sample mean ($\rho(t) = t^2$), (ii) the sample median ($\rho(t) = |t|$), and more generally, (iii) all maximum likelihood estimators ($\rho(t) = -\log f(t)$, where f is the assumed density of the untranslated distribution).

These (M)-estimators, as I shall call them for short, have rather pleasant asymptotic properties; sufficient conditions for asymptotic normality and an explicit expression for their asymptotic variance will be given.

How should one judge the robustness of an estimator $T_n(x) = T_n(x_1, \ldots, x_n)$? Since ill effects from contamination are mainly felt for large sample sizes, it seems that one should primarily optimize large sample robustness properties. Therefore, a convenient measure of robustness for asymptotically normal estimators seems to be the supremum of the asymptotic variance ($n \to \infty$) when F ranges over some suitable set of underlying distributions, in particular over the set of all $F = (1 - \varepsilon)\Phi + \varepsilon H$ for fixed ε and symmetric H.

On second thought, it turns out that the asymptotic variance is not only easier to handle, but that even for moderate values of n it is a better measure of performance than the actual variance, because (i) the actual variance of an estimator depends very much on the behavior of the tails of H, and the supremum of the actual variance is infinite for any estimator whose value is always contained in the convex hull of the observations. (ii) If an estimator is asymptotically normal, then the important central part of its distribution and confidence intervals for moderate confidence levels can better be approximated in terms of the asymptotic variance than in terms of the actual variance.

If we adopt this measure of robustness, and if we restrict attention to (M)-estimators, then it will be shown that the most robust estimator is uniquely determined and corresponds to the following $\rho : \rho(t) = \frac{1}{2}t^2$ for $|t| < k$, $\rho(t) = k|t| - \frac{1}{2}k^2$ for $|t| \geqq k$, with k depending on ε. This estimator is most robust even among all translation invariant estimators. Sample mean ($k = \infty$) and sample median ($k = 0$) are limiting cases corresponding to $\varepsilon = 0$ and $\varepsilon = 1$, respectively, and the estimator is closely related and asymptotically equivalent to Winsorizing.

I recall the definition of Winsorizing: assume that the observations have been ordered, $x_1 \leqq x_2 \leqq \cdots \leqq x_n$, then the statistic $T = n^{-1}(gx_{g+1} + x_{g+1} + x_{g+2} + \cdots + x_{n-h} + hx_{n-h})$ is called the *Winsorized mean*, obtained by *Winsorizing* the g leftmost and the h rightmost observations. The above most robust (M)-estimators can be described by the same formula, except that in the first and in the last summand, the factors x_{g+1} and x_{n-h} have to be replaced by some numbers u, v satisfying $x_g \leqq u \leqq x_{g+1}$ and $x_{n-h} \leqq v \leqq x_{n-h+1}$, respectively; g, h, u and v depend on the sample.

In fact, this (M)-estimator is the maximum likelihood estimator corresponding to a unique least favorable distribution F_0 with density $f_0(t) =$

$(1-\varepsilon)(2\pi)^{-1/2}e^{-\rho(t)}$. This f_0 behaves like a normal density for small t, like an exponential density for large t. At least for me, this was rather surprising—I would have expected an f_0 with much heavier tails.

This result is a particular case of a more general one that can be stated roughly as follows: Assume that F belongs to some convex set C of distribution functions. Then the most robust (M)-estimator for the set C coincides with the maximum likelihood estimator for the unique $F_0 \in C$ which has the smallest Fisher information number $I(F) = \int (f'/f)^2 f\, dt$ among all $F \in C$.

Miscellaneous related problems will also be treated: the case of non-symmetric contaminating distributions; the most robust estimator for the model of indeterminacy $\sup_t |F(t) - \Phi(t)| \leqq \varepsilon$; robust estimation of a scale parameter; how to estimate location, if scale and ε are unknown; numerical computation of the estimators; more general estimators, e.g., minimizing $\sum_{i<j} \rho(x_i - T, x_j - T)$, where ρ is a function of two arguments.

Questions of small sample size theory will not be touched in this paper.

2. Asymptotic Normality of (M)-Estimators for Convex ρ

It will be assumed throughout this section that ρ is a continuous convex real-valued function of a real variable t, tending to $+\infty$ as $t \to \pm\infty$.

Let $x_1, \ldots, x_n$ be independent identically distributed random variables with common distribution function F. Let $[T_n(x)]$ be the set of all those ξ for which $Q(\xi) = \sum_{i=1}^n \rho(x_i - \xi)$ reaches its infimum Q_{inf}. Obviously, $[T_n(x)]$ is invariant under translations: $[T_n(x + c)] = [T_n(x)] + c$. By $T_n(x)$ we shall denote any representation of the set valued function $(x_1, \ldots, x_n) \to [T_n(x)]$ by a single-valued function $(x_1, \ldots, x_n) \to T_n(x) \in [T_n(x)]$, e.g., $T_n(x) =$ midpoint of $[T_n(x)]$.

Lemma 1. *$Q(\xi)$ is a convex function of ξ, and $[T_n(x)]$ is non-empty, convex and compact. If ρ is strictly convex, then $[T_n(x)]$ is reduced to a single point.*

Proof. (Strict) convexity of Q follows immediately from (strict) convexity of ρ. The sets $\{\xi \mid Q(\xi) \leqq Q_{\text{inf}} + m^{-1}\}$ form a decreasing sequence of non-empty convex compact sets as $m \to \infty$, hence their intersection $[T_n(x)]$ is non-empty convex compact. If Q is strictly convex, and if ξ', ξ'' were distinct points of $[T_n(x)]$, then we would have $Q_{\text{inf}} = Q(\frac{1}{2}\xi' + \frac{1}{2}\xi'') < \frac{1}{2}Q(\xi') + \frac{1}{2}Q(\xi'') = Q_{\text{inf}}$, which is a contradiction. □

Let $\psi = \rho'$ be the derivative of ρ, normalized such that $\psi(t) = \frac{1}{2}\psi(t-0) + \frac{1}{2}\psi(t+0)$. ψ is monotone increasing and strictly negative (positive) for large negative (positive) values of t.

If ψ is continuous, $T_n(x)$ may equivalently be defined by the equation $\sum_{i=1}^n \psi(x_i - T_n(x)) = 0$.

Define $\lambda(\xi) = \int \psi(t - \xi)F(dt) = E\psi(t - \xi)$.

Lemma 2. *If there is a ξ_0 such that $\lambda(\xi_0)$ exists and is finite, then $\lambda(\xi)$ exists for all ξ (possibly $\lambda(\xi) = \pm\infty$), is monotone decreasing and strictly positive (negative) for large negative (positive) values of ξ.*

(*Editor's note*: The proof has been omitted.)

Lemma 3. ("Consistency of T_n"). *Assume that there is a c such that $\lambda(\xi) > 0$ for $\xi < c$ and $\lambda(\xi) < 0$ for $\xi > c$. Then $T_n \to c$ almost sure and in probability.*

Proof. Let $\varepsilon > 0$. Then, by the law of large numbers,

$$(1/n) \sum_{i=1}^n \psi(x_i - c - \varepsilon) \to \lambda(c + \varepsilon) < 0,$$

$$(1/n) \sum_{i=1}^n \psi(x_i - c + \varepsilon) \to \lambda(c - \varepsilon) > 0,$$

a.s. and in probability. Hence, by monotonicity of ψ, for a.a. sample sequences, $c - \varepsilon < [T_n(x)] < c + \varepsilon$ holds from some n on, and similarly, $P[c - \varepsilon < [T_n(x)] < c + \varepsilon] \to 1$. □

Remark. The assumption of Lemma 3 could have been replaced—in view of Lemma 2—by the assumption that $\lambda(\xi)$ exists and is finite for some ξ_0, and that it does not identically vanish on a nondegenerate interval.

Lemma 4. ("Asymptotic normality"). *Assume that* (i) $\lambda(c) = 0$, (ii) $\lambda(\xi)$ *is differentiable at $\xi = c$, and $\lambda'(c) < 0$,* (iii) $\int \psi^2(t - \xi)F(dt)$ *is finite and continuous at $\xi = c$. Then $n^{1/2}(T_n(x) - c)$ is asymptotically normal with asymptotic mean 0 and asymptotic variance $V(\psi, F) = \int \psi^2(t - c)F(dt)/(\lambda'(c))^2$.*

Proof. Without loss of generality assume $c = 0$. We have to show that for every fixed real number g, $P[n^{1/2}T_n < g\sigma] \to \Phi(g)$, where $\sigma = V(\psi, F)^{1/2}$. Since

$$[T_n < g\sigma n^{-1/2}] \subset \left[\sum_{i=1}^n \psi(x_i - g\sigma n^{-1/2}) < 0\right] \subset [T_n \leqq g\sigma n^{-1/2}],$$

it suffices to show that

$$p_n = \left[\sum_{i=1}^n \psi(x_i - g\sigma n^{-1/2}) < 0\right] \to \Phi(g).$$

Let $s^2 = \int (\psi(t - g\sigma n^{-1/2}) - \lambda(g\sigma n^{-1/2}))^2 F(dt)$, then the $y_i = (\psi(x_i - g\sigma n^{-1/2}) - \lambda(g\sigma n^{-1/2}))/s$ are independent random variables with mean 0 and variance 1. We have

$$p_n = P\left[n^{-1/2} \sum_{i=1}^n y_i < -n^{1/2}\lambda(g\sigma n^{-1/2})/s\right]$$

and $-n^{1/2}\lambda(g\sigma n^{-1/2})/s \to g$. We shall see presently that $n^{-1/2}\sum y_i$ is asymptotically normal with mean 0 and variance 1, hence $p_n \to \Phi(g)$.

There is a slight complication: although the y_i are independent identically distributed summands, they are different for different values of n, and the usual formulations of normal convergence theorems do not apply. But by the normal convergence criterion, as given in Loève (1960), p. 295, the distribution of $\sum n^{-1/2} y_i$ converges toward the standard normal iff for every $\varepsilon > 0$, as $n \to \infty$, $\int_E y_1^2 F(dx_1) \to 0$, the integration being extended over the set $E = [|y_1| \geqq n^{1/2}\varepsilon]$. Since $\lambda(g\sigma n^{-1/2}) \to 0$, this holds iff for every $\varepsilon > 0$, $\int_{E'} \psi^2(x_1 - n^{-1/2}g\sigma)F(dx_1) \to 0$, the integration being extended over the set $E' = [|\psi(x_1 - n^{-1/2}g\sigma)| \geqq n^{1/2}\varepsilon]$. Now, let $\eta > 0$ be given and let n_0 be such that $|n^{-1/2}g\sigma| < \eta$ for $n \geqq n_0$. Then, since ψ is monotone, we have $\psi^2(x_1 - n^{-1/2}g\sigma) \leqq u^2(x_1)$ for $n \geqq n_0$ with $u(x_1) = \max\{|\psi(x_1 - \eta)|, |\psi(x_1 + \eta)|\}$. Since $E\psi^2(x_1 \pm \eta)$ exists, we have $\int_{E''} u^2(x_1)F(dx_1) \to 0$, the integration being extended over the set $E'' = [|u| \geqq n^{1/2}\varepsilon] \supset E'$. This proves the assertion. □

Remark. Lemma 4 admits a slight generalization, which is sometimes useful: If λ has different left and right derivatives at c, then the asymptotic distribution of $n^{1/2}(T_n(x) - c)$ is pieced together from two half-normal distributions, with variances determined by the one-sided derivatives of λ respectively in the same way as in Lemma 4.

For the remainder of this section, assume for simplicity $c = 0$. By formally interchanging the order of integration and differentiation one obtains $\lambda'(0) = [(d/d\xi)\int \psi(t - \xi)F(dt)]_{\xi=0} = -\int \psi'(t)F(dt) = -E\psi'$. This makes sense either when ψ is sufficiently regular or when F is sufficiently regular so that ψ' may be interpreted as a Schwartz distribution. Moreover, if F has an absolutely continuous density f, we find by partial integration $\lambda'(0) = \int \psi(t)f'(t)\,dt$. The Schwarz inequality now yields the inequality

$$V(\psi, F) = \frac{E\psi^2}{(E\psi')^2} = \frac{\int \psi^2 f\,dt}{\left[\int \psi(f'/f)f\,dt\right]^2} \geqq \frac{1}{\int (f'/f)^2 f\,dt}. \tag{1}$$

We have strict inequality unless $\psi = -pf'/f$, for some constant p, that is, unless $f(t) = \text{const.}\exp(-\rho(t)/p)$ and then the (M)-estimator is the maximum likelihood estimator.

3. The Case of Non-Convex ρ

If ρ is not convex, the estimators T_n will, in general, no longer converge toward some constant c. Apparently, one has to impose not only local but also some global conditions on ρ in order to have consistency. Compare also Wald (1949).

Asymptotic normality is easier to handle. The following is a simple proof of asymptotic normality under somewhat too stringent regularity conditions, but without assuming monotonicity of ψ.

Lemma 5. *Assume that* (i) $T_n(x) \to 0$ *in probability*; (ii) ψ *is continuous and has a uniformly continuous derivative* ψ'; (iii) $E\psi^2 < \infty$; (iv) $0 < E\psi' < \infty$. *Then* $n^{1/2}T_n$ *is asymptotically normal, with asymptotic mean* 0 *and asymptotic variance* $E\psi^2/(E\psi')^2$.

PROOF. T_n satisfies $\sum \psi(x_i - T_n) = 0$, hence, by the mean value theorem, there is some ϑ, $0 \leqq \vartheta \leqq 1$, such that $\sum \psi(x_i) - T_n \sum \psi'(x_i - \vartheta T_n) = 0$, or $n^{1/2}T_n = n^{-1/2}\sum \psi(x_i)/(n^{-1}\sum \psi'(x_i - \vartheta T_n))$. The numerator is asymptotically normal with mean 0 and variance $E\psi^2$, the denominator tends in probability toward the constant $E\psi'$, hence $n^{1/2}T_n$ is asymptotically normal with mean and variance as given above (see Cramér (1946), 20.6). □

4. Examples

In all examples below we shall choose the origin such that $E\psi(x) = 0$, that is, $T_n \to 0$.

(i) $\rho(t) = t^2$. The corresponding estimator is the sample mean $T_n = \Sigma x_i/n$, $T_n \to Ex = 0$, and $n^{1/2}T_n$ is asymptotically normal with mean 0 and variance $Ex^2 = \int t^2 F(dt)$.

(ii) $\rho(t) = |t|$. The corresponding estimator is the sample median, and $T_n \to \mu = 0$, where μ is the median of F. If F has a non-zero derivative at the origin, then $n^{1/2}T_n$ is asymptotically normal with asymptotic mean 0 and asymptotic variance $1/(2F'(0))^2$.

(iii) $\rho(t) = \frac{1}{2}t^2$ for $|t| \leqq k$, $\rho(t) = k|t| - \frac{1}{2}k^2$ for $|t| > k$. The corresponding estimator is closely related to Winsorizing, since it has the following property: if we put for short $t_0 = T_n(x)$, then all observations x_l such that $|x_i - t_0| > k$ may be replaced by $t_0 \pm k$, whichever is nearer, without changing the estimate $T_n(x)$, and t_0 equals the arithmetic mean of the modified set of observation. *Proof*: since ρ is convex, it suffices to look at the first derivative of $Q(\xi) = \sum \rho(x_i - \xi)$, which still vanishes at $\xi = t_0$ if it did before the change:

$$Q'(t_0) = \sum_{|x_i - t_0| \leqq k} (x_i - t_0) + \sum_{x_i > t_0 + k} ((t_0 + k) - t_0) + \sum_{x_i < t_0 - k} ((t_0 - k) - t_0).$$

The asymptotic variance of $n^{1/2}T_n$ is

$$V = \left(\int_{-k}^{+k} t^2 F(dt) + k^2 \int_{-\infty}^{-k} F(dt) + k^2 \int_{k}^{\infty} F(dt) \right) \Big/ \left(\int_{-k}^{+k} F(dt) \right)^2.$$

(iv) $\rho(t) = \frac{1}{2}t^2$ for $|t| \leqq k$, $\rho(t) = \frac{1}{2}k^2$ for $|t| > k$. The corresponding estimator is a trimmed mean: let $t_0 = T_n(x)$, and assume for simplicity that for no observation $|x_i - t_0| = k$. Then $T_n(x)$ equals the sample mean of those observations for which $|x_i - t_0| < k$, and remains unchanged if some or all of the x_i for which $|x_i - t_0| > k$ are removed. *Proof*: Compute the derivative of $Q(\xi) = \sum \rho(x_i - \xi)$ at t_0

$$Q'(t_0) = \sum_{|x_i - t_0| < k} (x_i - t_0).$$

Thus, if $Q(\xi)$ attained its infimum at t_0 before the outliers were removed, it must still be stationary there when the outliers are exluded. But removing an outlier decreases $Q(t_0)$ by $\frac{1}{2}k^2$ and cannot decrease $Q(\xi)$ by more than that for any ξ, hence Q still attains its infimum at t_0.

Assume that T_n is consistent $T_n \to 0$, then the asymptotic variance of $n^{1/2} T_n$ is, formally,

$$V = \int_{-k}^{+k} t^2 F(dt) \Big/ \left(\int_{-k}^{+k} F(dt) - kF'(k) - kF'(-k) \right)^2.$$

(However, one should be reminded that we did not prove the formula in this case, ρ not being convex. I conjecture that it is valid if and only if the expression inside the parentheses of the denominator is strictly positive.)

A glance at the asymptotic variances computed in these four cases shows plainly that

(i) The sample mean is very sensitive to the tails of F.
(ii) The sample median is very sensitive to the behavior of F at its median, and neglects its behavior elsewhere.
(iii) "Winsorizing" avoids these shortcomings and seems to be practically foolproof. Apparently, this is connected with the fact that the corresponding ψ is monotone, bounded and absolutely continuous.
(iv) The "trimming procedure" is rather sensitive to the behavior of F at the rejection points $\pm k$; a high density at these points will play havoc with the estimate. This shortcoming seems to be common also to other rejection procedures; here one might avoid it by smoothing ρ at $\pm k$.

5. Minimax Questions; a Special Case

At first glance, it might seem absurd to look for asymptotic minimax solutions, since, asymptotically, one could do better by estimating the true underlying distribution (see also Stein (1956), Hájek (1962)). However, this seems to require an exorbitant number of observations. On the other hand, it is hoped—and preliminary numerical experiments seem to substantiate this hope—that (M)-estimators approach their asymptotic behavior rather fast, provided ψ is bounded. So an asymptotic minimax theory should be useful in those frequent cases where the sample size is perhaps large enough to indicate deviations from the assumed model but not yet large enough to establish their nature. In the case of the contaminated normal distribution $F =$

$(1-\varepsilon)\Phi + \varepsilon H$, this means that asymptotic minimax theory would be appropriate whenever the sample size n is fairly large, but εn, the average number of outliers, is still rather small. We shall discuss this point in a later section.

First we shall treat a special case that can be solved explicitly by a direct verification of the saddlepoint property.

Let C be the set of all distributions of the form $F = (1-\varepsilon)G + \varepsilon H$, where $0 \leqq \varepsilon < 1$ is a fixed number, G is a fixed and H is a variable distribution function. Assume that G has a convex support and a twice continuously differentiable density g such that $-\log g$ is convex on the support of G. Let T_n be an (M)-estimator belonging to a certain ρ, let $\psi = \rho'$ be the derivative of ρ, and let $c = c(F)$ be such that $\int \psi(t-c)F(dt) = 0$. The asymptotic variance of $n^{1/2}(T_n - c)$ will be

$$V(\psi, F) = E_F\psi^2(t-c)/(E_F\psi'(t-c))^2,$$

provided ψ is "nice." For the moment we shall not bother about c—it might be interpreted as the bias of T_n—and shall try to minimize the supremum $\sup_F V(\psi, F)$ of the asymptotic variance only for those pairs (ψ, F) for which $c(F) = 0$.

Theorem 1. *The asymptotic variance $V(\psi, F)$ has a saddlepoint: there is an $F_0 = (1-\varepsilon)G + \varepsilon H_0$ and a ψ_0 such that*

$$\sup_F V(\psi_0, F) = V(\psi_0, F_0) = \inf_\psi V(\psi, F_0)$$

where F ranges over those distributions in C for which $E_F\psi_0 = 0$. Let $t_0 < t_1$ be the endpoints of the interval where $|g'/g| \leqq k$ (either or both of these endpoints may be at infinity), and k is related to ε by

$$(1-\varepsilon)^{-1} = \int_{t_0}^{t_1} g(t)\,dt + (g(t_0) + g(t_1))/k.$$

Then the density f_0 of F_0 is given explicitly by

$$\begin{aligned} f_0(t) &= (1-\varepsilon)g(t_0)e^{k(t-t_0)} && \text{for } t \leqq t_0, \\ &= (1-\varepsilon)g(t) && \text{for } t_0 < t < t_1, \\ &= (1-\varepsilon)g(t_1)e^{-k(t-t_1)} && \text{for } t \geqq t_1. \end{aligned}$$

$\psi_0 = -f_0'/f_0$ is monotone and bounded and corresponds to the maximum likelihood estimator of the location parameter when F_0 is the underlying distribution.

Remark. The statement of this theorem is unsatisfactory insofar as the class over which H ranges depends on ψ_0. This could be avoided by restricting G to be symmetric, and letting H range over all symmetric distributions. However, I preferred to give the stronger statement above.

PROOF. If it easy to check that F_0 has total mass 1, hence also $H_0 = (F_0 - (1-\varepsilon)G)/\varepsilon$ has total mass 1, and it remains to check that its density h_0 is nonnegative. But

$$\begin{aligned}\varepsilon h_0(t) &= (1-\varepsilon)[g(t_0)e^{k(t-t_0)} - g(t)] && \text{for } t \leqq t_0,\\ &= 0 && \text{for } t_0 < t < t_1,\\ &= (1-\varepsilon)[g(t_1)e^{-k(t-t_1)} - g(t)] && \text{for } t \geqq t_1.\end{aligned}$$

Because the function $-\log g(t)$ is convex, it lies above its tangents at t_0 and t_1, i.e., $-\log g(t) \geqq -\log g(t_0) - k(t - t_0)$, thus $g(t) \leqq g(t_0)e^{k(t-t_0)}$, etc., which implies non-negativity of h_0.

$\psi_0 = -f_0'/f_0$ is bounded and monotone, so we have

$$V(\psi_0, F_0) = \frac{E_{F_0}\psi_0^2}{(E_{F_0}\psi_0')^2} = \frac{(1-\varepsilon)E_G\psi_0^2 + \varepsilon k^2}{((1-\varepsilon)E_G\psi_0')^2}.$$

The right side is an obvious upper bound for $V(\psi_0, F)$, provided $E_F\psi_0 = 0$, so we have $V(\psi_0, F) \leqq V(\psi_0, F_0)$.

Note that λ_F has left and right derivatives, so that asymptotic distribution of $n^{1/2}T_n$ must be either normal or pieced together from two half normal distributions, and the latter case can only occur if F puts positive mass on the points t_0, t_1.

The inequality $V(\psi_0, F_0) \leqq V(\psi, F_0)$ follows directly from inequality (1) of Section 2, noticing that $V(\psi_0, F_0) = (\int (f_0'/f_0)^2 f_0\, dt)^{-1}$. □

More generally, using results of LeCam (1953), (1958), one may prove that the maximum likelihood estimator for F_0 is indeed efficient among all translation invariant estimators, in the sense that for any translation invariant measurable estimator $T_n = T_n(x_1, \ldots, x_n)$ one has

$$V(\psi_0, F_0) \leqq \lim_{c\to\infty} \liminf_{n\to\infty} \int \min(nT_n^2, c^2) F_0(dx_1)\ldots F_0(dx_n).$$

In other words, one may strengthen the theorem: the (M)-estimator corresponding to ψ_0 minimizes the maximal asymptotic variance not only among (M)-estimators, but even among all translation invariant estimators. (Instead of translation invariance, any weaker condition still excluding superefficient estimators would suffice in this context.)

6. The Contaminated Normal Distribution

The assumptions of the preceding section are satisfied if $G = \Phi$ is the standard normal cumulative with density $\varphi(t) = (2\pi)^{-1/2}\exp\{-\frac{1}{2}t^2\}$. Because of the importance of this case, we summarize the results in a

Corollary. *Define the estimator $T_n = T_n(x_1, \ldots, x_n)$ by the property that it minimizes $\sum_i \rho(x_i - T_n)$, where $\rho(t) = \frac{1}{2}t^2$ for $|t| < k$, $\rho(t) = k|t| - \frac{1}{2}k^2$ for*

Table I. Upper Bounds for the Asymptotic Variance of $n^{1/2}T_n$ for $F = (1 - \varepsilon)\Phi + \varepsilon H$, H Symmetric.

k	$E_\Phi\psi'$	$E_\Phi\psi^2$	$\varepsilon_{\min}$	$\varepsilon = 0$	0.001	0.002	0.005	0.010	0.020	0.050	0.100	0.200	0.500
0.0	0.0000	0.0000	1.0000	1.571	1.574	1.577	1.587	1.603	1.636	1.741	1.939	2.454	6.283
0.1	0.0797	0.0095	0.8753	1.492	1.495	1.498	1.508	1.523	1.556	1.658	1.853	2.358	6.137
0.2	0.1585	0.0358	0.7542	1.423	1.426	1.429	1.438	1.454	1.485	1.586	1.778	2.276	6.030
0.3	0.2358	0.0758	0.6401	1.362	1.365	1.368	1.377	1.393	1.424	1.524	1.714	2.209	5.962
0.4	0.3108	0.1265	0.5354	1.309	1.312	1.315	1.324	1.339	1.370	1.470	1.659	2.154	5.930
0.5	0.3829	0.1851	0.4417	1.263	1.266	1.269	1.278	1.293	1.324	1.423	1.613	2.111	5.935
0.6	0.4515	0.2491	0.3599	1.222	1.225	1.228	1.237	1.252	1.284	1.384	1.576	2.079	5.976
0.7	0.5161	0.3160	0.2899	1.187	1.190	1.193	1.202	1.217	1.249	1.351	1.546	2.058	6.053
0.8	0.5763	0.3840	0.2311	1.156	1.159	1.162	1.172	1.188	1.220	1.324	1.523	2.047	6.166
0.9	0.6319	0.4511	0.1825	1.130	1.133	1.136	1.146	1.162	1.195	1.302	1.506	2.046	6.317
1.0	0.6827	0.5161	0.1428	1.107	1.111	1.114	1.124	1.140	1.175	1.284	1.495	2.055	6.506
1.1	0.7287	0.5777	0.1109	1.088	1.091	1.095	1.105	1.122	1.158	1.272	1.490	2.072	6.734
1.2	0.7699	0.6352	0.0855	1.072	1.075	1.079	1.089	1.107	1.144	1.263	1.491	2.099	7.003
1.3	0.8064	0.6880	0.0655	1.058	1.062	1.065	1.077	1.095	1.134	1.258	1.497	2.135	7.314
1.4	0.8385	0.7358	0.0498	1.047	1.050	1.054	1.066	1.086	1.126	1.256	1.507	2.180	7.669
1.5	0.8664	0.7785	0.0376	1.037	1.041	1.045	1.057	1.078	1.121	1.258	1.522	2.233	8.069
1.6	0.8904	0.8160	0.0282	1.029	1.034	1.038	1.051	1.073	1.118	1.262	1.542	2.296	8.517
1.7	0.9109	0.8487	0.0211	1.023	1.027	1.032	1.046	1.069	1.116	1.270	1.567	2.367	9.012
1.8	0.9281	0.8767	0.0156	1.018	1.023	1.027	1.042	1.066	1.117	1.280	1.595	2.448	9.558
1.9	0.9426	0.9006	0.0115	1.014	1.019	1.024	1.039	1.065	1.119	1.292	1.628	2.537	10.154

T_n is defined by $\sum \psi(x_i - T_n) = 0$, where $\psi(t) = t$ for $|t| < k$, $\psi(t) = k \operatorname{sign}(t)$ for $|t| \geqq k$.

Editor's note: The part of the table for $k = 2.0(1)3.0$ has been omitted.

$|t| \geqq k$. Let $\psi = \rho'$. For symmetric H, $n^{1/2}T_n$ is asymptotically normal with asymptotic mean 0, *and attains the maximal asymptotic variance* $\sup_F V(\psi, F) = \{(1-\varepsilon)E_\Phi\psi^2 + \varepsilon k^2\}/\{(1-\varepsilon)E_\Phi\psi'\}^2$, *whenever H puts all its mass outside the interval $[-k, +k]$. If ε and k are related by $(1-\varepsilon)^{-1} = \int_{-k}^{+k}\varphi(t)\,dt + 2\varphi(k)/k$, then T_n minimizes the maximal asymptotic variance among all translation invariant estimators, the maximum being taken over the set of all symmetric ε-contaminated distributions $F = (1-\varepsilon)\Phi + \varepsilon H$. There is a unique asymptotically least favorable F_0 having the density $f_0(t) = (1-\varepsilon)(2\pi)^{-1/2}e^{-\rho(t)}$.*

Table I gives the value of $\sup_F V(\psi, F)$ in dependence of k and ε. Notice that $E_\Phi\psi' = P_\Phi[|x| < k]$ is the probability that an observation falls inside the interval $(-k, +k)$ if Φ is the true underlying distribution. $\varepsilon_{\min}$ is the value of ε for which the corresponding k yields the asymptotic minimax solution.

It can be seen from the table that the performance of T_n is not very sensitive to the choice of k—any value between 1.0 and 2.0 will give acceptable results for all $\varepsilon \leqq 0.2$. The limiting case $k = 0$ is the sample median.

7. The Question of Bias

Assume that we have the situation of Section 5, the prototype distribution being $F = (1-\varepsilon)G + \varepsilon H$, with a symmetric G. What happens if the contaminating distribution H is not symmetric? (Suggested interpretation: most observations are distributed according to $G(x - \vartheta)$, but a small fraction ε of them is distributed according to $H(x)$, and is unrelated to the parameter ϑ to be estimated. This might happen if the observations are made with an instrument that jams with probability ε.)

Let ψ be any of the minimax procedures described in Theorem 1, or any other sufficiently regular bounded monotone skew symmetric function. Let $c = c(F)$ be such that $\lambda_F(c) = E_F\psi(t - c) = 0$; for reasons of symmetry, we have $c(G) = 0$. Thus, $c(F)$ may be considered as the bias caused by the contamination εH. The mean value theorem implies that $\lambda_F(0) + c(F)\lambda_F'(\vartheta c) = 0$ for some ϑ, $0 < \vartheta < 1$, if λ_F is differentiable. Hence

$$c(F) = -\varepsilon\lambda_H(0)/[(1-\varepsilon)\lambda_G'(\vartheta c) + \varepsilon\lambda_H'(\vartheta c)].$$

Thus, since ψ and λ are monotone,

$$|c(F)| \leqq [\varepsilon/(1-\varepsilon)]\sup|\psi|/(-\lambda_G'(\vartheta c)). \tag{2}$$

If $\lambda_G'(c)$ is continuous at $c = 0$, and if ε and hence c are small, we may replace ϑc by 0 to obtain an approximation to the right side of (2).

The upper bound of the asymptotic variance calculated above for symmetric H will have to be increased if also asymmetric H are admitted. But under mild regularity conditions this correction will be of the order $o(c)$ and may thus be neglected in comparison with the bias c, if the latter is small.

In the particular case where $G = \Phi$ is the standard normal cumulative, and ψ is the minimax solution for ε (minimizing the maximal asymptotic variance for symmetric H), we obtain the following approximate upper bounds for the asymptotic bias c:

$$|c| \leqq 0.17 \qquad \text{for } \varepsilon = 0.1, k = 1.1.4,$$

$$|c| \leqq 0.021 \qquad \text{for } \varepsilon = 0.01, k = 1.95,$$

$$|c| \leqq 0.0027 \qquad \text{for } \varepsilon = 0.001, k = 2.64.$$

These bounds will be reached by c whenever H puts all its mass to the right of $k + c$, or to the left of $-k - c$.

It does not seem that one can make the bias much smaller without doing serious damage to the variance of the estimate. The sample median gives the smallest bias, namely $|c| \leqq \varepsilon(1 - \varepsilon)^{-1}(2\varphi(0))^{-1} \approx 1.25\varepsilon(1 - \varepsilon)^{-1}$. It is easy to see that this is indeed the best possible bound for translation invariant estimators: let

$$f(t) = (1 - \varepsilon)\varphi(0) \qquad \text{for } |t| < c,$$

$$= (1 - \varepsilon)\varphi(|t| - c) \qquad \text{for } |t| \geqq c,$$

f and ε being related by $\int f(t)\,dt = 1$, i.e., $c = \varepsilon(1 - \varepsilon)^{-1}(2\varphi(0))^{-1}$. Then both $f(t - c)$ and $f(t + c)$ may be considered as ε-contaminated standard normal densities, one having the contamination to the right, and the other to the left of the origin. No translation invariant estimator can have a bias of absolute value less than c for both of these densities simultaneously, since the difference of the two biases must be $2c$.

As a consequence of these considerations, if somebody wants to control the asymptotic bias and keep it below $\frac{1}{2}n^{-1/2}$, that is, below one half of the asymptotic standard deviation of T_n, he should not increase the sample size above $n = 9,600$ and 35,000, respectively, for the above-mentioned values of ε and k.

Of course, such sample sizes are not large enough to estimate H to a sufficient degree of accuracy, since, on the average, less than $\varepsilon n = 1$, 6 or 35 observations, respectively, would come from H, and one would not even know which ones. This may be taken as a justification for using minimax procedures, in the sense that sample sizes reasonable for a given amount of unknown contamination will not allow to determine the nature of this contamination, except in rather extreme cases.

8. Minimax Theory

The most robust estimator constructed in Section 5 coincided with the maximum likelihood estimator for some least favorable distribution. The present section shows that this is quite a general feature of most robust (M)-estimators.

Consider the following game against Nature:

Nature chooses a distribution F from some set C of distributions on the real line.

The statistician chooses a function ψ from some set Ψ of functions.

The payoff to the statistician is

$$K(\psi, F) = \left(\int \psi' \, dF\right)^2 \Big/ \int \psi^2 \, dF$$

where ψ' denotes the derivative (in measure) of ψ; it shall be tacitly understood that the statistician chooses not only ψ, but also a particular representative of ψ'.

On purpose, we shall remain vague about the set Ψ and shall change it according to convenience. For instance, if F has an absolutely continuous density f, the payoff function may be transformed by partial integration into $K(\psi, F) = (\int \psi f' \, dt)^2 / \int \psi^2 f \, dt$, which makes sense for a broader class of functions than the original definition and allows to use a larger class Ψ.

If C and Ψ are restricted in a suitable way, for instance to symmetric distributions and sufficiently regular skew symmetric monotone functions ψ, we will have $K(\psi, F) =$ (asymptotic variance of $n^{1/2} T_n)^{-1}$, hence $K(\psi, F)$ is a measure of the asymptotic efficiency of the estimator. For the moment, we shall not bother about this aspect and shall just try to establish properties of the abstractly defined game.

The reason for preferring the utility function $K(\psi, F)$ over the loss function $V(\psi, F) = K(\psi, F)^{-1}$ used in earlier sections is that K has more convenient convexity properties.

The following convexity inequality is hardly new, but since it will be used several times, it is given the status of a lemma.

Lemma 6. *Let* $v_1 > 0$, $v_2 > 0$, $0 \leqq \alpha \leqq 1$. *Then*

$$\frac{(\alpha u_1 + (1 - \alpha) u_2)^2}{\alpha v_1 + (1 - \alpha) v_2} \leqq \alpha \frac{u_1^2}{v_1} + (1 - \alpha) \frac{u_2^2}{v_2}.$$

PROOF. Put $v = \alpha v_1 + (1 - \alpha) v_2$, $\beta = \alpha v_1 / v$. Then

$$\frac{(\alpha u_1 + (1 - \alpha) u_2)^2}{\alpha v_1 + (1 - \alpha) v_2} = v \left\{\beta \frac{u_1}{v_1} + (1 - \beta) \frac{u_2}{v_2}\right\}^2 \leqq v \left\{\beta \left(\frac{u_1}{v_1}\right)^2 + (1 - \beta) \left(\frac{u_2}{v_2}\right)^2\right\}$$

$$= \alpha \frac{u_1^2}{v_1} + (1 - \alpha) \frac{u_2^2}{v_2}. \qquad \square$$

In particular, if we put $u_i = \int \psi' \, dF_i$, $v_i = \int \psi^2 \, dF_i$, $(i = 1, 2)$, we obtain that $K(\psi, \cdot)$ is a convex function: $K(\psi, \alpha F_1 + (1 - \alpha) F_2) \leqq \alpha K(\psi, F_1) + (1 - \alpha) K(\psi, F_2)$. Hence, if μ is a mixed strategy of Nature, that is, a probability distribution on C, it follows that $K(\psi, F_\mu) \leqq \int K(\psi, F) \, d\mu$, where $F_\mu = \int F \, d\mu$. In other words, every mixed strategy of Nature is dominated by a pure one, provided C is convex in the sense that it contains all averages F_μ of elements of C. However, we shall not use this result, so we do not give it a formal proof.

Theorem 2. *Let C be a convex set of distribution functions such that every $F \in C$ has an absolutely continuous density f satisfying $I(F) = \int (f'/f)^2 f\,dt < \infty$.* (i) *If there is an $F_0 \in C$ such that $I(F_0) \leqq I(F)$ for all $F \in C$, and if $\psi_0 = -f_0'/f_0$ is contained in Ψ, then (ψ_0, F_0) is a saddlepoint of the game, that is,*

$$K(\psi, F_0) \leqq K(\psi_0, F_0) = I(F_0) \leqq K(\psi_0, F)$$

for all $\psi \in \Psi$ and all $F \in C$. (ii) *Conversely, if (ψ_0, F_0) is a saddlepoint and Ψ contains a nonzero multiple of $-f_0'/f_0$, then $I(F_0) \leqq I(F)$ for all $F \in C$, F_0 is uniquely determined, and ψ_0 is $[F_0]$-equivalent to a multiple $-f_0'/f_0$.* (iii) *Necessary and sufficient for F_0 to minimize $I(F)$ is that $\int (-2\psi_0 g' - \psi_0^2 g)\,dt \geqq 0$ for all functions $g = f_1 - f_0$, $F_1 \in C$.*

(*Editor's note*: The proof has been omitted.)

Theorem 2 does not answer the question whether such an F_0 exists, and it does not tell us what happens for distributions F that do not possess an absolutely continuous density. To settle these questions, we have to generalize a little bit further.

From now on, let C be a vaguely compact convex set of substochastic distributions on the real line. (Under the vague topology we understand the weakest topology such that the maps $F \to \int \psi\,dF$ are continuous for all continuous functions ψ with compact support.) We no longer require that the elements of C have a density, and we do not exclude the possibility that some might have a total mass < 1. This latter case is not so far-fetched as it might seem. It corresponds loosely to the usual practice of screening the data and excluding extremely wild observations from further processing: If we formalize this by assuming that Nature chooses a probability distribution on the extended real line and that the statistician excludes infinite-valued observations, we end up with substochastic distributions on the ordinary real line. Theorem 2 remains valid for substochastic distributions.

On the other hand, vague compactness would be unduly restrictive if we did not admit substochastic distributions.

First we shall extend the definition of the Fisher information number, essentially by putting it equal to ∞ whenever the classical expression $\int (f'/f)^2 f\,dt$ does not work. However, a more devious approach is more useful, where the above statement turns up as the conclusion of a theorem.

Definition. *Let*

$$I(F) = \sup_\psi \left(\int \psi'\,dF\right)^2 \Big/ \int \psi^2\,dF$$

where ψ ranges over the set of continuous and continuously differentiable functions with compact support satisfying $\int \psi^2\,dF \neq 0$.

Theorem 3. *$I(F) < \infty$ if and only if F has an absolutely continuous density f such that $\int (f'/f)^2 f\, dt < \infty$, f' being the derivative in measure of f, and then $I(F) = \int (f'/f)^2 f\, dt$.*

(*Editor's note*: The proof has been omitted.)

Theorem 4. *Assume that C is vaguely compact and convex. If $\inf_{F \in C} I(F) = a < \infty$, then there exists a unique $F_0 \in C$ such that $I(F_0) = a$.*

PROOF. For $b > a$, the set $K_b = \{F \in C \mid I(F) \leqq b\}$ is nonempty and convex. Being the intersection of the vaguely closed sets $K_{b,\psi} = \{F \in C \mid (\int \psi'\, dF)^2 \leqq b \int \psi^2\, dF\}$, it is vaguely closed and hence compact. By the finite intersection property, $K_a = \bigcap_{b>a} K_b$ is nonempty, convex and compact, and Theorem 2 implies that it is reduced to a single element F_0. □

Theorem 5. *Assume that C is vaguely compact and convex, and that the set $C_1 = \{F \in C \mid I(F) < \infty\}$ is dense in C. Let F_0 be the unique distribution minimizing $I(F)$. If F_0 happens to be such that $\psi_0 = -f_0'/f_0$ is in Ψ, is differentiable in measure and ψ_0' is either continuous or nonnegative upper semi-continuous, and vanishes at infinity, then (ψ_0, F_0) is not only a saddlepoint with respect to C_1, but also with respect to C.*

PROOF. Assume that $F \in C$, and let (F_ν) be a net of elements of C_1 converging toward F. We have to prove that $K(\psi_0, F) \geqq a$. If ψ_0' is continuous and vanishes at infinity, then $\int \psi_0'\, dF_\nu \to \int \psi_0'\, dF$. If $\psi_0' \geqq 0$ is upper semi-continuous and vanishes at infinity, then $\int \psi_0'\, dF = \inf_\chi \int \chi\, dF = \inf_\chi \operatorname{limsup}_\nu \int \chi\, dF_\nu \geqq \operatorname{limsup}_\nu \int \psi_0'\, dF$, where χ ranges over the continuous functions $\chi \geqq \psi_0'$ vanishing at infinity. On the other hand, we have

$$\int \psi_0^2\, dF = \sup_\chi \int \chi\, dF = \sup_\chi \operatorname{liminf}_\nu \int \chi\, \mathrm{d}F_\nu \leqq \operatorname{liminf}_\nu \int \psi_0^2\, dF_\nu,$$

where χ ranges over the set of continuous functions $0 \leqq \chi \leqq \psi_0^2$ vanishing at infinity. It follows that $K(\psi_0, F) \geqq \operatorname{limsup}_\nu K(\psi_0, F_\nu) \geqq a$. □

Remark. This theorem solves the abstract game; however, for the original statistical problem some further considerations are needed. Let $\lambda_F(\xi) = \int \psi_0(t - \xi)\, dF$. According to Lemma 4, the asymptotic variance of $n^{1/2} T_n$ is $V(\psi_0, F) = \int \psi_0^2\, \mathrm{d}F/(\lambda_F'(0))^2$, provided ψ_0 is bounded and monotone increasing, $\lambda_F(0) = 0$ and $\lambda_F'(0) < 0$ exists. If ψ_0' is uniformly continuous, then $-\lambda_F'(0) = \int \psi_0'\, dF$, thus $V(\psi_0, F) = K(\psi_0, F)^{-1}$ and (ψ_0, F_0) is also a saddlepoint of $V(\psi, F)$ if it is one of $K(\psi, F)$ and if $\lambda_F(0) = 0$ for all $F \in C$. If ψ_0' is discontinuous, then $-\lambda_F'(0) = \int \psi_0'\, dF$ still holds for $F \in C_1$. Under the assumptions (i) ψ_0 is continuous, bounded and monotone increasing, (ii) for all $F \in C$, $\lambda_F(0) = 0$ and $\lambda_F'(0) < 0$ exists, (iii) for every $F \in C$, there is a net $F_\nu \to F$, $F_\nu \in C_1$, such that $\lim \int \psi_0'\, dF_\nu \leqq -\lambda_F'(0)$, we may still conclude that (ψ_0, F_0) is a saddlepoint of $V(\psi, F)$. The proof is a simplified version of that of Theorem 5.

9. The Minimax Solution for ε-Normal Distributions

We say that a distribution F is ε-normal with approximate mean ξ and approximate variance σ^2 if it satisfies $\sup_t|F(t) - \Phi((t - \xi)/\sigma)| \leqq \varepsilon$. Of course, ξ and σ are in general not uniquely determined, but the indeterminacy is small for small ε. We shall now determine the saddlepoint of the game, when Nature chooses an F from the set C of symmetric ε-standard-normal distributions:

$$C = \{F|\ \sup_t|F(t) - \Phi(t)| \leqq \varepsilon,\ F(t) + F(-t) = 1\}.$$

This set C is vaguely closed in the set of all substochastic distributions, hence it is compact; it is convex and $C_1 = \{F \in C\,|\,I(F) < \infty\}$ is dense in C. So the theory of the preceding section can be applied.

The saddlepoint (ψ_0, F_0) was found by some kind of enlightened guesswork, using variational techniques; instead of going through the cumbersome heuristic arguments, I shall state the solution explicitly and shall then verify it.

The solution is valid only for values of ε between 0 and (approximately) 0.03, but it seems that this range is just the range important in practice.

F_0 has the symmetric density

$$\begin{aligned} f_0(t) = f_0(-t) &= \varphi(a)(\cos \tfrac{1}{2}ca)^{-2}(\cos \tfrac{1}{2}ct)^2 && \text{for } 0 \leqq t \leqq a \\ &= \varphi(t) && \text{for } a < t < b \\ &= \varphi(b)e^{-b(t-b)} && \text{for } t \geqq b \end{aligned}$$

where $\varphi(t) = (2\pi)^{-1/2}e^{-t^2/2}$, and a, b, c are three constants determined through the relations

(i) $c\tan(\frac{1}{2}ca) = a\ (0 \leqq ca < \pi)$,
(ii) $\int_0^a f_0\,dt = \int_0^a \varphi\,dt - \varepsilon$,
(iii) $\int_b^\infty f_0\,dt = \int_b^\infty \varphi\,dt + \varepsilon$.

It is easy to check that the F_0 thus defined belongs to C, provided (i), (ii), and (iii) can be satisfied with $0 \leqq a \leqq b$; this is the case for values of ε between 0 and (approximately) 0.03. For numerical results, see Table II.

Table II.

ε	a	b	c
0.001	0.65	2.44	1.37
0.002	0.75	2.23	1.35
0.005	0.91	1.95	1.32
0.01	1.06	1.72	1.29
0.02	1.24	1.49	1.26
0.03	1.34	1.36	1.23

Putting $\psi_0 = -f_0'/f_0$, we have

$$\begin{aligned}\psi_0(t) = -\psi_0(-t) = c\tan\tfrac{1}{2}ct, &\qquad \text{for } 0 \leqq t \leqq a\\ = t &\qquad \text{for } a < t < b\\ = b &\qquad \text{for } t \geqq b,\end{aligned}$$

which is continuous and has a piecewise continuous bounded derivative.

We have to prove that (ψ_0, F_0) is a saddlepoint. In view of Theorem 2, we have to show that $J'(0) = \int(-2\psi_0 g' - \psi_0^2 g)\,dt \geqq 0$ for all functions $g = f_1 - f_0$, $F_1 \in C_1$. By partial integration, the condition becomes

$$J'(0) = \int (2\psi_0' - \psi_0^2) g\, dt \geqq 0 \qquad \text{for all } g = f_1 - f_0.$$

Let $\hat{g}(t) = \int_0^t g(s)\, ds = F_1(t) - F_0(t)$, then $F_1 \in C$ implies the necessary feasibility condition $\hat{g}(t) \geqq 0$ for $a \leqq t \leqq b$, since in this interval $F_0(t) = \Phi(t) - \varepsilon$.

We have

$$J'(0) = \int_0^a c^2 g\, dt + \int_a^b (2 - t^2) g\, dt + \int_b^\infty (-b)^2 g\, dt,$$

thus, if we transform the middle integral by partial integration,

$$\begin{aligned}J'(0) &= c^2\hat{g}(a) + (2 - b^2)\hat{g}(b) - (2 - a^2)\hat{g}(a) + \int_a^b 2t\hat{g}(t)\, dt + b^2\hat{g}(b)\\ &\quad + b^2(1 - F_1(\infty))\\ &= (c^2 + a^2 - 2)\hat{g}(a) + 2\hat{g}(b) + \int_a^b 2t\hat{g}(t)\, dt + b^2(1 - F_1(\infty)).\end{aligned}$$

We shall show presently that $c^2 + a^2 - 2 \geqq 0$, hence all summands are nonnegative and it follows that $J'(0) \geqq 0$. To prove that remaining inequality, notice that in the interval $(0, a)$, ψ_0' increases monotonely from $\psi_0'(0) = \frac{1}{2}c^2$ to $\psi_0'(a) = \frac{1}{2}(c^2 + a^2)$. Since $\int_0^a \psi_0'\, dt = \psi_0(a) = a$, it follows from the mean value theorem that $\psi_0'(t) = 1$ for some $0 \leqq t \leqq a$, hence $\psi_0'(a) = \frac{1}{2}(c^2 + a^2) \geqq 1$, which establishes the inequality in question.

This proves the saddlepoint property for distributions from C_1. To prove it for the whole of C, one has to check the points (i), (ii) and (iii) of the remark at the end of Section 8. (i) and (ii) are immediate. Instead of doing (iii), it is probably easier to go back to Theorem 5, to show that $-\lambda_F'(0) = \int \psi_0'\, dF$ unless F puts a positive mass on $\{\pm a, \pm b\}$; and that in this latter case one may decrease $-\lambda_F'(0)$ by shifting this mass away from $\{\pm a, \pm b\}$.

10. Estimation of a Scale Parameter

The theory of estimating a scale parameter is less satisfactory than that of estimating a location parameter. Perhaps the main source of trouble is that there is no natural "canonical" parameter to be estimated. In the case of a location parameter, it was convenient to restrict attention to symmetric distributions; then there is a natural location parameter, namely the location of the center of symmetry, and we could separate difficulties by optimizing the estimator for symmetric distributions (where we know what we are estimating) and then investigate the properties of this optimal estimator for nonstandard conditions, e.g., for nonsymmetric distributions. In the case of scale parameters, we meet, typically, highly asymmetric distributions, and the above device to ensure unicity of the parameter to be estimated fails. Moreover, it becomes questionable, whether one should minimize bias or variance of the estimator.

So we shall just go ahead and shall construct estimators that are invariant under scale transformations and that estimate their own asymptotic values as accurately as possible. Of course, one has to check afterward in a few typical cases what these estimators really do estimate.

The problem of estimating a scale parameter for the random variable X can be reduced to that of estimating a location parameter for the random variable $Y = \log X^2$. This amounts to estimating the parameter $\tau = \log \sigma^2$, where σ is a scale parameter for X. The change of parameters here is made for technical reasons, but it might also be justified on purely philosophical grounds. Compare also Tukey (1960), especially p. 461 ff.

We shall again be concerned with the contaminated normal case, that is, we shall assume that the distribution of Y is of the form $F = (1 - \varepsilon)\, G + \varepsilon H$, where G is the distribution of Y corresponding to a standard normal distribution for X, ε is a known number and H is an unknown contaminating distribution.

If the distribution of X is the standard normal, then the distribution function of $Y = \log X^2$ is $G(u) = P[Y < u] = P[X^2 < e^u] = \Phi(e^{(1/2)u}) - \Phi(-e^{(1/2)u})$, having the density $g(u) = G'(u) = (2\pi)^{-(1/2)} \exp(-\frac{1}{2}e^u + \frac{1}{2}u)$.

Then, $-\log g$ is a convex function, and $-g'(u)/g(u) = \frac{1}{2}(e^u - 1)$ is monotone and differentiable. Hence, we may apply the theory of Section 5, especially Theorem 1. To avoid confusion later on, we replace the letters ψ, k of Theorem 1 by χ, c and define

$$\begin{aligned} \chi_0(u) &= \tfrac{1}{2}(e^u - 1) && \text{for } \tfrac{1}{2}|e^u - 1| < c, \\ &= c \operatorname{sign}(e^u - 1) && \text{for } \tfrac{1}{2}|e^u - 1| \geqq c. \end{aligned}$$

Then, if c is chosen appropriately, the estimator T_n defined by $\sum_i \chi_0(y_i - T_n) = 0$ will minimize the maximal asymptotic variance among all estimators of $\tau = \log \sigma^2$ that are invariant under a change of scale of the x_i (resulting in a translation of the $y_i = \log x_i^2$), the maximum being taken over those H for which $E_F \chi_0 = 0$.

The minimax solution shows a different behavior for $c < \frac{1}{2}$ than for $c \geqq \frac{1}{2}$; for the following, we shall assume that $c \geqq \frac{1}{2}$; thus, with $q^2 = 2c + 1 \geqq 2$,

$$\begin{aligned} \chi_0(u) &= \tfrac{1}{2}(e^u - 1) && \text{for } e^u < q^2 \\ &= \tfrac{1}{2}(q^2 - 1) && \text{for } e^u \geqq q^2. \end{aligned}$$

Unfortunately, the condition $E_F\chi_0 = 0$ now becomes rather restrictive: it means that only those H are admitted for competition in $\sup_F V(\chi_0, F)$ which put all their mass on the set $[|x| > q]$, and for this class of distributions we have identically $V(\chi_0, F) = \sup_F V(\chi_0, F)$. Nevertheless, it seems intuitively plausible that an estimator behaving satisfactorily against this type of contamination should not fare too badly against other types.

The parameters q and ε are related by

$$(1-\varepsilon)^{-1} = \int_{-\infty}^{t_1} g(t)\,dt + g(t_1)/c$$

where $t_1 = \log q^2$. Hence

$$(1-\varepsilon)^{-1} = \int_{-q}^{+q} \varphi(t)\,dt + 2q(q^2-1)^{-1}\varphi(q). \tag{3}$$

It will be convenient to express the results again in terms of the x_i and of the estimate $S_n = e^{(1/2)T_n}$ of the scale parameter of the x_i. If we introduce the function ψ_0(already used in Section 6), displaying explicitly the parameter q occurring in it

$$\begin{aligned} \psi_0(q, t) &= t && \text{for } |t| < q \\ &= q\,\text{sign}(t) && \text{for } |t| \geqq q, \end{aligned} \tag{4}$$

then S_n satisfies

$$\sum_{i=1}^{n} \psi_0^2(q, x_i/S_n) = n,$$

as an elementary calculation shows, and may be defined by this equation.

Summarizing, we may describe the properties of S_n as follows. Let C be the class of all distributions of the form $F = (1-\varepsilon)\Phi + \varepsilon H$, where ε is given by (3), and H puts all its mass outside the interval $[-q, +q]$. The problem is to estimate the variance σ^2 of the normal component of F, where F has been obtained by a scale transformation from some (unknown) element of C. Then, $\log S_n^2$ is an estimate of $\log \sigma^2$, invariant under scale transformations, asymptotically unbiased and asymptotically normal; it minimizes the maximal asymptotic variance among all scale invariant estimators of $\log \sigma^2$, and has a constant asymptotic variance over the whole of C.

In other words, S_n is gauged for the class C and the class of distributions that can be generated from C by scale transformations and has there about the most pleasant asymptotic properties one can reasonably expect. The flaw is that S_n may have quite a considerable bias for distributions not in C, for instance for the normal distribution Φ itself.

In most cases it will be preferable to gauge the bias such that it is zero at Φ. In order to do this, let S_∞ be the asymptotic value of S_n under Φ; then define $S_n' = S_n/S_\infty$. Obviously, S_n' has the same pleasant properties as S_n, except that it is biased for the class C, but asymptotically unbiased for Φ.

We have, by the definition of S_n,

$$1 - n^{-1} \sum \psi_0^2(q, x_i/S_n) = n^{-1} S_\infty^{-2} \sum \psi_0^2(qS_\infty, x_i/S_n'),$$

and, going over to the limit,

$$1 = E_\Phi \psi_0^2(q, t/S_\infty) = S_\infty^{-2} E_\Phi \psi_0^2(qS_\infty, t).$$

Put $q' = qS_\infty$, then S_n' can be defined by the equation

$$n^{-1} \sum_i \psi_0^2(q', x_i/S_n') = E_\Phi \psi_0^2(q', t). \tag{5}$$

Table III gives some numerical results. $\varepsilon_{\min}$ is the value of ε for which the corresponding q' yields the asymptotic minimax solution.

Table III.

q'	S_∞	$q = q'/S_\infty$	$\varepsilon_{\min}$
1.1	0.760	1.45	0.182
1.5	0.882	1.70	0.074
2.0	0.960	2.08	0.019
2.5	0.989	2.53	0.0038
3.0	0.997	3.01	0.0006

11. Estimation of a Location Parameter, if Scale and ε Are Unknown

Now consider the problem of estimating a location parameter if neither the scale σ of the normal component nor the amount ε of contamination are known. We are now not interested in estimating scale or ε, but only in estimating location by some estimator T_n, and in addition, in estimating the (asymptotic) variance of T_n.

The following are three heuristic proposals on how one can apply the theory of (M)-estimators to this problem. I conjecture that the third proposal is asymptotically minimax, in the sense that independently of scale σ and ε, the supremum over symmetric H of the asymptotic variance of T_n has the least possible value. The first two proposals are asymptotically minimax if k is chosen in relation to the true ε.

All three proposals are asymptotically equivalent to Winsorizing; they differ among each other in the determination of the number of observations to be Winsorized. From now on I shall omit the indices from ψ_0, T_n and S_n; $\psi = \psi_0$ is defined by (4) in the preceding section.

Proposal 1. Choose beforehand a fixed number k (either by intuition or by using previous experience). Then determine T and S such that

$$n^{-1}\sum\psi(k,(x_i-T)/S)=0$$
$$n^{-1}\sum\psi'(k,(x_i-T)/S)=E_\Phi\psi'(k,x).$$

The second equation can only be fulfilled approximately, since $\sum\psi'(k,(x_i-T)/S)$ is an integer, namely the number of observations contained in the interval $[T-kS, T+kS]$. Hence this proposal is similar (and asymptotically equivalent) to Winsorizing a fixed, predetermined number of observations; it is asymmetric in the sense that it tends to Winsorize more observations on the side with the heavier contamination. k might be compared to a kind of insurance against ill effects from contamination: if one insures against high contamination by choosing a small k, one has to pay by losing some efficiency if the actual contamination is low, and vice versa; compare also Anscombe (1960), especially p. 127.

Proposal 2. Choose beforehand a fixed number k. Then determine T and S such that

$$n^{-1}\sum\psi(k,(x_i-T)/S)=0$$
$$n^{-1}\sum\psi^2(k,(x_i-T)/S)=E_\Phi\psi^2(k,x).$$

This proposal is of course suggested by the most robust estimator of scale as determined in Equation (5) of the preceding section. It corresponds to Winsorizing a variable number of observations: slightly more if F has heavier tails, slightly less if F has lighter tails, and if H is asymmetric, more on the side with heavier contamination. Certainly for large n, but probably also for small n, this proposal is better than the preceding one, since it is less sensitive to a "wrong" choice of k (i.e., to a choice not adapted to the true value of ε). Moreover, the S of Proposal 2 will be more accurate in general ($\log S$ having the smaller asymptotic variance), which also improves accuracy of T.

For both proposal, T and S can be determined relatively easily by iterative procedures (see below), and the (asymptotic) variance of T might be estimated by

$$\frac{n}{n-1}\frac{\sum\psi^2(k,(x_i-T)/S)}{[\sum\psi'(k,(x_i-T)/S)]^2}S^2.$$

The correction term $n/(n-1)$ is suggested by the classical theory ($k=\infty$); one will have to revise it in the light of later knowledge, but it seems to agree quite well with exploratory Monte Carlo computations for sample sizes 5 and 10.

Proposal 3. Determine k such that the estimated (asymptotic) variance of T is minimized, i.e., minimize

$$\sum\psi^2(k,x_i-T)/[\sum\psi'(k,x_i-T)]^2$$

subject to the side condition (determining T)

$$\sum\psi(k,x_i-T)=0.$$

To avoid trouble at $k = 0$, one should safeguard by requiring that enough observations (say more than $n^{1/2}$) are contained in the interval $[T - k, T + k]$. This proposal seems to give the best asymptotic properties of all three, but numerical computation of k and T apparently is a very difficult task.

For practical application, I personally would favor Proposal 2, since it seems to fit best into the framework of conventional least squares techniques. For nonlinear least squares problems, which have to be solved by iterative methods anyway, this proposal probably could be incorporated into existing computer programs with an only marginal increase of the amount of computation.

First we have to show that Proposal 2 constitutes a legitimate definition of T and S; that is, we have to show existence and uniqueness of the solution.

Let $x_1, \ldots, x_n$ be a fixed sample of size n and consider the equations

$$\begin{aligned} &\sum_{i=1}^{n} \psi(k, (x_i - T)/S) = 0 \\ &\sum_{i=1}^{n} \psi^2(k, (x_i - T)/S = n\beta. \end{aligned} \tag{6}$$

Proposal 2 corresponds to $\beta = E_\Phi \psi^2(k, x)$, and for this choice we have, obviously, $\beta < k^2$ whenever $k > 0$. I do not want to exclude the possibility that small sample size considerations might lead to a different choice of β.

Proposition. *Assume that $\beta < k^2$ and that the sample is such that $x_{i_1} = x_{i_2} = \cdots = x_{i_m}$, $i_1 < i_2 < \cdots < i_m$, implies $m < n(1 - \beta/k^2)$. Then the system* (6) *has a unique solution T, S.*

(*Editor's note*: The proof has been omitted.)

Let m_1, m_2, m_3 be the number of observations satisfying $x_i \leqq T - kS$, $T - kS < x_i < T + kS$ and $T + kS \leqq x_i$, respectively. Then (6) may be written

$$\sum{}' x_i - m_2 T + (m_3 - m_1)kS = 0$$

$$\sum{}' (x_i - T)^2 + (m_1 + m_3)k^2 S^2 - n\beta S^2 = 0.$$

By determining T from the first equation and inserting it in the second, one obtains the equivalent system

$$\begin{aligned} T &= \bar{x}' + kS(m_3 - m_1)/m_2 \\ S^2 &= \sum{}' (x_i - \bar{x}')^2/(n\beta - (m_1 + m_3 + (m_3 - m_1)^2/m_2)k^2) \end{aligned} \tag{6'}$$

where $\bar{x}' = m_2^{-1} \sum{}' x_i$ is the trimmed mean.

This may be used to compute T and S by iterations: start with some initial values T_0, S_0. Compute

$$\bar{x}' = m_2^{-1} \sum{}' x_i$$

Table IV. Sharp Upper Bounds for the Asymptotic Variance of $n^{1/2}T$, if T and S Are Computed According to Proposal 2, for Symmetric Contamination.

	$\varepsilon = 0.000$	0.010	0.020	0.050	0.100	0.200
$k = 1.0$	1.107	1.138	1.172	1.276	1.490	2.15
1.1	1.088	1.120	1.155	1.265	1.495	2.23
1.2	1.072	1.105	1.140	1.258	1.506	2.35
1.3	1.058	1.093	1.131	1.256	1.526	2.50
1.4	1.047	1.084	1.124	1.257	1.557	2.70
1.5	1.037	1.077	1.120	1.264	1.592	2.97
1.6	1.029	1.072	1.117	1.275	1.639	3.36
1.7	1.023	1.068	1.117	1.288	1.697	3.92
1.8	1.018	1.066	1.118	1.306	1.769	>4
1.9	1.014	1.065	1.121	1.324	1.859	>4
2.0	1.010	1.065	1.127	1.353	2.029	>4

$$S_1^2 = \sum{}' (x_i - \bar{x}')^2/(n\beta - (m_1 + m_3 + (m_3 - m_1)^2/m_2)k^2)$$

$$T_1 = \bar{x}' + kS_1(m_3 - m_1)/m_2,$$

where $\sum'$ denotes that the summation is extended over those indices for which $|(x_i - T_0)/S_0| < k$, and m_1, m_2, m_3 are the numbers of observations satisfying $x_i \leqq T_0 - kS_0$, $T_0 - kS_0 < x_i < T_0 + kS_0$ and $T_0 + kS_0 \leqq x_i$, respectively. If $T_1 = T_0$ and $S_1 = S_0$, one has found the solution, otherwise do the same again, with T_1, S_1 in place of T_0, S_0, etc. For automatic computation, it might be wise to safeguard against too small values of m_2 and of the denominator in the expression defining S_1^2.

Since, for a fixed sample, the values of T_1 and S_1 are uniquely determined by the pair of integers m_1, m_3, the process must either stop after a finite number of steps, or it will repeat itself periodically. I do not know whether and when the latter case occurs.

Ordinarily, the procedure converges rather fast, as an exploratory study with Monte Carlo methods shows. For $k = 1.5$, sample sizes up to 100 and distributions that do not have too heavy tails, the stationary value is on the average reached after 1–2 steps, starting from sample mean and sample variance. For distributions with heavier tails (Cauchy), somewhat more steps are needed (about 2 for sample size 10, 4 for sample size 100).

Table IV gives sharp upper bounds for the asymptotic variance of $n^{1/2}T$, if T and S are computed according to Proposal 2, and $F = (1 - \varepsilon)\Phi + \varepsilon H$, H symmetric. These upper bounds can be determined as follows. The asymptotic variance of $n^{1/2}T$ is $E_F\psi^2(kS, x)/(E_F\psi'(kS, x))^2$, where S satisfies $E_F\psi^2(kS, x) = S^2\beta(k)$, with $\beta(k) = E_\Phi\psi^2(k, x)$. Hence we have, with $q = kS$,

$$q^2\beta(k)/k^2 = (1 - \varepsilon)\beta(q) + \varepsilon E_H\psi^2(q, x) \leqq (1 - \varepsilon)\beta(q) + \varepsilon q^2$$

or

$$\beta(q)/q^2 \geqq ((\beta(k)/k^2 - \varepsilon))/(1 - \varepsilon). \tag{7}$$

Since $\psi^2(q, x)/q^2$ is monotone decreasing for increasing q, $\beta(q)/q^2$ is also monotone decreasing, and equality in (7) determines a sharp upper bound $q_{\max}$ for q.

Now we have to find a sharp upper bound for the asymptotic variance

$$\frac{E_F\psi^2(q, x)}{(E_F\psi'(q, x))^2} = \frac{q^2\beta(k)/k^2}{((1 - \varepsilon)\alpha(q) + \varepsilon E_H\psi'(q, x))^2}$$

where $\alpha(q) = E_\Phi\psi'(q, x)$. This is solved if we can find a sharp lower bound for $(1 - \varepsilon)\alpha(q)/q + \varepsilon E_H\psi'(q, x)/q$. But $\alpha(q)/q$ is monotone decreasing for increasing q (notice that $\alpha(q)/q$ is the average of the normal density in the interval $(0, q)$), hence this expression is bounded from below by $(1 - \varepsilon)\alpha(q_{\max})/q_{\max}$. Hence it follows that the asymptotic variance of $n^{1/2}T$ is bounded from above by

$$\frac{q_{\max}^2\beta(k)/k^2}{((1 - \varepsilon)\alpha(q_{\max}))^2} = \frac{(1 - \varepsilon)\beta(q_{\max}) + \varepsilon q_{\max}^2}{((1 - \varepsilon)\alpha(q_{\max}))^2}.$$

This bound is sharp: it is attained for contaminating distributions H that put all their mass outside the interval $[-q_{\max}, q_{\max}]$. Hence, one may determine the upper bound $q_{\max}$ from (7) and enter Table 1 with $k = q_{\max}$ and ε to determine the upper bound for the variance. This was done, using linear interpolation, to obtain Table 4.

12. A Generalization

(M)-estimators $T_n = T_n(x_1, \ldots, x_n)$ have been defined by the property that they minimize an expression of the form $\sum_{i=1}^n \rho(x_i - T_n)$. More generally, one may define (M_r)-estimators by the property that they minimize an expression of the form $\sum_I \rho_r(x_I - T_n)$ where ρ_r is a symmetric function of r arguments, the summation is extended over the $\binom{n}{r}$ subsets $I = \{i_1, \ldots, i_r\} \subset \{1, 2, \ldots, n\}$ containing r distinct elements, and $(x_I - T_n)$ stands short for $(x_{i_1} - T_n, \ldots, x_{i_r} - T_n)$.

Let $\psi_r(x_I - T) = -(\partial/\partial T)\rho_r(x_I - T)$, and assume that $E\psi_r = 0$. Then define $\psi(x_1) = \int \psi_r(x_1, \ldots, x_r)F(dx_2)\ldots F(dx_r)$.

It follows from Hoeffding's work on U-statistics (1948), that the suitably normed sum $\sum_I \psi_r(x_I - T)$ behaves asymptotically like $\sum_i \psi(x_i - T)$. In particular, one shows in much the same way as in Section 2, and under analogous regularity conditions, that $T_n \to 0$ in probability, and that $n^{1/2}T_n$ is asymptotically normal, with asymptotic mean 0 and asymptotic variance $V(\psi_r, F) = E\psi^2/(E\psi')^2$.

Since the amount of computation needed for determining the value of an (M_r)-estimator increases with the rth power of the sample size, they will hardly be of practical use for $r > 2$.

EXAMPLE. Take $r = 2$, and $\rho_2(t_1, t_2) = |t_1 + t_2|$. It is easy to see that $\sum_I \rho_2(x_I - T)$ is minimized by $T(x) = \{\text{sample median of the } z_{ij} = \frac{1}{2}(x_i + x_j), i < j\}$, i.e., by the estimator proposed by Hodges and Lehmann (1963). If the underlying distribution has a symmetric density f, we end up with an asymptotic variance $V(\psi_2, F) = 1/(12(\int f^2\,dt)^2)$. In the contaminated normal case, $F = (1 - \varepsilon)\Phi + \varepsilon H$, H symmetric, we have the sharp upper bound $V(\psi_2, F) < (1 - \varepsilon)^{-4}\pi/3$ for this particular ψ_2; the bound is approached when H spreads its mass toward infinity. Numerically:

$\varepsilon =$ 0.000	0.001	0.002	0.005	0.010	0.020	0.050	0.100	0.200	0.500
$V \leqq$ 1.047	1.051	1.056	1.068	1.090	1.135	1.286	1.596	2.557	16.76

The reader is invited to compare this with Tables 1 and 4; apparently, the Hodges-Lehmann estimator and the estimators proposed in the preceding sections of the present paper are close competitors.

For the conventional model of contamination $F(t) = (1 - \varepsilon)\Phi(t) + \varepsilon\Phi(t/3)$, these estimators have practically equivalent performances also:

	$\varepsilon =$ 0.0	0.01	0.02	0.05	0.10
Hodges-Lehmann:	$V =$ 1.047	1.07	1.09	1.17	1.31
Proposal 2, $k = 1.5$:	$V =$ 1.037	1.06	1.08	1.16	1.30

I would like to express my thanks for a number of stimulating discussions with Professor E.L. Lehmann on robust estimation and with Professor L. LeCam on asymptotic efficiency.

References

Anscombe, F.J. (1960). Rejection of outliers. *Technometrics* **2** 123–147.

Cramér, H. (1946). *Mathematical Methods of Statistics*. Princeton Univ. Press.

Hajek, J. (1962). Asymptotically most powerful rank order tests. *Ann. Math. Statist.* **33** 1124–1147.

Hodges, J.L., Jr. and Lehmann, E. (1963). Estimates of location based on rank tests. *Ann. Math. Statist.* **34** 598–611.

Hoeffding, W. (1948). A class of statistics with asymptotically normal distributions. *Ann. Math. Statist.* **19** 293–325.

LeCam, L. (1953). On some asymptotic properties of maximum likelihood estimates and related Bayes estimates. *Univ. California Publ. Statist.* **1** 277–330.

LeCam, L. (1958). Les propriétés asymptotiques des solutions de Bayes. *Publications de l'Institut de Statistique de l'Université de Paris.* **7** 17–35.

Loève, M. (1960). *Probability Theory*, (2nd ed.). Van Nostrand, New York.

Stein, C. (1956). Efficient nonparametric testing and estimation. *Proc. Third Berkeley Symp. Math. Statist. and Prob.* **I** 187–195.

Tukey, J.W. (1960). A survey of sampling from contaminated distributions. In *Contributions to Probability and Statistics* (ed. Olkin). Stanford Univ. Press.

Wald, A. (1949). Note on the consistency of the M.L. estimate. *Ann. Math. Statist.* **20** 595–601.

Introduction to Cox (1972) Regression Models and Life-Tables

Ross L. Prentice
Fred Hutchinson Cancer Research Center

1. General Remarks on the Paper

In this paper, Sir David Cox proposed a stimulating and pioneering procedure for the regression analysis of censored failure time data. Within a few years of publication, this procedure became a data analytic standard in a number of application areas, most notably in the biomedical sciences. The procedure has also stimulated considerable related methodologic development.

Earlier regression methods for right-censored data focused on parametric models for failure time, or log-failure time, whereas Cox's paper specified a semiparametric model for the instantaneous failure rate, or hazard function. An emphasis on failure rate modeling is natural in many application areas; for example, much information and intuition often accompanies age-specific disease rates in epidemiology, whereas models for failure time, or functions thereof, are typically more abstract. Cox's model specifies a parametric regression function, of exponential form, for the ratio of failure rates between study subjects, while allowing the shape of the "baseline" hazard function to be an arbitrary function of time. This hazard ratio, or relative risk, provides a very natural estimation target, while the arbitrariness of the dependence of the hazard function on time, at a given regression vector, circumvents certain problems with "guarantee times," failure-free periods, and competing risks that are not well handled by parametric models for failure time. Cox's class of models becomes very general upon relaxing the relative risk to be a function of time. Such a relaxation is typically impractical for failure time models outside the Cox model family.

In spite of the flexibility of this regression model, Cox provided a very simple and intuitive procedure for estimating relative risk parameters. Specifically, the discrete conditional probability that a particular study subject fails at a given time is simply the ratio of the relative risk for the failing individual to the sum of the relative risks (at the same time) for all subjects at risk. Moreover the "scores" corresponding to such conditional probabilities are uncorrelated, so that under regularity, ordinary likelihood formulae apply to the product of these conditional probabilities over all failure times.

The publication of Cox's paper stimulated a substantial statistical effort that continues today. This effort has focused on properties, extensions, and applications of the proposed hazard ratio estimation procedure. As a result, the basic relative risk estimation procedure for univariate failure time data is now fairly well understood. A range of other topics discussed in Cox's paper, for example, the modeling and estimation of multivariate failure time data and the use of accelerated failure time regression models, are less well understood or standardized and merit further investigation.

2. Some Specific Contents of the Paper

The proposed regression procedure is variably referred to as Cox regression, proportional hazards' regression, or perhaps best as relative risk regression. It is based on a model

$$\lambda(t; \mathbf{z}) = \lambda_0(t) \exp(\mathbf{z}\boldsymbol{\beta}) \tag{1}$$

for the hazard rate $\lambda(t; \mathbf{z})$ at time t for an individual with regression vector $\mathbf{z}' = (z_1, \ldots, z_p)$. Note that $\lambda_0(\cdot)$ is an arbitrary baseline hazard function, $\exp(\mathbf{z}'\beta)$ is the relative risk of failure at a general covariate value z as compared to a standard value $\mathbf{z}' = \mathbf{0}$, and $\boldsymbol{\beta}$ is a column p-vector of parameters to be estimated. For example, in a clinical trial z may include indicator variables for treatment assignment, as well as other qualitative or quantitative study subject characteristics. This model generalized earlier work on exponential and Weibull regression models, as well as the little-known work of Sheehe (1962) that proposed a similar model for matched-pair data. The major assumption in (1) is the lack of dependence of the relative risk on time. Cox presented the novel notion that $\mathbf{z}$ on the right side of (1) could be replaced by $\mathbf{z}(t)$, a vector that may include product terms between elements of $\mathbf{z}$ and functions of time, in order to test this proportional hazards' assumption and relax it as necessary.

It was proposed that the estimation of the regression parameter β in (1) be carried out by applylng standard asymptotic likelihood procedures to

$$L(\beta) = \prod_{i=1}^{k} \left[\frac{\exp\{\mathbf{z}_i(t_i)\beta}{\sum_{l \in R(t_i)} \exp\{\mathbf{z}_l(t_i)\beta\}} \right], \tag{2}$$

where $t_1 \leq \cdots \leq t_k$ are the observed failure times, assumed distinct, in the sample; $\mathbf{z}_i(t_i)$ is the covariate vector at time t_i for the subject failing at t_i; and

$\mathbf{z}_l(t_i)$ is the corresponding covariate vector for the lth member of $R(t_i)$, the set of subjects at risk at time t_i. Under standard independent failure time and independent censorship assumptions, the ith factor in (2) is precisely the probability that the subject with covariate vector $\mathbf{z}_i(t_i)$ fails at t_i given the failure, censoring, and covariate information prior to t_i and given that a failure occurs at t_i. Hence, the score statistic $U(\boldsymbol{\beta}) = \partial \log L(\boldsymbol{\beta})/\partial\boldsymbol{\beta}$ is the sum of variates that have a conditional, and hence, an unconditional mean of zero so that $U(\boldsymbol{\beta}) = 0$ provides an unbiased estimating equation for $\boldsymbol{\beta}$. It was pointed out that in the special case of the two-sample problem, $U(0)$ reduces to the log-rank test [Mantel (1966); Peto and Peto (1972)].

In order to accommodate tied failure times, Cox proposed the replacement of (1) and (2) by more complicated corresponding expressions based on a discrete logistic model. Some discussants of this paper [e.g., Peto (1972)] pointed out that (2) continues to provide a sufficiently accurate basis for inference on β even if failure times from (1) are moderately grouped. Accordingly, the simple expression (2) forms the basis for virtually all available relative-risk regression software. Similarly, the simple estimator

$$\hat{\Lambda}_0(t) = \sum_{t_i \leqslant t} \left(\sum_{l \in R(t_i)} e^{\mathbf{z}_l(t_i)\hat{\boldsymbol{\beta}}} \right)^{-1}$$

[e.g., Oakes (1972) and Breslow (1972)] of the cumulative hazard process $\Lambda_0(t) = \int_0^t \lambda_0(u)\, du$, where $\hat{\boldsymbol{\beta}}$ solves $U(\boldsymbol{\beta}) = 0$, is included in most software packages, rather than the iterative estimator based on the discrete logistic model. A discrete version of (1) is available [Kalbfleisch and Prentice (1973)] for the analysis of severely grouped failure times.

3. Impact of the Paper

One measure of the impact of this remarkable paper is provided by the amount of related methodologic work that it has stimulated. Several contributors to the published discussion noted that (2) is generally not a conditional likelihood function, and hence, it was not obvious that standard asymptotic likelihood procedures should apply. Cox clarified the interpretation of (2) with his subsequent definition of a partial likelihood function [Cox (1975)]. He showed that the contributions to $U(\boldsymbol{\beta})$ from distinct failure times are uncorrelated, owing to the nestedness of the conditioning events mentioned above, from which it follows that $-\partial^2 \log L(\boldsymbol{\beta})/\partial\boldsymbol{\beta}^2$ is an estimator of the variance of $U(\boldsymbol{\beta})$. He argued further that standard asymptotic formulae can be expected to apply to $L(\boldsymbol{\beta})$ under suitable regularity and stability conditions. Explicit conditions for these asymptotic results and for an asymptotic Gaussian distribution for $\hat{\Lambda}_0(t)$ were proposed by a number of subsequent authors and provided in a quite general form some years later using martingale convergence results [Andersen and Gill (1982)] assuming continuous failure times. In fact, the prominence of counting process representations and martingale procedures in recent failure time developments is attributable, in part, to their successful application to relative-risk regression models.

Since (2) is not, in general, a likelihood function corresponding to any marginal or conditional distribution, standard results on estimating efficiency do not apply. A number of authors examined aspects of the asymptotic and small-sample efficiency of $\hat{\boldsymbol{\beta}}$ that maximize (2) as compared to the maximum likelihood estimate under a parametric specification of the baseline hazard function $\lambda_0(\cdot)$ [e.g., Efron (1977); Oakes (1977); Kalbfleisch and McIntosh (1977); Kay (1979)]. It was found that the loss of efficiency in using the partial likelihood function (2) was typically modest, even if $\lambda_0(\cdot)$ was assumed to be known up to a single scale parameter, and that the asymptotic efficiency could be made arbitrarily close to unity relative to the maximum likelihood estimates under a sequence of models having increasingly more flexible parametrizations of $\lambda_0(\cdot)$. This latter property is sometimes summarized by stating that the maximum partial likelihood estimate $\hat{\boldsymbol{\beta}}$ is fully efficient in the absence of assumption about $\lambda_0(\cdot)$. This type of result has stimulated interesting related work on efficiency and parameter orthogonality in semiparametric models [e.g., Begun et al. (1983)].

Another offshoot of attempts to understand (2) was the systematic development of classes of censored-data rank tests. Kalbfleisch and Prentice (1973) noted that (2) could be derived as the likelihood corresponding to the marginal distribution of a generalized rank vector if $z(t)$ is time-independent. This generalized rank statistic notion, when applied to models that are linear in log-failure time, generated a class of censored-data rank tests that included the log-rank statistic and a generalization of the Wilcoxon statistic as special cases [Prentice (1978)]. In fact, this class of statistics was subsequently shown to have an equivalent representation as a partial likelihood score statistic $U(0)$ under suitable specifications of the time-dependent covariate $\mathbf{z}(t)$ in (2) [e.g., Lustbader (1980); Mehrotra et al. (1982)]. Hence, the estimation procedure emanating from (1) and (2) served to unify such fundamental failure time techniques as cumulative hazard and Kaplan–Meier estimates, and censored-data rank tests, and to extend these into a full and flexible regression methodology.

The novel notion of using time-dependent covariates in (1) much enhances the power and flexibility of the relative-risk regression procedure. Of particular interest is the use of stochastic covariates that are evolving over time, in which case, $\mathbf{z}$ on the left side of (1) is replaced by the history $\{\mathbf{z}(u); u < t\}$ a covariate process up to time t, while the modeled regression vector on the right side of (1) consists of functions of this history and t. Many practical and theoretical issues arise in relation to the use of such evolving covariates [e.g., Crowley and Hu (1977); Kalbfleisch and Prentice (1980, Chap. 5); Cox and Oakes (1984, Chap. 8); Andersen (1986)], and quite a number of applications have appeared in the literature. For example, selected applications have used time-dependent covariates in attempts to explain treatment effects using post-randomization measurements in clinical trials [e.g., Crowley and Hu (1977)] to relate disease risk to changes over time in covariate values in epidemiologic cohort studies [e.g., Prentice et al. (1982a)], and to relate relative risk shape to the administered dose in animal carcinogenesis experiments [e.g., Prentice et al. (1982b)].

Cox's paper also included a section on bivariate failure times, which, in terms of the preceding paragraph, involved modeling the hazard rate for each component using (1), while regarding the counting process for the other component as an evolving covariate. A rather large number of subsequent papers have used this formulation in more general multivariate and multistate failure problems [e.g., Prentice et al. (1981)]. Other authors have considered models of the form (1) for the marginal hazard rates, rather than the counting process intensity [e.g., Clayton and Cuzick (1985); Wei et al. (1989)]. This latter approach is still in the early stages of development, but promises results for marginal relative-risk parameters that are analogous to those recently developed using generalized estimating equations [e.g., Liang and Zeger (1986)] for the estimation of parameters in the mean of other types of response variables.

The relative risk in (1) is an instantaneous version of the ratio of disease probabilities and the closely related odds' ratio that have long played a central role in epidemiologic modeling. Expressions (1) and (2) stimulated the development of time-matched case-control sampling procedures [e.g., Prentice and Breslow (1978)], which allow relative-risk estimation from case-control studies without making a rare disease assumption. These expressions also stimulated the development of case-control and case-cohort procedures for sampling within a cohort [e.g., Liddell et al. (1977); Prentice (1986)]. Such sampling procedures are now widely used in epidemiologic research.

The above discussion mentions just a few of the methodologic developments that were directly stimulated by this remarkable paper. A considerable literature, too extensive to mention here, has also arisen on procedures for model criticism in relation to (1). These include graphic procedures, regression diagnostics, residual specifications, test-of-fit proposals, robustness results under misspecified models, and studies of the effects of mismeasured covariates.

The applications of (1) and (2) are also too numerous to document here, but a scan of recent issues of prominent clinical and epidemiologic journals will convince the reader of the impact of this work in these settings. There have also been frequent applications in a number of other research areas including industrial life testing, toxicology, demography, sociology, and econometrics.

One of the most lasting impacts of Cox's paper may be the effect it has had on the manner in which statistical researchers and practitioners, and their collaborators, have learned to think about failure time problems. Prior to 1972, failure time regression problems in clinical trial and industrial life-testing contexts were evidently conceived primarily as ordinary regression problems in which the response variable might be right-censored. While this formulation is useful and persists today, the concept of comparing failure rates at common points in time has now become more intuitive and fundamental. Perhaps Richard Peto (1972) said it best when he seconded the vote of thanks to Professor Cox "because he has opened up new territories to common sense." Certainly, the arsenal of sensible analytic tools available to statistical and biostatistical researchers today has been much enhanced by the technical and conceptual advances of this pioneering paper.

4. About the Author

Sir David Cox, knighted in 1985, is Warden at Nuffield College at Oxford, a post he has held since 1988. Born in 1924, Professor Cox received a master's degree from Cambridge and a Ph.D. from Leeds in 1949. Early in his scientific career, he held posts at the Royal Aircraft Establishment (1944–46) and the Wool Industries Research Association (1946–50), settings in which the topics of reliability and life testing are of obvious importance. He worked in the Statistical Laboratory at Cambridge from 1950–55 and became reader (1956–60) and then professor of statistics (1961–66) at Birkbeck College, following a year as visiting professor at the University of North Carolina. From 1966–88 he served as professor of statistics at the Imperial College of Science and Technology and was head of the math department from 1970–74. Professor Cox also served as president of the Bernouilli Society (1979–81) and the Royal Statistical Society (1980–82).

The national and international esteem accorded Professor Cox is indicated by his election to the Royal Danish Academy of Sciences and Letters (1983), the American Academy of Arts and Sciences (1974), and the U.S. National Academy of Sciences (1989). He has also seen the recipient of the Weldon Memorial Prize (1984) Oxford University and the recent recipient of the Kettering Prize for Cancer Research (1990) awarded by the General Motors Corporation. In fact the impact of Cox's (1972) paper on the analysis and interpretation of cancer treatment trials was a major reason for his receipt of this award.

The paper discussed above is but one of the statistical breakthroughs that has earned Sir David Cox the respect and admiration of a broad scientific community. In addition to such breakthroughs, he has authored or coauthored a series of lucid books and monographs including *Statistical Methods in the Textile Industry* (1949), *Planning of Experiments* (1958), *Queues* (1961), *Renewal Theory* (1962), *Theory of Stochastic Processes* (1965), *Statistical Analysis of a Series of Events* (1966), *Analysis of Binary Data* (1970), *Theoretical Statistics* (1974), *Problems and Solutions in Theoretical Statistics* (1978), *Point Processes* (1980), *Applied Statistics* (1981), *Analysis of Survival Data* (1984), and *Asymptotic Techniques* (1989). He is also well known for major scientific and industrial consulting activities, and for many years of editorial leadership, including service as editor of *Biometrika* since 1966.

References

Andersen, P.K., and Gill, R.D. (1982). Cox's regression model for counting processes: A large sample study, *Ann. Statist.*, **10**, 1100–1120.

Anderson, P.K. (1986). Time-dependent covariates and Markov processes, in *Modern Statistical Methods in Chronic Disease Epidemiology* (S.H. Moolgavkar, and R.L. Prentice, eds.). Wiley, New York, pp. 82–103.

Begun, J.M., Hall, W.J., Huang, W., and Wellner, J.A. (1983). Information and asymptotic efficiency in parametric-nonparametric models, *Ann. Statist.*, **11**, 432–452.

Breslow, N. (1972). Contribution to the discussion of paper by D.R. Cox, *J. Roy. Statist. Soc., Ser. B*, **34**, 216–217.

Clayton, D., and Cuzick, J. (1985). Multivariate generalizations of the proportional hazards model, *J. Roy. Statist. Soc., Ser, A*, **148**, 82–117 (with discussion).

Cox, D.R. (1972). Regression models and life-tables (with discussion). *J. R. Statist. Soc. Ser. B*, **34**, 187–202.

Cox, D.R. (1975). Partial likelihood, *Biometrika*, **62**, 269–276.

Cox, D.R., and Oakes, D. (1984). *Analysis of Survival Data.* Chapman and Hall, New York.

Crowley, J., and Hu, M. (1977). Covariance analysis of heart transplant survival data, *J. Amer. Statist. Assoc.*, **73**, 27–36.

Efron, B. (1977). The efficiency of Cox's likelihood function for censored data, *J. Amer. Statist. Assoc.*, **72**, 557–565.

Kalbfleisch, J.D., and McIntosh, A.A. (1977). Efficiency in survival distributions with time-dependent covariables, *Biometrika*, **64**, 47–50.

Kalbfleisch, J.D., and Prentice, R.L. (1973). Marginal likelihoods based on Cox's regression and life model, *Biometrika*, **60**, 267–278.

Kalbfleisch, J.D., and Prentice, R.L. (1980). *The Statistical Analysis of Failure Time Data.* Wiley, New York.

Kay, R. (1979). Some further asymptotic efficiency calculations for survival data regression models, *Biometrika* **66**, 91–96.

Liang, K.Y., and Zeger, S.L. (1986). Longitudinal data analysis using generalized linear models, *Biometrika*, **73**, 13–22.

Liddell, F.D.K., McDonald, J.C. and Thomas, D.C. (1977). Methods of cohort analysis. Appraisal by application to asbestos mining, *J. Roy. Statist. Soc., Ser. A*, **140**, 469–491 (with discussion).

Lustbader, E.D. (1980). Time-dependent covariates in survival analysis, *Biometrika*, **67**, 697–698.

Mantel, N. (1966). Evaluation of survival data and two new rank order statistics arising in its consideration, *Cancer Chemother. Rept.*, **50**, 163–170.

Mehrotra, K.D., Michalek, J.E., and Mihalko, D. (1982). A relationship between two forms of linear rank procedures for censored data, *Biometrika*, **69**, 674–676.

Oakes, D. (1972). Contribution to discussion of paper by D.R. Cox, *J. Roy. Statist. Soc., Ser. B*, **34**, 208.

Oakes, D. (1977). The asymptotic information in censored survival data, *Biometrika*, **64**, 441–448.

Peto, R. (1972). Contribution to discussion of paper by D.R. Cox, *J. Roy. Statist. Soc., Ser. B*, **34**, 205–207.

Peto, R., and Peto, J. (1972). Asymptotically efficient rank invariant test procedures, *J. Roy. Statist. Soc., Ser. A*, **135**, 185–206 (with discussion).

Prentice, R.L. (1978). Linear rank tests with right censored data, *Biometrika*, **65**, 167–179.

Prentice, R.L. (1986). A case-cohort design for epidemiologic cohort studies and disease prevention trials, *Biometrika*, **73**, 1–11.

Prentice, R.L., and Breslow, N.E. (1978). Retrospective studies and failure time models, *Biometrika*, **65**, 153–158.

Prentice, R.L., Williams, B.J. and Peterson A.V. On the regression analysis of multivariate failure time data (1981). *Biometrika* **68**, 373–379.
Prentice, R.L., Shimizu, Y., Lin, C.H., Peterson, A.V., Kato, H., Mason, M.W., and Szatrowski, T.P. (1982a). Serial blood pressure measurements and cardiovascular disease in a Japanese cohort, *Amer. J. Epid.*, **116**, 1–28.
Prentice, R.L., Peterson, A.V., and Marek, P. (1982b). Dose mortality relationship in RFM mice following 136Cs gamma irradiation, *Radiation Res.*, **90**, 57–76.
Sheehe, P.R. (1962). Dynamic risk analysis in retrospective matched pair studies of disease, *Biometrics*, **18**, 323–341.
Wei, L.J., Lin, D.Y., and Weissfield, L. (1989). Regression analysis of multivariate incomplete failure time data by modelling marginal distributions, *J. Amer. Statist. Assoc.*, **84**, 1065–1073.

Regression Models and Life-Tables

David R. Cox
Imperial College, London

[Read before the Royal Statistical Society, at a meeting organized by the Research Section, on Wednesday, March 8th, 1972, Mr. M.J.R. Healy in the Chair]

Summary

The analysis of censored failure times is considered. It is assumed that on each individual are available values of one or more explanatory variables. The hazard function (age-specific failure rate) is taken to be a function of the explanatory variables and unknown regression coefficients multiplied by an arbitrary and unknown function of time. A conditional likelihood is obtained, leading to inferences about the unknown regression coefficients. Some generalizations are outlined.*

1. Introduction

Life tables are one of the oldest statistical techniques and are extensively used by medical statisticians and by actuaries. Yet relatively little has been written about their more formal statistical theory. Kaplan and Meier (1958) gave a comprehensive review of earlier work and many new results. Chiang in a series of papers has, in particular, explored the connection with birth–death processes; see, for example, Chiang (1968). The present paper is largely concerned with the extension of the results of Kaplan and Meier to the comparison of life tables and more generally to the incorporation of regression-like arguments into life-table analysis. The arguments are asymptotic but are relevant to situations where the sampling fluctuations are large enough to be

* *Editor's note*: This paper could be read in conjunction with Kaplan and Meier (1958) which is also reproduced in this volume.

of practical importance. In other words, the applications are more likely to be in industrial reliability studies and in medical statistics than in actuarial science. The procedures proposed are, especially for the two-sample problem, closely related to procedures for combining contingency tables; see Mantel and Haenzel (1959), Mantel (1963) and, especially for the application to life tables, Mantel (1966). There is also a strong connection with a paper read recently to the Society by R. and J. Peto (1972).

We consider a population of individuals; for each individual we observe either the time to "failure" or the time to "loss" or censoring. That is, for the censored individuals we know only that the time to failure is greater than the censoring time.

Denote by T a random variable representing failure time; it may be discrete or continuous. Let $\mathscr{F}(t)$ be the survivor function,

$$\mathscr{F}(t) = \text{pr}(T \geqslant t)$$

and let $\lambda(t)$ be the hazard or age-specific failure rate. That is,

$$\lambda(t) = \lim_{\Delta t \to 0+} \frac{\text{pr}(t \leqslant T < t + \Delta t \mid t \leqslant T)}{\Delta t}. \tag{1}$$

Note that if T is discrete, then

$$\lambda(t) = \sum \lambda_{u_j} \delta(t - u_j), \tag{2}$$

where $\delta(t)$ denotes the Dirac delta function and $\lambda_t = \text{pr}(T = t \mid T \geq t)$. By the product law of probability $\mathscr{F}(t)$ is given by the product integral

$$\mathscr{F}(t) = \mathop{\mathscr{P}}_{u=0}^{t-0} \{1 - \lambda(u)\, du\} = \lim \prod_{k=0}^{r-1} \{1 - \lambda(\tau_k)(\tau_{k+1} - \tau_k)\}, \tag{3}$$

the limit being taken as all $\tau_{k+1} - \tau_k$ tend to zero with $0 = \tau_0 < \tau_1 < \cdots < \tau_{r-1} < \tau_r = t$. If $\lambda(t)$ is integrable this is

$$\exp\left\{-\int_0^t \lambda(u)\, du\right\}, \tag{4}$$

whereas if $\lambda(t)$ is given by (2), the product integral is

$$\prod_{u_j < t} (1 - \lambda_{u_j}). \tag{5}$$

If the distribution has both discrete and continuous components the product integral is a product of factors (4) and (5).

2. The Product-Limit Method

Suppose observations are available on n_0 independent individuals and, to begin with, that the failure times are identically distributed in the form specified in Section 1. Let n individuals be observed to failure and the rest be censored. The rather strong assumption will be made throughout that the only information available about the failure time of a censored individual is that it exceeds the censoring time. This assumption is testable only if suitable supplementary information is available. Denote the distinct failure times by

$$t_{(1)} < t_{(2)} < \cdots < t_{(k)}. \tag{6}$$

Further let $m_{(i)}$ be the number of failure times equal to $t_{(i)}$ the multiplicity of $t_{(i)}$; of course $\sum m_{(i)} = n$, and in the continuous case $k = n$, $m_{(i)} = 1$.

The set of individuals at risk at time $t - 0$ is called the *risk set* at time t and denoted $\mathscr{R}(t)$; this consists of those individuals whose failure or censoring time is at least t. Let $r_{(i)}$ be the number of such individuals for $t = t_{(i)}$. The product-limit estimate of the underlying distribution is obtained by taking estimated conditional probabilities that agree exactly with the observed conditional frequencies. That is,

$$\hat{\lambda}(t) = \sum_{i=1}^{k} \frac{m_{(i)}}{r_{(i)}} \delta(t - t_{(i)}). \tag{7}$$

Correspondingly,

$$\hat{\mathscr{F}}(t) = \mathop{\mathscr{P}}_{u=0}^{t-0} \{1 - \hat{\lambda}(u)\, du\} = \prod_{t_{(i)}<t} \left\{1 - \frac{m_{(i)}}{r_{(i)}}\right\}. \tag{8}$$

For uncensored data this is the usual sample survivor function; some of the asymptotic properties of (8) are given by Kaplan and Meier (1958) and by Efron (1967) and can be used to adapt to the censored case tests based on sample cumulative distribution function.

The functions (7) and (8) are maximum-likelihood estimates in the family of all possible distributions (Kaplan and Meier, 1958). However, as in the uncensored case, this property is of limited importance and the best justification is essentially (7). The estimates probably also have a Bayesian interpretation involving a very "irregular" prior.

If the class of distributions is restricted, either parametrically or by some such condition as requiring $\lambda(t)$ to be monotonic or smooth, the maximum-likelihood estimates will be changed. For the monotone hazard case with uncensored data, see Grenander (1956). The smoothing of estimated hazard functions has been considered by Watson and Leadbetter (1964a, b) for the uncensored case.

3. Regression Models

Suppose now that on each individual one or more further measurements are available, say on variables $z_1, \ldots, z_p$. We deal first with the notationally simpler case when the failure-times are continuously distributed and the possibility of ties can be ignored. For the jth individual let the values of $\mathbf{z}$ be

$\mathbf{z}_j = (z_{1j}, \ldots, z_{pj})$. The z's may be functions of time. The main problem considered in this paper is that of assessing the relation between the distribution of failure time and $\mathbf{z}$. This will be done in terms of a model in which the hazard is

$$\lambda(t; \mathbf{z}) = \exp(\mathbf{z}\boldsymbol{\beta})\lambda_0(t), \tag{9}$$

where $\boldsymbol{\beta}$ is a $p \times 1$ vector of unknown parameters and $\lambda_0(t)$ is an unknown function giving the hazard function for the standard set of conditions $\mathbf{z} = \mathbf{0}$. In fact $(\mathbf{z}\boldsymbol{\beta})$ can be replaced by any known function $h(\mathbf{z}, \boldsymbol{\beta})$, but this extra generality is not needed at this stage. The following examples illustrate just a few possibilities.

Example 1 (Two-Sample Problem). Suppose that there is just one z variable, $p = 1$, and that this takes values 0 and 1, being an indicator variable for the two samples. Then according to (9) the hazards in samples 0 and 1 are respectively $\lambda_0(t)$ and $\psi\lambda_0(t)$, where $\psi = e^{\beta}$. In the continuous case the survivor functions are related (Lehmann, 1953) by $\mathscr{F}_1(t) = \{\mathscr{F}_0(t)\}^{\psi}$. There is an obvious extension for the k sample problem.

Example 2 (The Two-Sample Problem; Extended Treatment). We can deal with more complicated relationships between the two samples than are contemplated in Example 1 by introducing additional time-dependent components into $\mathbf{z}$. Thus if $z_2 = tz_1$, where z_1 is the binary variable of Example 1, the hazard in the second sample is

$$\psi e^{\beta_2 t}\lambda_0(t). \tag{10}$$

Of course in defining z_2, t could be replaced by any known function of t; further, several new variables could be introduced involving different functions of t. This provides one way of examining consistency with a simple model of proportional hazards. In fitting the model and often also in interpretation it is convenient to reparametrize (10) in the form

$$\rho \exp\{\beta_2(t - t^*)\}, \tag{11}$$

where t^* is any convenient constant time somewhere near the overall mean. This will avoid the more extreme non-orthogonalities of fitting. All the points connected with this example extend to the comparison of several samples.

Example 3 (Two-Sample Problem with Covariate). By introducing into the models of Examples 1 and 2 one or more further z variables representing concomitant variables, it is possible to examine the relation between two samples adjusting for the presence of concomitant variables.

Example 4 (Regression). The connection between failure-time and regressor variables can be explored in an obvious way. Note especially that by introducing functions of t, effects other than constant multiplication of the hazard can be included.

4. Analysis of Regression Models

There are several approaches to the analysis of the above models. The simplest is to assume $\lambda_0(t)$ constant, i.e. to assume an underlying exponential distribution; see, for example, Chernoff (1962) for some models of this type in the context of accelerated life tests. The next simplest is to take a two-parameter family of hazard functions, such as the power law associated with the Weibull distribution or the exponential of a linear function of t. Then standard methods such as maximum likelihood can be used; to be rigorous extension of the usual conditions for maximum-likelihood formulae and theory would be involved to cover censoring, but there is little doubt that some such justification could be given. This is in many ways the most natural approach but will not be explored further in the present paper. In this approach a computationally desirable feature is that both probability density and survivor function are fairly easily found. A simple form for the hazard is not by itself particularly advantageous, and models other than (9) may be more natural. For a normal theory maximum-likelihood analysis of factorial experiments with censored observations, see Sampford and Taylor (1959), and for the parametric analysis of response times in bioassay, see, Sampford (1954).

Alternatively we may restrict $\lambda_0(t)$ qualitatively, for example by assuming it to be monotonic or to be a step function (a suggestion of Professor J.W. Tukey). The latter possibility is related to a simple spline approximation to the log survivor function.

In the present paper we shall, however, concentrate on exploring the consequence of allowing $\lambda_0(t)$ to be arbitrary, main interest being in the regression parameters. That is, we require our method of analysis to have sensible properties whatever the form of the nuisance function $\lambda_0(t)$. Now this is a severe requirement and unnecessary in the sense that an assumption of some smoothness in the distribution $\mathscr{F}_0(t)$ would be reasonable. The situation is parallel to that arising in simpler problems when a nuisance parameter is regarded as completely unknown. It seems plausible in the present case that the loss of information about $\boldsymbol{\beta}$ arising from leaving $\lambda_0(t)$ arbitrary is usually slight; if this is indeed so the procedure discussed here is justifiable as a reasonably cautious approach to the study of $\boldsymbol{\beta}$. A major outstanding problem is the analysis of the relative efficiency of inferences about $\boldsymbol{\beta}$ under various assumptions about $\lambda_0(t)$.

The general attitude taken is that parametrization of the dependence on $\mathbf{z}$ is required so that our conclusions about that dependence are expressed concisely; of course any form taken is provisional and needs examination in the light of the data. So far as the secondary features of the system are concerned, however, it is sensible to make a minimum of assumptions leading to a convenient analysis, provided that no major loss of efficiency is involved.

5. A Conditional Likelihood

Suppose then that $\lambda_0(t)$ is arbitrary. No information can be contributed about $\boldsymbol{\beta}$ by time intervals in which no failures occur because the component $\lambda_0(t)$ might conceivably be identically zero in such intervals. We therefore argue conditionally on the set $\{t_{(i)}\}$ of instants at which failures occur; in discrete time we shall condition also on the observed multiplicities $\{m_{(i)}\}$. Once we require a method of analysis holding for all $\lambda_0(t)$, consideration of this conditional distribution seems inevitable.

For the particular failure at time $t_{(i)}$, conditionally on the risk set $\mathscr{R}(t_{(i)})$, the probability that the failure is on the individual as observed is

$$\exp\{\mathbf{z}_{(i)}\boldsymbol{\beta}\} \Big/ \sum_{l \in \mathscr{R}(t_{(i)})} \exp\{\mathbf{z}_{(l)}\boldsymbol{\beta}\}. \tag{12}$$

Each failure contributes a factor of this nature and hence the required conditional log likelihood is

$$L(\boldsymbol{\beta}) = \sum_{i=1}^{k} \mathbf{z}_{(i)}\boldsymbol{\beta} - \sum_{i=1}^{k} \log\left[\sum_{l \in \mathscr{R}(t_{(i)})} \exp\{\mathbf{z}_{(l)}\boldsymbol{\beta}\}\right]. \tag{13}$$

Direct calculation from (13) gives for $\xi, \eta = 1, \ldots, p$

$$U_\xi(\boldsymbol{\beta}) = \frac{\partial L(\boldsymbol{\beta})}{\partial \beta_\xi} = \sum_{i=1}^{k} \{z_{(\xi i)} - A_{(\xi i)}(\boldsymbol{\beta})\}, \tag{14}$$

where

$$A_{(\xi i)}(\boldsymbol{\beta}) = \frac{\sum z_{\xi l} \exp(\mathbf{z}_l \boldsymbol{\beta})}{\sum \exp(\mathbf{z}_l \boldsymbol{\beta})}, \tag{15}$$

the sum being over $l \in \mathscr{R}(t_{(i)})$. That is, $A_{(\xi i)}(\boldsymbol{\beta})$ is the average of z_ξ over the finite population $\mathscr{R}(t_{(i)})$, using an "exponentially weighted" form of sampling. Similarly

$$\mathscr{I}_{\xi\eta}(\boldsymbol{\beta}) = -\frac{\partial^2 L(\boldsymbol{\beta})}{\partial \beta_\xi \partial \beta_\eta} = \sum_{i=1}^{k} C_{(\xi\eta i)}(\boldsymbol{\beta}), \tag{16}$$

where

$$C_{(\xi\eta i)}(\boldsymbol{\beta}) = \{\textstyle\sum z_{\xi l} z_{\eta l} \exp(\mathbf{z}_l \boldsymbol{\beta}) / \sum \exp(\mathbf{z}_l \boldsymbol{\beta})\} - A_{(\xi i)}(\boldsymbol{\beta}) A_{(\eta i)}(\boldsymbol{\beta}) \tag{17}$$

is the covariance of z_ξ and z_η in this form of weighted sampling.

To calculate the expected value of (16) it would be necessary to know the times at which individuals who failed would have been censored had they not failed. This information would often not be available and in any case might well be thought irrelevant; this point is connected with difficulties of conditionality at the basis of a sampling theory approach to statistics (Pratt, 1962). Here we shall use asymptotic arguments in which (16) can be used directly for the estimation of variances, $\boldsymbol{\beta}$ being replaced by a suitable estimate. For a rigorous justification, assumptions about the censoring times generalizing those of Breslow (1970) would be required. It would not be satisfactory to assume that the censoring times are random variables distributed independently of the $\mathbf{z}$'s. For instance in the two-sample problem censoring might be much more severe in one sample than in the other.

Maximum-likelihood estimates of $\boldsymbol{\beta}$ can be obtained by iterative use of (14) and (16) in the usual way. Significance tests about subsets of parameters can be derived in various ways, for example by comparison of the maximum log likelihoods achieved. Relatively simple results can, however, be obtained for testing the global null hypothesis, $\boldsymbol{\beta} = \mathbf{0}$. For this we treat $\mathbf{U}(\mathbf{0})$ as asymptotically normal with zero mean vector and with covariance matrix $\mathscr{I}(\mathbf{0})$. That is, the statistic

$$\{\mathbf{U}(\mathbf{0})\}^T\{\mathscr{I}(\mathbf{0})\}^{-1}\{\mathbf{U}(\mathbf{0})\} \tag{18}$$

has, under the null hypothesis, an asymptotic chi-squared distribution with p degrees of freedom.

We have from (14) and (15) that

$$\mathbf{U}_\xi(\mathbf{0}) = \sum_{i=1}^{k} (z_{(\xi i)} - A_{(\xi i)}), \tag{19}$$

where $A_{(\xi i)} = A_{(\xi i)}(\mathbf{0})$ is the mean of z_ξ over $\mathscr{R}(t_{(i)})$. Further, from (16),

$$\mathscr{I}_{\xi\eta}(\mathbf{0}) = \sum_{i=1}^{k} C_{(\xi\eta i)}, \tag{20}$$

where $C_{(\xi\eta i)} = C_{(\xi\eta i)}(\mathbf{0})$ is the covariance of z_ξ and z_η in the finite population $\mathscr{R}(t_{(i)})$. The form of weighted sampling associated with general $\boldsymbol{\beta}$ has reduced to random sampling without replacement.

6. Analysis in Discrete Time

Unfortunately it is quite likely in applications that the data will be recorded in a form involving ties. If these are small in number a relatively *ad hoc* modification of the above procedures will be satisfactory. To cover the possibility of an appreciable number of ties, we generalize (9) formally to discrete time by

$$\frac{\lambda(t; \mathbf{z})\,dt}{1 - \lambda(t; \mathbf{z})\,dt} = \exp(\mathbf{z}\boldsymbol{\beta})\frac{\lambda_0(t)\,dt}{1 - \lambda_0(t)\,dt}. \tag{21}$$

In the continuous case this reduces to (9); in discrete time $\lambda(t; \mathbf{z})\, dt$ is a non-zero probability and (21) is a logistic model.

The typical contribution (12) to the likelihood now becomes

$$\exp\{\mathbf{s}_{(i)}\boldsymbol{\beta}\} \Big/ \sum_{l \in \mathscr{R}(t_{(i)}, m_{(i)})} \exp\{\mathbf{s}_{(l)}\boldsymbol{\beta}\}, \tag{22}$$

where $\mathbf{s}_{(i)}$ is the sum of $\mathbf{z}$ over the individuals failing at $t_{(i)}$ and the notation in the denominator means that the sum is taken over all distinct sets of $m_{(i)}$ individuals drawn from $\mathscr{R}(t_{(i)})$.

Thus the full conditional log likelihood is

$$\sum_{i=1}^{k} \mathbf{s}_{(i)}\boldsymbol{\beta} - \sum_{i=1}^{k} \log\left[\sum_{l \in \mathscr{R}(t_{(i)}, m_{(i)})} \exp\{\mathbf{s}_{(l)}\boldsymbol{\beta}\}\right]. \tag{22}$$

The derivatives can be calculated as before. In particular,

$$U_{\xi}(\mathbf{0}) = \sum_{i=1}^{k} \{s_{(\xi i)} - m_{(i)} A_{(\xi i)}\}, \tag{23}$$

$$\mathscr{I}_{\xi\eta}(\mathbf{0}) = \sum_{i=1}^{k} \frac{m_{(i)}\{r_{(i)} - m_{(i)}\}}{\{r_{(i)} - 1\}} C_{(\xi\eta i)}. \tag{24}$$

Note that (24) gives the exact covariance matrix when the observations $z_{(\xi i)}$ and the totals $s_{(\xi i)}$ are drawn randomly without replacement from the *fixed* finite populations $\mathscr{R}(t_{(1)}), \ldots, \mathscr{R}(t_{(k)})$. In fact, however, the population at one time is influenced by the outcomes of the "trials" at previous times.

7. The Two-Sample Problem

As an illustration, consider the two-sample problem with the proportional hazard model of Section 3, Example 1. Here $p = 1$ and we omit the first suffix on the indicator variable. Then

$$U(0) = n_1 - \sum_{i=1}^{k} m_{(i)} A_{(i)}, \tag{25}$$

$$\mathscr{I}(0) = \sum_{i=1}^{k} \frac{m_{(i)}\{r_{(i)} - m_{(i)}\}}{\{r_{(i)} - 1\}} A_{(i)}\{1 - A_{(i)}\}, \tag{26}$$

where $A_{(i)}$ is the proportion of the risk population $\mathscr{R}(t_{(i)})$ that have $z = 1$, i.e. belong to sample 1, and n_1 is the total number of failures in sample 1. An asymptotic two-sample test is thus obtained by treating

$$U(0)/\sqrt{\mathscr{I}(0)} \tag{27}$$

as having a standard normal distribution under the null hypothesis. This is different from the procedure of Gehan who adapted the Wilcoxon test to censored data (Gehan, 1965; Efron, 1967; Breslow, 1970). The test has been considered in some detail by Peto and Peto (1972).

The test (27) is formally identical with that obtained by setting up at each failure point a 2×2 contingency table (sample 1, sample 2) (failed, survived). To test for the presence of a difference between the two samples the information from the separate tables can then be combined (Cochran, 1954; Mantel and Haenzel, 1959; Mantel, 1963). The application of this to life tables is discussed especially by Mantel (1966). Note, however, that whereas the test in the contingency table situation is, at least in principle, exact, the test here is only asymptotic, because of the difficulties associated with specification of the stopping rule. Formally the same test was given by Cox (1959) for a different life-table problem where there is a single sample with two types of failure and the hypothesis under test concerns the proportionality of the hazard function for the two types.

When there is a non-zero value of β, the "weighted" average of a single observation from the risk population $\mathscr{R}(t_{(i)})$ is

$$A_{(i)}(\beta) = \frac{e^{\beta} A_{(i)}}{1 - A_{(i)} + e^{\beta} A_{(i)}} \tag{28}$$

and the maximum-likelihood equation $U(\hat{\beta}) = 0$ gives, when all failure times are distinct,

$$\sum_{i=1}^{k} \frac{e^{\hat{\beta}} A_{(i)}}{1 - A_{(i)} + e^{\hat{\beta}} A_{(i)}} = n_1. \tag{29}$$

If $\hat{\beta}$ is thought to be close to some known constant, it may be useful to linearize (29). In particular, if $\hat{\beta}$ is small, we have as an approximation to the maximum-likelihood estimate

$$\hat{\beta}_0 = (n_1 - \sum A_{(i)}) / \sum A_{(i)}\{1 - A_{(i)}\}.$$

The procedures of this section involve only the ranked data, i.e. are unaffected by an arbitrary monotonic transformation of the time scale. Indeed the same is true for any of the results in Section 4 provided that the **z**'s are not functions of time. While the connection with the theory of rank tests will not be explored in detail, it is worth examining the form of the test (27) for uncensored data with all failure times distinct. For this, let the failure times in sample 1 have ranks $c_1 < c_2 < \cdots < c_{n_1}$ in the ranking of the full data. At the ith largest observed failure time, individuals with ranks $n, n-1, \ldots, i$ are at risk, so that

$$A_{(i)} = \frac{1}{n - i + 1} \sum_{l=1}^{n_1} H(c_l - i), \tag{30}$$

where $H(x)$ is the unit Heaviside function,

$$H(x) = \begin{cases} 1 & (x \geqslant 0), \\ 0 & (x < 0). \end{cases} \tag{31}$$

Thus, by (25),

$$U(0) = n_1 - \sum_{l=1}^{n_1} \sum_{i=1}^{c_l} \frac{1}{n-i+1}$$

$$= n_1 - \sum_{l=1}^{n} e_{nc_l}, \tag{32}$$

where e's are the expected values of the order statistics in a random sample of size n from a unit exponential distribution. The test based on (32) is asymptotically fully efficient for the comparison of two exponential distributions (Savage, 1956; Cox, 1964). Further, by (26),

$$\mathscr{I}(0) = \sum_{l=1}^{n_1} e_{nc_l} - \sum_{l=1}^{n_1} (1 + 2n_1 - 2l) v_{nc_l}, \tag{33}$$

where

$$v_{nc_l} = \sum_{i=1}^{c_l} \frac{1}{(n-i+1)^2} \tag{34}$$

is the variance of an exponential order statistic.

Here the test statistic is, under the null hypothesis, a constant minus the total of a random sample of size n_1 drawn without replacement from the finite population $\{e_{n1}, \ldots, e_{nn}\}$. The exact distribution can in principle be obtained and in particular it can be shown that

$$E\{U(0)\} = 0, \qquad \operatorname{var}\{U(0)\} = \frac{n_1(n-n_1)(n-e_{nn})}{n(n-1)}. \tag{35}$$

There is not much point in this case in using the more complicated asymptotic formula (33), especially as fairly simple more refined approximations to the distribution of the test statistic are available (Cox, 1964). It can easily be verified that

$$E\{\mathscr{I}(0)\} \sim \operatorname{var}\{U(0)\}. \tag{36}$$

8. Estimation of Distribution of Failure-Time

Once we have obtained the maximum-likelihood estimate of β, we can consider the estimation of the distribution associated with the hazard (10) either for $\mathbf{z} = \mathbf{0}$, or for some other given value of $\mathbf{z}$. Thus to estimate $\lambda_0(t)$ we need to generalize (7). To do this we take $\lambda_0(t)$ to be identically zero, except at the points where failures have occurred, and carry out a separate maximum likelihood estimation at each such failure point. For the latter it is convenient to write the contribution to $\lambda_0(t)$ at $t_{(i)}$ in the form

$$\frac{\pi_{(i)} \exp(-\boldsymbol{\beta}\tilde{\mathbf{z}}_{(i)})}{1 - \pi_{(i)} + \pi_{(i)} \exp(-\boldsymbol{\beta}\tilde{\mathbf{z}}_{(i)})} \delta(t - t_{(i)}),$$

where $\tilde{\mathbf{z}}_{(i)}$ is an arbitrary constant to be chosen; it is useful to take $\tilde{\mathbf{z}}_{(i)}$ as approximately the mean in the relevant risk set. The maximum-likelihood estimate of $\pi_{(i)}$ can then be shown to satisfy

$$\hat{\pi}_{(i)} = \frac{m_{(i)}}{r_{(i)}} - \frac{\hat{\pi}_{(i)}(1 - \hat{\pi}_{(i)})}{r_{(i)}} \sum_{j \in R(t_{(i)})} \frac{\exp\{\hat{\boldsymbol{\beta}}(\mathbf{z}_j - \tilde{\mathbf{z}}_{(i)})\} - 1}{1 - \hat{\pi}_{(i)} + \pi_{(i)} \exp\{\hat{\boldsymbol{\beta}}(\mathbf{z}_j - \tilde{\mathbf{z}}_{(i)})\}}, \tag{37}$$

which can be solved by iteration. The suggested choice of $\tilde{\mathbf{z}}_{(i)}$ is designed to make the second term in (37) small. Note that in the single-sample case, the second term is identically zero. Once (37) is solved for all i, we have by the product integral formula

$$\hat{\mathscr{F}}_0(t) = \prod_{t_{(i)} < t} \left\{1 - \frac{\hat{\pi}_{(i)} \exp(-\hat{\boldsymbol{\beta}}\tilde{\mathbf{z}}_{(i)})}{1 - \hat{\pi}_{(i)} + \hat{\pi}_{(i)} \exp(-\hat{\boldsymbol{\beta}}\tilde{\mathbf{z}}_{(i)})}\right\}. \tag{38}$$

For an estimate at a given non-zero $\mathbf{z}$, replace $\exp(-\hat{\boldsymbol{\beta}}\tilde{\mathbf{z}}_{(i)})$ by $\exp\{\hat{\boldsymbol{\beta}}(\mathbf{z} - \tilde{\mathbf{z}}_{(i)})\}$. Alternative simpler procedures would be worth having (Mantel, 1966).

9. Bivariate Life Tables

We now consider briefly the extension of life-table arguments to multivariate data. Suppose for simplicity that there are two types of failure time for each individual represented by random variables T_1 and T_2. For instance, these might be the failure-times of two different but associated components; observations may be censored on neither, one or both components. For analogous problems in bioassay, see Sampford (1952).

The joint distribution can be described in terms of hazard functions $\lambda_{10}(t)$, $\lambda_{20}(t)$, $\lambda_{21}(t|u)$, $\lambda_{12}(t|u)$, where

$$\left.\begin{aligned} \lambda_{p0}(t) &= \lim_{\Delta t \to 0+} \frac{\mathrm{pr}(t \leqslant T_p < t + \Delta t | t \leqslant T_1, t \leqslant T_2)}{\Delta t} \quad (p = 1, 2), \\ \lambda_{21}(t|u) &= \lim_{\Delta t \to 0+} \frac{\mathrm{pr}(t \leqslant T_2 < t + \Delta t | t \leqslant T_2, T_1 = u)}{\Delta t} \quad (u < t), \end{aligned}\right\} \tag{39}$$

with a similar definition for $\lambda_{12}(t|u)$. It is easily shown that the bivariate probability density function $f(t_2, t_2)$ is given by

$$f(t_1, t_2) = \exp\left[-\int_0^{t_1 - 0} \{\lambda_{10}(u) + \lambda_{20}(u)\}\, du - \int_{t_1 + 0}^{t_2 - 0} \lambda_{21}(u|t_1)\, du\right] \lambda_{10}(t_1)\lambda_{21}(t_2|t_1), \tag{40}$$

for $t_2 \geq t_1$, with again an analogous expression for $t_2 \leq t_1$. It is fairly easy to show formally that a necessary and sufficient condition for the independence of T_1 and T_2 is

$$\lambda_{12}(t\,|\,u) = \lambda_{10}(t), \qquad \lambda_{21}(t\,|\,u) = \lambda_{20}(t), \tag{41}$$

as is obvious on general grounds. Note also that if $\mathscr{F}(t_1, t_2)$ is the joint survivor function

$$\lambda_{10}(t) = -\frac{1}{\mathscr{F}(t, t)}\left[\frac{\partial \mathscr{F}(t, u)}{\partial t}\right]_{u=t}, \qquad \lambda_{12}(t\,|\,u) = -\frac{\partial^2 \mathscr{F}(t, u)}{\partial t \partial u} \bigg/ \frac{\partial \mathscr{F}(t, u)}{\partial u}. \tag{42}$$

Dependence on further variables $\mathbf{z}$ can be indicated in the same way as for (11). The simplest model would have the same function of $\mathbf{z}$ multiplying all four hazard functions, although this restriction is not essential.

Estimation and testing would in principle proceed as before, although grouping of the conditioning u variable seems necessary in the parts of the analysis concerning the function $\lambda_{12}(t|u)$ and $\lambda_{21}(t|u)$.

Further generalizations which will not, however, be explored here are to problems in multidimensional time and to problems connected with point processes (Cox and Lewis, 1972; Cox, 1972).

(*Editor's note*: Section 10, An Example, has been omitted.)

11. Physical Interpretation of Model

The model (9), which is the basis of this paper, is intended as a representation of the behaviour of failure-time that is convenient, flexible and yet entirely empirical. One of the referees has, however, suggested adding some discussion of the physical meaning of the model and in particular of its possible relevance to accelerated life testing. Suppose in fact that there is a variable s, called "stress", and that life tests are carried out at various levels of s. For simplicity we suppose that s is one-dimensional and that each individual is tested at a fixed level of s. The usual idea is that we are really interested in some standard stress, say $s = 1$, and which to use other values of s to get quick laboratory results as a substitute for a predictor of the expensive results of user trials.

Now in order that the distribution of failure-time at one level of stress should be related to that at some other level, the relationship being stable under a wide range of conditions, it seems necessary that the basic physical process of failure should be common at the different stress levels; and this is likely to happen only when there is a single predominant mode of failure. One difficulty of the problem is that of knowing enough about the physical process to be able to define a stress variable, i.e. a set of test conditions, with the right properties.

One of the simplest models proposed for the effect of stress on the distribution of failure-time is to assume that the mechanism of failure is identical at the various levels of s but takes place on a time-scale that depends on s. Thus if $\mathscr{F}(t; s)$ denotes the survivor function at stress s, this model implies that

$$\mathscr{F}(t; s) = \mathscr{F}\{g(s)t; 1\}, \tag{45}$$

where $g(s)$ is some function of s with $g(1) = 1$. Thus the hazard function at stress s is

$$g(s)\lambda_0\{g(s)t\}, \tag{46}$$

where $\lambda_0(\cdot)$ is the hazard at $s = 1$. In particular if $g(s) = s^\beta$ and if $z = \log s$ this gives

$$e^{\beta z}\lambda_0(e^{\beta z}t). \tag{47}$$

This is similar to but different from the model (9) of this paper. A special set of conditions where (47) applies is where the individual is subject to a stream of shocks of randomly varying magnitudes until the cumulative shock exceeds some time-independent tolerance. If, for instance, all aspects of the process except the rate of incidence of shocks are independent of s, then (45) will apply.

If, however, the shocks are non-cumulative and failure occurs when a rather high threshold is first exceeded, failures occur in a Poisson process with a rate depending on s. A special model of this kind often used for thermal stress is to suppose that failure corresponds to the excedence of the activation energy of some process; then by the theory of rate processes (47) can be used with $\lambda_0(\cdot) = 1$ and z equal to the reciprocal of absolute temperature.

As a quite different model suppose that some process of ageing goes on independently of stress. Suppose further that the conditional probability of failure at any time is the product of an instantaneous time-dependent term arising from the ageing process and a stress-dependent term; the model is non-cumulative. Then the hazard is

$$h(s)\lambda_0(t), \tag{48}$$

where $h(s)$ is some function of stress. Again if $h(s) = s^\beta$, the model becomes

$$e^{\beta z}\lambda_0(t) \tag{49}$$

exactly that of (9), where again $\lambda_0(t)$ is the hazard function at $s = 1, z = 0$. One special example of this model is rather similar to that suggested for (46), except that the critical tolerance varies in a fixed way with time and the shocks are non-cumulative, the rate of incidence of shocks depending on s. For another possibility, see Shooman (1968).

If hazard or survivor functions are available at various levels of s we might attempt an empirical discrimination between (46) and (48). Note, however, that if we have a Weibull distribution at $s = 1$, $\lambda_0(\cdot)$ is a power function and (46) and (48) are identical. Then the models cannot be discriminated from failure-time distributions alone. That is, if we did want to make such a discrimination we must look for situations in which the distributions are far from the Weibull form. Of course the models outlined here can be made much more specific by introducing explicit stochastic processes or physical models. The wide variety of possibilities serves to emphasize the difficulty of inferring an underlying mechanism indirectly from failure times alone rather than from direct study of the controlling physical processes.

As a basis for rather empirical data reduction (9), possibly with time-dependent exponent, seems flexible and satisfactory.

Acknowledgments

I am grateful to the referees for helpful comments and to Professor P. Armitage, Mr P. Fisk, Dr N. Mantel, Professors J.W. Tukey and M. Zelen for references and constructive suggestions.

References

Breslow, N. (1970). A generalized Kruskal–Wallis test for comparing samples subject to unequal patterns of censoring. *Biometrika*, **57**, 579–594.

Chernoff, H. (1962). Optimal accelerated life designs for estimation. *Technometrics*, **4**, 381–408.

Chiang, C.L. (1968). *Introduction to Stochastic Processes in Biostatistics*. New York: Wiley.

Cochran, W.G. (1954). Some methods for strengthening the common χ^2 tests. *Biometrics*, **10**, 417–451.

Cox, D.R. (1959). The analysis of exponentially distributed life-times with two types of failure *J. R. Statist. Soc.* **B**, **21**, 411–421.

——— (1964). Some applications of exponential ordered scores. *J. R. Statist. Soc.* **B**, **26**, 103–110.

——— (1972). The statistical analysis of dependencies in point processes. In *Symposium on Point Processes* (P.A.W. Lewis, ed.). New York: Wiley.

Cox, D.R., and Lewis, P.A.W. (1972). Multivariate point processes. *Proc. 6th Berkeley Symp.*

Efron, B. (1967). The two sample problem with censored data. *Proc. 5th Berkeley Symp.*, **4**, 831–853.

Gehan, E.A. (1965). A generalized Wilcoxon test for comparing arbitrarily single-censored samples. *Biometrika*, **52**, 203–224.

Grenander, U. (1956). On the theory of mortality measurement, I and II. *Skand Akt.*, **39**, 90–96, 125–153.

Kaplan, E.L., and Meier, P. (1958). Nonparametric estimation from incomplete observations. *J. Am. Statist. Assoc.*, **53**, 457–481.

Lehmann, E.L. (1953). The power of rank tests. *Ann. Math. Statist.*, **24**, 23–43.

Mantel, N. (1963). Chi-square tests with one degree of freedom: extensions of the Mantel-Haenzel procedure. *J. Am. Statist. Assoc.*, **58**, 690–700.

——— (1966). Evaluation of survival data and two new rank order statistics arising in its consideration. *Cancer Chemotherapy Reports*, **50**, 163–170.

Mantel, N., and Haenzel, W. (1959). Statistical aspects of the analysis of data from retrospective studies of disease. *J. Nat. Cancer Inst.*, **22**, 719–748.

Peto, R., and Peto, J. (1972). Asymptotically efficient rank invariant test procedures. *J. R. Statist. Soc.*, A **135**, 185–206.

Pratt, J.W. (1962). Contribution to discussion of paper by A. Birnbaum. *J. Am. Statist. Assoc.*, **57**, 314–316.

Sampford, M.R. (1952). The estimation of response-time distributions, II: Multi-stimulus distributions. *Biometrics*, **8**, 307–369.

——— (1954). The estimation of response-time distribution, III: Truncation and survival. *Biometrics*, **10**, 531–561.

Sampford, M.R., and Taylor, J. (1959). Censored observations in randomized block experiments. *J. R. Statist. Soc.* **B**, **21**, 214–237.
Savage, I.R. (1956). Contributions to the theory of rank order statistics—the two-sample case. *Ann. Math. Statist.*, **27**, 590–615.
Shooman, M.L. (1968). Reliability physics models. *IEEE Trans. on Reliability*, **17**, 14–20.
Watson, G.S., and Leadbetter, M.R. (1964a). Hazard analysis, I. *Biometrika*, **51**, 175–184.
Watson, G.S., and Leadbetter, M.R. (1964b). Hazard analysis, II. *Sankhyā*, A **26**, 101–116.

Sampford, M.R. and Taylor, J. (1959). Censored observations in randomized [illegible] experiments. J. R. Statist. Soc. B, 21, 214–2[illegible].

Sa[illegible] (19[illegible]). Contributions to the theory of order statistics—[illegible] samples [illegible] Ann. Math. Statist. 27, [illegible].

Shooman, M.L. (1968). Reliability [illegible] models. IEEE Trans. on Rel. [illegible]

Watson, G.S. and Leadbetter, M.R. (1964). Hazard analysis. I. Biometrika, 51, [illegible]

Watson, G.S. and Leadbetter, M.R. (1964). Hazard analysis. II. Sankhyā, [illegible] 101–1[illegible].

Introduction to Nelder and Wedderburn (1972) Generalized Linear Models

P. McCullagh
University of Chicago

In this paper, the authors show that maximum likelihood estimates for a large class of commonly used regression models can be obtained by the method of iteratively weighted least squares, in which both the weights and the response are adjusted from one iteration to the next. The proposed algorithm, sometimes linown as "Fisher-scoring," is an extension of Fisller's (1935) method for computing maximum likelihood estimates in linear probit models. The same result was obtained independently by Bradley (1973) and Jennrich and Moore (1975), though not exploited to its full extent.

The notation used in this introduction, taken from McCullagh and Nelder (1989), differs from that used in the original paper in several ways as shown below.

Nelder and Wedderburn		McCullagh and Nelder	
Terminology	Notation	Terminology	Notation
Observation	z_i	Observation	y_i
—	$g(\theta)$	Cumulant function	$b(\theta)$
Log likelihood	L	Log likelihood	$l(\boldsymbol{\mu}; \mathbf{y}) = \sum \{y_i\theta_i - b(\theta_i)\}/\phi$
Scale factor	$1/\alpha(\phi)$	Dispersion parameter	ϕ or σ^2
—	$Y_i = \sum x_{ij}\beta_j$	Linear predictor	$\eta_i = \sum x_{ij}\beta_j$
Linking function	$\theta_i = f(Y_i)$	Link function	$\eta_i = g(\mu_i)$
—	$2(L_{\max} - L)/\alpha(\phi)$	Deviance	$D(\mathbf{y}; \boldsymbol{\mu}) = 2\phi\{l(\mathbf{y}; \mathbf{y}) - l(\boldsymbol{\mu}; \mathbf{y})\}$
Deviance	$2(L_{\max} - L)$	Scaled deviance	$D^*(\mathbf{y}; \boldsymbol{\mu}) = D(\mathbf{y}; \boldsymbol{\mu})/\phi$

It is assumed that the observations, $y_1, \ldots, y_n$ are independent and that y_i is a realization of the random variable Y_i wllose distribution has the form

$$\exp\left\{\frac{y_i\theta_i - b(\theta_i)}{\phi} + c(y_i, \phi)\right\},$$

depending on parameters θ_i related to the mean of Y_i, and ϕ related to the variance of Y_i. For immediate practical purposes, this family includes the normal, exponential, gamma, Poisson, and binomial. In all cases, the first two moments satisfy

$$\mu_i = E(Y_i) = b'(\theta_i), \qquad \operatorname{var}(Y_i) = \phi b''(\theta_i) \stackrel{\text{def}}{=} \phi V(\mu_i).$$

For obvious reasons, ϕ is called the dispersion parameter and $V(\mu)$ the variance function. Generalized linear models are formulated in two additional parts. First, the link function $\eta_i = g(\mu_i)$ transforms the means to a scale on which effects are assumed to be additive. To complete the specification, we require a model matrix $\mathbf{X}$ such that the effect of the covariates is given by $\boldsymbol{\eta} = \mathbf{X}\boldsymbol{\beta}$ in the usual notation for linear models, where $\boldsymbol{\beta}$ is a p-dimensional parameter vector to be estimated. Usually, p is much smaller than n. In the computer program GLIM (Baker and Nelder, 1978), for example, $\mathbf{X}$ is specified by means of the model formula notation of Wilkinson and Rogers (1973).

In all generalized linear models, the dispersion parameter ϕ is regarded as either known or else an unknown constant independent of the covariates. The two cases are indistinguishable insofar as the iterative algorithm is concerned because the estimate of $\boldsymbol{\beta}$ is independent of ϕ. If ϕ is unknown, an estimate may be obtained from the residuals in much the same way that σ^2 is estimated by the residual mean square in linear regression models. Ordinary maximum likelihood is usually not the method of choice for this purpose.

If the log likelihood function at $\boldsymbol{\beta}$ for fixed ϕ is denoted by

$$l(\boldsymbol{\mu}(\boldsymbol{\beta}); \mathbf{y}) = \sum \frac{y_i\theta_i(\boldsymbol{\beta}) - b(\theta_i)}{\phi},$$

the deviance function, measuring the discrepancy between the data vector $\mathbf{y}$ and the mean vector $\boldsymbol{\mu}$, is defined by

$$D(\mathbf{y}; \boldsymbol{\mu}) = 2\phi\{l(\mathbf{y}; \mathbf{y}) - l(\boldsymbol{\mu}; \mathbf{y})\},$$

which does not depend on ϕ. For normal models, $D(\mathbf{y}; \hat{\boldsymbol{\mu}})$ is the residual sum of squares $\sum(y_i - \hat{\mu}_i)^2$. For Poisson models, we obtain

$$D(\mathbf{y}; \boldsymbol{\mu}) = 2\sum\left\{y_i \log\left(\frac{y_i}{\mu_i}\right) - (y_i - \mu_i)\right\}.$$

Corresponding to any sequence of nested models is an additive decomposition of the deviance function into approximately independent components whose null distribution is $\phi\chi^2_p$. This decomposition gives rise to an analysis-of-deviance table by analogy with the analysis-of-variance table for normal theory linear models.

The special subclass with a canonical link function, $\eta_i = \theta_i$, had been studied by Dempster (1971). Binomial models had been extensively studied by Cox (1958, 1970).

In retrospect, one remarkable aspect of the paper is that the generalization given is not nearly so general as it might have been. It extends the boundaries of linear models just enough to encompass the most commonly used statistical methods such as log linear models for counted data and linear logistic or probit models for quantal responses. This is an exceedingly useful extension, not simply because it unites into a single framework a large number of models from diverse fields, but also because it is achieved at so little cost. Apart from the usual model formula, all that is required to complete the specification is a link function and a variance function, the latter implicitly specified via the error distribution.

More recent work by Green (1984), Jørgensen (1987), and others shows that the method of iteratively ueighted least squares can be used to obtain maximum likelihood estimates and fitted values for a large class of models, including but not restricted to distributions from the family (1). Linearity, even after link transformation, is irrelevant in this more general framework. For practical purposes, however, this added generality comes at a substantial cost in terms of model specification, parameter interpretation, data organization, and computational overhead. To a large extent, the ready acceptance by practicing statisticians of the original generalization is explained by the availability of a suitable computer program (GLIM), the ease of model specification, and the familiarity of most users with some subset of the models that could be fitted.

Wedderburn died suddenly and unexpectedly in June 1975 at the age of 28. An obituary appears in *Series A* of the *Journal of the Royal Statistical Society* (Vol. 138, 1975, p. 587). His two other papers, one of which was published posthumously, are listed below.

References

Baker, R.J., and Nelder, J.A. (1978). *The GLIM System, Release 3: Generalized linear Interactive Modelling.* Numerical Algorithms Group, Oxford.

Bradley, E.L. (1973). The equivalence of maximum likelihood and weighted least squares in the exponential family, *J. Statist. Assoc.*, **68**, 199–200.

Cox, D.R. (1958). The regression analysis of binary sequences, *J. Roy. Statist. Soc., Ser. B*, **20**, 215–242 (with discussion).

Cox, D.R. (1970). *The Analysis of Binary Data.* Chapman and Hall, London.

Dempster, A.P. (1971). An overview of multivariate data analysis, *J. Multiv. Anal.*, **1**, 316–346.

Fisher, R.A. (1935). The case of zero survivors [appendix to Bliss (1935)], *Ann. Appl. Biol.*, **22**, 164–165.

Green, P.J. (1984). Iteratively reweighted least squares for maximum likelihood estimation, and some robust and resistant alternatives, *J. Roy. Statist. Soc., Ser. B*, **46**, 149–192 (with discussion).

Jennrich, R.I., and Moore, R.H. (1975). Maximum likelihood estimation by means of nonlinear least squares, *Amer. Statist. Assoc. Proc. Stalist. Computing Section*, 57–65.

Jørgensen, B. (1987). Exponential dispersion models, *J. Roy. Statist. Soc., Ser. B*, **49**, 127–162 (with discussion).

McCullagh, P., and Nelder, J.A. (1989). *Generalized Linear Models*. Chapman and Hall, London.

Wedderburn, R.W.M. (1974). Quasilikelihood functions, generalized linear models and the Gauss–Newton method, *Biometrika*, **61**, 439–447.

Wedderburn, R.W.M. (1976). On the existence and uniqueness of the maximum likelihood estimates for certain generalized linear models, *Biometrika*, **63**, 27–32.

Wilkinson, G.N., and Rogers, C.E. (1973). Symbolic description of factorial models for analysis of variance, *Appl. Statist.*, **22**, 392–399.

Generalized Linear Models

J.A. Nelder and R.W.M. Wedderburn
Rothamsted Experimental Station

Summary

The technique of iterative weighted linear regression can be used to obtain maximum likelihood estimates of the parameters with observations distributed according to some exponential family and systematic effects that can be made linear by a suitable transformation. A generalization of the analysis of variance is given for these models using log-likelihoods. These generalized linear models are illustrated by examples relating to four distributions; the Normal, Binomial (probit analysis, etc.), Poisson (contingency tables) and gamma (variance components).

The implications of the approach in designing statistics courses are discussed.

Introduction

Linear models customarily embody both systematic and random (error) components, with the errors usually assumed to have normal distributions. The associated analytic technique is least-squares theory, which in its classical form assumed just one error component; extensions for multiple errors have been developed primarily for analysis of designed experiments and survey data. Techniques developed for non-normal data include probit analysis, where a binomial variate has a parameter related to an assumed underlying tolerance distribution, and contingency tables, where the distribution is multinomial and the systematic part of the model usually multiplicative. In both these examples there is a linear aspect to the model; thus in probit analysis the parameter p is a function of tolerance Y which is itself linear on the dose (or some function thereof), and in a contingency table with a multiplicative model thc logarithm of the expected probability is assumed linear on classifying factors defining the table. Thus for both, the systematic part of the model has a linear basis. In another extension (Nelder, 1968) a certain transformation is used to produce normal errors, and a different transformation of the expected values is used to produce linearity.

So far we have mentioned models associated with the normal, binomial and multinomial distributions (this last can be thought of as a set of Poisson distributions with constraints). A further class is based on the χ^2 or gamma distribution and arises in the estimation of variance components from independent quadratic forms derived from the original observations. Again the systematic component of the model has a linear structure.

In this paper we develop a class of *generalized linear models*, which includes all the above examples, and we give a unified procedure for fitting them based on likelihood. This procedure is a generalization of the well-known one described by Finney (1952) for maximum likelihood estimation in probit analysis. Section 1 defines the models, and Section 2 develops the fitting process and generalizes the analysis of variance. Section 3 gives examples with four special distributions for the random components. In Section 4 we consider the usefulness of the models for courses of instruction in statistics.

1.1. The Random Component

Suppose our observations z come from a distribution with density function

$$\pi(z; \theta, \phi) = \exp[\alpha(\phi)\{z\theta - g(\theta) + h(z)\} + \beta(\phi, z)],$$

where $\alpha(\phi) > 0$ so that for fixed ϕ we have an exponential family. The parameter ϕ could stand for a certain type of nuisancc parameter such as the variance σ^2 of a normal distribution or the parameter p of a gamma distribution (see Section 3.4). We denote the mean of z by μ.

We require expressions for the first and second derivatives of the log-likelihood in terms of the mean and variance of z and the scale factor $\alpha(\phi)$. We use the results (see, for example, Kendall and Stuart, 1967, p. 9)

$$E(\partial L/\partial\theta) = 0 \tag{1}$$

and

$$E(\partial^2 L/\partial\theta^2) = -E(\partial L/\partial\theta)^2, \tag{2}$$

where the differentiation under the sign of integration used in their derivation can be justified by Theorem 9 of Lehmann (1959).

We have

$$\partial L/\partial\theta = \alpha(\phi)\{z - g'(\theta)\}.$$

Then (1) implies that

$$\mu = E(z) = g'(\theta);$$

hence

$$\partial L/\partial\theta = \alpha(\phi)(z - \mu). \tag{3}$$

From (2) we obtain $\alpha(\phi)g''(\theta) = [\alpha(\phi)]^2\text{var}(z)$, whence

$$g''(\theta) = \alpha(\phi)\text{var}(z) = V, \quad \text{say}, \tag{4}$$

so that V is the variance of z when the scale factor is unity. Then

$$\partial^2 L/\partial\theta^2 = -\alpha(\phi)V. \tag{5}$$

We note also that

$$V = d\mu/d\theta. \tag{6}$$

For a one-parameter exponential family in which $\alpha(\phi) = 1$, we can write

$$\pi(z; \theta) = \exp\{z\theta - g(\theta) + h(z)\},$$

so that

$$\partial L/\partial\theta = z - \mu$$

and

$$-\partial^2 L/\partial\theta^2 = V = \text{var}(z).$$

1.2. The Linear Model for Systematic Effects

The term "linear model" usually encompasses both systematic and random components in a statistical model, but we shall restrict the term to include only the systematic components. We write

$$Y = \sum_{i=1}^{m} \beta_i x_i$$

when the x_i are independent variates whose values are supposed known and β_i are parameters. The β_i may have fixed (known) values or be unknown and require estimation. An independent variate may be *quantitative* and produce a single x-variate in the model, or *qualitative* and produce a set of x-variates whose values are 0 and 1, or *mixed*. Consider the model

$$Y_{ij} = \alpha_i + \beta u_{ij} + \gamma_j v_{ij} \qquad (i = 1, \ldots, n, j = 1, \ldots, p),$$

where the data are indexed by factors whose levels are denoted by i and j. The term α_i includes n parameters associated with a qualitative variate represented by n dummy x-variate components taking values 1 for one level and 0 for the rest; βu_{ij} represents a quantitative variate, namely u with single parameter β, and $\gamma_j v_{ij}$ shows p parameters γ_j associated with a mixed independent variate whose p components take the values of v_{ij} for one level of j and zero for the rest. A notation suitable for computer use has been developed by C.E. Rogers and G.N. Wilkinson and is to be published in *Applied Statistics.**

1.3. The Generalized Linear Model

We now combine the systematic and random components in our model to produce the generalized linear model. This is characterized by

(i) A dependent variable z whose distribution with parameter θ is one of the class in Section 1.1.
(ii) A set of independent variables $x_1, \ldots, x_m$ and predicted $Y = \sum \beta_i x_i$ as in Section 1.2.
(iii) A linking function $\theta = f(Y)$ connecting the parameter θ of the distribution of z with the Y's of the linear model.

When z is normally distributed with mean θ and variance σ^2 and when $\theta = Y$, we have ordinary linear models with Normal errors. Other examples of these models will be described in Section 3 under the various distributions of the exponential type. We now consider the solution of the maximum likelihood equations for the parameters of the generalized linear models and show its equivalence to a procedure of iterative weighted least squares.

2. Fitting the Models

2.1. The Maximum Likelihood Equations

The solution of the maximum likelihood equations is equivalent to an iterative weighted least-squares procedure with a weight function

$$w = (d\mu/dY)^2/V$$

and a modified dependent variable (the working probit of probit analysis)

$$y = Y + (z - \mu)/(d\mu/dY),$$

where μ, Y and V are based on current estimates. This generalizes the results of Nelder (1968).

Proof. Writing L for the log-likelihood from one observation we have, from (3) and (6),

* *Applied Statistics* (1973), **22** 392–399.

$$\frac{\partial L}{\partial \beta_i} = \alpha(\theta)(z-\mu)\frac{d\theta}{d\mu}\frac{d\mu}{dY}x_i$$

$$= \alpha(\theta)\frac{z-\mu}{V}\frac{d\mu}{dY}x_i \tag{7}$$

and

$$\frac{\partial^2 L}{\partial \beta_i \partial \beta_j} = \frac{\partial^2 L}{\partial Y^2}x_i x_j,$$

where

$$\frac{\partial^2 L}{\partial Y^2} = \frac{\partial^2 L}{\partial \theta^2}\left(\frac{d\theta}{dY}\right)^2 + \frac{\partial L}{\partial \theta}\frac{d^2\theta}{dY^2}$$

$$= \alpha(\phi)\left\{-V\left(\frac{d\theta}{d\mu}\right)^2\left(\frac{d\mu}{dY}\right)^2 + (z-\mu)\frac{d^2\theta}{dY^2}\right\}$$

$$= \alpha(\phi)\left\{-\left(\frac{d\mu}{dY}\right)^2\Big/V + (z-\mu)\frac{d^2\theta}{dY^2}\right\}. \tag{8}$$

The expected second derivative with negative sign is thus

$$\alpha(\phi)\left\{\left(\frac{d\mu}{dY}\right)^2\Big/V\right\}x_i x_j \quad \text{from (8).}$$

Writing w for the weight function $(d\mu/dy)^2/V$, (7) gives

$$\partial L/\partial \beta_i = \alpha(\phi)wx_i(z-\mu)/(d\mu/dY).$$

Thus the Newton-Raphson process with expected second derivatives (equivalent to Fisher's scoring technique) for a sample of n gives

$$\mathbf{A}\delta\boldsymbol{\beta} = \mathbf{C}, \tag{9}$$

where $\mathbf{A}$ is a $m \times m$ matrix with

$$A_{ij} = \sum_{k=1}^{n} w_k x_{ik} x_{jk}$$

and $\mathbf{C}$ is a $m \times 1$ vector with

$$C_i = \sum w_k x_{ik}(z-\mu)/(d\mu/dY).$$

Finally we have

$$(A\beta)_i = \sum A_{ij}\beta_j = \sum w_k x_{ik} Y_k$$

so that (9) may be written in the form

$$\mathbf{A}\boldsymbol{\beta}^* = \mathbf{r}$$

where $r_i = \sum w_k x_{ik} y_k$, $y_k = Y_k + (z_k - \mu_k)/(d\mu_k/dY_k)$ and $\boldsymbol{\beta}^* = \boldsymbol{\beta} + \delta\boldsymbol{\beta}$. □

Starting Method

In practice we can obtain a good starting procedure for iteration as follows: take as a first approximation $\mu = z$ and calculate Y from it; then calculate w as before and set $y = Y$. Then obtain the first approximation to the β's by regression. The method may need slight modification to deal with extreme values of z. For instance, with the binomial distribution it will probably be adequate to replace instances of $z = 0$ or $z = n$ with $z = \frac{1}{2}$ and $z = n - \frac{1}{2}$, where, e.g. with the probit and logit transformations, $\mu = 0$ or $\mu = n$ would lead to infinite values for Y.

2.2. Sufficient Statistics

An important special case occurs when θ, the parameter of the distribution of the random element, and Y the predicted value of the linear model, coincide. Then

$$L = zY - g(Y) + h(z),$$

and, using (3),

$$\partial L/\partial \beta_i = \alpha(\phi)(z - \mu)x_i.$$

The maximum likelihood equations are then of the form $\sum_k (z - \hat{\mu})x_{ik} = 0$, the summation being over the observations. Hence we have

$$\sum_k z_k x_{ik} = \sum_k \hat{\mu}_k x_{ik}. \tag{10}$$

For a qualitative independent variate, this implies that the fitted marginal totals with respect to that variate will be equal to the observed ones.

From the expression for L we see that the quantities $\sum_k z_k x_{ik}$ are a set of sufficient statistics. Also, in (8) $d^2\theta/dY^2 = 0$ and so

$$\frac{\partial^2 L}{\partial \beta_i \partial \beta_j} = E\left(\frac{\partial^2 L}{\partial \beta_i \partial \beta_j}\right) = -\alpha(\phi)\left\{\left(\frac{d\mu}{dY}\right)^2 \Big/ V\right\} x_i x_j. \tag{11}$$

When θ is also the mean of the distribution, i.e. $\mu = \theta = Y$, we have the usual linear model with normal errors, for $g'(\theta) = \theta$ gives

$$g(\theta) = \tfrac{1}{2}\theta^2 + \text{const}$$

which uniquely determines the distribution as Normal with variance $1/\alpha(\phi)$ (using Theorem 1 of Patil and Shorrock, 1965). The sub-class of models for which there are sufficient statistics was noted by Cox (1968), and Dempster (1971) has extended it to include many dependent variates.

2.3. The Analysis of Deviance

A linear model is said to be *ordered* if the fitting of the β's is to be done in the same sequence as their declaration in the model. Ordering (or partial ordering) may be implied by the structure of the model; for instance it makes no sense to fit an interaction term $(ab)_{ij}$ before fitting the corresponding main effects a_i and b_j. It may also be implied by the objectives of the fitting, i.e. if a trend must be removed first before the fitting of further effects. More commonly, however, the ordering is to some extent arbitrary, and this gives rise to difficult problems of inference which we shall not try to tackle here. For ease of exposition of the basic ideas we shall assume that the model under consideration is ordered, and will be fitted sequentially a term at a time. The objectives of the fitting will be to assess how many terms are required for an adequate description of the data, and to derive the associated estimates of the parameters and their information matrix.

Two extreme models are conceivable for any set of data, the *minimal model* which contains the smallest set of terms that the problem allows, and the *complete model* in which all the Ys are different and match the data completely so that $\hat{\mu} = z$. An extreme case of the minimal model is the null model, which is equivalent to fitting the grand mean only and effectively consigns all the variation in the data to the random component of the model, while the complete model fits exactly and so consigns all the variation in the data to the systematic part. The model-fitting process with an ordered model thus consists of proceeding a suitable distance from the minimal model towards the complete model. At each stage we trade increasing goodness-of-fit to the current set of data against increasing complexity of the model. The fitting of the parameters at each stage is done by maximizing the likelihood for the current model and the matching of the model to the data will be measured quantitatively by the quantity $-2L_{\max}$ which we propose to call the *deviance*. For the four special distributions the deviance takes the form:

Normal	$\sum (z - \hat{\mu})^2/\sigma^2$,
Poisson	$2\{\sum z \ln(z/\hat{\mu}) - \sum (z - \hat{\mu})\}$,
Binomial	$2[\sum z \ln(z/\hat{\mu}) + \sum \{(n - z) \ln(n - \hat{\mu})\}]$,
Gamma	$2p\{-\sum \ln(z/\hat{\mu}) + \sum (z - \hat{\mu})/\hat{\mu}\}$.

Note that the deviance is measured from that of the complete model, so that terms involving constants, the data alone, or the scale factor alone are omitted. The second term in the expressions for the Poisson and gamma distribution is commonly identically zero (see Appendix for conditions and proof).

Associated with each model is a quantity r termed *the degrees of freedom* which is given by the rank of the X matrix, or equivalently the number of linearly independent parameters to be estimated. For a sample of n independent observations, the deviance for the model has residual degrees of freedom $(n - r)$. The degrees of freedom, multiplied where necessary by a scale factor, form a scale for a set of sequential models with which deviances can be compared; when (residual degrees of freedom $\times$ scale factor) is approximately equal to the deviance of the current model then it is unlikely that further

Table 1 Deviances and Their Differences.

Model term	Deviance	Difference	Component
Minimal	d_m		
		$d_m - d_A$	A
A	d_A		
		$d_A - d_{AB}$	B eliminating A
B	d_{AB}		
		$d_{AB} - d_{ABC}$	C eliminating A and B
C	d_{ABC}		
		$d_{ABC} - d_0$	Residual
Complete	d_0		

fitting of systematic components is worth while. The scale factor may be known (e.g. unity for the Poisson distribution) or unknown (e.g. for the normal distribution with unknown variance). If unknown it may be estimable directly, e.g. by replicate observations, or indirectly from the deviance after an adequate model has been fitted. The adequacy of the model may be determined by plotting successive deviances against their degrees of freedom, and accepting as a measure of the scale factor the linear portion through the origin determined by those points with fewest degrees of freedom.

2.4. The Generalization of Analysis of Variance

The first differences of the deviances for the normal distribution are (apart from a scale factor) the sums of squares in the analysis of variance for a sequential fit as shown for a three-term model in Table 1.

The generalized analysis of variance for a sequential model is now defined to have components given by the first differences of the deviance, with degrees of freedom defined as above. These components have distributions proportional to χ^2, exactly for normal errors, approximately for others. Such a generalization of the analysis of variance was suggested by Good (1967).

3. Special Distributions

3.1. The Normal Distribution

Here, we have

$$L = \sigma^{-2}(z\mu - \tfrac{1}{2}\mu^2 - \tfrac{1}{2}z^2) - \tfrac{1}{2}\ln\sigma^2$$

and in the notation of Section 1.2

$$\mu = \theta, \qquad V = 1.$$

Inverse polynomials provide an example where we assume that the observations u are normal on the log scale and the systematic effects additive on

Table 2. Fisher's Tuberculin-Test Data.

Sites	Cow class				Treatments			
	I	III	II	IV				
3 + 6	454	249	349	249	A	B	C	D
4 + 5	408	322	312	347	B	A	D	C
1 + 8	523	268	411	285	C	D	A	B
2 + 7	364	283	266	290	D	C	B	A

the inverse scale. Then

$$z = \ln u \qquad \text{and} \qquad Y = e^{\mu}.$$

Nelder (1966) gives examples of inverse polynomials calculated using the first approximation of the method in this paper.

More generally, as shown in Nelder (1968), we can consider models in which there is a linearizing transformation f and a normalizing transformation g. This means that if the observations are denoted by u, then $g(u)$ is normally distributed with mean μ and constant variance σ^2 and $f\{g^{-1}(\mu)\} = \sum \beta_i x_i$.

Then we have $V = 1$ and $Y = f\{g^{-1}(\mu)\}$ so that

$$w = [(d/dY)g\{f^{-1}(Y)\}]^2$$

and

$$y = Y + \{g(u) - \mu\}/[(d/dY)g\{f^{-1}(Y)\}].$$

Example (*Fisher's Tuberculin-Test Data*). Fisher (1949) published 16 measurements of tuberculin response which were classified by three four-level factors in a Latin-square type of arrangement as in Table 2.

Fisher gave reasons for believing that the variances of the observations were proportional to their expectation, and that the systematic part of the model was linear on the log-scale. The treatments were:

B Standard single

A Standard double

D Weybridge half

C Weybridge single

and these were treated as a 2×2 factorial arrangement, no interaction being fitted.

If the data had been Poisson observations the maximum likelihood estimates would have had the property that the marginal totals of the fitted values (on the untransformed scale) would be equal to the marginal totals of the observations. Although the observations were not, in fact, counts but measure-

Table 3. Tuberculin Effects.

	Fisher's result	Our result
B	0.0000	0.0000
A	0.2089	0.2092
D	0.0019	0.0023
C	0.2108	0.2115

ments in millimetres, Fisher decided to estimate the effects as if they were Poisson observations. He produced approximations to the effects by a method which made use of features of the particular Latin square, and then verified that these gave fitted values with margins approximately equal to the observed ones.

Another approach would be to treat the square roots of the observations as normally distributed with variance $\sigma^2/4$. Then we have $z = \sqrt{u}$ where u is an observation and

$$Y = 2 \ln \mu.$$

Fisher gave estimates of effects on the log scale relative to B. We produced estimates by the square-root/logarithmic method just described, and the two sets of estimates are given in Table 3.

The method of this paper can also be used to analyse the data as if they were Poisson observations. The estimates of effects obtained by this method agree with our other estimates to about four decimal places.

3.2. The Poisson Distribution

Here $L = z \ln \mu - \mu$ so we have $\theta = \ln \mu$ and $V = \mu$.

When $Y = \ln \mu$ there are sufficient statistics and a unique maximum likelihood estimate of β, provided it is finite. It will always be finite if there are no zero observations.

If $Y = \mu^\lambda$ $(0 < \lambda < 1)$, $L \to -\infty$ as $|\boldsymbol{\beta}|$ and hence L must have a maximum for finite $\boldsymbol{\beta}$. Also

$$\frac{\partial^2 L}{\partial \beta_i \partial \beta_j} = \sum \frac{\partial^2 L}{\partial Y^2} x_i x_j.$$

It is easily verified that $\partial^2 L/\partial Y^2 < 0$ and hence L is negative definite. It follows that $\hat{\boldsymbol{\beta}}$ is uniquely determined. When $Y = \mu$ the same result holds provided that the x's are linearly independent when units with $z = 0$ are excluded.

The main application of generalized linear models with Poisson errors is to contingency tables. These arise from data on counts classified by two or more factors, and the literature on them is enormous (see, for example,

Table 4. Disturbed Dreams Data.

Age in years	Rating 4	3	2	1	Total
5–7	7	3	4	7	21
8–9	13	11	15	10	49
10–11	7	11	9	23	50
12–13	10	12	9	28	59
14–15	3	4	5	32	44
Total	40	41	42	100	223

Simpson, 1951; Ireland and Kullback, 1968; Chapter 8 of Kullback 1968; Ku *et al.*, 1971).

Probabilistic models for contingency tables are built on assumptions of a multi-nomial distribution or a set of multinomial distributions. As Birch (1963) has snown, the estimation of a set of independent multinomial distributions is equivalent to the estimation of a set of independent Poisson distributions, and in what follows we shall regard a contingency table as a set of independent Poisson distributions.

The systematic part of models of contingency tables is usually multiplicative, and thus gives sufficient statistics with Poisson errors. The model terms usually correspond to qualitative x's, the equivalent of constant fitting, but quantitative terms occur naturally when the classifying factors have an underlying quantitative basis.

Example (A Contingency Table). Maxwell (1961, pp. 70–72) discusses the analysis of a 5×4 contingency table giving the number of boys with four different ratings for disturbed dreams in five different age groups. The data are given in Table 4. The higher the rating the more the boy suffers from disturbed dreams.

Here we can fit main effects and a linear × linear interaction using an x-variate of the form uv where $u = -2, -1, 0, 1$ or 2 according to the age group and v is the rating for disturbed dreams.

The estimated linear × linear interaction is -0.205.

We can form an analysis of deviance thus (regarding main effects as fixed):

Model term(s)	Deviance	Difference	Degrees of freedom
Minimal (i.e. main effects only)	32.46		
		18.38	1
Linear × linear	14.08		
		14.08	11
Complete	0		

Treating 18.38 and 14.08 as χ^2 variates with 1 and 11 degrees of freedom respectively we find that 18.38 is large while 14.08 is close to expectation. We conclude that the data are adequately described by a negative linear × linear interaction (indicating that the dream rating tends to decrease with age).

Maxwell, using the method of Yates (1948), obtained a decomposition of a Pearson χ^2 as follows:

Source of variation	Degrees of freedom	χ^2
Due to linear regression	1	17.94
Due to departure from linear regression	11	13.73
Total	12	31.67

Maxwell's values of χ^2 are clearly quite close to ours and his conclusions are essentially the same.

3.3. The Binomial Distribution

We re-write the usual form

$$L = r \ln p + (n - r) \ln q$$

as

$$\begin{aligned} L &= z \ln(\mu/n) + (n - z) \ln\{(1 - (\mu/n)\} \\ &= z \ln\{\mu/(n - \mu)\} + n \ln(n - \mu) + \text{terms in } n, \end{aligned}$$

i.e. we put z for r, and $\mu = E(z) = np$. Thus

$$\theta = \ln\{\mu/(n - \mu)\}$$

and

$$V = \mu(n - \mu)/n.$$

Sufficient statistics are provided by the logit transformation giving

$$\mu = ne^y/(1 + e^y).$$

Probit Analysis

We put

$$\mu = n\Phi(Y),$$

where Φ is the cumulative normal distribution function. There are no sufficient statistics here; the analysis via iterative weighted least-squares is well known (Finney, 1952).

Fitting Constants on a Logit Scale

This technique was introduced by Dyke and Patterson (1952) and is applied to multiway tables of proportions. It is a special case of the logistic transformation when all the x's are qualitative, and yields as sufficient statistics the total responding ($\sum z$) for each relevant margin.

The logit analogue of probit analysis is, of course, formally identical, with quantitative rather than qualitative x's. Again arbitrary mixtures of x types do not introduce anything new. Models based on the logistic transform have been extensively developed by Cox (1970).

From the results of Birch (1963) mentioned above, it follows that models with independent binomial data are equivalent to models with independent Poisson data. Bishop (1969) showed that a binomial model that is additive on the logit scale can be treated as a Poisson model additive on the log scale.

3.4. The Gamma Distribution

For the gamma distribution

$$L = -p(z/\mu + \ln \mu - \ln z) - \ln z.$$

We have $\theta = -1/\mu$ and $V = d\mu/d\theta = \mu^2$. There are sufficient statistics when $Y = 1/\mu$.

Thus the inverse transformation to linearity is related to the gamma distribution, as the logarithmic to the Poisson, or the identity transformation to the normal. The corresponding models seem not to have been explored.

One application of linear models involving the gamma distribution is the estimation of variance components. Here we have sums of squares which are proportional to χ^2 variates and the expectation of each is a linear combination of several variances which are to be estimated. It is better to write

$$L = -(zv/2\mu) - (v/2) \ln \mu + \{(v/2) + 1\} \ln z,$$

where v is the degrees of freedom of z; then putting $\theta = -v/2\mu$ we have $V = d\mu/d\theta = 2\mu^2/v$ and $y = z$, $Y = \mu$ and $w = v/2\mu^2$.

The deviance takes the form

$$\sum v[-\ln(z/\hat{\mu}) + \{(z - \hat{\mu})/\hat{\mu}\}]$$

and the result of the Appendix gives $\sum\{v(z - \hat{\mu})/\hat{\mu}\} = 0$; hence the deviance simplifies to

$$\sum v \ln(\hat{\mu}/z).$$

(*Editor's note*: The Example has been omitted.)

4. The Models in the Teaching of Statistics

We believe that the generalized linear models here developed could form a useful basis for courses in statistics. They give a consistent way of linking together the systematic elements in a model with the random elements. Too often the student meets complex systematic linear models only in connection with normal errors, and if he encounters probit analysis this may seem to have little to do with the linear regression theory he has learnt. By isolating the systematic linear component the student can be introduced to multiway tables and their margins, additivity, weighting, quantitative and qualitative independent variates, and transformations, quite independently of the added complications of errors and associated probability distributions. The essential unity of the linear model, encompassing qualitative, quantitative and mixed independent variates can be brought out and the introduction of qualitative variates brings in naturally the ideas of singularity of matrices and of constraints.

The complementary set of probability distributions would be introduced in the usual way, including the use of transformations of data to attain desirable properties of the errors. The difficult problem of discussing how far transformations can produce both linearity and normality simultaneously now disappears because the models allow two different transformations to be used, one to induce the linearity of the systematic component and one to induce the desired distribution in the error component. (Note that this distribution need not necessarily be equal-variance normal.)

The systematic use of log-likelihood-ratios (or, equivalently, differences in deviance) extends the ideas of analysis of variance to other distributions and produces an additive decomposition for the sequential fit of the model. To appreciate the simplicity that this can produce it is only necessary to look at the algebraic complexities arising from the attempts to analyse contingency tables by extensions of the Pearson χ^2 approach. Irwin (1949), Lancaster (1949, 1950) and Kimball (1954) have given modifications of the usual formulae for the Pearson χ^2 in contingency tables, to produce an additive decomposition of χ^2 when the table is partitioned. These formulae are much more complicated than the usual one for Pearson χ^2. However, the equivalent likelihood statistic, namely

$$2\left\{\sum_{ij} n_{ij} \ln n_{ij} - \sum_{i} n_{i} \ln n_{i.} - \sum_{i} n_{.j} \ln n_{.j} + n_{..} \ln n_{..}\right\}$$

does not need any modification to make it additive and is easy to compute as tables of $n \ln n$ are available (Kullback, 1968).

Finally, the fact that a single algorithm can be used to fit any of the models implies that quite a small set of routines can provide the basic computing facility to allow students to fit models to a wide range of data. They can get experience of programming by writing special routines to deal with special forms of output required by particular models, such as the LD50 of probit analysis. In this way the distinction can be made between the model-fitting part of an analysis and the subsequent derived quantities (with estimates of their uncertainties) which particular problems require.

We hope that the approach developed in this paper will prove to be a useful way of unifying what are often presented as unrelated statistical procedures, and that this unification will simplify the teaching of the subject to both specialists and non-specialists.

References

Birch, M.W. (1963). Maximum likelihood in three-way contingency tables. *J.R. Statist. Soc.* B, **25**, 220–233.

Bishop, Y.M.M. (1969). Full contingency tables, logits, and split contingency tables. *Biometrics*, **25**, 383–399.

Cox, D.R. (1968). Notes on some aspects of regression analysis. *J.R. Statist. Soc.* A, **131**, 265–279.

——— (1970). *Analysis of Binary Data*. London: Methuen.

Dempster, A.P. (1971). An overview of multivariate data analysis. *J. Multiuar. Anal.*, **1**, 316–346.

Dyke, G.V. and Patterson, H.D. (1952). Analysis of factorial arrangements when the data are proportions. *Biometrics*, **8**, 1–12.

Finney, D.J. (1952). *Probit Analysis*. 2nd ed. Cambridge: University Press.

Fisher, R.A. (1949). A biological assay of tuberculosis. *Biometrics*, **5**, 300–316.

Good, I.J. (1967). Analysis of log-likelihood ratios, "ANOΛ". (A contribution to the discussion of a paper on least squares by F.J. Anscombe.) *J.R. Statist. Soc.* B, **29**, 39–42.

Ireland, C.T. and Kullback, S. (1968). Contingency tables with given marginals. *Biometrika*, **55**, 179–188.

Irwin, J.O. (1949). A note on the subdivision of χ^2 into components. *Biometrika*, **36**, 130–134.

Kendall, M.G. and Stuart, A. (1967). *The Advanced Theory of Statistics*, 2nd ed, Vol. II. London: Griffin.

Kimball, A.W. (1954). Short-cut formulas for the exact partition of χ^2 in contingency tables. *Biometrics*, **10**, 452–458.

Ku, H.H., Varner, R.N. and Kullback, S. (1971). On the analysis of multidimensional contingency tables. *J. Amer. Statist. Ass.*, **66**, 55–64.

Kullback, S. (1968). *Information Theory and Statistics*. 2nd ed. New York: Dover.

Lancaster, H.O. (1949). The derivation and partition of χ^2 in certain discrete distributions. *Biometrika*, **36**, 117–128.

——— (1950). The exact partition of χ^2 and its application to the problem of the pooling of small expectations. *Biometrika*, **37**, 267–270.

Lehmann, E.L. (1959). *Testing Statistical Hypotheses*. London: Wiley.

Maxwell, A.E. (1961). *Analysing Qualitative Data*. London: Methuen.

Nelder, J.A. (1966). Inverse polynomials, a useful group of multi-factor response functions. *Biometrics*, **22**, 128–141.

——— (1968). Weighted regression, quantal response data and inverse polynomials. *Biometrics*, **24**, 979–985.

Patil, G.P. and Shorrock, R. (1965). On certain properties of the exponential-type families. *J. R. Statist. Soc.* B, **27**, 94–99.

Simpson, C.H. (1951). The interpretation of interaction in contingency tables. *J.R. Statist. Soc.* B, **13**, 238–241.

Yates, F. (1940). The recovery of inter-block information in balanced incomplete block designs. *Ann. Eugen.*, **10**, 317–325.
——— (1948). The analysis of contingency tables with groupings based on quantitative characters. *Biometrika*, **35**, 176–181.
——— (1970). *Experimental Design: Selected Papers*. London: Griffin.

Appendix

We can sometimes simplify the expression for deviance. In what follows, summation is to be taken to be over the observations unless otherwise stated.

Theorem. *If either* (a) $Y = \mu^\alpha$ *or* (b) $Y = \log \mu$ *and a constant term is being fitted in the model* (*say* $Y = \sum \beta_i x_i$ *where* x_0 *takes the constant value* 1) *then* $\sum\{(z - \hat{\mu})\hat{\mu}/\hat{V}\} = 0$, *where* $\hat{\mu}$ *and* V *are the maximum likelihood estimates of* μ *and* V.

Proof. For case (b)

$$\frac{\partial L}{\partial \beta_0} = \sum \frac{(z - \mu)\mu}{V} = 0.$$

In case (a), we have

$$\frac{\partial L}{\partial \beta_i} = \frac{z - \mu}{V} \frac{d\mu}{dY} x_i = \sum \frac{z - \mu}{V} \frac{\mu}{\alpha Y} x_i.$$

Hence

$$\sum \frac{(z - \hat{\mu})\hat{\mu}}{\hat{V}\hat{Y}} x_i = 0.$$

Then

$$\sum \frac{(z - \hat{\mu})\hat{\mu}}{\hat{V}} = \sum \frac{\hat{Y}(z - \hat{\mu})\hat{\mu}}{\hat{V}\hat{Y}}$$

$$= \sum \sum_i \beta_i x_i \frac{(z - \hat{\mu})\hat{\mu}}{\hat{V}\hat{Y}} = \sum_i \left\{\beta_i \sum \frac{(z - \hat{\mu})\hat{\mu}}{\hat{V}\hat{Y}} x_i\right\} = 0.$$

This completes the proof. □

When the conditions of this theorem are satisfied, we have, for the Poisson distribution

$$\sum (z - \hat{\mu}) = 0$$

and for the gamma distribution

$$\sum \frac{(z - \hat{\mu})}{\hat{\mu}} = 0.$$

Then the deviances for these distributions simplify as follows:

$$\text{Poisson} \quad 2\sum z \ln(z/\hat{\mu}),$$

$$\text{Gamma} \quad -2\sum \ln(z/\hat{\mu}).$$

Introduction to Efron (1979) Bootstrap Methods: Another Look at the Jackknife

Rudolf J. Beran
University of California at Berkeley

It is not unusual, in the history of statistics, that an important paper go scarcely noticed for a decade or longer. Examples from the past half-century include von Mises' (1937, 1947) papers on statistical functionals, Quenouille's (1949) paper, Tukey's (1956) abstract on the jackknife, and Wald's (1943) paper on the asymptotic optimality of likelihood ratio tests. Each of these pioneering works was well ahead of its time. Brad Efron's (1979) paper on the bootstrap sparked immediate interest among his peers. A decade after its publication, the bootstrap literature is large and still growing, with no immediate end in sight. Surely, the timing and formulation of Efron's paper were just right. But what were the yearnings in the statistical world of 1979 that the paper touched so well? Why did development of the bootstrap idea follow so swiftly?

I would suggest that statistical perceptions in 1979 were influenced by four historical developments. First, by the late 1970s, the revolution in computing, and subsequently in data analysis, had put theoretical statistics on the defensive. It was becoming increasingly clear that the classical formulations of statistical theory, whether frequentist or Bayesian, did not provide a realistic paradigm for the analysis of large data sets. One response was growing theoretical interest in the jackknife, cross-validation, and certain other resampling schemes [see references in Efron (1982)]. These were all methods that seemed to rely on direct internal examination of the data, rather than on fitting an externally conceived statistical model.

Second, some data analysts, not all professional statisticians, had been experimenting in the 1960s and 1970s with Monte Carlo simulations from fitted models as a means of generating plausible critical values for confidence statements or tests. Examples include Williams (1970) and two astrophysical papers from 1976 cited in Press et al. (1986, Sec. 14.5). Such direct simulation approaches were a natural response to the increased availability of inexpensive computing. However, they were supported more by intuition than by logical analysis and appeared mostly in papers published outside the mainstream statistical journals.

Third, another response to the problems of data analysis was the theory of robust estimation. By the late 1970s, theoretical workers in robust statistics and nonparametrics were familiar with estimates as statistical functionals—that is, as functions of the empirical cdf. The first derivatives of statistical functionals were being calculated to assess their robustness and to find their asymptotic distributions [see Huber (1981, Chap. 1)]. This background prepared the way for subsequent interpretation and analysis of a bootstrap distribution as a statistical functional.

Fourth, the probability theory of weak convergence had continued to develop through the 1970s. Studying optimality or robustness in large samples of various parametric and nonparametric procedures had required statisticians to use weak convergence results for triangular arrays [see the references in Ibragimov and Has'minskii (1981) and LeCam (1986)]. Studying second-order asymptotic efficiency had brought new life and insight to the theory of Edgeworth expansions [refer to Bickel (1974) and Bhattacharya and Ghosh (1978)]. These statistically motivated developments in probability theory became crucial tools in the analysis of bootstrap procedures that started immediately after the publication of Efron's paper.

Bradley Efron was born in St. Paul, Minnesota in 1938. After undergraduate work at the California Institute of Technology from 1956–60, he entered graduate school at Stanford University, eventually being offered a post in the department of statistics, where he has remained until the present day. The late 1960s and 1970s were a period of intense activity in the department, on the exploitation of newly available and increasingly powerful computing facilities. The present paper is among the more widely recognized of the resulting developments.

In the introduction to his paper, Efron wrote, "The paper proceeds by a series of examples, with little offered in the way of general theory." This statement errs on the side of modesty: The paper contains a wealth of insight, derivation and correct conjecture. In particular, the paper formulated the bootstrap idea as an intellectual object accessible to theoretical study and gave it a name (Sec. 2). It introduced the interpretation of a bootstrap distribution as a conditional distribution given the sample (Sec. 2 and Remark G in Sec. 8). It described the natural Monte Carlo approach to approximating such conditional distributions (Sec. 2). It pointed to the validity of the nonparametric bootstrap for the sample median, in contrast to the jackknife (Sec. 3). It indicated that jackknife procedures might often be seen as approximations to conceptually simpler bootstrap methods (Sec. 5). In specific two-sample situations, it illustrated the possible validity of nonparametric bootstrap estimates for bias and variance (Sec. 6) and proposed a bootstrap competitor to the cross-validation estimate of error rate in discriminant analysis (Sec. 4). It made a good start on bootstrap procedures for regression models, mentioning the key idea of recentering regression residuals before bootstrapping (Sec. 7). It proved, for finite sample spaces, the first theorem on consistency of the nonparametric bootstrap distribution as an estimate of an unknown sampling distribution (Remark G in section 8). It suggested, without quite pinning down the idea, that a nearly pivotal quantity would bootstrap more successfully than a less pivotal quantity (Sec. 2 and Remark F in Sec. 8). And it made provocative comments about the construction of bootstrap confidence intervals (Remark D in Sec. 8). Formal development of Efron's intuitive insights in this paper took nearly a decade and required substantial technical innovation.

As I write these words in early 1990, it seems that the first era of bootstrap exploration is over. Theoretical statisticians have proved the consistency of bootstrap estimates and derived their asymptotic distributions for a large variety of interesting models. In classical settings, we have Edgeworth expansion approximations to bootstrap distributions. We understand several situations where straightforward bootstrapping fails to yield trustworthy estimates. We are coming close to probabilistic characterizations of when the straightforward bootstrap can succeed. We understand clearly that the bootstrap, the jackknife, cross-validation, and other resampling schemes are, in fact, model-dependent, just like any other statistical procedure. A merit of the bootstrap is that the model enters explicitly into the definition of the bootstrap algorithm, thus encouraging thought on this point. In many situations, properly designed bootstrap estimates are asymptotically efficient, whether in the local asymptotic minimax sense or in the sense of the Hájek convolution representation. This efficiency property is retained by suitable jackknife or cross-validation approximations to bootstrap estimates.

What major insights have statisticians gained beyond those proved or conjectured in Efron's paper? Of the questions raised in the paper and now better understood, the problem of constructing confidence sets with accurate coverage probabilities has been most challenging. In asymptotically normal situations, Efron's (1982) percentile method, foreshadowed by the example in Remark D of Sec. 8, turns out not to be best. His later BC_a method makes the bias and skewness adjustments required by the asymptotics, at the cost of mixing some complicated algebra with the elegance of the pure bootstrap approach. Other workers have shown that the bootstrap-t method has the same asymptotic behavior as the BC_a method [see Hall (1988) for a comparative study of various bootstrap confidence intervals]. Interestingly, normal model bootstrap-t is asymptotically equivalent and numerically close, in the Behrens–Fisher problem, to Welch's famous analytical solution [Beran (1988)]. For small samples, one can find examples where either bootstrap-t or BC_a seems to dominate the other. Double bootstrap methods offer another new route to improving coverage probability of confidence sets, particularly those for multivariate parameters.

The most alluring aspect of the bootstrap is its substitution of feasible computer experiments for analytically difficult distribution theory. Of the questions not raised explicitly in Efron's paper but now reasonably well-understood, the problem of bootstrapping density estimates or nonparametric regression estimates has been especially fruitful. For example, recent research has established the validity of certain bootstrap methods for asymptotically optimal bandwidth selection in density estimation. This theoretical advance neatly confirms the conjectural findings of a few earlier numerical experiments. It seems likely now that valid bootstrap methods exist for other inverse problems where the sample size is much smaller than the dimension of the parameter space.

The distribution theory behind the chi-squared test, the t-test, and the F-test, all developed early in this century, created modern statistics and gave probability theory an important role in our discipline. The computational need to stay with statistical procedures that use tabulated distributions surely shaped research for several decades thereafter. Efron's 1979 paper on the bootstrap, the research it has inspired, and recent advances in the study of chaotic systems have revealed intriguing new possibilities for statistical thought in the computer age. Our exploration of this brave new world has scarcely begun.*

References

Beran, R. (1988). Prepivoting test statistics: a bootstrap view of asymptotic refinements, *J. Amer. Statist. Assoc.*, **83**, 687–697.

Bhattacharya, R.N., and Ghosh J.K. (1978). On the validity of the formal Edgeworth expansion, *Ann. Statist.*, **6**, 435–451.

Bickel, P.J. (1974). Edgeworth expansions in nonparametric statistics, *Ann. Statist.*, **2**, 1–20.

Efron, B. (1982). *The Jackknife, the Bootstrap and Other Resampling Plans.* SIAM, Philadelphia, Pa.

Hall, P. (1988). Theoretical comparison of bootstrap confidence intervals (with discussion), *Ann. Statist.*, **16**, 927–985.

Huber, P.J. (1981). *Robust Statistics.* Wiley, New York.

Ibragimov, I.A., and Has'minskii, R.Z. (1981). *Statistical Estimation: Asymptotic Theory.* Springer-Verlag, New York.

LeCam, L. (1986). *Asymptotic Methods in Statistical Decision Theory.* Springer-Verlag, New York.

Press, W.H., Flannery, B.P., Teukolsky, S.A., and Vetterling, W.T. (1986). *Numerical Recipes: The Art of Scientific Computing.* Cambridge University Press, New York.

Quenouille, M. (1949). Approximate tests of correlation in time series, *J. Roy. Statist. Soc., Ser. B.*, **11**, 18–84.

Tukey, J. (1956). Bias and confidence in not quite large samples, abstract, *Ann. Math. Statist.*, **29**, 614.

von Mises, R. (1937). Sur les fonctions statistiques, in *Conférence de la Réunion Internationale des Mathématiciens.* Gauthiers-Villars, Paris.

von Mises, R. (1947). On the asymptotic distribution of differentiable statistical functions, *Ann. Math. Statist.*, **18**, 309–348.

Wald, A. (1943). Tests of statistical hypotheses concerning several parameters when the number of observations is large, *Trans. Amer. Math. Soc.*, **54**, 426–482.

Williams, D.A. (1970). Discrimination between regression models to determine the pattern of enzyme synthesis in synchronous cell cultures, *Biometrics*, **28**, 23–32.

* *Editor's note*: A Russian translation of some of Efron's papers on Bootstrap appeared in 1987.

Bootstrap Methods: Another Look at the Jackknife

Bradley Efron
Stanford University

Abstract

We discuss the following problem given a random sample $\mathbf{X} = (X_1, X_2, \ldots, X_n)$ from an unknown probability distribution F, estimate the sampling distribution of some prespecified random variable $R(\mathbf{X}, F)$, on the basis of the observed data $\mathbf{x}$. (Standard jackknife theory gives an approximate mean and variance in the case $R(\mathbf{X}, F) = \theta(\hat{F}) - \theta(F)$, θ some parameter of interest.) A general method, called the "bootstrap", is introduced, and shown to work satisfactorily on a variety of estimation problems. The jackknife is shown to be a linear approximation method for the bootstrap. The exposition proceeds by a series of examples: variance of the sample median, error rates in a linear discriminant analysis, ratio estimation, estimating regression parameters, etc.

1. Introduction

The Quenouille-Tukey jackknife is an intriguing nonparametric method for estimating the bias and variance of a statistic of interest, and also for testing the null hypothesis that the distribution of a statistic is centered at some prespecified point. Miller [14] gives an excellent review of the subject.

This article attempts to explain the jackknife in terms of a more primitive method, named the "bootstrap" for reasons which will become obvious. In

Received June 1977; revised December 1977.

principle, bootstrap methods are more widely applicable than the jackknife, and also more dependable. In Section 3, for example, the bootstrap is shown to (asymptotically) correctly estimate the variance of the sample median, a case where the jackknife is known to fail. Section 4 shows the bootstrap doing well at estimating the error rates in a linear discrimination problem, outperforming "cross-validation," another nonparametric estimation method.

We will show that the jackknife can be thought of as a linear expansion method (i.e., a "delta method") for approximating the bootstrap. This helps clarify the theoretical basis of the jackknife, and suggests improvements and variations likely to be successful in various special situations. Section 3, for example, discusses jackknifing (or bootstrapping) when one is willing to assume symmetry or smoothness of the underlying probability distribution. This point reappears more emphatically in Section 7, which discusses bootstrap and jackknife methods for regression models.

The paper proceeds by a series of examples, with little offered in the way of general theory. Most of the examples concern estimation problems, except for Remark F of Section 8, which discusses Tukey's original idea for t-testing using the jackknife. The bootstrap results on this point are mixed (and won't be reported here), offering only slight encouragement for the usual jackknife t tests.

John Hartigan, in an important series of papers [5, 6, 7], has explored ideas closely related to what is called bootstrap "Method 2" in the next section, see Remark I of Section 8. Maritz and Jarrett [13] have independently used bootstrap "Method 1" for estimating the variance of the sample median, deriving equation (3.4) of this paper and applying it to the variance calculation. Bootstrap "Method 3," the connection to the jackknife via linear expansions, relates closely to Jaeckel's work on the infinitesimal jackknife [10]. If we work in a parametric framework, this approach to the bootstrap gives Fisher's information bound for the asymptotic variance of the maximum likelihood estimator, see Remark K of Section 8.

2. Bootstrap Methods

We discuss first the one-sample situation in which a random sample of size n is observed from a completely unspecified probability distribution F,

$$X_i = x_i, \qquad X_i \sim_{\text{ind}} F \qquad i = 1, 2, \ldots, n. \tag{2.1}$$

In all of our examples F will be a distribution on either the real line or the plane, but that plays no role in the theory. We let $\mathbf{X} = (X_1, X_2, \ldots, X_n)$ and $\mathbf{x} = (x_1, x_2, \ldots, x_n)$ denote the random sample and its observed realization, respectively.

The problem we wish to solve is the following. Given a specified random variable $R(\mathbf{X}, F)$, possibly depending on both $\mathbf{X}$ and the unknown distribution F, estimate the sampling distribution of R on the basis of the observed data $\mathbf{x}$.

Traditional jackknife theory focuses on two particular choices of R. Let $\theta(F)$ be some parameter of interest such as the mean, correlation, or standard deviation of F, and $t(\mathbf{X})$ be an estimator of $\theta(F)$, such as the sample mean, sample correlation, or a multiple of the sample range. Then the sampling distribution of

$$R(\mathbf{X}, F) = t(\mathbf{X}) - \theta(F), \tag{2.2}$$

or more exactly its mean (the bias of t) and variance, is estimated using the standard jackknife theory, as described in Section 5. The bias and variance estimates say $\widehat{\text{Bias}}(t)$ and $\widehat{\text{Var}}(t)$, are cleverly constructed functions of $\mathbf{X}$ obtained by recomputing $t(\cdot)$ n times, each time removing one component of $\mathbf{X}$ from consideration. The second traditional choice of R is

$$R(\mathbf{X}, F) = \frac{t(\mathbf{X}) - \widehat{\text{Bias}}(t) - \theta(F)}{(\widehat{\text{Var}}(t))^{1/2}}. \tag{2.3}$$

Tukey's original suggestion was to treat (2.3) as having a standard Student's t distribution with $n - 1$ degress of freedom. (See Remark F, Section 8.) Random variables (2.2), (2.3) play no special role in the bootstrap theory, and, as a matter of fact, some of our examples concern other choices of R.

The bootstrap method for the one-sample problem is extremely simple, at least in principle:

1. Construct the simple probability distribution $\hat{F}$, putting mass $1/n$ at each point $x_1, x_2, x_3, \ldots, x_n$.
2. With $\hat{F}$ fixed, draw a random sample of size n from $\hat{F}$, say

   $$X_i^* = x_i^*, X_i^* \sim_{\text{ind}} \hat{F} \qquad i = 1, 2, \ldots, n.$$

   Call this the *bootstrap sample*, $\mathbf{X}^* = (X_1^*, X_2^*, \ldots, X_n^*)$, $\mathbf{x}^* = (x_1^*, x_2^*, \ldots, x_n^*)$. Notice that we are not getting a permutation distribution since the values of $\mathbf{X}^*$ are selected *with* replacement from the set $\{x_1, x_2, \ldots, x_n\}$. As a point of comparison, the ordinary jackknife can be thought of as drawing samples of size $n - 1$ *without* replacement.
3. Approximate the sampling distribution of $R(\mathbf{X}, F)$ by the *bootstrap distribution* of

   $$R^* = R(\mathbf{X}^*, \hat{F}), \tag{2.5}$$

   i.e., the distribution of R^* induced by the random mechanism (2.4), with $\hat{F}$ held fixed at its observed value.

The point is that the distribution of R^*, which in theory can be calculated exactly once the data $\mathbf{x}$ is observed, equals the desired distribution of R if

$F = \hat{F}$. Any nonparametric estimator of R's distribution, i.e., one that does a reasonably good estimation job without prior restrictions on the form of F, must give close to the right answer when $F = \hat{F}$, since $\hat{F}$ is a central point amongst the class of likely F's, having observed $\mathbf{X} = \mathbf{x}$. Making the answer exactly right for $F = \hat{F}$ is *Fisher consistency* applied to our particular estimation problem.

Just how well the distribution of R^* approximates that of R depends upon the form of R. For example, $R(\mathbf{X}, F) = t(\mathbf{X})$ might be expected to bootstrap less successfully than $R(\mathbf{X}, F) = [t(\mathbf{X}) - E_F t]/(\mathrm{Var}_F\, t)^{1/2}$. This is an important question, related to the concept of pivotal quantities, Barnard [2], but is discussed only briefly here, in Section 8. Mostly we will be content to let the varying degrees of success of the examples speak for themselves.

As the simplest possible example of the bootstrap method, consider a probability distribution F putting all of its mass at zero or one, and let the parameter of interest be $\theta(F) = \mathrm{Prob}_F\{X = 1\}$. The most obvious random variable of interest is

$$R(\mathbf{X}, F) = \bar{X} - \theta(F) \qquad \bar{X} = (\Sigma_{i=1}^{n} X_i/n). \tag{2.6}$$

Having observed $\mathbf{X} = \mathbf{x}$, the bootstrap sample $\mathbf{X}^* = (X_1^*, X_2^*, \ldots, X_n^*)$ has each component independently equal to one with probability $\bar{x} = \theta(\hat{F})$, zero with probability $1 - \bar{x}$. Standard binomial results show that

$$R^* = R(\mathbf{X}^*, \hat{F}) = \bar{X}^* - \bar{x} \tag{2.7}$$

has mean and variance

$$E_*(\bar{X}^* - \bar{x}) = 0, \qquad \mathrm{Var}_*(\bar{X}^* - \bar{x}) = \bar{x}(1 - \bar{x})/n. \tag{2.8}$$

(Notations such as "E_*," "Var_*," "Prob_*," etc. indicate probability calculations relating to the bootstrap distribution of $\mathbf{X}^*$, with $\mathbf{x}$ and $\hat{F}$ fixed.) The implication that $\bar{X}$ is unbiased for θ, with variance approximately equal to $\bar{x}(1 - \bar{x})/n$, is universally familiar.

As a second example, consider estimating $\theta(F) = \mathrm{Var}_F\, X$, the variance of an arbitrary distribution on the real line, using the estimator $t(\mathbf{X}) = \Sigma_{i=1}^{n} (X_i - \bar{X})^2/(n - 1)$. Perhaps we wish to know the sampling distribution of

$$R(\mathbf{X}, F) = t(\mathbf{X}) - \theta(F). \tag{2.9}$$

Let $\mu_k(F)$ indicate the kth central moment of F, $\mu_k(F) = E_F(X - E_F X)^k$, and $\hat{\mu}_k = \mu_k(\hat{F})$ the kth central moment of $\hat{F}$. Standard sampling theory results, as in Cramér [3], Section 27.4, show that

$$R^* = R(\mathbf{X}^*, \hat{F}) = t(\mathbf{X}^*) - \theta(\hat{F})$$

has

$$E_* R^* = 0, \qquad \mathrm{Var}_* R^* = \frac{\hat{\mu}_4 - ((n - 3)/(n - 1))\hat{\mu}_2^2}{n}. \tag{2.10}$$

The approximation $\mathrm{Var}_F\, t(\mathbf{X}) \approx \mathrm{Var}_* R^*$ is (almost) the jackknife estimate for $\mathrm{Var}_F\, t$.

The difficult part of the bootstrap procedure is the actual calculation of the bootstrap distribution. Three methods of calculation are possible:

Method 1. Direct theoretical calculation, as in the two examples above and the example of the next section.

Method 2. Monte Carlo approximation to the bootstrap distribution. Repeated realizations of $\mathbf{X}^*$ are generated by taking random samples of size n from $\hat{F}$, say $\mathbf{x}^{*1}, \mathbf{x}^{*2}, \ldots, \mathbf{x}^{*N}$, and the histogram of the corresponding values $R(\mathbf{x}^{*1}, \hat{F}), R(\mathbf{x}^{*2}, \hat{F}), \ldots, R(\mathbf{x}^{*N}, \hat{F})$ is taken as an approximation to the actual bootstrap distribution. This approach is illustrated in Sections 3, 4 and 8.

Method 3. Taylor series expansion methods can be used to obtain the approximate mean and variance of the bootstrap distribution of R^*. This turns out to be the same as using some form of the jackknife, as shown in Section 5.

In Section 4 we consider a two sample problem where the data consists of a random sample $\mathbf{X} = (X_1, X_2, \ldots, X_n)$ from F and an independent random sample $\mathbf{Y} = (Y_1, Y_2, \ldots, Y_n)$ from G, F and G arbitrary probability distributions on a given space. In order to estimate the sampling distribution of a random variable $R((\mathbf{X}, \mathbf{Y}), (F, G))$, having observed $\mathbf{X} = \mathbf{x}$, $\mathbf{Y} = \mathbf{y}$, the one-sample bootstrap method can be extended in the obvious way: $\hat{F}$ and $\hat{G}$, the sample probability distributions corresponding to F and G, are constructed; bootstrap samples $X_i^* \sim \hat{F}$, $i = 1, 2, \ldots, m$, $Y_j^* \sim \hat{G}$, $j = 1, 2, \ldots, n$, are independently drawn; and finally the bootstrap distribution of $R^* = R((\mathbf{X}^*, Y^*), (\hat{F}, \hat{G}))$ is calculated, or use as an approximation to the actual distribution of R. The calculation of the bootstrap distribution proceeds by any of the three methods listed above. (The third method makes clear the correct analogue of the jackknife procedure for nonsymmetric situations, such as the two sample problem; see the remarks of Section 6.)

So far we have only used nonparametric maximum likelihood estimators, $\hat{F}$ and $(\hat{F}, \hat{G})$, to begin the bootstrap procedure. This isn't crucial, and as the examples of Sections 3 and 7 show, it is sometimes more convenient to use other estimates of the underlying distributions.

3. Estimating the Median

Suppose we are in the one-sample situation (2.1), with F a distribution on the real line, and we wish to estimate the median of F using the sample median. Let $\theta(F)$ indicate the median of F, and let $t(\mathbf{X})$ be the sample median,

$$t(\mathbf{X}) = X_{(m)}, \tag{3.1}$$

where $X_{(1)} \leqslant X_{(2)} \leqslant \cdots \leqslant X_{(n)}$ is the order statistic, and we have assumed an odd sample size $n = 2m - 1$ for convenience. Once again we take $R(\mathbf{X}, F) = t(\mathbf{X}) - \theta(F)$, and hope to say something about the sampling distribution of R on the basis of the observed random sample.

Having observed $\mathbf{X} = \mathbf{x}$, we construct the bootstrap sample $\mathbf{X}^* = \mathbf{x}^*$ as in (2.4). Let

$$N_i^* = \#\{X_i^* = x_i\}, \tag{3.2}$$

the number of times x_i is selected in the bootstrap sampling procedure. The vector $\mathbf{N}^* = (N_1^*, N_2^*, \ldots, N_n^*)$ has a multinomial distribution with expectation one in each of the n cells.

Denote the observed order statistic $x_{(1)} \leqslant x_{(2)} \leqslant x_{(3)} \leqslant \cdots \leqslant x_{(n)}$, and the corresponding N^* values $N_{(1)}^*, N_{(2)}^*, \ldots, N_{(n)}^*$. (Ties $x_i = x_{i'}$ can be broken by assigning the lower value of i, i' to the lower position in the order statistic.) The bootstrap value of R is

$$R^* = R(\mathbf{X}^*, \hat{F}) = X_{(m)}^* - x_{(m)}. \tag{3.3}$$

We notice that for any integer value l, $1 \leqslant l < n$,

$$\begin{aligned} \text{Prob}_*\{X_{(m)}^* > x_{(l)}\} &= \text{Prob}_*\{N_{(1)}^* + N_{(2)}^* + \cdots + N_{(l)}^* \leqslant m - 1\} \\ &= \text{Prob}\left\{\text{Binomial}\left(n, \frac{l}{n}\right) \leqslant m - 1\right\} \\ &= \Sigma_{j=0}^{m-1} \binom{n}{j} \left(\frac{l}{n}\right)^j \left(\frac{n-l}{n}\right)^{n-j}. \end{aligned} \tag{3.4}$$

Therefore

$$\begin{aligned} \text{Prob}_*\{R^* = x_{(l)} - x_{(m)}\} &= \text{Prob}\left\{\text{Binomial}\left(n, \frac{l-1}{n}\right) \leqslant m - 1\right\} \\ &\quad - \text{Prob}\left\{\text{Binomial}\left(n, \frac{l}{n}\right) \leqslant m - 1\right\}, \end{aligned} \tag{3.5}$$

a result derived independently by Maritz and Jarrett [13].

The case $n = 13$ $(m = 7)$ gives the following bootstrap distribution for R^*:

$l =$	2 or 12	3 or 11	4 or 10	5 or 9	6 or 8	7
(3.5) =	.0015	.0142	.0050	.1242	.1936	.2230.

(3.6)

For any given random sample of size 13 we can compute

$$E_*(R^*)^2 = \Sigma_{l=1}^{13} [x_{(l)} - x_{(7)}]^2 \,\text{Prob}_*\{R^* = x_{(l)} - x_{(7)}\}, \tag{3.7}$$

and use this number as an estimate of $E_F R^2 = E_F[t(\mathbf{X}) - \theta(F)]^2$, the expected squared error of estimation for the sample median. Standard asymptotic theory, applied to the case where F has a bounded continuous density $f(x)$, shows that as the sample size n goes to infinity, the quantity $nE_*(R^*)^2$ approaches $1/4f^2(\theta)$, where $f(\theta)$ is the density evaluated at the median $\theta(F)$. This is the correct asymptotic value, see Kendall and Stuart [11], page 237. The standard jackknife applied to the sample median gives a variance estimate which is not even asymptotically consistent (Miller [14], page 8, is incorrect on this point): $n\,\widehat{\text{Var}}(R) \to (1/4f^2(\theta))[\chi_2^2/2]^2$. The random variable $[\chi_2^2/2]^2$ has mean 2 and variance 20.

Suppose we happened to know that the probability distribution F was symmetric. In that case we could replace $\hat{F}$ by the symmetric probability distribution obtained from $\hat{F}$ by reflection about the median,

$$\begin{aligned}\hat{F}_{\text{SYM}}:\quad &\text{probability mass } \frac{1}{2n-1} \text{ at } x_{(1)}, x_{(2)}, \ldots, x_{(n)} \quad\text{and}\\ &2x_{(m)} - x_{(1)}, \ldots, 2x_{(m)} - x_{(n)}.\end{aligned} \tag{3.8}$$

This is not the nonparametric maximum likelihood estimator for F, but has similar asymptotic properties, see Hinkley [8]. Let $z_{(1)} \leqslant z_{(2)} \leqslant \cdots \leqslant z_{(2n-1)}$ be the ordered values appearing in the distribution of $\hat{F}_{\text{SYM}}$. The bootstrap procedure starting from $\hat{F}_{\text{SYM}}$ gives

$$\begin{aligned}\text{Prob}_*\{R^* = z_{(l)} - x_{(m)}\} &= \text{Prob}\left\{\text{Binomial}\left(n, \frac{l-1}{2n-1}\right) \leqslant m-1\right\}\\ &\quad - \text{Prob}\left\{\text{Binomial}\left(n, \frac{l}{2n-1}\right) \leqslant m-1\right\},\end{aligned} \tag{3.9}$$

by the same argument leading to (3.5). For $n = 13$ the bootstrap probabilities (3.9) equal

$l =$	4 or 22	5 or 21	6 or 20	7 or 19	8 or 18
(3.9) =	.0016	.0051	.0125	.0245	.0414

$l =$	9 or 17	10 or 16	11 or 15	12 or 14	13
(3.9) =	.0614	.0820	.1002	.1125	.1170.

(3.10)

The corresponding estimate of $E_F R^2$ would be

$$\Sigma_{l=1}^{25} [z_{(l)} - x_{(7)}]^2 \,\text{Prob}_*\{R^* = z_{(l)} - x_{(7)}\}.$$

Usually we would not be willing to assume F symmetric in a nonparametric estimation situation. However in dealing with continuous variables we might be willing to attribute a moderate amount of smoothness to F. This can be incorporated into the bootstrap procedure at step (2.4). Instead of choosing each X_i^* randomly from the set $\{x_1, x_2, \ldots, x_n\}$, we can take

$$X_i^* = \bar{x} + c[x_{I_i} - \bar{x} + \hat{\sigma} Z_i] \tag{3.11}$$

where the I_i are chosen independently and randomly from the set $\{1, 2, \ldots, n\}$, and the Z_i are a random sample from some fixed distribution having mean 0 and variance σ_Z^2, for example the uniform distribution on $[-\frac{1}{2}, \frac{1}{2}]$, which has $\sigma_Z^2 = 1/12$. The most obvious choice is a normal distribution for the Z_i, but this would be self-serving in the Monte Carlo experiment which follows, where the X_i themselves are normally distributed. The quantities $\bar{x}$, $\hat{\sigma}$, and c appearing in (3.11) are the sample mean, sample standard deviation $(=(\hat{\mu}_2)^{1/2})$, and $[1 + \sigma_Z^2]^{-1/2}$, respectively, so that X_i^* has mean $\bar{x}$ and variance $\hat{\sigma}^2$ under the bootstrap sampling procedure. In using (3.11) in place of (2.4), we are replacing $\hat{F}$ with a smoothed "window" estimator, having the same mean and variance as $\hat{F}$.

A small Monte Carlo experiment was run to compare the various bootstrap methods suggested above. Instead of comparing the squared error of the sample median, the quantity bootstrapped was

$$R(\mathbf{X}, F) = \frac{|t(\mathbf{X}) - \theta(F)|}{\sigma(F)}, \tag{3.12}$$

the absolute error of the sample median relative to the population standard deviation. (This quantity is more stable numerically, because the absolute value is less sensitive than the square and also because $R^* = |t(\mathbf{X}^*) - \theta(\hat{F})|/\hat{\sigma}$ is scale invariant, which eliminates the component of variation due to $\hat{\sigma}$ differing from $\sigma(F)$. The stability of (3.12) greatly increased the effectiveness of the Monte Carlo trial.)

The Monte Carlo experiment was run with $n = 13$, $X_i \sim_{\text{ind}} N(0,1)$, $i = 1, 2, \ldots, n$. In this situation the true expectation of R is

$$E_F R = 0.95. \tag{3.13}$$

The first two columns of Table 1 show $E_F R^*$ for each trial, using the bootstrap probabilities (3.6), and then (3.10) for the symmetrized version. It is not possible to theoretically calculate $E_* R^*$ for the smoothed bootstrap (3.11), so these entries of Table 1 were obtained by a secondary Monte Carlo simulation, as described in "Method 2" of Section 2. A total of $N = 50$ replications $\mathbf{x}^{*j}$ were generated for each trial. This means that the values in the table are only unbiased estimates of the actual bootstrap expectations $E_* R^*$ (which could be obtained by letting $N \to \infty$); the standard error being about .15 for each entry. The effect of this approximation is seen in the column "$d = 0$," which would exactly equal column "(3.6)" if $N \to \infty$. (Within each trial, the same set of random numbers was used to generate the four different uniform distributions for Z_i, $d = 0, .25, .5, 1$.)

The most notable feature of Table 1 is that the simplest form of the bootstrap, "(3.6)," seems to do just as well as the symmetrical or smoothed versions. A larger Monte Carlo investigation of the same situation as in Table 1, 200 trials, 100 bootstrap replications per trial, was just a little more favorable to the smoothed bootstrap methods:

	(3.6)	(3.10)	$d = 0$	$d = .25$	$d = .5$	$d = 1$	$d = 2$
AVE.:	1.01	1.00	1.00	1.01	1.00	.99	.93
S.D.:	.31	.33	.32 [31]	.32 [.30]	.32 [.30]	.30 [.29]	.26 [.25].

(The figures in square brackets are estimated standard deviations if N were increased from 100 to ∞, obtained by a components of variance calculation.)

Table 1.*

Trial #	Unsmoothed Bootstrap (3.6)	(3.10)	Smoothed Boostrap (3.11) Z_i uniform dist. on $[-d/2, d/2]$ $d = 0$	$d = .25$	$d = .5$	$d = 1$	Z_i triangular dist., $\sigma_Z^2 = 1/12$
1	1.07	1.18	1.09	1.10	1.12	1.11	1.16
2	.96	.74	1.10	1.10	1.08	1.09	1.15
3	1.22	.74	1.36	1.35	1.33	1.43	1.52
4	1.38	1.51	1.44	1.41	1.38	1.28	1.30
5	1.00	.83	1.03	1.05	1.09	1.14	1.17
6	1.13	1.21	1.27	1.26	1.23	1.20	1.26
7	1.07	.98	1.01	.94	.83	.79	.92
8	1.51	1.40	1.40	1.45	1.47	1.51	1.50
9	.56	.64	.69	.71	.74	.80	.81
10	1.05	.86	1.14	1.17	1.20	1.13	1.22
Ave.	1.09	1.01	1.15	1.15	1.15	1.15	1.20
S.D.	.26	.30	.23	.23	.23	.23	.22

* Ten Monte Carlo trials of $X_i \sim_{\text{ind}} N(0, 1)$, $i = 1, 2, \ldots, 13$ were used to compare different bootstrap methods for estimating the expected value of random variable (3.12). The true expectation is 0.95. The quantities tabled are $E_* R^*$, the bootstrap expectation for that trial. The values in the first two columns are for the bootstrap as described originally, and for the symmetrized version (3.8)–(3.10). The smoothed bootstrap expectations were approximated using a secondary Monte Carlo simulation for each trial, $N = 50$, as described in "Method 2," Section 2. Each of these entries estimates the actual value of $E_* R^*$ unbiasedly with a standard error of about .15. The column "$d = 0$" would exactly equal column "(3.6)" if $N \to \infty$.

Remembering that we are trying to estimate the true value $E_F R = .95$, these seem like good performances for a nonparametric method based on a sample size of just 13.

The symmetrized version of the bootstrap might be expected to do relatively better than the unsymmetrized version if R itself was of a less symmetric form than (3.12), e.g., $R(\mathbf{X}, F) = \exp\{X_{(m)} - \theta(F)\}$. Likewise, the smoothed versions of the bootstrap might be expected to do relatively better if R itself were less smooth, e.g., $R(\mathbf{X}, F) = \text{Prob}\{X_{(m)} > \theta(F) + \sigma(F)\}$. However no evidence to support these guesses is available at present.

4. Error Rate Estimation in Discriminant Analysis

This section discusses the estimation of error rates in a standard linear discriminant analysis problem. There is a tremendous literature on this problem, nicely summarized in Toussaint [17]. In the two examples considered below, bootstrap methods outperform the commonly used "leave-one-out," or *cross-validation*, approach (Lachenbruch and Mickey [12]).

The data in the discriminant problem consists of independent random samples from two unknown continuous probability distributions F and G on some k-dimensional space R^k,

$$\begin{aligned} X_i &= x_i, \quad X_i \sim_{\text{ind}} F \qquad i = 1, 2, \ldots, m \\ Y_j &= y_j, \quad Y_j \sim_{\text{ind}} G \qquad j = 1, 2, \ldots, n. \end{aligned} \tag{4.1}$$

On the basis of the observed data $\mathbf{X} = \mathbf{x}$, $\mathbf{Y} = \mathbf{y}$ we use some method (linear discriminant analysis in the examples below) to partition R^k into two complementary regions A and B, the intent being to ascribe a future observation z to the F distribution if $z \in A$, or to the G distribution if $z \in B$.

The obvious estimate of the error rate, for the F distribution, associated with the partition (A, B) is

$$\widehat{\text{error}_F} = \frac{\#\{x_i \in B\}}{m}, \tag{4.2}$$

which will tend to underestimate the true error rate

$$\text{error}_F = \text{Prob}_F\{X \in B\}. \tag{4.3}$$

(In probability calculation (4.3), B is considered fixed, at its observed value, even though it is originally determined by a random mechanism.) We will be interested in the distribution of the difference

$$R((\mathbf{X}, \mathbf{Y}), (F, G)) = \text{error}_F - \widehat{\text{error}_F}, \tag{4.4}$$

and the corresponding quantity for the distribution G. We could directly consider the distribution of $\widehat{\text{error}_F}$, but concentrating on the difference (4.4) is much more efficient for comparing different estimation methods. This point is discussed briefly at the end of the section.

Given $\mathbf{x}$ and $\mathbf{y}$, we define the region B by

$$B = \left\{z : (\bar{y} - \bar{x})' S^{-1}\left(z - \frac{\bar{x} + \bar{y}}{2}\right) > \log \frac{m}{n}\right\}, \tag{4.5}$$

where $\bar{x} = \Sigma x_i/m$, $\bar{y} = \Sigma y_j/n$, and $S = [\Sigma(x_i - \bar{x})(x_i - \bar{x})' + \Sigma(y_j - \bar{y})(y_j - \bar{y})']/(m + n)$. This is the maximum likelihood estimate of the optimum division under multivariate normal theory, and differs just slightly (in the definition of S) from the estimated version of the Fisher linear discriminant function discussed in Chapter 6 of Anderson [1].

"Method 2," the brute force application of the bootstrap via simulation, is implemented as follows: given the data $\mathbf{x}$, $\mathbf{y}$, bootstrap random samples

$$\begin{aligned} X_i^* &= x_i^*, \quad X_i^* \sim_{\text{ind}} \hat{F} \qquad i = 1, 2, \ldots, m \\ Y_j^* &= y_j^*, \quad Y_j^* \sim_{\text{ind}} \hat{G} \qquad j = 1, 2, \ldots, n \end{aligned} \tag{4.6}$$

are generated, $\hat{F}$ and $\hat{G}$ being the sample probability distributions corresponding to F and G. This yields a region B^* defined by (4.5) with $\bar{x}^*$, $\bar{y}^*$, S^*

Table 2.*

Random Variable	$m = n = 10$ Mean	(S.E.)	S.D.	$m = n = 20$ Mean	(S.E.)	S.D.	Remarks
Error Rate Diff. (4.4) R	**.062**	(.003)	**.143**	**.028**	(.002)	**.103**	Based on 1000 trials
Bootstrap Expectation $E_* R^*$	**.057**	(.002)	**.026** [.023]	**.029**	(.001)	**.015** [.011]	Based on 100 trials; $N = 100$ Bootstrap Replications per trial. (Figure in brackets is S.D. if $N = \infty$.)
Bootstrap Standard Deviation $SD_*(R^*)$	**.131**	(.0013)	**.016**	**.097**	(.002)	**.010**	
Cross-Validation Diff. $\tilde{R}$	**.054**	(.009)	**.078**	**.032**	(.002)	**.043**	Based on 40 trials

* The error rate difference (4.4) for linear discriminant analysis, investigated for bivariate normal samples (4.8). Sample sizes are $m = n = 10$ and $m = n = 20$. The values for the bootstrap method were obtained by Method 2, $N = 100$ bootstrap replications per trial. The bootstrap method gives useful estimates of both the mean and standard deviation of R. The cross-validation method was nearly unbiased for the expectation of R, but had about three times as large a standard deviation. All of the quantities in this table were estimated by repeated Monte Carlo trials; standard errors are given for the means.

replacing $\bar{x}$, $\bar{y}$, S. The bootstrap random variable in this case is

$$R^* = R((\mathbf{X}^*, \mathbf{Y}^*), (\hat{F}, \hat{G}) = \frac{\#\{x_i \in B^*\}}{m} - \frac{\#\{x_i^* \in B^*\}}{m}. \tag{4.7}$$

In other words, (4.7) is the difference between the actual error rate, actual now being defined with respect to the "true" distribution $\hat{F}$, and the apparent error rate obtained by counting errors in the bootstrap sample.

Repeated independent generation of $(\mathbf{X}^*, \mathbf{Y}^*)$ yields a sequence of independent realizations of R^*, say $R^{*1}, R^{*2}, \ldots, R^{*N}$, which are then used to approximate the actual bootstrap distribution of R^*, this hopefully being a reasonable estimate of the unknown distribution of R. For example, the bootstrap expectation $E_* R^* = \Sigma_{j=1}^N R^{*j}/N$ can be used as an estimate of the true expectation $E_{F,G} R$.

To test out this theory, bivariate normal choices of F and G were investigated,

$$F: X \sim N_2\left(\begin{pmatrix} -\frac{1}{2} \\ 0 \end{pmatrix}, \mathbf{I}\right) G: Y \sim N_2\left(\begin{pmatrix} \frac{1}{2} \\ 0 \end{pmatrix}, \mathbf{I}\right). \tag{4.8}$$

Two sets of sample sizes, $m = n = 10$ and $m = n = 20$, were looked at, with the results shown in Table 2. (The entries of Table 2 were themselves estimated by averaging over repeated Monte Carlo trials, which should not be confused with the Monto Carlo replications used in the bootstrap process. "Replications" will always refer to the bootstrap process, "trials" to repetitions of the basic situation.) Because situation (4.8) is symmetric, only random variable (4.4), and not the corresponding error rate for G, need be considered.

Table 2 shows that with $m = n = 10$, the random variable (4.4) has mean and standard deviation approximately (.062, .143). The apparent error rate underestimates the true error rate by about 6%, on the average, but the standard deviation of the difference is 14% from trial to trial, so bias is less troublesome than variability in this situation. The bootstrap method gave an average of .057 for $E_* R^*$, which, allowing for sampling error, shows that the statistic $E_* R^*$ is nearly an unbiased estimator for $E_{F,G} R$. Unbiasedness is not enough, of course; we want $E_* R^*$ to have a small standard deviation, ideally zero, so that we can rely on it as an estimate. The actual value of its standard deviation, .026, is not wonderful, but does indicate that most of the trials yielded $E_* R^*$ in the range [.02, .09], which means that the statistician would have obtained a reasonably informative estimate of the true bias $E_{F,G} R =$ .062.

As a point of comparison, consider the cross-validation estimate of R, say $\tilde{R}$, obtained by: deleting one x value at a time from the vector $\mathbf{x}$; recomputing B using (4.5), to get a new region $\tilde{B}$ (it is important *not* to change m to $m - 1$ in recomputing B—doing so results in badly biased estimation of R); seeing if the deleted x value is correctly classified by $\tilde{B}$; counting the proportion of x values misclassified in this way to get a cross-validated error rate $\widetilde{\text{error}}_F$; and finally, defining $\tilde{R} = \widetilde{\text{error}}_F - \widehat{\text{error}}_F$. The last row of Table 2 shows that $\tilde{R}$ has mean and standard deviation approximately (.054, .078). That is, *$\tilde{R}$ is three times as variable as $E_* R^*$ as an estimator of $E_{F,G} R$.*

The bootstrap standard deviation of R^*, $SD_*(R^*) = \{\Sigma_{j=1}^{N} [R^{*j} - E_* R^*]^2 /(N-1)\}^{1/2}$, can be used as an estimate of $SD_{F,G}(R)$, the actual standard deviation of R. Table 2 shows that $SD_*(R^*)$ had mean and standard deviation (.131, .016) across the 100 trials. Remembering that $SD_{F,G}(R) = .143$, the bootstrap estimate $SD_*(R^*)$ is seen to be a quite useful estimator of the actual standard deviation of R.

How much better would the bootstrap estimator $E_* R^*$ perform if the number of bootstrap replications N were increased from 100 to, say, 10,000? A components of variance analysis of all the data going into Table 2 showed that only moderate further improvement is possible. As $N \to \infty$, the trial-to-trial standard deviation of $E_* R^*$ would decrease from .026 to about .023 (from .015 to .011 in the case $m = n = 20$).

The reader may wonder which is the best estimator of the error rate error_F itself, rather than of the difference R. In terms of expected squared error, the order of preference is $\widehat{\text{error}}_F + E_* R^*$ (the bias-corrected value based on the bootstrap), $\widehat{\text{error}}_F$, and lastly $\widetilde{\text{error}}_F$, but the differences are quite small in the two situations of Table 2. The large variability of $\widehat{\text{error}}_F$, compared to its relatively small bias, makes bias correction an almost fruitless chore in these two situations. (Of course, this might not be so in more difficult discriminant problems.) *The bootstrap estimates of $E_{F,G} R$ and $SD_{F,G}(R)$ considered together* *make it clear that this is the case*, which is a good recommedation for the bootstrap approach.

5. Relationship with the Jackknife

This section concerns "Method 3" of approximating the bootstrap distribution, Taylor series expansion (or the *delta method*), which turns out to be the same as the usual jackknife theory. To be precise, it is the same as Jaeckel's *infinitesimal jackknife* [10, 14], a useful mathematical device which differs only in detail from the standard jackknife.

Returning to the one-sample situation, define $P_i^* = N_i^*/n$, where $N_i^* = \#\{X_i^* = x_i\}$ as at (3.2), and the corresponding vector

$$\mathbf{P}^* = (P_1^*, P_2^*, \ldots, P_n^*). \tag{5.1}$$

By the properties of the multinomial distribution, $\mathbf{P}^*$ has mean vector and covariance matrix

$$E_*\mathbf{P}^* = \mathbf{e}/n, \qquad \mathrm{Cov}_*\mathbf{P}^* = \mathbf{I}/n^2 - \mathbf{e}'\mathbf{e}/n^3 \tag{5.2}$$

under the bootstrap sampling procedure, where $\mathbf{I}$ is the identity matrix and $\mathbf{e} = (1, 1, 1, \ldots, 1)$.

Given the observed data vector $\mathbf{X} = \mathbf{x}$, and therefore $\hat{F}$, we can use the abbreviated notation

$$R(\mathbf{P}^*) = R(\mathbf{X}^*, \hat{F}) \tag{5.3}$$

for the bootstrap realization of R corresponding to $\mathbf{P}^*$. In making this definition we assume that the random variable of interest, $R(\mathbf{X}, F)$, is symmetrically defined in the sense that its value is invariant under any permutation of the components of X, so that it is sufficient to know $\mathbf{N}^* = n\mathbf{P}^*$ in order to evalute $R(\mathbf{X}^*, \hat{F})$. This is always the case in standard applications of the jackknife.

We can approximate the bootstrap distribution of $R(\mathbf{X}^*, \hat{F})$ by expanding $R(\mathbf{P}^*)$ in a Taylor series about the value $\mathbf{P}^* = \mathbf{e}/n$, say

$$R(\mathbf{P}^*) = R(\mathbf{e}/n) + (\mathbf{P}^* - \mathbf{e}/n)\mathbf{U} + \tfrac{1}{2}(\mathbf{P}^* - \mathbf{e}/n)\mathbf{V}(\mathbf{P}^* - \mathbf{e}/n)'. \tag{5.4}$$

Here

$$\mathbf{U} = \left.\begin{pmatrix} \vdots \\ \dfrac{\partial R(\mathbf{P}^*)}{\partial P_i^*} \\ \vdots \end{pmatrix}\right|_{P^*=e/n} \qquad \mathbf{V} = \left.\begin{pmatrix} & \vdots & \\ \cdots & \dfrac{\partial^2 R(\mathbf{P}^*)}{\partial P_i^* \partial P_j^*} & \cdots \\ & \vdots & \end{pmatrix}\right|_{P^*=e/n}. \tag{5.5}$$

Expansion (5.4), and definitions (5.5), assume that the definition of $R(\mathbf{P}^*)$ can be smoothly interpolated between the lattice point values originally contemplated for $\mathbf{P}^*$. How to do so will be obvious in most specific cases, but a general recipe is difficult to provide. See Remarks G and H of Section 8.

The restriction $\Sigma P_i^* = 1$ has been ignored in (5.4), (5.5). This computational convenience is justified by extending the definition of $R(\mathbf{P}^*)$ to all vectors $\mathbf{P}^*$ having nonnegative components, at least one positive, by the homogeneous extension

$$R(\mathbf{P}^*) = R\left(\frac{\mathbf{P}^*}{\Sigma_{i=1}^n P_i^*}\right). \tag{5.6}$$

It is easily shown that the homogeneity of definition (5.6) implies

$$\mathbf{eU} = 0, \qquad \mathbf{eV} = -n\mathbf{U}', \qquad \mathbf{eVe}' = 0. \tag{5.7}$$

From (5.2) and (5.4) we get the approximation to the bootstrap expectation

$$E_* R(\mathbf{P}^*) \doteq R(\mathbf{e}/n) + \frac{1}{2}\operatorname{trace}\mathbf{V}[\mathbf{I}/n^2 - \mathbf{e}'\mathbf{e}/n^3] = R(\mathbf{e}/n) + \frac{1}{2n}\bar{V}, \tag{5.8}$$

where

$$\bar{V} = \Sigma_{i=1}^n V_{ii}/n. \tag{5.9}$$

Ignoring the last term in (5.4) gives a cruder approximation for the bootstrap variance,

$$\operatorname{Var}_* R(\mathbf{P}^*) \doteq \mathbf{U}'[\mathbf{I}/n^2 - \mathbf{e}'\mathbf{e}/n^3]\mathbf{U} = \Sigma_{i=1}^n U_i^2/n^2. \tag{5.10}$$

(Both (5.8) and (5.10) involve the use of (5.7).)

Results (5.8) and (5.10) are essentially the jackknife expressions for bias and variance. The usual jackknife theory considers $R(\mathbf{X}, F) = \theta(\hat{F}) - \theta(F)$, the difference between the obvious nonparametric estimator $\theta(F)$ and $\theta(F)$ itself. In this case $R(\mathbf{X}^*, F) = \theta(\hat{F}^*) - \theta(\hat{F})$, $\hat{F}^*$ being the empirical distribution of the bootstrap sample, so that $R(\mathbf{e}/n) = \theta(\hat{F}) - \theta(\hat{F}) = 0$. Then (5.8) becomes $E_*[\theta(\hat{F}^*) - \theta(\hat{F})] \doteq (1/2n)\bar{V}$, suggesting $E_F[\theta(\hat{F}) - \theta(F)] \approx (1/2n)\bar{V}$; likewise (5.10) becomes $\operatorname{Var}_*[\theta(\hat{F}^*) - \theta(\hat{F})] \doteq \Sigma U_i^2/n^2$, suggesting $\operatorname{Var}_F \theta(\hat{F}) \approx \Sigma U_i^2/n^2$.

The approximations

$$\operatorname{Bias}_F \theta(\hat{F}) \approx \frac{1}{2n}\bar{V}, \qquad \operatorname{Var}_F \theta(\hat{F}) \approx \Sigma_{i=1}^n U_i^2/n^2 \tag{5.11}$$

exactly agree with those given by Jaeckel's infinitesimal jackknife [10], which themselves differ only slightly from the ordinary jackknife expressions. Without going into details, which are given in Jaeckel [10] and Miller [14], the ordinary jackknife replaces the derivatives $U_i = \partial R(\mathbf{P}^*)/\partial P_i$ with finite differences

$$\tilde{U}_i = (n-1)(R^* - R_{(i)}^*) \tag{5.12}$$

where $R_{(i)}^* = R(\mathbf{e}_{(i)}/(n-1))$, $\mathbf{e}_{(i)}$ being the vector with zero in the ith coordinate and ones elsewhere, and $R^* = \Sigma_{i=1}^n R_{(i)}^*/n$. Expansion (5.4) combines with (5.7) to give

$$\tilde{U}_i \doteq \frac{n-2}{n-1}U_i - \frac{1}{2(n-1)}(V_{ii} - \bar{V}), \tag{5.13}$$

so that $\tilde{U}_i/U_i = 1 + O(1/n)$. The ordinary jackknife estimate of variance is $\Sigma_{i=1}^n \tilde{U}_i^2/n\cdot(n-1)$, differing from the variance expression in (5.11) by a factor $1 + O(1/n)$, the same statement being true for the bias. (In the familiar case $R = \theta(\hat{F}) - \theta(F)$, definition (5.12) becomes $\tilde{U}_i = (n-1)(\hat{\theta} - \hat{\theta}_{(i)})$, where $\hat{\theta}_{(i)}$ is the estimate of θ with x_i removed from the sample, and $\hat{\theta} = \Sigma_i \hat{\theta}_{(i)}/n$; the jackknife estimate of θ is $\tilde{\theta} = \hat{\theta} + (n-1)(\hat{\theta} - \hat{\theta})$, and $\tilde{\theta}_i = \tilde{\theta} + \tilde{U}_i$ is the ith *pseudo-value*, to use the standard terminology.)

6. Wilcoxon's Statistic

We again consider the two-sample situation (4.1), this time with F and G being continuous probability distributions on the real line. The parameter of interest will be

$$\theta(F, G) = P_{F,G}(X < Y), \tag{6.1}$$

estimated by Wilcoxon's statistic

$$\hat{\theta} = \theta(\hat{F}, \hat{G}) = \frac{1}{mn}\Sigma_{i=1}^m \Sigma_{j=1}^n I(X_i, Y_j), \tag{6.2}$$

where

$$\begin{aligned} I(a, b) &= 1 \quad a < b \\ &= 0 \quad a \geqslant b. \end{aligned} \tag{6.3}$$

The bootstrap variance of $\hat{\theta}$ can be calculated directly by Method 1, and will turn out below to be the same as the standard variance approximation for Wilcoxon's statistic. The comparison with Method 3, the infinitesimal jackknife, illustrates how this theory works in a two-sample situation. More importantly, *it suggests the correct analogue of the ordinary jackknife for such situations.*

There has been considerable interest in extending the ordinary jackknife to "unbalanced" situations, i.e., those where it is not clear what the correct analogue of "leave one out" is, see Miller [15], Hinkley [9]. In the two-sample problem, for example, should we leave out one x_i at a time, then one y_j at a time, or should we leave out all mn pairs (x_i, y_j) one at a time? (The former turns out to be correct.) This problem gets more crucial in the next section, where we consider regression problems.

Let $R((\mathbf{X}, \mathbf{Y}), (F, G))$ be $\hat{\theta}$ itself, so that the bootstrap value of R corresponding to $(\mathbf{X}^*, \mathbf{Y}^*)$ is $R((\mathbf{X}^*, \mathbf{Y}^*), (\hat{F}, \hat{G})) = \hat{\theta}^*$,

$$\hat{\theta}^* = \frac{1}{mn}\Sigma_i\Sigma_j I(X_i^*, Y_j^*). \tag{6.4}$$

Letting $I_{ij}^* = I(X_i^*, Y_j^*)$, straightforward calculations familiar from standard nonparametric theory, give

$$E_* I_{ij}^* = \hat{\theta}, \qquad \text{Var}_* I_{ij}^* = \hat{\theta}(1-\hat{\theta}), \qquad E_* I_{ij}^* I_{i'j'}^* = \hat{\theta}^2 \quad i \neq i', j \neq j' \tag{6.5}$$

and

$$\begin{aligned} E_* I_{ij}^* I_{ij'}^* &= \int_{-\infty}^{\infty} [1-\hat{G}(z)]^2 \, d\hat{F}(z) \equiv \hat{\alpha}, \qquad j \neq j' \\ E_* I_{ij}^* I_{i'j}^* &= \int_{-\infty}^{\infty} \hat{F}^2(z) \, d\hat{G}(z) \equiv \hat{\beta}, \qquad i \neq i'. \end{aligned} \tag{6.6}$$

Using these results in (6.4) gives

$$\text{Var}_* \hat{\theta}^* = \frac{(n-1)(\hat{\alpha}-\hat{\theta}^2) + (m-1)(\hat{\beta}-\hat{\theta})^2 + \hat{\theta}(1-\hat{\theta})}{mn}, \tag{6.7}$$

which is the usual estimate for the variance of the Wilcoxon statistic, see Noether [16], page 32.

Method 3, the Taylor series or infinitesimal jackknife, proceeds as in Section 5, with obvious modifications for the two-sample situation. Let $\mathbf{N}_F^* = (N_{F1}^*, N_{F2}^*, \ldots, N_{Fm})$ be the numbers of times $x_1, x_2, \ldots, x_m$ occur in the bootstrap sample $\mathbf{X}^*$, likewise $\mathbf{N}_G^* = (N_{G1}^*, N_{G2}^*, \ldots, N_{Gn}^*)$ for $\mathbf{Y}^*$, and define and covariance as in (5.2). The expansion corresponding to (5.4) is

$$\begin{aligned} R(\mathbf{P}_F^*, \mathbf{P}_G^*) \doteq{}& R(\mathbf{e}/m, \mathbf{e}/n) + (\mathbf{P}_F^* - \mathbf{e}/m)\mathbf{U}_F + (\mathbf{P}_G^* - \mathbf{e}/n)\mathbf{U}_G \\ &+ \tfrac{1}{2}[(\mathbf{P}_F^* - \mathbf{e}/m) V_F (\mathbf{P}_F^* - \mathbf{e}/m)' \\ &+ 2(\mathbf{P}_F^* - \mathbf{e}/m) V_{FG} (\mathbf{P}_G^* - \mathbf{e}/n)' \\ &+ (\mathbf{P}_G^* - \mathbf{e}/n) V_G (\mathbf{P}_G^* - \mathbf{e}/n)'], \end{aligned} \tag{6.8}$$

where

$$U_{Fi} = \partial R / \partial P_{Fi}^*, \qquad V_{Fii'} = \partial^2 R / \partial P_{Fi}^* \partial P_{Fi'}^*, \qquad V_{FGij} = \partial^2 R / \partial P_{Fi}^* \partial P_{Gj}^*, \tag{6.9}$$

all the derivatives being evaluated at $(\mathbf{P}_F^*, \mathbf{P}_G^* = (\mathbf{e}/m, \mathbf{e}/n)$, analogous definitions applying to $\mathbf{U}_G$ and $\mathbf{V}_G$.

The results corresponding to (5.8) and (5.10) are

$$E_* R^* \doteq R(\mathbf{e}/m, \mathbf{e}/n) + \frac{1}{2}\left[\frac{\bar{V}_F}{m} + \frac{\bar{V}_G}{n}\right] \tag{6.10}$$

and

$$\text{Var}_* R^* \doteq \Sigma_{i=1}^m U_{Fi}^2/m^2 + \Sigma_{j=1}^n U_{Gj}^2/n^2, \tag{6.11}$$

$\bar{V}_F = \Sigma_i V_{Fii}/m$, $\bar{V}_G = \Sigma_j V_{Gjj}/n$. For $R = \theta(\hat{F}, \hat{G}) - \theta(F, G)$, the approximations corresponding to (5.11) are

$$\text{Bias}_{F,G}\, \theta(\hat{F}, \hat{G}) \approx \frac{1}{2}\left[\frac{\bar{V}_F}{m} + \frac{\bar{V}_G}{n}\right], \qquad \text{Var}_{F,G}\, \theta(F, G) \approx \frac{\Sigma_{i=1}^m U_{Fi}^2}{m^2} + \frac{\Sigma_{j=1}^n U_{Gj}^2}{n^2}. \tag{6.12}$$

For the case of the Wilcoxon statistic (6.11) (or (6.12)) gives

$$\text{Var}_* \, \theta^* \doteq \frac{n[\hat{\alpha} - \hat{\theta}^2] + m[\hat{\beta} - \hat{\theta}^2]}{mn}, \tag{6.13}$$

which should be compared with (6.7)

How can we use the ordinary jackknife to get results like (6.12)? A direct analogy of (5.12) can be carried through, but it is simpler to change definitions slightly, letting

$$\begin{aligned} D_{(i,)} &= R(\mathbf{e}/m, \mathbf{e}/n) - R(\mathbf{e}_{(i)}/(m-1), \mathbf{e}/n) \\ D_{(,j)} &= R(\mathbf{e}/m, \mathbf{e}/n) - R(\mathbf{e}/m, \mathbf{e}_{(j)}/(n-1)), \end{aligned} \tag{6.14}$$

the difference from $R((x, y), (\hat{F}, \hat{G}))$ obtained by deleting x_i from $\mathbf{x}$ y_j from $\mathbf{y}$. Expansion (6.8) gives

$$\begin{aligned} D_{(i,)} &\doteq \frac{m-2}{(m-1)^2} U_{Fi} - \frac{1}{2(m-1)^2} V_{Fii} \\ D_{(,j)} &\doteq \frac{(n-2)^2}{(n-1)^2} U_{Gj} - \frac{1}{2(n-1)^2} V_{Gjj}. \end{aligned} \tag{6.15}$$

From (6.15) it is easy to obtain approximations for the bias and variance expressions in terms of the D's

$$-\Sigma_{i=1}^{m} D_{(i,)} + \Sigma_{j=1}^{n} D_{(,j)}] \doteq \frac{1}{2}\left[\left(\frac{m}{m-1}\right)^2 \bar{V}_F + \left(\frac{n}{n-1}\right)^2 \bar{V}_G\right], \tag{6.16}$$

which, as m and n grow large, approaches the second term in (6.10). (For $R = \hat{\theta} - \theta$, this gives the bias-corrected estimate $\tilde{\theta} = (m + n - 1)\hat{\theta} - \Sigma_i \hat{\theta}_{(i,)} - \Sigma_j \hat{\theta}_{(,j)}$.) Likewise, just using the first line of (6.8) gives

$$\Sigma_{i=1}^{m} D_{(i,)}^2 + \Sigma_{j=1}^{n} D_{(,j)}^2 \doteq \frac{m^2(m-2)^2}{(m-1)^4} \frac{\Sigma_{i=1}^{m} U_{Fi}^2}{m^2} + \frac{n^2(n-2)^2}{(n-1)^2} \frac{\Sigma_{j=1}^{n} U_{Gj}^2}{n^2}, \tag{6.17}$$

which approaches (6.11) as $m, n \to \infty$.

The advantage of the D's over expressions like (5.12) is that no group averages, such as R^*, need be defined. Group averages are easy enough to define in the two-sample problem, but are less clear in more complicated situations such as regression. Expressions (6.16) and (6.17) are easy to extend to any situation (which doesn't necessarily mean they give good answers—see the remarks of the next section!).

7. Regression Modes

A reasonably general regression model is

$$X_i = g_i(\beta) + \varepsilon_i \qquad i = 1, 2, \ldots, n, \tag{7.1}$$

the $g(\cdot)$ being known functions of the unknown parameter vector β, and

$$\varepsilon_i \sim_{\text{ind}} F \qquad i = 1, 2, \ldots, n. \tag{7.2}$$

All that is assumed known about F is that it is centered at zero in some sense, perhaps $E_F\varepsilon = 0$ or $\text{Median}_F\varepsilon = 0$. Having observed $\mathbf{X} = \mathbf{x}$, we use some fitting technique to estimate β, perhaps least squares,

$$\hat{\beta} : \min_\beta \Sigma_{i=1}^n [x_i - g_i(\beta)]^2, \tag{7.3}$$

and wish to say something about the sampling distribution of $\hat{\beta}$.

Method 2, the brute force application of the bootstrap, can be carried out by defining $\hat{F}$ as the sample probability distribution of the residuals $\hat{\varepsilon}_i$,

$$\hat{F} : \text{mass} \frac{1}{n} \qquad \text{at } \hat{\varepsilon}_i = x_i - g_i(\hat{\beta}), \qquad i = 1, 2, \ldots, n. \tag{7.4}$$

(If one of the components of β is a translation parameter for the functions $g(\cdot)$, then $\hat{F}$ has mean zero. If not, and if the assumption $E_F\varepsilon = 0$ is very firm, one might still modify $\hat{F}$ by translation to achieve zero mean.) The bootstrap sample, given $(\hat{\beta}, \hat{F})$, is

$$X_i^* = g_i(\hat{\beta}) + \varepsilon_i^*, \qquad \varepsilon_i^* \sim_{\text{ind}} \hat{F} \qquad i = 1, 2, \ldots, n. \tag{7.5}$$

Each realization of (2.5) yields a realization of $\hat{\beta}^*$ by the same minimization process that gave $\hat{\beta}$,

$$\hat{\beta}^* : \min_\beta \Sigma_{i=1}^n [x_i^* - g_i(\beta)]^2. \tag{7.6}$$

Repeated independent bootstrap replications give a random sample $\hat{\beta}^{*1}, \hat{\beta}^{*2}, \hat{\beta}^{*3}, \ldots, \hat{\beta}^{*N}$ which can be used to estimate the bootstrap distribution of $\hat{\beta}^*$.

A handly test case is the familiar linear model, $g_i(\beta) = c_i\beta$, c_i a known $1 \times p$ vector, with first coordinate $c_{i1} = 1$ for convenience. Let $\mathbf{C}$ be the $n \times p$ matrix whose ith row is c_i, and $\mathbf{G}$ the $p \times p$ matrix $\mathbf{C}'\mathbf{C}$, assumed nonsingular. Then the least squares estimator $\hat{\beta} = \mathbf{G}^{-1}\mathbf{C}'\mathbf{X}$ has mean β and covariance matrix $\sigma_F^2\mathbf{G}^{-1}$ by the usual theory.

The bootstrap values ε_i^* used in (7.5) are independent with mean zero and variance $\hat{\sigma}^2 = \Sigma_{i=1}^n [x_i - g(\hat{\beta})]^2/n$. This implies that $\hat{\beta}^* = \mathbf{G}^{-1}\mathbf{C}'\mathbf{X}^*$ has bootstrap mean and variance

$$E_*\hat{\beta}^* = \hat{\beta}, \qquad \text{Cov}_*\hat{\beta}^* = \hat{\sigma}^2\mathbf{G}^{-1}. \tag{7.7}$$

The implication that $\hat{\beta}$ is unbiased for β, with covariance matrix approximately equal to $\hat{\sigma}^2\mathbf{G}^{-1}$, agrees with traditional theory, except perhaps in using the estimate $\hat{\sigma}^2$ for σ^2.

Miller [15] and Hinkley [9] have applied, respectively, the ordinary jackknife and infinitesimal jackknife to the linear regression problem. They formulate the situation as a one-sample problem, with (c_i, x_i) as the ith observed data point, essentially removing one row at a time from the model $\mathbf{X} = \mathbf{C}\beta + \boldsymbol{\varepsilon}$. The infinitesimal jackknife gives the approximation

$$\text{Cov}\,\hat{\beta} \approx \mathbf{G}^{-1}[\Sigma_{i=1}^n c_i'c_i\hat{\varepsilon}_i^2]\mathbf{G}^{-1}, \tag{7.8}$$

(and the ordinary jackknife a quite similar expression) for the estimated covariance matrix. This doesn't look at all like (7.7)!

The trouble lies in the fact that the jackknife methods as used above ignore an important aspect of the regression model, namely that the errors ε_i are assumed to have the same distribution for every value of i. To make (7.8) agree with (7.7) it is only necessary to "symmetrize" the data set by adding hypothetical data points, corresponding to all the possible values of the residual $\hat{\varepsilon}$, at each value of i, say

$$x_{ij} = c_i\hat{\beta} + \hat{\varepsilon}_j \qquad j = 1, 2, \ldots, n \qquad (i = 1, 2, \ldots, n). \tag{7.9}$$

Notice that the bootstrap implicitly does this at step (7.5). Applying the infinitesimal jackknife to data set (7.9), and remembering to take account of the artificially increased amount of data as at step (5.16), gives covariance estimate (7.7).

Returning to the nonlinear regression model (7.1), (7.2), where bootstrap-jackknife methods may really be necessary in order to get estimates of variability for $\hat{\beta}$, we now suspect that jackknife procedures like "leave out one row at a time" may be inefficient unless preceded by some form of data symmetrization such as (7.9). To put things the other way, as in Hinkley [9], such procedures tend to give consistent estimates of Cov $\hat{\beta}$ without assumption (7.2) that the residuals are identically distributed. The price of such complete generality is low efficiency. Usually assumption (7.2) can be roughly justified, perhaps after suitable transformations on X, in which case the bootstrap should give a better estimate of Cov $\hat{\beta}$.

8. Remarks

Remark A. Method 2, the straightforward calculation of the bootstrap distribution by repeated Monte Carlo sampling, is remarkably easy to implement on the computer. Given the original algorithm for computing R, only minor modifications are necessary to produce bootstrap replications $R^{*1}, R^{*2}, \ldots, R^{*N}$. The amount of computer time required is just about N times that for the original computations. For the discriminant analysis problem reported in Table 2, each trial of $N = 100$ replications, $m = n = 20$, took about 0.15 seconds and cost about 40 cents on Stanford's 370/168 computer. For a single real data set with $m = n = 20$, we might have taken $N = 1000$, at a cost of $4.00.

Remark B. Instead of estimating $\theta(F)$ with $t(\mathbf{X})$, we might make a transformation $\phi = g(\theta)$, $s = g(t)$, and estimate $\phi(F) = g(\theta(F))$ with $s(\mathbf{X}) = g(t(\mathbf{X}))$. That is, we might consider the random variable $S(\mathbf{X}, \mathbf{F}) = s(\mathbf{X}) - \phi(F)$ instead of $R(\mathbf{X}, \mathbf{F}) = t(\mathbf{X}) - \theta(F)$. The effect of such a transformation on the bootstrap is very simple: a bootstrap realization $R^* = R^*(\mathbf{X}^*, \hat{F}) = t(\mathbf{X}^*) - \theta(F)$ transforms into $S = S(\mathbf{X}^*, \hat{F}) = g(t(\mathbf{X}^*)) - g(\theta(\hat{F}))$, or more simply

$$S^* = g(R^* + \hat{\theta}) - g(\hat{\theta}). \tag{8.1}$$

so the bootstrap distribution of R^* transforms into that of S^* by (8.1).

Figure 1 illustrates a simple example. Miller [14], page 12, gives 9 pairs of numbers having sample Pearson correlation coefficient $\hat{\rho} = .945$. The top half of Figure 1 shows the histogram of $N = 1000$ bootstrap replications of $\hat{\rho}^* - \hat{\rho}$, the bottom half the corresponding histogram of $\tanh^{-1} \hat{\rho}^* - \tanh^{-1} \hat{\rho}$. The first distribution straggles off to the left, the second distribution to the right. The median is above zero, but only slightly so compared to the spread of the distributions, indicating that bias correction is not likely to be important in this example.

The purpose of making transformations is, presumably, to improve the inference process. In the example above we might be willing to believe, on the basis of normal theory, that $\tanh^{-1} \hat{\rho} - \tanh^{-1} \rho$ is more nearly *pivotal* than $\hat{\rho} - \rho$ (see Remark E) and so more worthwhile investigating by the bootstrap procedure. This does not mean that the bootstrap gives more accurate result, only that the results are more useful. Notice that if $g(\cdot)$ is monotone, then any quantile of the bootstrap distribution of R^* maps into the corresponding quantile of S^* via (8.1), and vice-versa. In particular, if we use the median (rather than the mean) to estimate the center of the bootstrap distribution, then we get the same answer working directly with $\hat{\theta}^* - \hat{\theta}$ ($\hat{\rho}^* - \hat{\rho}$ in the example), or first transforming to $\hat{\phi}^* - \phi$ ($\tanh^{-1} \hat{\rho}^* - \tanh^{-1} \hat{\rho}$), taking the median, and finally transforming back to the original scale.

Remark C. The bias and variance expressions (5.11) suggested by the infinitesimal jackknife transform exactly as in more familiar applications of the "delta method." That is, if $\phi = g(\theta)$, $\hat{\phi} = g(\hat{\theta})$ as above, and $\widehat{\text{Bias}}_F\, \hat{\theta}$, $\widehat{\text{Var}}_F\, \hat{\theta}$ are as given in formula (5.11), then it is easy to show that

$$\begin{aligned} \widehat{\text{Bias}}_F\, \hat{\phi} &= g'(\hat{\theta}) \widehat{\text{Bias}}_F\, \hat{\theta} + \frac{g''(\hat{\theta})}{2} \widehat{\text{Var}}_F\, \hat{\theta}, \\ \widehat{\text{Var}}_F\, \hat{\phi} &= [g'(\hat{\theta})]^2\, \widehat{\text{Var}}_F\, \hat{\theta}. \end{aligned} \tag{8.2}$$

In the context of this paper, the infinitesimal jackknife *is* the delta method; starting from a known distribution, that of $\mathbf{P}^*$, approximations to the moments of an arbitrary function $R(\mathbf{P}^*)$ are derived by Taylor series expansion. See Gray et al. [4] for a closely related result.

Remark D. A standard nonparametric confidence statement for the median $\theta(F)$, $n = 13$, is

$$\text{Prob}_F\{x_{(4)} < \theta < x_{(10)}\} = \text{Prob}\{4 \leqslant \text{Bi}(13, \tfrac{1}{2}) \leqslant 9\} = .908. \tag{8.3}$$

If we make the continuity correction of halving the end point probabilities,

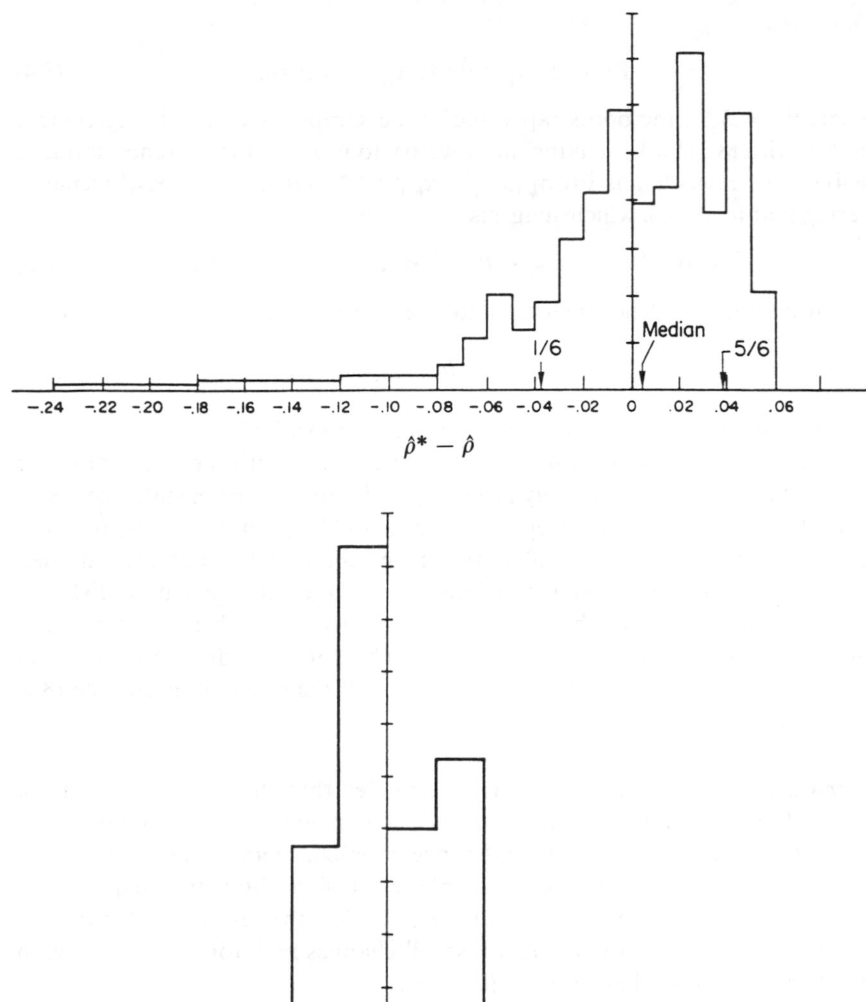

Figure 1. The top histogram shows $N = 1000$ bootstrap replications of $\hat{\rho}^* - \hat{\rho}$ for the nine data pairs from Miller [10]: (1.15, 1.38), (1.70, 1.72), (1.42, 1.59), (1.38, 1.47), (2.80, 1.66), (4.70, 3.45), (4.80, 3.87), (1.41, 1.31), (3.90, 3.75). The bottom histogram shows the corresponding replications for $\tanh^{-1} \hat{\rho}^* - \tanh^{-1} \hat{\rho}$. The 1/6, 1/2, and 5/6 quantiles are shown for both distributions. All quantiles transform according to equation (8.1).

(3.6) gives

$$\text{Prob}_*\{x_{(4)} < \hat{\theta}^* < x_{(10)}\} = .914, \tag{8.4}$$

where $\hat{\theta}^* = X^*_{(m)}$, the bootstrap value of the sample median. The agreement of (8.4) with (8.3) looks striking, until we try to use (8.4) for inference about θ; (8.4) can be rewritten as $\text{Prob}_*\{x_{(4)} - x_{(7)} < \hat{\theta}^* - \hat{\theta} < x_{(10)} - x_{(7)}\}$ (remembering that $\hat{\theta} = x_{(7)}$), which suggests

$$\text{Prob}_F\{x_{(4)} - x_{(7)} < \hat{\theta} - \theta < x_{(10)} - x_{(7)}\} \approx .914. \tag{8.5}$$

The resulting confidence interval statement for θ, again using $\hat{\theta} = x_{(7)}$, is

$$\text{Prob}_F\{2x_{(7)} - x_{(10)} < \theta < 2x_{(7)} - x_{(4)}\} \approx .914, \tag{8.6}$$

which is the reflection of interval (8.3) about the median!

The trouble here has nothing in particular to do with the bootstrap, and does not arise from the possibly large approximation error in statement (8.5), but rather in the inferential step from (8.5) to (8.6), which tries to use $\hat{\theta} - \theta$ as a *pivotal quantity*. The same difficulty can be exhibited in parametric families: suppose we know that F is a translation of a standard exponential distribution (density e^{-x}, $x > 0$). Then there exist two positive numbers a and b, $a < b$, such that $\text{Prob}_F\{-a < \hat{\theta} - \theta < b\} = .91$. The corresponding interval statement $\text{Prob}_F\{x_{(7)} - b < \theta < x_{(7)} + a\} = .91$ will tend to look more like (8.6) than (8.3).

Remark E. The difficulty above is a reminder that the bootstrap, and the jackknife, provide approximate *frequency* statements, not approximate *likelihood* statements. Fundamental inference problems remain, no matter how well the bootstrap works. For example, even if the bootstrap expectation $E_*(\hat{\theta}^* - \theta)^2$ very accurately estimates $E_F(\hat{\theta} - \theta)^2$, the resulting interval estimate for θ given $\hat{\theta}$ will be useless if small changes in F (or more exactly, in $\theta(F)$) result in large changes in $E_F(\hat{\theta} - \theta)^2$.

For the correlation coefficient, as discussed in Remark B, Fisher showed that $\tanh^{-1}\hat{\rho} - \tanh^{-1}\rho$ is nearly pivotal when sampling from bivariate normal populations. That is, its distribution is nearly the same for all bivariate normal populations, at least in the range $-.9 < \rho < .9$. This property tends to ameliorate inference difficulties, and is the principal reason for transforming variables, as in Remark B. The theory of pivotal quantities is well developed in parametric families, see Barnard [2], but not in the nonparametric context of this paper.

Remark F. The classic pivotal quantity is Student's t-statistic. Tukey has suggested using the analogous quantity (2.3) for hypothesis testing purposes, relying on the standard t tables for significance points. This amounts to treating (2.3) as a pivotal quantity for all choices of F, $\theta(F)$, and $t(\mathbf{X})$. The only theoretical justifications for this rather optimistic assumption apply to large samples, where the Student t effect rapidly becomes negligible, see Miller [14]. Given the current state of the theory, one is as well justified in comparing (2.3) to a $\mathcal{N}(0, 1)$ distribution as to a Student's t distribution (except when $t(\mathbf{X}) = \bar{X}$).

An alternative approach is to bootstrap (2.3) by Method 2 to obtain a direct estimate of its distribution, instead of relying on the t distribution, and then compare the observed value of (2.3) to the bootstrap distribution.

Remark G. The rationale for bootstrap methods becomes particularly clear when the sample space $\mathcal{X}$ of the X_i is a finite set, say

$$\mathcal{X} = \{1, 2, 3, \ldots, L\}. \tag{8.7}$$

The distribution F can now be represented by the vector of probabilities $\mathbf{f} = (f_1, f_2, \ldots, f_L)$, $f_l = \text{Prob}_F\{X_i = l\}$. For a given random sample $\mathbf{X}$ let $\hat{f}_l = \#\{X_i = l\}/n$ and $\hat{\mathbf{f}} = (\hat{f}_1, \hat{f}_2, \ldots, \hat{f}_L)$, so that if $R(\mathbf{X}, F)$ is symmetrically defined in the components of $\mathbf{X}$ we can write it as a function of $\hat{\mathbf{f}}$ and $\mathbf{f}$, say

$$R(\mathbf{X}, F) = Q(\hat{\mathbf{f}}, \mathbf{f}). \tag{8.8}$$

Likewise $R(\mathbf{X}^*, \hat{F}) = Q(\hat{\mathbf{f}}^*, \hat{\mathbf{f}})$, where $\hat{f}_l^* = \#\{X_i^* = l\}/n$ and $\hat{\mathbf{f}}^* = (\hat{f}_1^*, \hat{f}_2^*, \ldots, \hat{f}_L^*)$.

Bootstrap methods estimate the sampling distribution of $Q(\hat{\mathbf{f}}, \mathbf{f})$, given the true distribution $\mathbf{f}$, by the conditional distribution of $Q(\hat{\mathbf{f}}^*, \hat{\mathbf{f}})$ given the observed value of $\hat{\mathbf{f}}$. This is plausible because

$$\hat{\mathbf{f}} | \mathbf{f} \sim \mathcal{M}_L(n, \mathbf{f}) \qquad \text{and} \qquad \hat{\mathbf{f}}^* | \hat{\mathbf{f}} \sim \mathcal{M}_L(n, \hat{\mathbf{f}}), \tag{8.9}$$

where $\mathcal{M}_L(n, \mathbf{f})$ is the L-category multinomial distribution with sample size n, probability vector $\mathbf{f}$. In large samples we expect $\hat{\mathbf{f}}$ to be close to $\mathbf{f}$, so that for reasonable functions $Q(\cdot, \cdot)$ (8.9) should imply the approximate validity of the bootstrap method.

The *asymptotic* validity of the bootstrap is easy to verify in this framework, assuming some regularity conditions on $Q(\cdot, \cdot)$. Suppose that $Q(\mathbf{f}, \mathbf{f}) = 0$ for all $\mathbf{f}$ (as it does in the usual jackknife situation where $R(\mathbf{X}, F) = \theta(\hat{F}) - \theta(F)$); that the vector $\mathbf{u}(\hat{\mathbf{f}}^*, \hat{\mathbf{f}})$ with lth component equal to $\partial Q(\hat{\mathbf{f}}^*, \hat{\mathbf{f}})/\partial \hat{f}_l^*$ exits continuously for $(\hat{\mathbf{f}}^*, \mathbf{f})$ in an open neighborhood of $(\mathbf{f}, \mathbf{f})$; and that $\mathbf{u} = \mathbf{u}(\mathbf{f}, \mathbf{f})$ does not equal zero. By Taylor's theorem, and the fact that $\hat{\mathbf{f}}^*$ and $\hat{\mathbf{f}}$ converge to $\mathbf{f}$ with probability one,

$$Q(\hat{\mathbf{f}}, \mathbf{f}) = (\hat{\mathbf{f}} - \mathbf{f})(\mathbf{u} + \boldsymbol{\varepsilon}_n) \qquad \text{and} \qquad Q(\hat{\mathbf{f}}^*, \mathbf{f}) = \hat{\mathbf{f}}^* - \hat{\mathbf{f}})(\mathbf{u} + \hat{\boldsymbol{\varepsilon}}_n), \tag{8.10}$$

both $\boldsymbol{\varepsilon}_n$ and $\hat{\boldsymbol{\varepsilon}}_n$ converging to zero with probability one. From (8.9) and the fact that $\hat{\mathbf{f}}$ converges to $\mathbf{f}$ with probability one, we have

$$n^{1/2}(\hat{\mathbf{f}} - \mathbf{f}) | \mathbf{f} \to \mathcal{N}_L(\mathbf{0}, \boldsymbol{\Sigma}_f) \qquad \text{and} \qquad n^{1/2}(\hat{\mathbf{f}}^* - \hat{\mathbf{f}}) | \hat{\mathbf{f}} \to \mathcal{N}_L(\mathbf{0}, \boldsymbol{\Sigma}_f), \tag{8.11}$$

where $\boldsymbol{\Sigma}_f$ is the matrix with element (l, m) equal to $f_l(\delta_{lm} - f_m)$. Combining (8.10) and (8.12) shows that the bootstrap distribution of $n^{1/2}Q(\hat{\mathbf{f}}^*, \hat{\mathbf{f}})$, given

$\hat{\mathbf{f}}$, is asymptotically equivalent to the sampling distribution of $n^{1/2}Q(\hat{\mathbf{f}}, \mathbf{f})$, given the true probability distribution $\mathbf{f}$. Both have the limiting distribution $\mathcal{N}(0, \mathbf{u}'\Sigma_f\mathbf{u})$.

The argument above assumes that the form of $Q(\cdot, \cdot)$ does not depend upon n. More careful considerations are necessary in cases like (2.3) where $Q(\cdot, \cdot)$ does depend on n, but in a minor way. Some nondifferentiable functions such as the sample median (3.3) can also be handled by a smoothing argument, though direct calculation of the limiting distribution is easier in that particular case.

(*Editor's note*: Remarks H–K have been omitted.)

Acknowledgments

I am grateful to Professors Rupert Miller and David Hinkley for numerous discussions, suggestions and references, and to Joseph Verducci for help with the numerical computations. The referees contributed several helpful ideas, especially concerning the connection with Hartigan's work, and the large sample theory. I also wish to thank the many friends who suggested names more colorful than *Bootstrap*, including *Swiss Army Knife*, *Meat Axe*, *Swan-Dive*, *Jack-Rabbit*, and my personal favorite, the *Shotgun*, which, to paraphrase Tukey, "can blow the head off any problem if the statistician can stand the resulting mess."

References

[1] Anderson, T.W. (1958). *An Introduction to Multivariate Statistical Analysis.* Wiley, New York.

[2] Barnard, G. (1974). Conditionality, pivotals, and robust estimation. *Proceedings of the Conference on Foundational Questions in Statistical Inference.* Memoirs No. 1, Dept. of Theoretical Statist., Univ. of Aarhus, Denmark.

[3] Cramér, H. (1946). *Mathematical Methods in Statistics.* Princeton Univ. Press.

[4] Gray, H., Schucany, W. and Watkins, T. (1975). On the generalized jackknife and its relation to statistical differentials. *Biometrika* **62** 637–642.

[5] Hartigan, J.A. (1969). Using subsample values as typical values. *J. Amer. Statist. Assoc.* **64** 1303–1317.

[6] Hartigan, J.A. (1971). Error analysis by replaced samples. *J. Roy. Statist. Soc. Ser. B* **33** 98–110.

[7] Hartigan, J.A. (1975). Necessary and sufficient conditions for asymptotic joint normality of a statistic and its subsample values. *Ann. Statist.* **3** 573–580.

[8] Hinkley, D. (1976[a]). On estimating a symmetric distribution. *Biometrika* **63** 680.

[9] Hinkley, D. (1976[b]). On jackknifing in unbalanced situations. Technical Report No. 22, Division of Biostatistics, Stanford Univ.

[10] Jaeckel, L. (1972). The infinitesimal jackknife. Bell Laboratories Memorandum #MM 72-1215-11.

[11] Kendall, M. and Stuart, A. (1950). *The Advanced Theory of Statistics.* Hafner, New York.

[12] Lachenbruch, P. and Mickey, R. (1968). Estimation of error rates in discriminant analysis. *Technometrics* **10** 1–11.

[13] Maritz, J.S. and Jarrett, R.G. (1978). A note on estimating the variance of the sample median. *J. Amer. Statist. Assoc.* **73** 194–196.

[14] Miller, R.G. (1974[a]). The jackknife—a review. *Biometrika* **61** 1–15.

[15] Miller, R.G. (1974[b]). An unbalanced jackknife. *Ann. Statist.* **2** 880–891.

[16] Noether, G. (1967). *Elements of Nonparametric Statistics.* Wiley, New York.

[17] Toussaint, G. (1974). Bibliography on estimation of misclassification. *IEEE Trans. Information Theory* **20** 472–479.

Index